T0136292

ISOSURFACES

ISOSURFACES

GEOMETRY, TOPOLOGY, AND ALGORITHMS

REPHAEL WENGER

CRC Press
Taylor & Francis Group
Boca Raton London New York

CRC Press is an imprint of the
Taylor & Francis Group, an **informa** business

AN A K PETERS BOOK

CRC Press
Taylor & Francis Group
6000 Broken Sound Parkway NW, Suite 300
Boca Raton, FL 33487-2742

First issued in hardback 2019

ISBN-13: 978-1-4665-7097-9 (hbk)

Library of Congress Cataloging-in-Publication Data

Wenger, Rephael.
   Isosurfaces : geometry, topology, and algorithms / Rephael Wenger.
   pages cm
   "An A K Peters Book."
   Includes bibliographical references and index.
   ISBN 978-1-4665-7097-9 (hardback)
   1. Surfaces. 2. Isogeometric analysis. 3. Three-dimensional imaging--Mathematics. I. Title.

QA573.W46 2013
516.3'52--dc23
                               2012044434

To my wife, Shifra,
for her love, companionship, and support.

CONTENTS

PREFACE

Ever since Lorensen and Cline published their 1987 paper on the MARCHING CUBES algorithm, isosurfaces have been a standard technique for visualization of three-dimensional volumetric data. Nevertheless, there is no book specifically devoted to isosurfaces. Part of this is because of the elegance and simplicity of the MARCHING CUBES algorithm, which can easily be described in a few pages. Yet, extensive work has been done since 1987 on extensions and variations of the MARCHING CUBES algorithm, on other algorithms for isosurface construction, on isosurface simplification, and on isosurface topology.

This book is my attempt to give a clear presentation of the basic algorithms for isosurface construction. It is also an attempt at a more rigorous, mathematical perspective for some of the algorithms and results. My targeted audience is designers of visualization software who would like an organized overview of the various algorithms associated with isosurfaces; graduate students pursuing research in visualization who need a solid introduction to research in the areas; and visualization researchers for whom this can serve as a reference for the vast amount of literature on isosurfaces.

The mathematical proofs in this book are more rigorous and challenging than one might see in a typical graphics or visualization text. Despite the many readers who will skip the proofs, I have included them because they are "guarantors" of the correctness of the claims about the various algorithms. Starting with the paper by Lorensen and Cline, numerous papers on isosurfaces contain erroneous, obscure, or unsubstantiated claims. The proofs in this book are an attempt to remedy this deficiency. I have tried to place the proofs in separate sections so that readers who wish to avoid them can easily do so.

Of course, it is possible (and probable) that some of the claims and/or proofs in this book are incorrect. Providing the proofs will hopefully help others uncover and correct any erroneous claims.

Because some readers will be interested only in a subset of the topics in this book, I have tried to make the chapters as self-contained as possible. Unfortunately, this resulted in some redundancy in the text, for which I apologize.

Everyone should read Chapters 1 and 2, the introduction and the MARCHING CUBES algorithm. Chapters 5, 6, and 7 on isosurface patch construction, four-dimensional isosurfaces, and interval volumes are related and should be read in order. Chapter 9 on multiresolution tetrahedral meshes should be read before Section 10.1 on multiresolution convex polyhedral meshes. Chapter 3 on dual contouring should be read before Section 10.2 on multiresolution surface nets. The other chapters are relatively independent and can be read independently.

ACKNOWLEDGMENTS

I am indebted to many people for their support and assistance. First and foremost is my colleague Tamal Dey, who continually challenged me to apply rigorous methods to geometric modeling. He is a source of inspiration and a role model for research excellence. Roger Crawfis initiated my interest in isosurfaces during a graduate seminar on isosurfaces. That seminar led to a joint paper on generalizing the Marching Cubes algorithm to four dimensions and my ongoing interest and research in isosurfaces. Colleagues and collaborators, Han-Wei Shen and Yusu Wang, were also sources of encouragement and inspiration.

A special thanks to Josh Levine who carefully reviewed this book and suggested numerous corrections and improvements. Thanks also to my former and current students, Ramakrishnan Khaziyur-Mannar, Marc Khoury, and Arindam Bhattacharya. Thanks to Hamish Carr at University of Leeds and Carlos Scheidegger from ATT Labs for numerous conversations about isosurfaces and scalar data sets. Thanks also for support from the National Science Foundation.

Many of the images in this text were produced from data sets compiled by Michael Meissner at www.volvis.org and the data sets at the Volume Library, www.stereofx.org, compiled by Stefan Roettger. These data sets are an invaluable resource for research in volume graphics.

Finally, thanks to my wife, Shifra. Without her encouragement and support, I would never have completed this book.

CHAPTER 1

INTRODUCTION

1.1 What Are Isosurfaces?

A scalar field is a function ϕ which assigns a scalar value (a real number) to each point in $\mathbb{R}^d$. The value d is known as the dimension of the scalar field. Examples of three-dimensional scalar fields are densities, pressures, or temperatures associated with points in $\mathbb{R}^3$. If these values change with time, then the addition of time as a fourth dimension gives a four-dimensional scalar field.

Given a scalar field $\phi : \mathbb{R}^d \to \mathbb{R}$ and a constant $\sigma \in \mathbb{R}$, the set $\{x : \phi(x) = \sigma\}$ is called a level set[1] of ϕ. We use the notation $\phi^{-1}(\sigma)$ to represent the level set $\{x : \phi(x) = \sigma\}$. If ϕ is a continuous function, then the level set $\phi^{-1}(\sigma)$ separates $\mathbb{R}^d$ into two sets of points, those with scalar value above σ and those with scalar value below σ.

In two dimensions ($d = 2$), level sets are called isocontours or contour lines. Contour lines in topographic maps are a familiar example of isocontours. Each contour line on a topographic map represents a specific elevation. Walking along the contour line means walking along a level path that does not change elevation. Crossing contour lines means climbing up or down and changing elevations.

In three dimensions ($d = 3$), level sets are also called implicit surfaces or isosurfaces. In computer graphics, the term *implicit surface* is generally used to refer to surfaces defined by explicitly providing a function ϕ. Problems include rendering such a surface, converting the implicit representation to a parameterized one, and computing intersections of implicit surfaces.

---

[1]This mathematical formulation of level sets should not be confused with the level set method for segmentation. The level set method defines a continuous, smooth function g based on the input data and then uses the level sets from this function to segment the data.

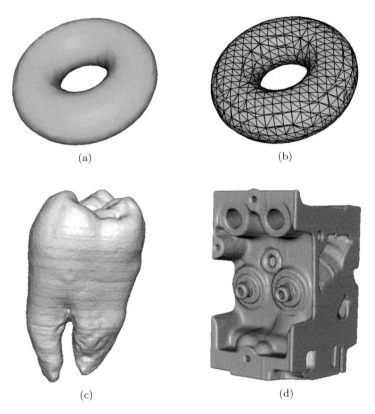

(a)   (b)

(c)   (d)

Figure 1.1. (a) Isosurface (isovalue 3) forming a torus. Scalar data set is a $20 \times 20 \times 20$ regular grid with origin $(0, 0, 0)$ measuring the distance to a circle with radius 6 centered at $(9.5, 9.5, 9.5)$. (b) Torus isosurface edges. (c) Isosurface (isovalue 600) of a micro CT scan of a tooth using a GE Industrial Micro CT scanner. Data set created by GE Aircraft Engines. (d) Isosurface (isovalue 80) of CT scan of an engine block. Data set created by General Electric.

Isosurface is the term for level sets used in volume visualization. Generally, it refers to a surface constructed from a finite set of input points, each associated with a scalar value. (See Figure 1.1.) This set of input points is a sampling of some continuous function ϕ and the isosurface is an approximation of the level set of ϕ. Of course, numerous functions take on the same value at a finite set of sample points, so the function ϕ and the isosurface are not uniquely defined. In addition, sample data often contains noise and so is not even a precise representation of ϕ at the sample points. Finally, the very idea that the input data represents a sampling of some continuous function ϕ is itself a modeling assumption and may be misleading.

Unfortunately, the term *isosurface* is sometimes used to represent the level set of a function ϕ and at other times is used to represent an approximation to a level set. In this book, we will always use level set to refer to the mathematically defined set $\phi^{-1}(\sigma)$. We use the term isosurface to refer to an approximation to a level set $\phi^{-1}(\sigma)$ where function ϕ is represented by a finite set of sample points. The value σ is called the isovalue of the isosurface.

1.2   Applications of Isosurfaces

Two well-established procedures in medical imaging produce extensive scalar field data. Computerized tomography (CT) scanners send beams of radiation through a person and measure the amount of radiation that arrives at various detectors. The radiation measurements are processed to produce a (radiation) density at various sample points within the person. Magnetic resonance imaging (MRI) scanners measure changes in a magnetic field caused by excited hydrogen nuclei in water. Mathematical transformations map these measurements to water density values at sample points within the person. The CT and MRI density measurements implicitly represent a scalar density field on the scanned person with each point associated with a density. Since CT and MRI scans are measuring different material properties, they have different relative strengths and weaknesses. CT scans excel at imaging solid organs while MRI scans are better at imaging subtle differences in soft tissue.

The output of a CT or MRI scan is simply a set of values associated with sample points, usually on a regular grid. Regions within this data represent individual objects such as skin, muscle, or bones or pathologies such as tumors, hemorrhages, or bowel obstructions. There are two approaches to visualizing objects within this data. One, called direct volume rendering, is to cast imaginary rays from a specified eye location through the data and integrate a color along the rays based on the density values. A transfer function determines how the color is constructed from the density values. By varying the transfer function, the user can view or highlight different objects within the data. Direct volume rendering can produce excellent images, but it is computationally expensive and produces only a visual image of a specific view of the data. In addition, the transfer function is difficult to set and adjust.

The other approach to visualizing data is to produce surfaces representing the boundaries of objects within the data. This approach is called surface reconstruction. Once such surfaces are produced, they can be rendered from any viewpoint using standard computer graphics techniques. Moreover, the surfaces model the object boundaries and can be used to measure object volume and surface area. The most direct way to produce surfaces from volumetric data is to construct an isosurface that approximates the level set of a scalar field implicitly represented by the data.

Computational fluid dynamics also produces extensive scalar field data. In computational fluid dynamics, the flow space is partitioned into small elements (polyhedra). Each element has a flow density that is derived by solving a set of finite difference equations. The flow density of an element can be thought of as the density of some point within the element, perhaps the barycenter. Usually this density varies with time. The objects of interest in fluid flow are high or low pressure regions, perhaps representing shock waves or turbulence. Again, either direct volume rendering or surface reconstruction can be used to visualize such regions at a fixed time.

1.3   Isosurface Properties

As previously mentioned, we use the term *isosurface* to refer to an approximation of a level set. There are infinitely many approximations to a level set. What properties are required or desired in such an approximation?

One obvious property is that the isosurface should be a surface. However, this is not as simple as it seems. For example, the level set is not necessarily a surface: the level set of the constant function, $\phi(x, y, z) = \sigma$, is all of $\mathbb{R}^3$ for isovalue σ and the empty set for all other isovalues. If $g : \mathbb{R}^3 \to \mathbb{R}$ is the distance from (x, y, z) to the origin, then the level set of isovalue 0 is a point.

Another problem is what exactly is meant by surface. Consider the union of two unit spheres in $\mathbb{R}^3$, one lying above and one below the $x - y$ plane, such that the two spheres touch at the origin. Is the union of these two spheres a surface? The union of two spheres tangent at the origin separates points inside the spheres from points outside the spheres. On the other hand, in the neighborhood of the origin this set of points looks like two surfaces glued together at a single point. In technical terms, the union of two spheres is not a 2-manifold. Should the isosurface be a 2-manifold?

An isosurface is an approximation to a level set of a continuous scalar field ϕ. However, only a finite sampling P of ϕ is given. There are numerous scalar fields ϕ with drastically different geometry and topology that have the same scalar values on P. These different scalar fields can give rise to very different isosurfaces. How do we choose among such isosurfaces?

One assumption we will make is that function ϕ is continuous. Under this assumption, it is possible to at least identify some line segments that are intersected by the level set $\phi^{-1}(\sigma)$.

Let p and p' be points in P where p has scalar values above $\sigma \in \mathbb{R}$ and p' has scalar values below σ. For any continuous field ϕ, the level set $\phi^{-1}(\sigma)$ intersects line segment (p, p'). Thus, any isosurface approximation of such a level set should intersect line segment (p, p').

On the other hand, if p and p' both have scalar values above or both have scalar values below σ, then the level set $\phi^{-1}(\sigma)$ may or may not intersect line segment (p, p'). In general, the isosurface should not intersect such an edge.

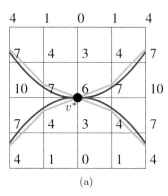

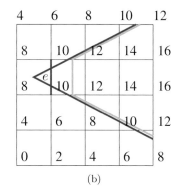

(a)   (b)

Figure 1.2. (a) Scalar grid sampling the function $\phi_a(x, y) = (x - 2)^2 - 3|y - 2| + 6$, the red level set $\phi_a^{-1}(6)$ and the green isocontour with isovalue 6. Four branches of the level set and four edges of the isocontour meet at the grid center, v^*. (b) Scalar grid sampling the function $\phi_b(x, y) = 2x - |4y - 10| + 10$, the red level set $\phi_b^{-1}(11)$ and the green isocontour with isovalue 11. The red level set intersects the blue grid edge e twice.

The boundary of many objects, particularly manufactured ones, is often piecewise smooth but with sharp edges or corners connecting the pieces. An isosurface representing such a boundary should not smooth over such edges or corners.

We summarize some of the desirable properties of an isosurface:

1. It separates sample points with scalar value above isovalue from scalar points with value below isovalue.

2. It does not intersect a grid edge more than once.

3. It does not intersect grid edges with both endpoint scalar values above or both endpoint scalar values below the isovalue.

4. It is a manifold.

5. It represents sharp edges and corners.

Not all of these properties are always desirable. For instance, Figure 1.2(a) displays a scalar grid sampling the function $\phi_a(x, y) = (x - 2)^2 - 3|y - 2| + 6$. The vertex v^* at the grid center has scalar value 6, so any isocontour with isovalue 6 should pass through this vertex. Each of the four squares surrounding v^* contains a grid edge with scalar values of 4 and 7 so the isocontour passes through each such grid edge. The result is that four isocontour edges meet at v^*, and the isocontour is not a manifold. However, the isocontour does faithfully represent the topology of $\phi_a^{-1}(6)$ which has four curves meeting at v^*.

Some of the properties listed above are mutually exclusive. Figure 1.2(b) displays a scalar grid sampling the function $\phi_b(x, y) = 2x - |4y - 10| + 10$. The red level set $\phi_b^{-1}(11)$ has a sharp corner at (0.5,2.5) and intersects the blue grid edge e twice. The green isocontour does not properly represent the sharp corner at (0.5,2.5) and does not intersect grid edge e. Any isocontour that reproduces the sharp corner, satisfying Property 5, would intersect grid edge e twice, violating Properties 2 and 3.

1.4   Isosurface Construction

There are four basic approaches to isosurface construction. The first and earliest approach is to partition volumetric data into two-dimensional (2D) slices, construct isocontours in each slice, and then "stitch" together the slices using triangles. This approach mimics the way early radiologists used CT and MRI data by examining slices of the data. The difficulty is in the stitching, which is both time-consuming and error-prone. This approach has been superseded by volumetric methods, which construct the isosurface directly in 3-space.

The second approach is to partition space into cubes and associate each cube with a scalar value. The isosurface is the boundary of all cubes with scalar values below a given value. This approach was motivated by pixel graphics, which represents images as a collection of square pixels. The obvious drawback is that the boundary of a set of cubes is extremely nonsmooth, with faces meeting at ninety-degree angles. In visualization, this problem can be mitigated by rendering the surface using "phony" surface normals constructed from the original data. Alternatively, smoothing techniques can be applied to the choppy surface but with potential loss of some detail.

The third and most popular approach is the MARCHING CUBES algorithm and its variants introduced by Lorensen and Cline [Lorensen and Cline, 1987a] in 1987. The MARCHING CUBES algorithm partitions the volume into cubes and then independently constructs surface patches within each cube. Each patch is a small triangulated surface with a boundary on the cube. Based on a comparison of the scalar values of the cube corners and the isovalue, a cube is classified into one of 256 cases. The surface patches are constructed using a precomputed table based on these 256 cases.

The original MARCHING CUBES algorithm sometimes created cube patches that did not properly meet the patches of adjacent cubes. A number of solutions were proposed, the simplest being a change to the precomputed table of 256 cases.

Variants of the MARCHING CUBES algorithm include using tetrahedra instead of cubes and extending the algorithm to higher dimensions.

The last and most recent approach is called dual contouring. The volume is partitioned into cubes and each cube is replaced by a single point. Points in adjacent cubes are connected to form a surface using quadrilaterals that are the dual of cube edges. Dual contouring has the nice property of producing surfaces that are tiled by quadrilaterals, not triangles. It can also be easily used with multiresolution techniques where the volume partitioning may not be uniform. On the other hand, the surfaces produced by dual contouring are usually not manifolds.

1.5   Limitations of Isosurfaces

Using isosurfaces to model object boundaries from volumetric data has some significant advantages. Isosurfaces encode basic, simple structures of the scalar field sampled in the input data. They are easy to define and understand. They correspond to a formal mathematical object, the level set of a scalar field, and so lend themselves to rigorous mathematical analysis. They can be constructed in time proportional to the size of the input data (linear time).

Unfortunately, isosurfaces have some significant deficiencies and limitations as models for object boundaries. These deficiencies are caused by problems of sampling and noise and by the lack of any global criterion in the isosurface definition. We list some below:

1. undersampling of the spatial domain,

2. high-frequency noise,

3. low-frequency noise,

4. overspecification of the scalar values,

5. lack of smoothness criterion,

6. choice of isosurface,

7. lack of global information,

8. lack of a priori information.

Undersampling and high-frequency noise generate adjacent samples with large variations in scalar value. These scalar variations create surfaces with complicated geometric and topological features that are not representative of the object. In regions where scalar values are constant or near-constant, using scalar values with precision beyond the range of the scanner creates isosurfaces which wind arbitrarily through the regions. Without any smoothness criterion, isosurfaces have no restrictions on their susceptibility to undersampling and noise, even though most objects are best represented by some smooth or piecewise smooth boundaries.

Applying smoothing and noise reduction filters to the raw data helps mitigate some of the problems described above but at the expense of losing some of the fine isosurface features and nonsmooth features that may be present in the data. On the other hand, one of the benefits of isosurfaces is their faithfulness to the data, including all the irregularities and noise in the data. The trade-off between smooth filtering versus exact data representation is data- and application-dependent and is best left to the individual researcher or clinician.

Low-frequency noise produces shifts in scalar values in different regions of the data. The boundary of the object or objects of interest may have different scalar values in different regions of the data. One isosurface will capture the objects in one region while a different isosurface with a different isovalue will bound the objects in the other region. Between the two regions, an isosurface may give object fragments, representing portions of the object. Normalizing the data across regions by adjusting scalar values may help, but it creates the danger of introducing normalization errors.

Isosurfaces depend upon a single parameter, the isovalue of the points on the isosurface. Choosing this parameter is itself a challenging task. Both visualization and data analysis tools exist to help in finding interesting or relevant isovalues.

Isosurfaces are intrinsically local with no global criteria about their shape or structure. In almost all applications such global criteria do exist and are known to researchers or clinicians. On the other hand, because isosurfaces make no application or data-specific assumptions, they are versatile structures that can be used in almost any geometric application. They are a basic tool for anyone visualizing or modeling data but only as the building blocks for more sophisticated data-specific tools.

1.6   Multivalued Functions and Vector Fields

Many applications produce more than a single scalar value at each point. The simplest example is color images that have an RGB (red, green, blue) value associated with each pixel. In fluid flow simulation, both a pressure and temperature could be associated with sample points in the flow. Combinations of scans from different instruments, such as a CT scan and an MRI scan of the same individual, can produce a radiation density and a water density at each sample point.

Visualizing and modeling multivalued data is much more difficult than analyzing scalar fields. Sometimes multiple values are combined into a single scalar value at each point producing a single scalar field. Isosurfaces can then be used to visualize and model objects in that scalar field. The resulting surface is highly sensitive to the function used to create the scalar field from the multivalued functions.

Vector fields are multivalued functions that map $\mathbb{R}^d$ to $\mathbb{R}^d$. In fluid flow simulation, they can represent direction and speed of the flow. Critical points in a vector field are points that are assigned the zero vector, $(0, 0, \ldots, 0)$. Visualization and modeling of vector fields usually relies upon identification of critical points and representation of the flow between critical points.

Various transformations can be used to transform a vector field into a scalar one—for instance, by replacing each vector by its length. Such transformations are usually too crude to extract all but the most rudimentary information.

1.7   Definitions and Basic Techniques

Before discussing isosurface construction, we need to review some basic definitions and techniques that are used throughout this book.

1.7.1   Definitions

Regular scalar grid. Isosurface construction algorithms take as input a sample set of points. This sample set is often represented by a regular grid.

In two dimensions, a regular grid is a partition of a large rectangle into small congruent rectangles. More generally, a regular grid in $\mathbb{R}^d$ is a partition of a large hyperrectangle into small congruent hyperrectangles. (See Figure 1.3.) The vertices and edges of the regular grid are the vertices and edges of the small hyperrectangles. A typical example of a regular grid is the partition of the region $[0, m_1] \times [0, m_2] \times [0, m_3]$ into $m_1 \times m_2 \times m_3$ cubes. Note that along each axis d this regular grid has m_d edges and $(m_d + 1)$ vertices. The grid has $(m_1 + 1) \times (m_2 + 1) \times (m_3 + 1)$ vertices.

(a) A 2D regular grid.

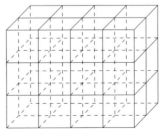

(b) A 3D regular grid.

Figure 1.3. (a) A 2D regular grid with vertex dimensions 5×4 and cube dimensions 4×3. (b) A 3D regular grid with vertex dimensions $5 \times 4 \times 3$ and cube dimensions $4 \times 3 \times 2$.

The vertex dimensions of a regular grid is the number of vertices along each axis. A regular grid of cubes with vertex dimensions $n_1 \times n_2 \times n_3$ has n_d vertices along each axis, $n_1 \times n_2 \times n_3$ vertices, and $(n_1 - 1) \times (n_2 - 1) \times (n_3 - 1)$ cubes.

The cube dimensions of a regular grid is the number of edges along each axis. A regular grid of cubes with cube dimensions $m_1 \times m_2 \times m_3$ has m_d edges along each axis, $(m_1 + 1) \times (m_2 + 1) \times (m_3 + 1)$ vertices, and $m_1 \times m_2 \times m_3$ cubes. A regular grid with cube dimensions $m_1 \times m_2 \times m_3$ has vertex dimensions $(m_1 + 1) \times (m_2 + 1) \times (m_3 + 1)$.

Unless otherwise noted, the dimensions of a grid refers to its vertex dimensions. Thus, an $n_1 \times n_2 \times n_3$ regular grid has vertex dimensions $n_1 \times n_2 \times n_3$ and cube dimensions $(n_1 - 1) \times (n_2 - 1) \times (n_3 - 1)$.

A regular scalar grid is a regular grid where each grid vertex v_i is associated with a scalar value $s_i \in \mathbb{R}$. A simple example is a grayscale image—for instance, a black-and-white picture.[2] The sample points are the pixel centers. The scalar value at each point is the grayscale value of the pixel containing the point.

Triangulation. Isosurfaces are often triangulations, sets of triangles or simplices with appropriate intersection conditions.

Definition 1.1. A triangulation τ is a set of simplices such that for every pair of simplices $\mathbf{t}, \mathbf{t}' \in \tau$, the intersection $\mathbf{t} \cap \mathbf{t}'$ is either empty or a face of each simplex.

For instance, if triangulation τ is a set of triangles, then the intersection $\mathbf{t} \cap \mathbf{t}'$ is either empty, a common vertex of $\mathbf{t}$ and $\mathbf{t}'$, or a common edge of $\mathbf{t}$ and $\mathbf{t}'$. (See Figure 1.4.)

Mathematics texts usually add a formal requirement that if simplex $\mathbf{t}$ is in τ, then every face of $\mathbf{t}$ is in τ. See Appendix B.4 for further discussion and definitions.

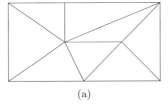
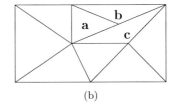

(a)   (b)

Figure 1.4. (a) A triangulation of a rectangle. (b) A partition of a rectangle into triangles, which is not a triangulation. The intersection of triangles **a** and **c** is a line segment that is not an edge of **c**. The intersection of triangles **b** and **c** is also not an edge of **c**.

---

[2]The term *black-and-white* is a bit misleading since black-and-white pictures generally contain all different shades of gray.

The notation $|\tau|$ represents the set of all points in all triangles of τ, i.e., $|\tau| = \bigcup_{\mathbf{t} \in \tau} \mathbf{t}$.

Definition 1.2. A set $\mathbb{X} \subseteq \mathbb{R}^d$ is piecewise linear if $\mathbb{X}$ equals $|\tau|$ for some triangulation τ.

Convex Polyhedral Mesh. In many instances, a scalar field is represented not by a regular scalar grid but by a mesh composed of triangles or convex polyhedra.

Definition 1.3. A convex polyhedral mesh Γ is a set of convex polyhedra in $\mathbb{R}^3$ such that for every pair of convex polyhedra $\mathbf{c}, \mathbf{c}' \in \Gamma$, the intersection $\mathbf{c} \cap \mathbf{c}'$ is either empty or a face of each convex polyhedron.

Mathematics texts usually add a formal requirement that if convex polyhedron $\mathbf{c}$ is in Γ, then every face of $\mathbf{c}$ is in Γ.

The notation $|\Gamma|$ represents the set of all points in all elements of Γ, i.e., $|\Gamma| = \bigcup_{\mathbf{c} \in \Gamma} \mathbf{c}$.

A tetrahedral mesh is a convex polyhedral mesh where every mesh element is a tetrahedron. A scalar mesh is a mesh where each mesh vertex v_i is associated with a scalar value $s_i \in \mathbb{R}$.

The generalization of a convex polyhedral mesh to $\mathbb{R}^d$ is called a convex polytopal mesh. The definition is given in Appendix B.5. A convex polytopal mesh where every mesh element is a simplex is a triangulation. It is also sometimes called a simplicial mesh.

Orientation. Let L be a line segment L with vertices $\{v_0, v_1\}$. The orientation of L is an ordering of the vertices of L, either (v_0, v_1) or (v_1, v_0).

Let $\mathbf{t}$ be a triangle with vertices $\{v_0, v_1, v_2\}$. The orientation of $\mathbf{t}$ is a cyclic order of the vertices of $\mathbf{t}$. (See Figure 1.5(a).) There are two possible cyclic

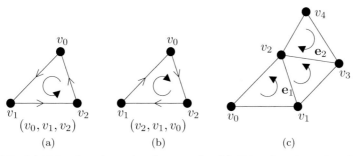

Figure 1.5. (a) Triangle orientation (v_0, v_1, v_2). (b) Triangle orientation (v_2, v_1, v_0). (c) Triangle orientations (v_0, v_1, v_2), (v_1, v_3, v_2), and (v_4, v_3, v_2). Orientations (v_0, v_1, v_2) and (v_1, v_3, v_2) are consistent. Orientation (v_0, v_1, v_2) induces the orientation (v_1, v_2) on edge $\mathbf{e}_1$ while (v_1, v_3, v_2) induces the opposite orientation (v_2, v_1) on edge $\mathbf{e}_1$. Orientations (v_1, v_3, v_2) and (v_4, v_3, v_2) are not consistent. Both orientations induce the same orientation (v_3, v_2) on edge $\mathbf{e}_2$.

orders, either (v_0, v_1, v_2) or (v_2, v_1, v_0). The sequences (v_1, v_2, v_0) and (v_2, v_0, v_1) represent the same cyclic order as (v_0, v_1, v_2). Only the starting vertex has changed. Similarly, the sequences (v_1, v_0, v_2) and (v_0, v_2, v_1) represent the same cyclic order as (v_2, v_1, v_0).

The cyclic order (v_0, v_1, v_2) induces orientations, (v_0, v_1), (v_1, v_2), and (v_2, v_0) of the edges of **t**. The reverse cyclic order (v_2, v_1, v_0) induces opposite orientations (v_1, v_0), (v_2, v_1), and (v_0, v_2) of the edges of **t**. Two oriented triangles, **t**$_1$ and **t**$_2$, which share an edge **e** have consistent orientations if the orientation of **e** induced by **t**$_1$ is the opposite of the orientation of **e** induced by **t**$_2$. (See Figure 1.5(c).)

The orientation (v_0, v_1, v_2) of a triangle **t** in $\mathbb{R}^3$ determines the vector

$$
\begin{aligned}
u & = & (v_1 - v_0) \times (v_2 - v_0) \\
& = & v_1 \times v_2 - v_0 \times v_2 - v_1 \times v_0 \\
& = & v_0 \times v_1 + v_1 \times v_2 + v_2 \times v_0,
\end{aligned}
$$

where $\times$ is the cross product. Vector u is orthogonal to **t**. The sequence (v_1, v_2, v_0) determines the vector $(v_2 - v_1) \times (v_0 - v_1) = v_0 \times v_1 + v_1 \times v_2 + v_2 \times v_0$ that is u. Similarly, (v_2, v_0, v_1) determines the vector $(v_0 - v_2) \times (v_1 - v_2)$, which equals u. Thus, the vector u is independent of the representation of the cycle (v_0, v_1, v_2). The orientation (v_2, v_1, v_0) determines the vector

$$
\begin{aligned}
(v_1 - v_2) \times (v_0 - v_2) & = & v_1 \times v_0 - v_1 \times v_2 - v_2 \times v_0 \\
& = & v_1 \times v_0 + v_2 \times v_1 + v_0 \times v_2 \\
& = & -u.
\end{aligned}
$$

Thus, the two orientations of **t** determine two opposite vectors, u and $-u$, which are both orthogonal to **t**.

In computer graphics, triangle orientations are used to determine the front and back faces of triangles. Triangle shading is often dependent on whether the viewer is seeing a front or back face. Thus, it is important that any two triangles that share a common edge have consistent orientations.

Orientations are defined for higher dimensional simplices, where they are also represented by sequences of simplex vertices. The orientation of a $(d-1)$-simplex in $\mathbb{R}^d$ determines a unique vector u orthogonal to the simplex. The opposite orientation determines the vector $-u$. Appendix B.6 contains the definition and discussion of orientations in higher dimensional simplices.

Separation. An important property of isosurfaces is that they "separate" those points with scalar value above the isovalue from those points with scalar value below the isovalue [Nielson et al., 2003]. We give the following formal definition of this concept.

Let $\mathbb{X}$ and $\mathbb{Y}$ be sets of points in $\mathbb{R}^d$. We first define what it means for $\mathbb{X}$ to separate two points in $\mathbb{Y}$.

Definition 1.4.

- Set $\mathbb{X}$ separates point $p \in \mathbb{Y}$ from point $q \in \mathbb{Y}$ if every path in $\mathbb{Y}$ connecting p to q intersects $\mathbb{X}$.

- Set $\mathbb{X}$ strictly separates p from q if $\mathbb{X}$ separates p from q and neither p nor q is in $\mathbb{X}$.

We next define what it means for $\mathbb{X}$ to separate two subsets of $\mathbb{Y}$.

Definition 1.5.

- Set $\mathbb{X}$ separates $\mathbb{Y}_1 \subseteq \mathbb{Y}$ from $\mathbb{Y}_2 \subseteq \mathbb{Y}$ if $\mathbb{X}$ separates every $p \in \mathbb{Y}_2$ from every $q \in \mathbb{Y}_2$.

- Set $\mathbb{X}$ strictly separates $\mathbb{Y}_1 \subseteq \mathbb{Y}$ from $\mathbb{Y}_2 \subseteq \mathbb{Y}$ if $\mathbb{X}$ separates $\mathbb{Y}_1$ from $\mathbb{Y}_2$ and $\mathbb{X}$ does not intersect $\mathbb{Y}_1$ or $\mathbb{Y}_2$.

(See Appendix B.9 for further discussion of separation and its properties.)

Manifolds. A manifold is a mathematical formalization of the intuitive concept of a surface.

Let $\mathbb{B}^k$ be the k-dimensional open ball with radius one centered at the origin. Ball $\mathbb{B}^1$ is an open line segment and $\mathbb{B}^2$ is an open disk. A k-dimensional manifold (k-manifold) is a set of points that locally resembles $\mathbb{B}^k$. Examples of 1-manifolds are circles or simple, closed curves. Every point of a 1-manifold has a small neighborhood that is topologically equivalent to an open line segment ($\mathbb{B}^1$). Examples of 2-manifolds are spheres, tori, or double tori. Every point of a 2-manifold has a small neighborhood that is topologically equivalent to an open disk ($\mathbb{B}^2$).

Let $\mathbb{B}^{k+}$ be the intersection of the open ball $\mathbb{B}^k$ and the closed half-space $\{(x_1, \ldots, x_k) : x_k \geq 0\}$. Note that $\mathbb{B}^{k+}$ is neither closed nor open. $\mathbb{B}^{1+}$ is a line segment open at one endpoint and closed at the other. $\mathbb{B}^{2+}$ is a half-disk, open along the disk and closed at the bounding edge. $\mathbb{B}^{3+}$ is a half-sphere, open along the sphere and closed at the bounding disk. A k-dimensional manifold with boundary (k-manifold with boundary) is a set of points which locally resembles either $\mathbb{B}^k$ or $\mathbb{B}^{k+}$. Examples of 1-manifolds with boundary are line segments or simple curves with two endpoints. Examples of 2-manifolds with boundary are disks or convex polygons (including the polygon interior.) Examples of 3-manifolds with boundary are closed balls or cubes (including the cube interior.) For more precise definitions of k-manifold and k-manifold with boundary, see Appendix B.3.

Piecewise linear manifold. A k-manifold (possibly with boundary) is piecewise linear if it is the union of a set of k-simplices that form a triangulation of the manifold. A piecewise linear manifold is orientable if every simplex in the manifold can be assigned an orientation and these orientations are consistent. The orientation of the manifold is the orientation of all its simplices. If a piecewise linear manifold is connected and orientable, then assigning an orientation to one simplex fixes the orientations of all the other manifold simplices.

1.7.2   Linear Interpolation

A basic step in MARCHING CUBES and its variants is approximating the intersection of a level set and a line segment. These algorithms use linear interpolation to find a point on the line segment that approximates the intersection.

Let $\phi : \mathbb{R}^d \to \mathbb{R}$ be a scalar field and let $\sigma \in \mathbb{R}$ be an isovalue defining the level set $\phi^{-1}(\sigma)$. Given two grid vertices p and q where $\phi(p) \neq \phi(q)$, if σ is between $\phi(p)$ and $\phi(q)$, then the level set intersects line segment $[p, q]$. We wish to approximate the intersection of $\phi^{-1}(\sigma)$ and line segment $[p, q]$. We do so by defining a linear function $\widehat{\phi}$ based on the two scalar values $\phi(p)$ and $\phi(q)$ and calculating the point $r \in [p, q]$ where $\widehat{\phi}(r) = \sigma$.

Every point on line segment $[p, q]$ can be described as a linear combination of p and q. More specifically, every point on line segment $[p, q]$ equals $(1 - \alpha)p + \alpha q$ for some α where $0 \leq \alpha \leq 1$. For example, in $\mathbb{R}^3$ where p equals (p_x, p_y, p_z) and q equals (q_x, q_y, q_z), the linear combination is

$$((1 - \alpha)p_x + \alpha q_x, \ (1 - \alpha)p_y + \alpha q_y, \ (1 - \alpha)p_z + \alpha q_z).$$

Define $\widehat{\phi} : [p, q] \to \mathbb{R}$ by

$$\widehat{\phi}((1 - \alpha)p + \alpha q) = (1 - \alpha)\phi(p) + \alpha\phi(q).$$

Note that $\widehat{\phi}(p) = \phi(p)$ $(\alpha = 0)$ and $\widehat{\phi}(q) = \phi(q)$ $(\alpha = 1)$. Values of $\widehat{\phi}$ vary linearly with α.

We approximate the intersection of $\phi^{-1}(\sigma)$ with $[p, q]$ as the point r where $\widehat{\phi}(r)$ equals σ. Since r is on line segment $[p, q]$, point r equals $(1 - \alpha_r)p + \alpha_r q$ for some α_r. Thus,

$$\sigma = \widehat{\phi}(r) = \widehat{\phi}((1 - \alpha_r)p + \alpha_r q) = (1 - \alpha_r)\phi(p) + \alpha_r\phi(q).$$

Solving for α_r gives

$$\alpha_r = \frac{\sigma - \phi(p)}{\phi(q) - \phi(p)}.$$

Note that since $\phi(p) \neq \phi(q)$, the denominator $\phi(q) - \phi(p)$ is nonzero.

In $\mathbb{R}^3$, the equations for the coordinates of $r = (r_x, r_y, r_z)$ are

$$
\begin{aligned}
r_x &= (1 - \alpha_r)p_x + \alpha_r q_x, \\
r_y &= (1 - \alpha_r)p_y + \alpha_r q_y, \\
r_z &= (1 - \alpha_r)p_z + \alpha_r q_z.
\end{aligned}
$$

Input   : Points $p, q \in \mathbb{R}^d$, scalar values s_p, s_q, and an isovalue σ.
Requires   : $s_p \neq s_q$ and either $s_p \leq \sigma \leq s_q$ or $s_p \geq \sigma \geq s_q$.
Output   : Point r lying on $[p, q]$.

LinearInterpolation(p, s_p, q, s_q, σ)

1 $\alpha \leftarrow \frac{\sigma - s_p}{s_q - s_p}$;
2 for $i = 1$ to d do
3 $\quad | \quad r_i \leftarrow (1 - \alpha)p_i + \alpha q_i$;
4 end
5 return (r);

Algorithm 1.1. Linear interpolation.

More generally, in $\mathbb{R}^d$ the equations for the coordinates of $r = (r_1, r_2, \ldots, r_d)$ are

$$
\begin{aligned}
r_1 &= (1 - \alpha_r)p_1 + \alpha_r q_1, \\
r_2 &= (1 - \alpha_r)p_2 + \alpha_r q_2, \\
&\cdots \\
r_d &= (1 - \alpha_r)p_d + \alpha_r q_d.
\end{aligned}
$$

Pseudocode is given in Algorithm 1.1.

The assumption for all these algorithms is that $\phi(p)$ does not equal $\phi(q)$. If both $\phi(p)$ and $\phi(q)$ equal σ, then the level set contains both p and q and there is no way to approximate the intersection of $\phi^{-1}(\sigma)$ and $[p, q]$ by a single point. Where or whether the isosurface approximation intersects line segment $[p, q]$ is dependent upon the specific isosurface construction algorithm.

1.7.3   Mesh Representation

The output of surface reconstruction algorithms is a mesh consisting of a set of small, simple surface elements. Typical surface elements are triangles or quadrilaterals. In curve reconstruction, the elements are line segments, while in higher dimensions the elements are simplices, cubes or hypercubes.

A mesh is represented by a list of mesh vertices, $\mathcal{L}_1$, followed by a list, $\mathcal{L}_2$, of surface elements. The list $\mathcal{L}_1$ of mesh vertices contains the mesh vertex coordinates, the location of each mesh vertex in $\mathbb{R}^d$. This representation is called an indexed mesh or a face-vertex mesh.

The list $\mathcal{L}_2$ of surface elements contains the element vertices, the mesh vertices determining the element. For instance, triangles are specified by three vertices, while quadrilaterals are specified by four vertices in order around the

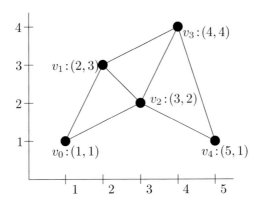

Figure 1.6. Triangle mesh. List of vertices (specified by vertex coordinates): $((1,1),(2,3),(3,2),(4,4),(5,1))$. List of triangles (specified by triangle vertices): $((v_0,v_2,v_1),(v_1,v_2,v_3),(v_2,v_4,v_3))$.

quadrilateral. Each mesh vertex stored in $\mathcal{L}_2$ is actually a reference to a mesh vertex in the list $\mathcal{L}_1$.

Figure 1.6 contains an example of a triangle mesh. The list $\mathcal{L}_1$ of mesh vertices for this mesh is $((1,1),(2,3),(3,2),(4,4),(5,1))$. The list $\mathcal{L}_2$ of mesh triangles is $((v_0,v_2,v_1),(v_1,v_2,v_3),(v_2,v_4,v_3))$.

CHAPTER 2

MARCHING CUBES AND VARIANTS

In the introduction, we mentioned four different approaches to isosurface construction. In this chapter, we describe one of those approaches to isosurface construction, the widely used MARCHING CUBES algorithm by Lorensen and Cline [Lorensen and Cline, 1987a].

The MARCHING CUBES algorithm is based on two ideas. First, the isosurface can be constructed piecewise within each cube of the grid without reference to other grid cubes. Second, the combinatorial structure of each isosurface patch in a grid cube can be retrieved from a lookup table. Since the main operation is retrieving this structure from the lookup table, the algorithm runs in time proportional to the number of grid cubes.

We first present a two-dimensional version of the algorithm, called MARCHING SQUARES, for constructing two-dimensional isocontours. Before discussing the MARCHING SQUARES algorithm, we define some terminology that will be used by the algorithms in this chapter.

2.1 Definitions

Given a regular scalar grid and an isovalue σ, it is convenient to assign "+" and "−" labels to each grid vertex based on the relationship between its scalar value and σ.

Definition 2.1.

- A grid vertex is positive, "+", if its scalar value is greater than or equal to σ.

- A grid vertex is negative, "−", if its scalar value is less than σ.

- A positive vertex is strictly positive if its scalar value does not equal σ.

Since the scalar value of a negative vertex never equals the isovalue, there is no point in defining a similar "strictly negative" term.

Grid edges can be characterized by the labels at their endpoints.

Definition 2.2.

- A grid edge is positive if both its endpoints are positive.

- A grid edge is negative if both its endpoints are negative.

- A positive grid edge is strictly positive if both its endpoints are strictly positive.

- A grid edge is bipolar if one endpoint is positive and one endpoint is negative.

Note that a grid vertex or edge is only positive or negative in relationship to some isovalue.

The definitions given above apply not just to regular scalar grids but also to curvilinear grids. They also apply to the vertices and edges of polyhedral meshes such as tetrahedral and simplicial meshes.

2.2   Marching Squares

2.2.1   Algorithm

Input to the MARCHING SQUARES algorithm is an isovalue and a set of scalar values at the vertices of a two-dimensional regular grid. The algorithm has three steps. (See Figure 2.1.) Read in the isocontour lookup table from a pre-constructed data file. For each square, retrieve from the lookup table a set of

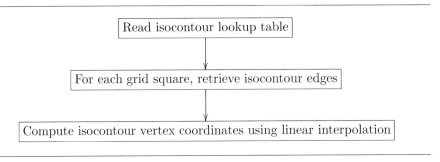

Figure 2.1. MARCHING SQUARES.

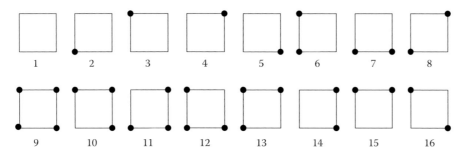

Figure 2.2. Square configurations. Black vertices are positive.

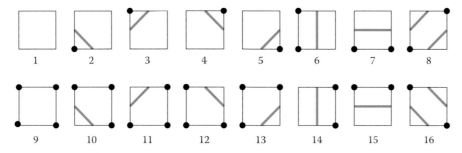

Figure 2.3. Square isocontours. Configurations 1 and 9 have no isocontour. Isocontours for configurations 2–7 and 10–15 are single line segments. Isocontours for configurations 8 and 16 are two line segments.

isocontour edges representing the combinatorial structure of the isocontour. The endpoints of these edges form the isocontour vertices. Assign geometric locations to the isocontour vertices based on the scalar values at the square edge endpoints. We explain the last two steps of the algorithm next.

Each grid vertex is labeled positive or negative as described in Section 2.1. (See Figure 2.4(b) for an example.) Since a square has four vertices, there are $2^4 = 16$ different configurations of square vertex labels. These configurations are listed in Figure 2.2.

The combinatorial structure of the isocontour within each square is determined from the configuration of the square's vertex labels. In order to separate the positive vertices from the negative ones, the isocontour must intersect any square edge that has one positive and one negative endpoint. An isocontour that intersects a minimal number of grid edges will not intersect any square edge whose endpoints are both strictly positive or whose endpoints are both negative.

For each square configuration κ, let $E_\kappa^{+/-}$ be the set of bipolar edges. Note that the size of $E_\kappa^{+/-}$ is either zero, two, or four. Pair the edges of $E_\kappa^{+/-}$. Each such pair represents an isocontour edge with endpoints on the two elements of the pair. Figure 2.3 contains the sixteen square configurations and their

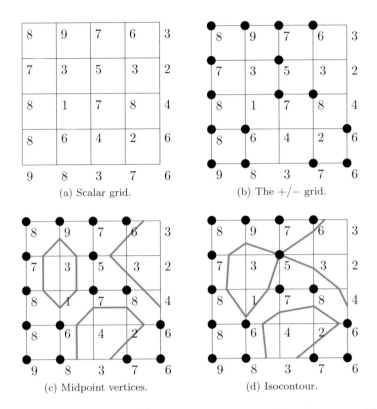

Figure 2.4. (a) 2D scalar grid. (b) Black vertices are positive. Vertex v with scalar value s_v is positive if $s_v >= 5$ and negative if $s_v < 5$. Note that $s_v = 5$ for one grid vertex v. (c) Isocontour with vertices at edge midpoints (before linear interpolation). (d) Isocontour with isovalue 5.

isocontours. The isocontour lookup table, Table, contains sixteen entries, one for each configuration. Each entry, Table[κ] is a list of the $E_\kappa^{+/-}$ pairs.

In Figure 2.3 the isocontour edges are drawn connecting the midpoints of each square edge. This is for illustration purposes only. The geometric locations of the isocontour vertices are not defined by the lookup table.

The isocontour lookup table is constructed on the unit square with vertices $(0,0), (1,0), (0,1), (1,1)$. To construct the isocontour in grid square (i,j), we have to map pairs of unit square edges to pairs of square (i,j) edges. Each vertex $v = (v_x, v_y)$ of the unit square maps to $v + (i,j) = (v_x, v_y) + (i,j) = (v_x + i, v_y + j)$. Each edge $\mathbf{e}$ of the unit square with endpoints (v, v') maps to edge $\mathbf{e} + (i,j) = (v + (i,j), v' + (i,j))$. Finally, each edge pair $(\mathbf{e}_1, \mathbf{e}_2)$ maps to $(\mathbf{e}_1 + (i,j), \mathbf{e}_2 + (i,j))$.

The endpoints of the isocontour edges are the isocontour vertices. To map each isocontour edge to a geometric line segment, we use linear interpolation to

> Input : F is a 2D array of scalar values.
>
> Coord is a 2D array of (x, y) coordinates.
>
> σ is an isovalue.
>
> Result : A set Υ of isocontour line segments.
>
> MarchingSquares(F, Coord, σ, Υ)
>
> 1 Read Marching Squares lookup table into Table;
>   /* Assign "+" or "−" signs to each vertex   */
> 2 foreach grid vertex (i, j) do
> 3 | if F$[i, j] < \sigma$ then Sign$[i, j] \leftarrow$ "−";
> 4 | else Sign$[i, j] \leftarrow$ "+";   /* F$[i, j] \geq \sigma$ */
> 5 end
> 6 S $\leftarrow \emptyset$;
>   /* For each grid square, retrieve isocontour edges   */
> 7 foreach grid square (i, j) do
> | /* Grid square vertices are $(i, j), (i{+}1, j), (i, j{+}1), (i{+}1, j{+}1)$ */
> 8 | $\kappa \leftarrow$ (Sign$[i, j]$, Sign$[i{+}1, j]$, Sign$[i, j{+}1]$, Sign$[i{+}1, j{+}1]$);
> 9 | foreach edge pair $(e_1, e_2) \in$ Table$[\kappa]$ do
> 10 | | Insert edge pair $(e_1 + (i, j), e_2 + (i, j))$ into S;
> 11 | end
> 12 end
>   /* Compute isocontour vertex coordinates using linear interpolation   */
> 13 foreach bipolar grid edge **e** with endpoints (i_1, j_1) and (i_2, j_2) do
> | /* Compute the isosurface vertex w_e on edge **e**   */
> 14 | $w_e \leftarrow$ LinearInterpolation
> 15 | (Coord$[i_1, j_1]$, F$[i_1, j_1]$, Coord$[i_2, j_2]$, F$[i_2, j_2]$, σ);
> 16 end
>   /* Convert S to set of line segments   */
> 17 $\Upsilon \leftarrow \emptyset$;
> 18 foreach pair of edges $(\mathbf{e}_1, \mathbf{e}_2) \in$ S do
> 19 | $\Upsilon \leftarrow \Upsilon \cup \{(w_{\mathbf{e}_1}, w_{\mathbf{e}_2})\}$;
> 20 end

Algorithm 2.1. Marching Squares.

position the isocontour vertices as described in Section 1.7.2. Each isocontour vertex v lies on a grid edge $[p, q]$. If s_p and s_q are the scalar values at p and q and σ is the isovalue, then map v to $(1 - \alpha)p + \alpha q$ where $\alpha = (\sigma - s_p)/(s_q - s_p)$. Note that since p and q have different signs, scalar s_p does not equal s_q and the denominator $(s_q - s_p)$ is never zero.

The Marching Squares algorithm is presented in Algorithm 2.1. Function LinearInterpolation, called by this algorithm, is defined in Algorithm 1.1 in Section 1.7.2.

Figure 2.4 contains an example of a scalar grid, an assignment of positive and negative labels to the grid vertices, the isocontour before linear interpolation, and the final isocontour after linear interpolation.

2.2.2   Running Time

The MARCHING SQUARES algorithm runs in linear time.

Proposition 2.3. *Let N be the total number of vertices of a 2D scalar grid. The running time of the* MARCHING SQUARES *algorithm on the scalar grid is $\Theta(N)$.*

Proof: Reading the MARCHING SQUARE lookup table takes constant time. Each grid square is processed once. At each grid square, at most two isocontour edges are retrieved from the lookup table. Since the number of grid squares is bounded by the number of grid vertices, determining the isocontour edges takes $O(N)$ time.

Computing the isocontour vertex on each grid edge takes time proportional to the number of isocontour vertices. Since each grid edge has at most one isocontour edge, the time to compute isocontour vertices is proportional the number of grid edges. The number of grid edges is less than twice the number of grid vertices, so the number of grid edges is at most $2N$. Thus computing the isocontour vertices takes $O(N)$ time.

The algorithm examines every grid square, so its running time has an $\Omega(N)$ lower bound. Thus, the running time of the MARCHING SQUARES algorithm is $\Theta(N)$. $\qquad\square$

2.2.3   Isocontour Properties

To properly discuss the output produced by the MARCHING SQUARES algorithm, we need to differentiate between two cases based on the isovalue. In the first case, the isovalue does not equal the scalar value of any grid vertex. In this case, the MARCHING SQUARES algorithm produces a piecewise linear 1-manifold with boundary. The boundary of the 1-manifold lies on the boundary of the grid. In the second case, the isovalue equals the scalar value of one or more grid vertices. In this case, the MARCHING SQUARES algorithm may not produce a 1-manifold with boundary or the boundary may not lie on the boundary of the grid. For instance, the MARCHING SQUARES algorithm applied to the 3×3 grids in Figures 2.5 and 2.6 produces non-manifold isocontours or isocontours with boundary not on the scalar grid. In Figure 2.5(a), four isocontour line segments intersect at a single point; in Figure 2.5(b), the isocontour is a single point, and in Figure 2.6, the boundary of the isocontour lies inside the grid.

The two cases also differ in the nature of the line segments produced by the algorithm. The isocontour produced by the MARCHING SQUARES algorithm is

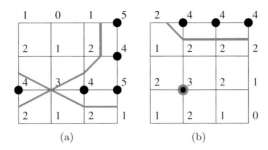

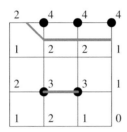

Figure 2.5. Examples of non-manifolds produced by MARCHING SQUARES (isovalue 3). Black vertices are positive. (a) Four curves joining at the grid vertex with isovalue 3. (b) Isosurface includes an isolated point at the grid vertex with isovalue 3.

Figure 2.6. Examples of a manifold produced by MARCHING SQUARES whose boundary does not lie on the grid boundary (isovalue 3). Black vertices are positive.

a set of line segments whose vertices lie on the grid edges. If the isovalue does not equal the scalar value of any grid vertex, then these line segments all have positive length. If the isovalue equals the scalar value of one or more grid vertices, then the isocontour may have zero-length edges. For instance, the MARCHING SQUARES algorithm applied to the three grids in Figure 2.7 produces isocontours for isovalue 3 with zero-length edges.

In Figure 2.7(a), the lower-left grid square has configuration 4, producing a single isocontour edge, but both endpoints of that edge map to the vertex in the middle of the grid. In Figure 2.7(b), each grid square produces an isocontour edge, but all four edges have zero length and collapse to a single point. In Figure 2.7(c), leftmost and rightmost grid squares produce zero-length isocontour edges and two middle grid squares produce two duplicate isocontour edges on a grid edge.

MARCHING SQUARES returns a finite set, Υ, of line segments. The isocontour is the union of those line segments. The vertices of the isocontour are the endpoints of the line segments.

The following properties apply to all isocontours produced by the MARCHING SQUARES algorithm.

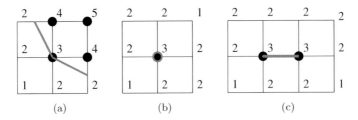

Figure 2.7. Examples of zero-length contour edges produced by MARCHING SQUARES (isovalue 3). Black vertices are positive. (a) Isocontour with one zero-length isocontour edge (from lower-left grid square). (b) Isocontour with four zero-length isocontour edges. (c) Another isocontour with four zero-length isocontour edges. Isocontour also has two duplicate nonzero isocontour edges (from the two middle grid squares).

Property 1.   *The isocontour is piecewise linear.*

Property 2.   *The vertices of the isocontour lie on grid edges.*

Property 3.   *The isocontour intersects every bipolar grid edge at exactly one point.*

Property 4.   *The isocontour does not intersect any negative or strictly positive grid edges.*

Property 5.   *The isocontour separates positive grid vertices from negative grid vertices and strictly separates strictly positive grid vertices from negative grid vertices.*

Set $\mathbb{Y} \subseteq \mathbb{X}$ separates point $p \in \mathbb{X}$ from point $q \in \mathbb{X}$ if every path in $\mathbb{X}$ connecting p to q intersects $\mathbb{Y}$. Set $\mathbb{Y}$ strictly separates p from q if $\mathbb{Y}$ separates p from q and neither p nor q is on $\mathbb{Y}$. (See Section 1.7.1 and Appendix B.9.)

Properties 3 and 4 imply that the isocontour intersects a minimum number of grid edges. If both endpoints of a grid edge have scalar value equal to the isovalue, then the isocontour may intersect the grid edge zero, one, or two times or may contain the grid edge. (See Figure 2.8.)

A grid vertex may have scalar value equal to the isovalue and yet no isocontour passes through any edge containing that grid vertex. For instance, the MARCHING SQUARES algorithm returns the empty set when run on the scalar grid in Figure 2.9 with isovalue 3. Each vertex, including the center vertex, is positive, so each grid square has configuration 9 (Figure 2.2) and has no isocontour edges.

By Property 3, the isocontour intersects every bipolar grid edge. However, the bipolar grid edge may be intersected by zero-length isocontour edges as in Figure 2.7(b).

The following properties apply to MARCHING SQUARES isocontours whose isovalues do not equal the scalar value of any grid vertex.

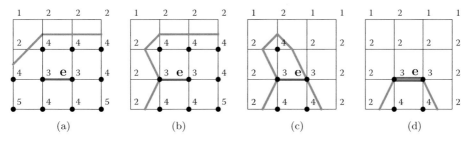

(a)   (b)   (c)   (d)

Figure 2.8. Examples of grid edges with both endpoint scalar values equal to the isovalue (3). Black vertices are positive. (a) Red grid edge **e** does not intersect the isocontour. (b) Red grid edge **e** intersects the isocontour at one endpoint. (c) Red grid edge **e** intersects the isocontour at both endpoints. (d) Red grid edge **e** is contained in the isocontour.

Figure 2.9. Example of a scalar grid whose MARCHING SQUARES isocontour is the empty set, even though the center grid vertex has scalar value equal to the isovalue 3. All vertices are positive.

Property 6. *The isocontour is a piecewise linear 1-manifold with boundary.*

Property 7. *The boundary of the isocontour lies on the boundary of the grid.*

Property 8. *Set Υ does not contain any zero-length line segments or duplicate line segments, and the line segments in Υ form a "triangulation" of the isocontour.*

The triangulation in Property 8 simply means that line segments in Υ intersect at their endpoints. The isocontour is one-dimensional and does not contain any triangles.

2.2.4   Proof of Isocontour Properties

We give a proof of each of the properties listed in the previous section.

Property 1. *The isocontour is piecewise linear.*

Property 2. *The vertices of the isocontour lie on grid edges.*

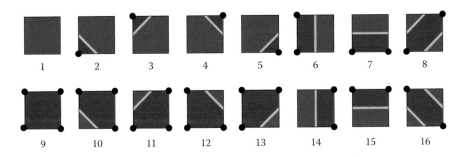

Figure 2.10. Red, positive regions and blue, negative regions for each square configuration. The green isocontour is part of the positive region. Black vertices are positive.

Proof of Properties 1 & 2: The MARCHING SQUARES isocontour consists of a finite set of line segments, so it is piecewise linear. These line segments intersect only at their endpoints and thus form a triangulation of the isocontour. The endpoints of these line segments lie on the grid edges, confirming Property 2.   □

Property 3.   *The isocontour intersects every bipolar grid edge at exactly one point.*

Property 4.   *The isocontour does not intersect any negative or strictly positive grid edges.*

Proof of Properties 3 & 4: Each isocontour edge is contained in a grid square. Since the grid squares are convex, only isocontour edges with endpoints (vertices) on the grid edge intersect the grid edge. If the grid edge has one positive and one negative endpoint, the unique location of the isocontour vertex on the grid edge is determined by linear interpolation. Thus the isocontour intersects a bipolar grid edge at only one point.

If the grid edge is negative or strictly positive, then no isocontour vertex lies on the grid edge. Thus the isocontour does not intersect negative or strictly positive grid edges.   □

Within each grid square the isocontour partitions the grid square into two regions. Let the positive region for a grid square c be the set of points which can be reached by a path ζ from a positive vertex. More precisely, a point p is in the positive region of c if there is some path $\zeta \subset c$ connecting p to a positive vertex of c such that the interior of ζ does not intersect the isocontour. A point p is in the negative region of c if there is some path $\zeta \subset c$ connecting p to a negative vertex of c such that ζ does not intersect the isocontour. Since any path $\zeta \subset c$ from a positive to a negative vertex must intersect the isocontour, the positive and negative regions form a partition of the square c. Figure 2.10 illustrates the positive and negative regions, colored red and blue, respectively, for each square configuration.

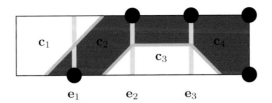

Figure 2.11. Adjacent grid squares, c_1, c_2, c_3, and c_4, and their positive (red) regions, $R^+_{c_1}$, $R^+_{c_2}$, $R^+_{c_3}$ and $R^+_{c_4}$, respectively. Yellow edges e_1, e_2 and e_3 separate the squares. Positive regions agree on the grid square boundaries, i.e., $R^+_{c_1} \cap e_1 = R^+_{c_2} \cap e_1$ and $R^+_{c_2} \cap e_2 = R^+_{c_3} \cap e_2$ and $R^+_{c_3} \cap e_3 = R^+_{c_4} \cap e_3$.

Note the asymmetry in the definitions of the positive and negative regions. For the positive region the *interior* of ζ does not intersect the isocontour, while for the negative region the entire path ζ must not intersect the isocontour. Thus, the positive region contains the isocontour while the negative region does not. The positive region is also closed. Any point within the positive region that does not lie in the isocontour has a neighborhood contained in the positive region.

Every negative vertex is contained in the negative region since the zero-length path connects the vertex to itself. Similarly, every positive vertex is contained in the positive region.

Let R^+_c be the positive region for a grid square c. We claim that positive and negative regions agree on the grid square boundaries. For instance, in Figure 2.11 $R^+_{c_1} \cap e_1$ equals $R^+_{c_2} \cap e_1$ where $R^+_{c_1}$ and $R^+_{c_2}$ are the positive regions for grid squares c_1 and c_2, respectively, and e_1 is the edge between c_1 and c_2. Similarly, $R^+_{c_2} \cap e_2$ equals $R^+_{c_3} \cap e_2$ and $R^+_{c_3} \cap e_3$ equals $R^+_{c_4} \cap e_3$.

Lemma 2.4. *Let c_1 and c_2 be adjacent grid squares where each vertex of c_1 and c_2 has a positive or a negative label. Let p be a point in $c_1 \cap c_2$. Point p is in $R^+_{c_1}$ if and only if p is in $R^+_{c_2}$.*

Proof: If p is a grid vertex, then p is in $R^+_{c_1}$ and $R^+_{c_2}$ if it is positive and not in $R^+_{c_1}$ or $R^+_{c_2}$ if it is negative. Otherwise, p must be in the interior of some grid edge e. If edge e is positive, then p is in $R^+_{c_1}$ and $R^+_{c_2}$. If edge e is negative, then p is not in $R^+_{c_1}$ or $R^+_{c_2}$. If one endpoint, v_1, is positive and the other endpoint, v_2, is negative, then the isocontour in both grid squares intersects the grid edge in the same interpolated point q. The closed segment $[v_1, q]$ is in both $R^+_{c_1}$ and $R^+_{c_2}$ while the segment $(q, v_2]$ (open at q and closed at v_2) is in neither. Thus if p is in $[v_1, q]$, then p is in both $R^+_{c_1}$ and $R^+_{c_2}$ and if p is in $(q, v_2]$, then p is in neither. $\qquad\square$

Using Lemma 2.4, we prove that the isocontour separates positive vertices from negative ones.

Property 5. *The isocontour separates positive grid vertices from negative grid vertices and strictly separates strictly positive grid vertices from negative grid vertices.*

Proof: For all the possible configurations, a path from a positive vertex to a negative one in a grid square must intersect the isocontour. We must show that this also holds true for paths through many grid squares.

Let R^+ be the union of the positive regions over all the grid squares. Consider a path ζ in the grid from a positive grid vertex to a negative one. The positive grid vertex lies in R^+ while the negative one does not. Thus ζ must intersect some point p on the boundary of R^+ where it crosses out of R^+. Every neighborhood of p must contain points that are not in R^+.

Since R^+ is closed, point p lies in R^+. Thus point p lies in $R^+_{\mathbf{c}'}$ for some grid square $\mathbf{c}'$. By Lemma 2.4, point p lies in $R^+_{\mathbf{c}}$ for every grid square $\mathbf{c}$ containing p. Assume p is not on the isocontour. Within each grid square containing p, some neighborhood of p is contained in the positive region for that grid square. The union of those neighborhoods is a neighborhood of p within the grid and is contained in R^+. Thus ζ does not cross out of R^+ at p. We conclude that p must lie on the isocontour and that ζ intersects the isocontour. Thus the isocontour separates positive from negative grid vertices.

If the scalar value of a grid vertex does not equal the isovalue, then the grid vertex does not lie on the isocontour. Thus the isocontour strictly separates strictly positive grid vertices from negative ones. (By definition, the scalar value of a negative vertex never equals the isovalue.) $\square$

To prove properties 6 and 7, we prove something slightly more general.

Proposition 2.5. *Let p be any point on the MARCHING SQUARES isocontour that is not a grid vertex with scalar value equal to the isovalue.*

  1. *If p is in the interior of the grid, then the isocontour restricted to some sufficiently small neighborhood of p is a 1-manifold.*

  2. *If p is on the boundary of the grid, then the isocontour restricted to some sufficiently small neighborhood of p is a 1-manifold with boundary.*

Proof: Let v be a grid vertex with scalar value s_v. If s_v is not the isovalue, then the isocontour does not contain v, so point p is not v. If s_v equals the isovalue, then, by assumption, point p is not v. Therefore, point p is not a grid vertex.

If p lies in the interior of a grid square, then it lies in the interior of some isocontour edge. The interior of this edge is a 1-manifold containing p.

Assume p lies on the boundary of a grid square but not on the boundary of the grid. Since p is not a grid vertex, point p must lie in the interior of some grid edge $\mathbf{e}$ with one positive and one negative vertex. The two grid squares containing $\mathbf{e}$ each contain a single contour edge with endpoint at p. The interior of these two contour edges and the point p form a 1-manifold containing p.

Finally, assume p lies on the boundary of the grid. Since p is not a grid vertex, point p is contained in a single grid square. This grid square contains a single contour edge with endpoint at p. This contour edge is a manifold with boundary containing p. □

Properties 6 and 7 apply to MARCHING SQUARES isocontours whose isovalues do not equal the scalar value of any grid vertex.

Property 6. *The isocontour is a piecewise linear 1-manifold with boundary.*

Property 7. *The boundary of the isocontour lies on the boundary of the grid.*

Proof of Properties 6 & 7: Consider a point p on the isocontour. Since the isovalue does not equal the scalar value of any grid vertex, point p is not a grid vertex. By Proposition 2.5, the isocontour restricted to some suitably small neighborhood of point p is either a 1-manifold or a 1-manifold with boundary. Thus the isocontour is a 1-manifold with boundary. Since the restricted isocontour is a 1-manifold whenever p is in the interior of the grid, the boundary of the isocontour must lie on the grid boundary. □

The last property is that Υ does not contain any zero-length or duplicate edges and forms a triangulation of the isocontour.

Property 8. *Set Υ does not contain any zero-length line segments or duplicate line segments, and the line segments in Υ form a "triangulation" of the isocontour.*

Proof: Since no grid vertex has scalar value equal to the isovalue, no isocontour vertex lies on a grid vertex. By Property 4, each bipolar grid edge contains only one isocontour vertex. Thus, the linear interpolation on isocontour vertices does not create any zero-length or duplicate isocontour edges. Since isocontour edges intersect only at their endpoints, Υ forms a triangulation of the isocontour. □

2.2.5 2D Ambiguity

Set $E_\kappa^{+/-}$ is the set of bipolar square edges for configuration κ. The combinatorial structure of the isocontour depends upon the matching of the elements of $E_\kappa^{+/-}$. If $E_\kappa^{+/-}$ has two elements, then there is no choice. However, if $E_\kappa^{+/-}$ has four bipolar edges, then there are two possible pairings and two possible isocontours that could be constructed for configuration κ. Configurations 8 and 16 from Figure 2.2 have four bipolar edges. They are called ambiguous configurations. These two ambiguous configurations are shown in Figure 2.12 along with the two combinatorially distinct isocontours for each ambiguous configuration.

Choosing different isocontours for the ambiguous configurations will change the topology of the overall isocontour. For instance, Figure 2.13 shows the same

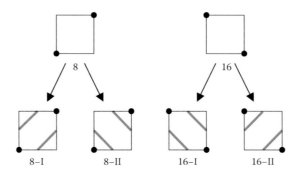

Figure 2.12. Ambiguous square configurations.

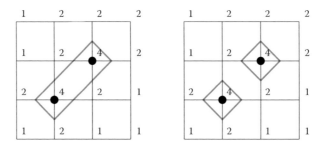

Figure 2.13. Topologically distinct isocontours created by using different isocontours for the ambiguous configuration in the central grid square.

scalar grid with two topologically distinct isocontours created by different resolutions of the ambiguous configurations. The first isocontour has two components while the second has one.

While the choice of isocontours for the ambiguous configurations changes the isocontour topology, any of the choices will produce isocontours that are 1-manifolds and strictly separate strictly positive vertices from negative vertices. As we shall see, this is not true in three dimensions.

2.3   Marching Cubes

2.3.1   Algorithm

The three-dimensional MARCHING CUBES algorithm follows precisely the steps in the two-dimensional MARCHING SQUARES algorithm. Input to the MARCH-

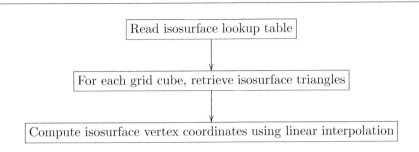

Figure 2.14. MARCHING CUBES.

ING CUBES algorithm is an isovalue and a set of scalar values at the vertices of a three-dimensional regular grid. The algorithm has three steps. (See Figure 2.14.) Read the isosurface lookup table from a preconstructed data file. For each cube, retrieve from the lookup table a set of isosurface triangles representing the combinatorial structure of the isosurface. The vertices of these triangles form the isosurface vertices. Assign geometric locations to the isosurface vertices based on the scalar values at the cube edge endpoints. We explain the last two steps below.

Grid vertices are labeled positive or negative as described in Section 2.1. Grid edges are labeled positive, negative, or bipolar.

The combinatorial structure of the isosurface within each cube is determined from the configuration of the cube's vertex labels. In order to separate the positive vertices from the negative ones, the isosurface must intersect any cube edge that has one positive and one negative endpoint. An isosurface that intersects a minimal number of grid edges will not intersect any edge whose endpoints are both strictly positive or whose endpoints are both negative.

Since each vertex is either positive or negative and a cube has eight vertices, there are $2^8 = 256$ different configurations of cube vertex labels. Many of these configurations are rotations or reflections of one another. By exploiting this symmetry, the number of distinct configurations can be reduced to twenty-two.[1] These distinct configurations are listed in Figure 2.15. All other configurations are rotations or reflections of these twenty-two.

For each cube configuration κ, let $E_\kappa^{+/-}$ be the set of edges with one positive and one negative endpoint. The isosurface lookup table contains 256 entries, one for each configuration κ. Each entry is a list of triples of edges of $E_\kappa^{+/-}$. Each triple $(\mathbf{e}_1, \mathbf{e}_2, \mathbf{e}_3)$ represents a triangle whose vertices lie on $\mathbf{e}_1$, $\mathbf{e}_2$, and $\mathbf{e}_3$. The list of triples define the combinatorial structure of the isosurface patch for

---

[1] Lorensen and Cline's original paper on MARCHING CUBES [Lorensen and Cline, 1987a] listed only fifteen configurations. For reasons discussed in Section 2.3.5, twenty-two configurations are preferable.

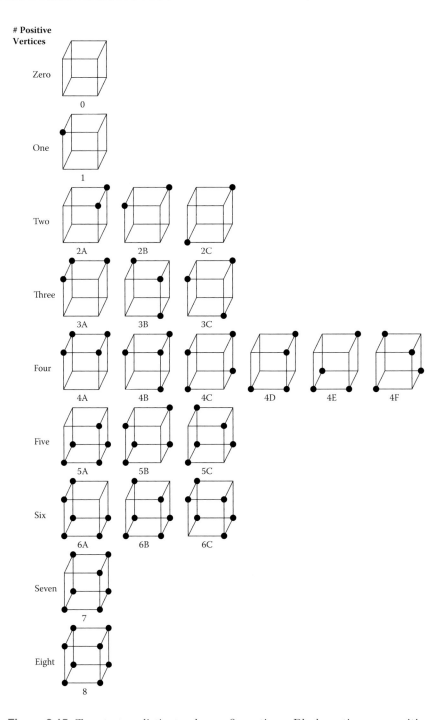

Figure 2.15. Twenty-two distinct cube configurations. Black vertices are positive.

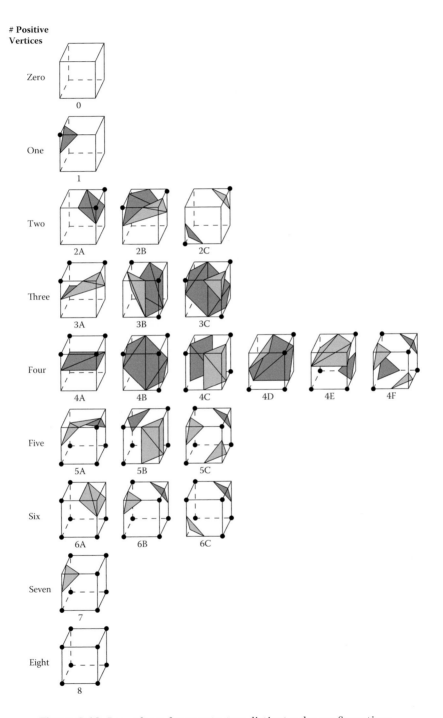

Figure 2.16. Isosurfaces for twenty-two distinct cube configurations.

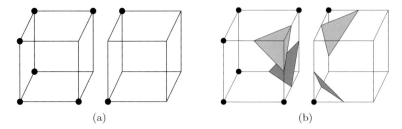

Figure 2.17. (a) Adjacent configurations sharing a common face. (b) Incompatible isosurface patches for the adjacent configurations.

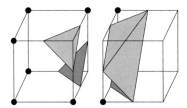

Figure 2.18. Compatible isosurface patches for adjacent configurations in Figure 2.17(a).

configuration κ. The isosurface patch intersects every edge of $E_\kappa^{+/-}$ exactly once and does not intersect any other grid cube edges.

To define the 256 entries in the table, it is only necessary to determine the table entries for the twenty-two distinct configurations. The table entries for the other configurations can be derived using rotation and reflection symmetry. Figure 2.16 contains the twenty-two distinct cube configurations and their isosurfaces.

The isosurface lookup table is constructed on the unit cube with vertices $(0,0,0),(1,0,0),(0,1,0),\ldots,(0,1,1),(1,1,1)$. To construct the isosurface in grid cube (i,j,k), we have to map unit cube edges to edges of cube (i,j,k). Each vertex $v = (v_x, v_y, v_z)$ of the unit cube maps to $v + (i,j,k) = (v_x, v_y, v_z) + (i,j,k) = (v_x + i, v_y + j, v_z + k)$. Each edge $\mathbf{e}$ of the unit square with endpoints (v, v') maps to edge $\mathbf{e} + (i,j,k) = (v + (i,j,k), v' + (i,j,k))$. Finally, each edge triple $(\mathbf{e}_1, \mathbf{e}_2, \mathbf{e}_3)$ maps to $(\mathbf{e}_1 + (i,j,k), \mathbf{e}_2 + (i,j,k), \mathbf{e}_3 + (i,j,k))$.

In Figure 2.16, the isosurface vertices lie on the midpoints of the grid edges. This is for illustration purposes only. The geometric locations of the isosurface vertices are not defined by the lookup table.

The vertices of the isosurface triangles are the isosurface vertices. To map each isosurface triangle to a geometric triangle, we use linear interpolation to position the isosurface vertices as described in Section 1.7.2. Each isosurface vertex v lies on a grid edge $[p, q]$. If s_p and s_q are the scalar values at p and q and σ is the isovalue, then map v to $(1-\alpha)p + q$ where $\alpha = (\sigma - s_p)/(s_q - s_p)$.

Input : F is a 3D array of scalar values.

Coord is a 3D array of (x, y, z) coordinates.

σ is an isovalue.

Result : A set Υ of isosurface triangles.

MarchingCubes(F, Coord, σ, Υ)

1 Read Marching Cubes lookup table into Table;

   /* Assign "+" or "−" signs to each vertex   */

2 foreach grid vertex (i, j, k) do

3 | if $F[i, j, k] < \sigma$ then $\mathsf{Sign}[i, j, k] \leftarrow$ "−";

4 | else $\mathsf{Sign}[i, j, k] \leftarrow$ "+";   /* $F[i, j, k] \geq \sigma$ */

5 end

6 $\mathsf{T} \leftarrow \emptyset$;

   /* For each grid cube, retrieve isosurface triangles   */

7 foreach grid cube (i, j, k) do

      /* Cube vertices are $(i, j, k), (i{+}1, j, k), \ldots, (i{+}1, j{+}1, k{+}1)$ */

8 | $\kappa \leftarrow (\mathsf{Sign}[i, j, k], \mathsf{Sign}[i{+}1, j, k], \mathsf{Sign}[i, j{+}1, k], \ldots, \mathsf{Sign}[i{+}1, j{+}1, k{+}1])$;

9 | foreach edge triple $(\mathbf{e}_1, \mathbf{e}_2, \mathbf{e}_3) \in \mathsf{Table}[\kappa]$ do

10 | | Insert edge triple $(\mathbf{e}_1 + (i, j, k), \mathbf{e}_2 + (i, j, k), \mathbf{e}_3 + (i, j, k))$ into T;

11 | end

12 end

   /* Compute isosurface vertex coordinates using linear interpolation   */

13 foreach bipolar grid edge e with endpoints (i_1, j_1, k_1) and (i_2, j_2, k_2) do

      /* Compute the isosurface vertex w_e on edge e */

14 | $w_e \leftarrow$ LinearInterpolation

15 | $(\mathsf{Coord}[i_1, j_1, k_1], \mathsf{F}[i_1, j_1, k_1], \mathsf{Coord}[i_2, j_2, k_2], \mathsf{F}[i_2, j_2, k_2], \sigma)$;

16 end

   /* Convert T to set of triangles   */

17 $\Upsilon \leftarrow \emptyset$;

18 foreach triple of edges $(e_1, e_2, e_3) \in \mathsf{T}$ do

19 | $\Upsilon \leftarrow \Upsilon \cup \{(w_{e_1}, w_{e_2}, w_{e_3})\}$;

20 end

Algorithm 2.2. MARCHING CUBES.

Note that since p and q have different sign, scalar s_p does not equal s_q and the denominator $(s_q - s_p)$ is never zero.

The MARCHING CUBES algorithm is presented in Algorithm 2.2. Function LinearInterpolation, called by this algorithm, is defined in Algorithm 1.1 in Section 1.7.2.

Two configurations can be adjacent to one another if the face they share in common has the same set of positive and negative vertices. Figure 2.17 contains

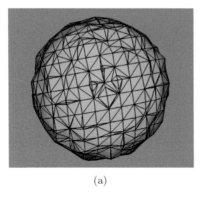 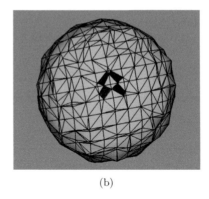

(a)   (b)

Figure 2.19. (a) Spherical isosurface of noisy point cloud. (b) Isosurface of noisy point cloud constructed using incorrect isosurface lookup table. The sphere interior is colored red and is visible through holes in the sphere. The holes are caused by incompatible isosurface patches in the isosurface lookup table.

an example of two such configurations. The isosurface patches for each configuration should be constructed so that their boundaries align on the common face. As shown in Figure 2.17, two reasonable isosurface patches for adjacent configurations can have boundaries that do not align on the common face and so are incompatible. Isosurfaces constructed using such incompatible isosurface patches for adjacent configurations may have "holes" and may not be a 2-manifold. (See Figure 2.19.) Compatible isosurface patches for the configuration in Figure 2.17(a) are given in Figure 2.18.

The isosurface patches for the twenty-two distinct configurations in Figure 2.16 were constructed so that they and all the 256 derived configurations are compatible. The isosurface patch boundaries align on the common face of any two adjacent configurations. As will be discussed in Section 2.3.5, Lorensen and Cline's original MARCHING CUBES algorithm [Lorensen and Cline, 1987a] lacked this property.

If the constructed isosurface is a manifold, then the manifold is orientable. After the construction of the isosurface, the isosurface can be assigned an orientation by assigning consistent orientations to all its triangles. However, a better approach is to properly orient the triangles in the lookup table. Each isosurface patch in a lookup table entry separates a positive region from a negative one. As noted in Section 1.7.1, a triangle orientation determines a vector orthogonal to the triangle. Orient each triangle in the isosurface patch so that the induced orthogonal vector points toward the positive region. Retrieve the oriented triangles from the lookup table to determine the orientation of the isosurface triangles. The orientations of the triangles are consistent and form an orientation of the isosurface.

2.3.2   Running Time

The MARCHING CUBES algorithm runs in linear time.

Proposition 2.6. *Let N be the total number of vertices of a 3D scalar grid. The running time of the* MARCHING CUBES *algorithm on the scalar grid is $\Theta(N)$.*

The proof is similar to the proof for the 2D MARCHING SQUARES algorithm in Section 2.2.2 and is omitted.

2.3.3   Isosurface Properties

The isosurface produced by the MARCHING CUBES algorithm has the same properties as the isocontour produced by the MARCHING SQUARES algorithm. As in the 2D version, we differentiate between the case where the isovalue equals the scalar value of one or more grid vertices and the case where it does not. If the isovalue does not equal the scalar value of any grid vertices, then the isosurface is a piecewise linear, orientable 2-manifold with boundary. If the isovalue equals the scalar value of some grid vertex, then the isosurface may not be a 2-manifold and the isosurface may have zero-length edges and zero-area triangles.

MARCHING CUBES returns a finite set, Υ, of oriented triangles. The isosurface is the union of these triangles. The vertices of the isosurface are the triangle vertices.

The following properties apply to all isosurfaces produced by the MARCHING CUBES algorithm.

Property 1.   *The isosurface is piecewise linear.*

Property 2.   *The vertices of the isosurface lie on grid edges.*

Property 3.   *The isosurface intersects every bipolar grid edge at exactly one point.*

Property 4.   *The isosurface does not intersect any negative or strictly positive grid edges.*

Property 5.   *The isosurface separates positive grid vertices from negative ones and strictly separates strictly positive grid vertices from negative grid vertices.*

Properties 3 and 4 imply that the isosurface intersects a minimum number of grid edges. As in two dimensions, if both endpoints of a grid edge have scalar value equal to the isovalue, then the isosurface may intersect the grid edge zero, one, or two times or may contain the grid edge.

By Property 3, the isosurface intersects every bipolar grid edge. However, the bipolar grid edge may be intersected by zero-area isosurface triangles.

The following properties apply to the MARCHING CUBES isosurfaces whose isovalues do not equal the scalar value of any grid vertex.

Property 6.   *The isosurface is a piecewise linear, orientable 2-manifold with boundary.*

Property 7.   *The boundary of the isosurface lies on the boundary of the grid.*

Property 8.   *Set Υ does not contain any zero-area triangles or duplicate triangles and the triangles in Υ form a triangulation of the isosurface.*

Scanning devices usually produce data sets whose scalar values are 8-bit, 12-bit, or 16-bit integers. For such data sets, a MARCHING CUBES isosurface with noninteger isovalues has no degenerate triangles and is always a manifold. Typically, the isovalue $(x + 0.5)$ is used, where x is some integer.

2.3.4   Proof of Isosurface Properties

We give a proof of each of the properties listed in Section 2.3.3. The proofs are the same as the proofs for MARCHING SQUARES isocontours.

Property 1.   *The isosurface is piecewise linear.*

Property 2.   *The vertices of the isosurface lie on grid edges.*

Proof of Properties 1 & 2: The MARCHING CUBES isosurface consists of a finite set of triangles, so it is piecewise linear. By construction, the vertices of these triangles lie on the grid edges.   □

Property 3.   *The isosurface intersects every bipolar grid edge at exactly one point.*

Property 4.   *The isosurface does not intersect any negative or strictly positive grid edges.*

Proof of Properties 3 & 4: Each isosurface triangle is contained in a grid cube. Since the grid cubes are convex, only isosurface triangles with vertices on the grid edge intersect the grid edge. If the grid edge has one positive and one negative endpoint, the unique location of the isosurface vertex on the grid edge is determined by linear interpolation. Thus the isosurface intersects a bipolar grid edge at exactly one point.

If the grid edge is negative or strictly positive, then no isosurface vertex lies on the grid edge. Thus the isosurface does not intersect negative or strictly positive grid edges.   □

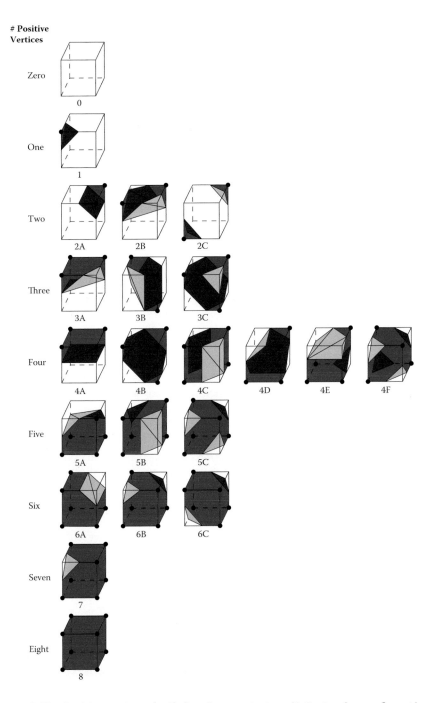

Figure 2.20. Positive regions (red) for the twenty-two distinct cube configurations. Visible portion of each isosurface is green. Portion of the isosurface behind each positive region is colored dark red.

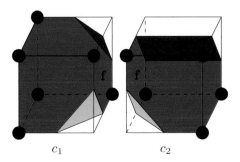

c_1 $\quad\quad\quad\quad\quad$ c_2

Figure 2.21. Adjacent cubes $\mathbf{c}_1$ and $\mathbf{c}_2$ and (red) positive regions $R_{\mathbf{c}_1}^+$ and $R_{\mathbf{c}_2}^+$. Black cube vertices are positive. Shared facet $\mathbf{f}$ lies between the two cubes. Note that $R_{\mathbf{c}_1}^+ \cap \mathbf{f}$ equals $R_{\mathbf{c}_2}^+ \cap \mathbf{f}$.

The proof of Property 5, the separation property, is similar to the proof of Property 5 for the MARCHING SQUARES isosurface (Section 2.2.4). We first extend the definition of positive and negative regions from grid squares to grid cubes.

Within each grid cube the isosurface partitions the grid cube into two regions. Define the positive region for a grid cube $\mathbf{c}$ to be points $p \in \mathbf{c}$ where some path $\zeta \subset \mathbf{c}$ connects p to a positive vertex of $\mathbf{c}$ and the interior of ζ does not intersect the isosurface (Figure 2.20). Define the negative region for a grid cube $\mathbf{c}$ to be points $p \in \mathbf{c}$ where some path $\zeta \subset \mathbf{c}$ connects p to a negative vertex of $\mathbf{c}$ and ζ does not intersect the isosurface. Since any path $\zeta \subset \mathbf{c}$ from a positive to a negative vertex must intersect the isosurface, the positive and negative regions form a partition of the cube $\mathbf{c}$.

The positive region is closed and contains the isosurface. Any point within the positive region that does not lie on the isosurface has a neighborhood contained in the positive region.

Every negative vertex is contained in the negative region since the zero-length path connects the vertex to itself. Every positive vertex is contained in the positive region since a path in a cube from any negative cube vertex to a positive one must intersect the isosurface.

We claim that positive and negative regions agree on the grid cube boundaries (Figure 2.21). Let $R_{\mathbf{c}}^+$ be the positive region for a grid cube $\mathbf{c}$.

Lemma 2.7. *Let $\mathbf{c}_1$ and $\mathbf{c}_2$ be adjacent grid cubes where each vertex of $\mathbf{c}_1$ and $\mathbf{c}_2$ has a positive or a negative label. Let p be a point in $\mathbf{c}_1 \cap \mathbf{c}_2$. Point p is in $R_{\mathbf{c}_1}^+$ if and only if p is in $R_{\mathbf{c}_2}^+$.*

The proof of Lemma 2.7 is based on an exhaustive examination of all possible adjacent configurations and is omitted. A more detailed and satisfying analysis is given in Chapter 5, which contains an algorithm for generating the lookup table for the MARCHING CUBES and similar algorithms.

Using Lemma 2.7, we prove that the isosurface separates positive vertices from negative ones.

Property 5.   *The isosurface separates positive grid vertices from negative ones and strictly separates strictly positive grid vertices from negative grid vertices.*

Proof: For all the possible configurations, a path from a positive vertex to a negative one in a grid cube must intersect the isosurface. We must show that this also holds true for paths through many grid cubes.

Consider a path ζ in the grid from a positive grid vertex to a negative one. The positive grid vertex lies in R^+ while the negative one does not. Thus ζ must intersect some point p on the boundary of R^+ where it crosses out of R^+. Every neighborhood of p must contain points that are not in R^+.

Since R^+ is closed, point p lies in R^+. Thus point p lies in $R^+_{\mathbf{c}'}$ for some grid cube $\mathbf{c}'$. By Lemma 2.7, point p lies in $R^+_{\mathbf{c}}$ for every grid cube $\mathbf{c}$ containing p. If p does not lie on the isosurface, then some neighborhood of p is contained in the positive region $R^+_{\mathbf{c}}$ of each grid cube $\mathbf{c}$ containing p. The union of those neighborhoods is a neighborhood of p within the grid and is contained in R^+. Thus ζ does not cross out of R^+ at p. We conclude that p must lie on the isosurface and that ζ intersects the isosurface. Thus the isosurface separates positive from negative grid vertices.

If the scalar value of a grid vertex does not equal the isovalue, then the grid vertex does not lie on the isosurface. Thus the isosurface strictly separates strictly positive grid vertices from negative ones. (By definition, the scalar value of a negative vertex never equals the isovalue.)   □

As in two dimensions, to prove Properties 6 and 7, we first prove something slightly more general.

Proposition 2.8. *Let p be any point on the* MARCHING CUBES *isosurface that is not a grid vertex with scalar value equal to the isovalue.*

1. *If p is in the interior of the grid, then the isosurface restricted to some sufficiently small neighborhood of p is a 2-manifold.*

2. *If p is on the boundary of the grid, then the isosurface restricted to some sufficiently small neighborhood of p is a 2-manifold with boundary.*

Proof: Let v be a grid vertex with scalar value s_v. If s_v is not the isovalue, then the isosurface does not contain v, so point p is not v. If s_v equals the isovalue, then, by assumption, point p is not v. Therefore, point p is not a grid vertex.

If p lies in the interior of a grid cube, then it lies in the interior of some isosurface patch. The interior of this patch is a 2-manifold containing p.

Assume p lies on the boundary of a grid cube but not on the boundary of the grid. Since p is not a grid vertex, point p either lies in the interior of a grid edge or the interior of a square grid facet. If point p lies in the interior of a square grid facet, then it is contained in two isosurface triangles lying in two adjacent grid cubes. The interior of the union of these two isosurface triangles forms a 2-manifold containing p. If point p lies in the interior of a grid edge $\mathbf{e}$, then p is contained in four isosurface patches lying in the four grid cubes containing $\mathbf{e}$. The interior of the union of these four patches forms a 2-manifold containing p.

Finally, assume p lies on the boundary of the grid. Since p is not a grid vertex, point p either lies in one or two grid cubes. If p lies in a single grid cube, then a single isosurface triangle in this grid cube contains p. This triangle is a 2-manifold with boundary containing p. If p lies in two grid cubes, then the union of the isosurface patches in these two grid cubes form a 2-manifold with boundary containing p. □

Properties 6 and 7 apply to MARCHING CUBES isosurfaces whose isovalues do not equal the scalar value of any grid vertex.

Property 6. *The isosurface is a piecewise linear, orientable 2-manifold with boundary.*

Property 7. *The boundary of the isosurface lies on the boundary of the grid.*

Proof of Properties 6 & 7: Consider a point p on the isosurface. Since the isovalue does not equal the scalar value of any grid vertex, point p is not a grid vertex. By Proposition 2.8, the isosurface restricted to some suitably small neighborhood of point p is either a 2-manifold or a 2-manifold with boundary. Thus the isosurface is a 2-manifold with boundary. All of the triangles in the isosurface are oriented so that the induced orthogonal vector points toward the positive vertices; thus, all the triangle orientations are consistent and the manifold is orientable.

Since the restricted isosurface is a 2-manifold whenever p is in the interior of the grid, the boundary of the isosurface must lie on the grid boundary. □

The last property is that Υ does not contain any zero-area or duplicate triangles and forms a triangulation of the isosurface.

Property 8. *Set Υ does not contain any zero-area triangles or duplicate triangles and the triangles in Υ form a triangulation of the isosurface.*

Proof: Since no grid vertex has scalar value equal to the isovalue, no isosurface vertex lies on a grid vertex. By Property 3, each bipolar grid edge contains only one isosurface vertex. Thus, the linear interpolation on isosurface vertices does not create any zero-area or duplicate isosurface triangles. The isosurface triangles within a grid cube are a subset of the triangulation of an isosurface patch and so the nonempty intersection of any two such triangles is either a vertex or an edge of both. The isosurface patches in Figure 2.16 were constructed so that the intersection of any two isosurface triangles in adjacent grid cubes is either a vertex or an edge of both. Since the nonempty intersection of any two triangles in Υ is a face of both triangles, Υ forms a triangulation of the isosurface. □

2.3.5 3D Ambiguity

The 256 configurations of positive and negative vertices are generated from the twenty-two configurations in Figure 2.15 using rotational and reflection symmetry. There is another kind of symmetry that is not being used. Configuration κ_1 is the complement of configuration κ_2 if κ_1 can be derived from κ_2 by switching κ_2's positive and negative vertex labels. The configurations with five, six, seven, or eight positive vertices are all complements of configurations with three, two, one, and zero positive vertices. Complementary symmetry reduces the number of unique configurations from twenty-two to fourteen.

Unfortunately, complementary symmetry creates incompatible isosurface patches in the isosurface table. For instance, the configuration on the left in Figure 2.17 is a rotation of configuration 6B from Figure 2.15 while the configuration on the right is a rotation of configuration 2B. Configuration 2B is the complement of configuration 6B. Both isosurface patches on the left and on the right come from a rotation of the isosurface patch for 6B in Figure 2.16. The boundaries of the isosurface patches clearly do not align on the square between the two cubes.

Lorensen and Cline's paper [Lorensen and Cline, 1987a] used complementary symmetry and so created incompatible isosurface patches in the isosurface table [Dürst, 1988].[2] The problem configurations were the ones that had ambiguous configurations on their square facets. These are configurations 2B, 3B, 3C, 4C, 4E, 4F, 5B, 5C and 6B. Numerous papers were written on how to handle these problematic configurations. The most widely adopted approach, proposed in [Montani et al., 1994] and in [Zhou et al., 1994], is to simply drop complementary symmetry. This is the approach adopted in the previous section.

As discussed in Section 2.2.5, the two-dimensional ambiguous configurations have two isocontour edges and two combinatorially different ways to position those edges. One of these edge positions separates the two negative vertices by isocontour edges (8-I and 16-I in Figure 2.12) while the other separates the positive vertices (8-II and 16-II in Figure 2.12).

The border of a cube's three-dimensional isosurface patch defines an isocontour on each of the cube's square facets. If some configuration's isosurface patch separates the negative vertices on the facet while an adjacent configuration's isosurface patch separates the positive ones, then the isosurface edges on the common facet will not align. The isosurface patches in Figure 2.16 do not separate the positive vertices on any facet.[3] Moreover, the derived isosurface surface patches in any rotation or reflection of the configurations also do not separate positive vertices on any facet. Thus the isosurface patches in any two adjacent

---

[2]Lorensen and Cline's paper did not use reflection symmetry and so they had fifteen, not fourteen, distinct configurations.

[3]Note that in configuration 2C, the two positive vertices do not share a facet. The isosurface patch separates the two positive vertices, but not on any facet. Since these vertices do not share a facet, the isosurface patch aligns with all possible adjacent configurations.

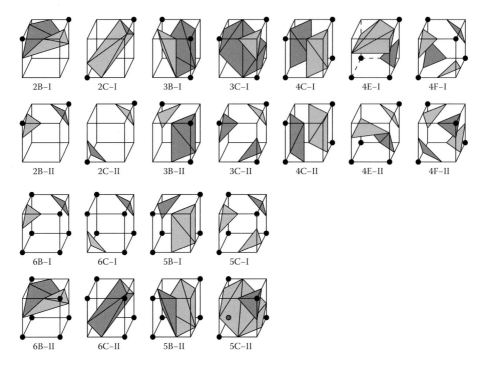

Figure 2.22. Two "natural" isosurface patches for the ambiguous 3D configurations. Isosurface patches in the first and third rows separate negative vertices. Isosurface patches in the second and fourth rows separate positive vertices.

cubes are properly aligned on their boundaries. An equally valid, but combinatorially distinct, isosurface table could be generated by using isosurface patches that do not separate the negative vertices on any square facet.

The ambiguous configurations in $\mathbb{R}^3$ are 2B, 2C, 3B, 3C, 4C, 4E, 4F, 5B, 5C, 6B and 6C. These configurations have topologically distinct piecewise linear isosurfaces with all the properties listed in the previous section. Ambiguous configurations admit two "natural" isosurface patches, one separating positive vertices and one separating negative vertices. (See Figure 2.22.)

As we have already discussed, configurations 2B, 3B, 3C, 4C, 4E, 4F, 5B, 5C and 6B have square facets with ambiguous 2D configurations. Configurations 2C and 6C are the exceptions.[4] None of their square facets have an ambiguous configuration, yet they admit two topologically distinct piecewise linear isosurfaces. (See Figure 2.22.) Usually, isosurface patches 2C-II and 6C-I are used since they each require only two isosurface triangles and do not create the "tunnel" in 2C-I and 6C-II.

---

[4]The numerous papers on resolving the ambiguous 3D configurations do not discuss these two, but they do admit topologically distinct isosurfaces and they are ambiguous.

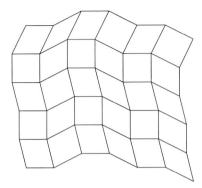

Figure 2.23. Curvilinear grid. Grid cells are quadrilaterals.

2.3.6 Curvilinear Grids

Sections 2.2 and 2.3 describe the MARCHING SQUARES and MARCHING CUBES algorithms for regular grids where each grid element is a geometric square or cube. These algorithms work equally well for curvilinear grids that have the same combinatorial structure as regular grids but whose elements are convex quadrilaterals or hexahedra. (See Figure 2.23.) The grid vertices are assigned positive and negative labels just as in the case of the regular grid. The same lookup table is used to determine the combinatorial structure of each isocontour or isosurface patch within a grid element. The isosurface geometry is determined by using linear interpolation to position the grid vertices along each grid edge.

2.4 Marching Tetrahedra

The MARCHING CUBES algorithm applies to regular and curvilinear grids. However, many applications decompose space into unstructured tetrahedral meshes composed of tetrahedral mesh elements. MARCHING TETRAHEDRA is a modified version of the MARCHING CUBES algorithm.

2.4.1 Algorithm

Input to the MARCHING TETRAHEDRA algorithm is an isovalue, a tetrahedral mesh, and a set of scalar values at the vertices of the tetrahedra. The mesh tetrahedra must form a triangulation of some three-dimensional region, i.e., the intersection of any two tetrahedra is a vertex, edge, or facet of both tetrahedra or is empty.

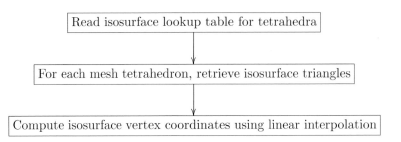

Figure 2.24. MARCHING TETRAHEDRA.

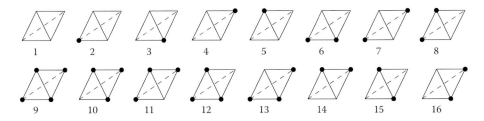

Figure 2.25. Tetrahedral configurations. Black vertices are positive.

The algorithm has three steps. (See Figure 2.24.) Read in the isosurface lookup table from a preconstructed data file. For each tetrahedron, retrieve from the lookup table a set of isosurface triangles representing the combinatorial structure of the isosurface. The endpoints of these edges form the isosurface vertices. Assign geometric locations to the isosurface vertices based on the scalar values at the triangle edge endpoints. We explain the last two steps next.

Mesh vertices are labeled positive or negative, as described in Section 2.1. Mesh edges are labeled positive, negative, or bipolar.

Since each vertex is either positive or negative and a tetrahedron has four vertices, there are $2^4 = 16$ different configurations of tetrahedron vertex labels. These configurations are listed in Figure 2.25. Applying affine transformations, only five of these configurations are distinct (Figure 2.26).

The combinatorial structure of the isosurface within each tetrahedron is determined from the configuration of the tetrahedron's vertex labels. In order to separate the positive vertices from the negative ones, the isosurface must intersect a tetrahedron edge that has one positive and one negative endpoint. An isosurface which intersects a minimal number of mesh edges will not intersect any edge whose endpoints are both strictly positive or both negative.

Each isosurface patch separates the tetrahedron into two regions, a positive one containing the positive vertices and a negative one containing the negative

Figure 2.26. Distinct tetrahedral configurations. Configuration k has exactly k positive (black) vertices.

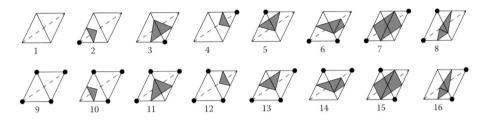

Figure 2.27. Tetrahedral isosurface patches.

vertices. Each tetrahedron on the isosurface patch is oriented so that the induced normal points to the positive region.

In Figure 2.27, the isosurface vertices lie on the midpoints of the grid edges. This is for illustration purposes only. The geometric locations of the isosurface vertices are not defined by the lookup table.

For each tetrahedral configuration κ, let $E_\kappa^{+/-}$ be the set of edges with one positive and one negative endpoint. The isosurface lookup table contains 16 entries, one for each configuration κ. Each entry is a list of triples of edges of $E_\kappa^{+/-}$. Each triple $(\mathbf{e}_1, \mathbf{e}_2, \mathbf{e}_3)$ represents an oriented triangle whose vertices lie on $\mathbf{e}_1$ and $\mathbf{e}_2$ and $\mathbf{e}_3$. The orientation of the triangle is given by the order, $\mathbf{e}_1$, $\mathbf{e}_2$, $\mathbf{e}_3$. The list of triples defines the combinatorial structure of the isosurface patch for configuration κ. The isosurface patch intersects every edge of $E_\kappa^{+/-}$ exactly once and does not intersect any other grid tetrahedron edges.

To determine the isosurface patch for a specific mesh tetrahedron, match the vertices of the mesh tetrahedron to the vertices of the tetrahedron in the lookup table by numbering the vertices of the mesh tetrahedron from 1 to 4. The vertex numbering can be arbitrary and will not affect the isosurface topology, although it may alter slightly the isosurface triangulation and geometry. Determine the lookup table configuration that matches the configuration of positive and negative vertices in the mesh tetrahedron and retrieve the isosurface patch from the lookup table.

To map each isosurface triangle to a geometric triangle, we use linear interpolation to position the isosurface vertices as described in Section 1.7.2. Each isosurface vertex v lies on a grid edge $[p, q]$. If s_p and s_q are the scalar values at p and q and σ is the isovalue, then map v to $(1-\alpha)p + \alpha q$ where $\alpha = (\sigma - s_p)/(s_q - s_p)$.

2.4.2   Isosurface Properties

The isosurface produced by MARCHING TETRAHEDRA has the same properties as the isosurface produced by the MARCHING CUBES algorithm (Section 2.3.3). The proofs are the same as the proofs in Section 2.3.4 and are omitted.

MARCHING TETRAHEDRA returns a finite set, Υ, of oriented triangles. The isosurface is the union of these triangles. The vertices of the isosurface are the triangle vertices.

The following properties apply to all isosurfaces produced by the MARCHING TETRAHEDRA algorithm.

Property 1.   *The isosurface is piecewise linear.*

Property 2.   *The vertices of the isosurface lie on grid edges.*

Property 3.   *The isosurface intersects every bipolar grid edge at exactly one point.*

Property 4.   *The isosurface does not intersect any negative or strictly positive grid edges.*

Property 5.   *The isosurface separates positive grid vertices from negative ones and strictly separates strictly positive grid vertices from negative grid vertices.*

Properties 3 and 4 imply that the isosurface intersects a minimum number of grid edges. As with MARCHING CUBES, if both endpoints of a grid edge have scalar value equal to the isovalue, then the isosurface may intersect the grid edge zero, one, or two times or may contain the grid edge.

By Property 3, the isosurface intersects every bipolar grid edge. However, the bipolar grid edge may be intersected by zero-area isosurface triangles.

Under appropriate conditions, the isosurface produced by MARCHING TETRA-HEDRA is a 2-manifold with boundary. One condition is that the isovalue does not equal the scalar value of any grid vertex. MARCHING TETRAHEDRA can be applied to any tetrahedral mesh, as long as the tetrahedra form a triangulation. As shown in Figure 2.28, the isosurface produced by MARCHING TETRAHEDRA may not be a manifold, even if the isovalue does not equal the scalar value. The problem is that the underlying space is not a manifold. A similar problem could face an isosurface constructed by MARCHING CUBES from an arbitrary mesh of cubes. In Section 2.3, we applied MARCHING CUBES to a regular grid whose underlying space is a manifold.

Consider a tetrahedral mesh that has the following two conditions:

- The isovalue does not equal the scalar value of any mesh vertex.

- The tetrahedral mesh is a partition of a 3-manifold with boundary.

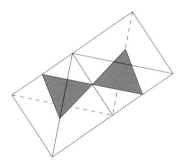

Figure 2.28. Non-manifold isosurface produced by MARCHING TETRAHEDRA. The tetrahedral mesh consists of two tetradedra sharing an edge. The region covered by the tetrahedral mesh is not a manifold with boundary.

Under these conditions, the isosurface produced by MARCHING TETRAHEDRA has the following two properties:

Property 6. *The isosurface is a piecewise linear, orientable 2-manifold with boundary.*

Property 7. *The boundary of the isosurface lies on the boundary of the grid.*

Property 8. *Set Υ does not contain any zero-area triangles or duplicate triangles and the triangles in Υ form a triangulation of the isosurface.*

One significant difference between MARCHING CUBES and MARCHING TETRAHEDRA is that MARCHING TETRAHEDRA has no ambiguous configurations. The four vertices of each tetrahedron are connected by tetrahedron edges and each such edge is intersected at most once by the isosurface, so there is no possible ambiguity in the isosurface construction.

2.4.3 Cube Tetrahedralization

MARCHING TETRAHEDRA can be used for isosurface construction in regular grids by tetrahedralizing the grid cubes. There are two distinct ways to tetrahedralize a grid cube, one that breaks the cube into five tetrahedra and one that breaks it into six. (See Figure 2.29.) The decomposition into five tetrahedra consists of a single large tetrahedra in the cube center and four smaller tetrahedra cutting off four corners of the cube. The decomposition into six tetrahedra consists of six congruent tetrahedra that all share a diagonal edge between two opposing corners of the cube.

The decomposition of a grid cube into tetrahedra creates diagonal edges on the square facets of the cube. These diagonals must match the diagonals created on adjacent cubes for the decomposition to be a triangulation. If all cubes

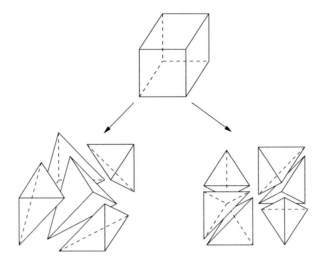

Figure 2.29. Subdivision of a cube into five or six tetrahedra.

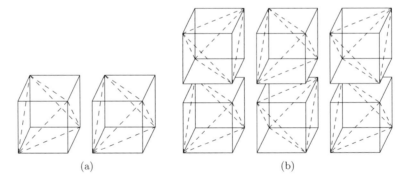

(a)   (b)

Figure 2.30. (a) Duplicate tetrahedralizations of two cubes into five tetrahedra with mismatching diagonals on the common face between the two cubes. (b) Alternating tetrahedralization of cube into five tetrahedra with matching diagonals on common faces between cubes.

are decomposed in the same manner into six tetrahedra, then these diagonals do match. However, if the cubes are decomposed in the same manner into five tetrahedra, then the diagonals of adjacent cubes will fail to match (Figure 2.30(a)). Two decompositions into five tetrahedra must be used in an alternating manner for the diagonals to match (Figure 2.30(b)).

The regular grid cubes have ambiguous configurations while the tetrahedral decomposition does not. What happened to the ambiguous configurations? These configurations are resolved by the choice of triangulation. For instance, in Figure 2.31, the first triangulation gives an isosurface patch with two compo-

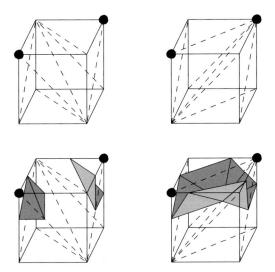

Figure 2.31. Different tetrahedralizations of a cube into six tetrahedra.

nents corresponding to 2B-II in Figure 2.22 while the second gives an isosurface patch with one component corresponding to 2B-I.

Using MARCHING TETRAHEDRA on a triangulated regular grid in place of MARCHING CUBES increases the number of triangles in resulting isosurfaces. The size of that increase depends upon the different configurations and the manner in which the tetrahedral decomposition intersects those configurations.

2.4.4   Isocontouring a Triangular Mesh

A two-dimensional simplified version of MARCHING TETRAHEDRA can be used to isocontour a triangular mesh.[5] The simplified version does not require a table lookup since each mesh triangle contains at most one isocontour edge.

Input to the isocontouring algorithm is an isovalue, a triangular mesh, and a set of scalar values at the vertices of the mesh. The mesh triangles must form a triangulation of some two-dimensional region, i.e., the intersection of any two triangles must be a vertex or edge of both triangles or must be empty.

For each bipolar mesh edge $[p, q]$, use linear interpolation to position the isosurface vertex on $[p, q]$ as described in Section 1.7.2. If s_p and s_q are the scalar values at p and q and σ is the isovalue, then map the isosurface vertex to $\alpha p + (1 - \alpha)q$ where $\alpha = (s_q - \sigma)/(s_q - s_p)$.

---

[5]While the name "Marching Triangles" might be appropriate for the two-dimensional version of MARCHING TETRAHEDRA, the name is already used by Hilton et al. [Hilton et al., 1996] for a completely different algorithm that constructs a triangular mesh on an implicit surface.

A triangle contains zero, one, two, or three positive vertices. If a triangle contains one or two positive vertices, then it contains two mesh edges with one positive and one negative endpoint. Connect the isocontour vertices lying on each of these mesh edges by an isocontour edge. If the isovalue equals the isovalue of exactly one of the triangle vertices v and is less than the isovalue of the other two vertices, then the isocontour vertices for each of these mesh edges will both be located at v. In that case, either connect them by a zero-length isocontour edge or merge them into a single vertex.

The resulting isocontour has all the same properties listed in Section 2.2.3 as the isocontour produced by MARCHING SQUARES. As with tetrahedral meshes, triangular meshes have no ambiguous configurations of positive and negative vertices.

2.5   Notes and Comments

Lorensen and Cline published the MARCHING CUBES algorithm in 1987 [Lorensen and Cline, 1987a]. A year earlier, Wyvill, McPheeters, and Wyvill [Wyvill et al., 1986] published a somewhat similar isosurface extraction algorithm, but without the use of isosurface patch lookup tables. A US patent [Lorensen and Cline, 1987b] for the MARCHING CUBES algorithm was granted in 1987 and the patent expired in 2005.

Dürst [Dürst, 1988] noted the problem discussed in Section 2.3.5 of incompatible surface patches in Lorensen and Cline's isosurface lookup table. Montani, Scateni, and Scopigno [Montani et al., 1994] and Zhou, Shu, and Kankanhalli [Zhou et al., 1994] proposed the corrected isosurface lookup tables described in this chapter. Wilhelms and Gelder [Gelder and Wilhelms, 1990] explored different techniques for consistently resolving ambiguous facets. Nielson and Hamann [Nielson and Hamann, 1991] used bilinear interpolation to resolve ambiguous facets. Their algorithm, the ASYMPTOTIC DECIDER, is described in Chapter 4.

The MARCHING TETRAHEDRA algorithm was proposed by a number of people [Bloomenthal, 1988, Bloomenthal, 1994, Carneiro et al., 1996, Payne and Toga, 1990, Shirley and Tuchman, 1990] although not necessarily under that name. Chan and Purisima [Chan and Purisima, 1998] apply MARCHING TETRAHEDRA to a tetrahedralization based on the body-centered cubic lattice. Theußl, Möller, and Gröller [Theußl et al., 2001] show that the body-centered cube lattice provides more accurate sampling than the regular grid. Carr, Möller, and Snoeyink [Carr et al., 2001, Carr et al., 2006b] discuss the effects of different tetrahedral subdivisions on isosurfaces generation.

The cube configurations in Figure 2.15 are distinct under the transformations of rotation and reflection. In [Banks et al., 2004], Banks, Linton, and Stockmeyer discuss counting the number of distinct configurations of cubes, hypercubes, and other convex polyhedra using computational group theory software.

The MARCHING CUBES algorithm does not produce good representations of sharp edges or corners. Kobbelt et al. [Kobbelt et al., 2001] modified the MARCHING CUBES algorithm to more accurately reconstruct sharp edges and corners when input is the distance to a given surface. Other algorithms for reproducing sharp edges and corners are presented in [Ho et al., 2005] and [Schaefer and Warren, 2004]. Dual contouring algorithms for reconstructing sharp edges and corners are referenced in the notes and comments section of Chapter 3.

The algorithms in [Ho et al., 2005, Schaefer and Warren, 2004] and the dual contouring algorithms can also create multiresolution isosurfaces (multiresolution isosurfaces are defined and discussed in Chapters 9 and 10). All the algorithms for reconstructing sharp edges and corners require accurate measurements of the gradients or surface normals at isosurface vertices.

The MARCHING CUBES and MARCHING TETRAHEDRA algorithms produce triangulated isosurface meshes with many small, thin triangles. There are numerous geometric algorithms for improving the quality and reducing the size of a triangle surface mesh. (See the surveys [Cignoni et al., 1998, Luebke, 2001, van Kaick and Pedrini, 2006, Alliez et al., 2007].) We will only mention the work specific to isosurface generation.

The MARCHING CUBES and MARCHING TETRAHEDRA algorithms produce poor quality triangles when the isosurface passes close to a grid or mesh vertex. Tzeng [Tzeng, 2004] improves triangle quality by warping the underlying regular grid so that isosurface vertices are near grid edge midpoints and far from the grid vertices. Improvements on MARCHING TETRAHEDRA triangle quality based on "snapping" tetrahedral mesh vertices to the isosurface are given in [Hall and Warren, 1990, Treece et al., 1999, Labelle and Shewchuk, 2007]. Labelle and Schewchuk [Labelle and Shewchuk, 2007] provide guaranteed bounds on isosurface triangle quality using a "snapping"-based algorithm. Raman and Wenger [Raman and Wenger, 2008] provide guaranteed bounds on MARCHING CUBES isosurface triangle quality using the "snapping" of regular grid vertices to the isosurface.

Various triangulations can be used for isosurface patches in the MARCHING CUBES lookup table. Dietrich et al. [Dietrich et al., 2008, Dietrich et al., 2009a, Dietrich et al., 2009b] improve the quality of the isosurface triangles by modifying and selecting the "best" triangulation. Nielson [Nielson, 2003a] gives triangulations for all cases in the lookup table such that the projection to some coordinate plane is one-to-one. This allows local representation of the isosurface as a function of the two axes of the coordinate plane.

Alternative approaches have been developed for generating isosurface meshes with quality elements. Schreiner et al. [Schreiner et al., 2006] use an advancing front technique to generate high-quality isosurfaces. Dey and Levine [Dey and Levine, 2007] present a quality isosurface mesh generation algorithm using the Voronoi diagram of a set of sample points on the isosurface. Both methods are adaptive, with more triangles inserted in areas of high curvature. Based on a related paper [Cheng et al., 2004], Dey and Levine's algorithm provides guarantees of triangle quality but at the expense of potentially adding large numbers of triangles in high curvature areas. Both the advancing front technique and the Voronoi-based technique take time that is an order of magnitude longer than the running time of the MARCHING CUBES algorithm.

Isosurface patches in the isosurface lookup table can be constructed of smooth elements such as bicubic or Bezier surfaces instead of piecewise linear triangles. Smooth isosurface patches are described in [Gallagher and Nagtegaal, 1989, Hall and Warren, 1990, Hamann et al., 1997, Theisel, 2002].

Nielson et al. [Nielson et al., 2003] present an algorithm to improve smoothness of the MARCHING CUBES isosurface by moving isosurface vertices along grid edges.

The SPIDERWEB algorithm by Karron [Karron, 1992] constructs an isosurface by placing an isosurface vertex on each bipolar grid edge and connecting those vertices to another isosurface vertex within the grid cube. It produces substantially more isosurface triangles than MARCHING CUBES or dual contouring and creates non-manifold "bubbles" whenever a grid facet has four bipolar edges (i.e., the grid facet is ambiguous). Properties and proofs of SPIDERWEB isosurfaces are discussed in [Karron and Cox, 1992] and [Karron et al., 1993].

Dual contouring algorithms, isosurface construction in four dimensions, and interval volumes are in the notes and comments sections of Chapters 3, 6, and 7.

Approximating isosurface normals from volumetric data is discussed in [Cline et al., 1988, Möller et al., 1997a, Möller et al., 1997b, Nielson et al., 2002, Lee et al., 2008, Hossain et al., 2011]. Approximating Gaussian and mean curvature is discussed in [Nielson, 2003a, Kindlmann et al., 2003].

Patera and Skala [Patera and Skala, 2004] compare the accuracy of the MARCHING CUBES isosurface with the accuracy of isosurfaces extracted from tetrahedralizations of the regular grid and with isosurfaces extracted from the body-centered cubic lattice.

Etiene et al. [Etiene et al., 2009] describe techniques for assessing the correctness of isosurface extraction programs. Pöthkow and Hege [Pöthkow and Hege, 2011] give a method for computing the positional uncertainy of isosurface vertices. They represent this uncertainty by coloring the isosurface.

Nooruddin and Turk [Nooruddin and Turk, 2003] use isosurfaces to repair polygonal models. They convert the polygonal model to a volumetric data set. Grid vertices with scalar value 1 lie inside the model while grid vertices with scalar value 0 lie outside. They extract a MARCHING CUBES isosurface with scalar value between 0 and 1. As noted in Section 2.3.3, this isosurface is a manifold. Ju [Ju, 2004] gives a variation of this algorithm using dual contouring. Ju's approach detects and patches hole boundaries. Bischoff, Pavic, and Kobbelt [Bischoff et al., 2005] repair polygonal models by using morphological operators to close the holes. They also use dual contouring to reconstruct the polygonal surface. Ju in [Ju, 2009] surveys techniques for repairing polynomial models.

Newman and Yi in [Newman and Yi, 2006] provide an excellent survey of the MARCHING CUBES algorithm and its variants.

CHAPTER 3

DUAL CONTOURING

Dual contouring[1] is an isosurface construction technique that places isosurface vertices inside mesh elements instead of on mesh edges. Isosurface vertices in adjacent mesh elements are connected by isosurface edges and facets.

In the simplest form, each mesh element intersected by the isosurface contains exactly one isosurface vertex. In more complicated versions, multiple isosurface vertices can appear within a single mesh element. Each such isosurface vertex lies on a different connected component of the mesh element's isosurface patch.

Dual contouring on regular grids produces quadrilateral patches instead of triangles. The ability to place a vertex anywhere within a mesh element allows better representation of sharp vertices and edges. The simpler versions of dual contouring, with one vertex per mesh element, extend easily to higher dimensional grids and multiresolution grids, as discussed in Sections 6.7 and 10.2. On the other hand, the placement of isosurface vertices is less precise in dual contouring. When dual contouring uses only one isosurface vertex per mesh element, the isosurface is usually not a manifold.

We present here two dual contouring algorithms, SURFACE NETS and DUAL MARCHING CUBES. SURFACE NETS is the first and simplest dual contouring algorithm, where each mesh element contains at most one isosurface vertex. DUAL MARCHING CUBES is a more sophisticated version, where an isosurface vertex is generated for each connected component in a mesh element's isosurface patch.

---

[1] Dual contouring constructs an isosurface that is a subset of the dual mesh of the original mesh, hence the name "dual contouring."

3.1   Definitions

Positive and negative vertices and faces. The definitions of positive, strictly positive, and negative vertices and edges given here are the same as in Section 2.1. However, the definitions are extended to grid facets and grid cubes.

Definition 3.1.

- A grid vertex is positive, "+", if its scalar value is greater than or equal to the isovalue σ.

- A grid vertex is negative, "$-$", if its scalar value is less than σ.

- A positive vertex is strictly positive if its scalar value does not equal to σ.

Definition 3.2.

- A grid face (vertex, edge, facet, or cube) is positive if all of its vertices are positive.

- A grid face is negative if all of its vertices are negative.

- A positive grid face is strictly positive if all of its vertices are strictly positive.

The definitions given in the beginning of this section apply not just to regular scalar grids but also to curvilinear grids. They also apply to the vertices and faces of polyhedral meshes such as tetrahedral and simplicial meshes.

Bipolar edges and active faces. The definitions of bipolar edges is the same as in Section 2.1. We add a similar term for grid squares, cubes, and their faces that contain one negative and one positive vertex.

Definition 3.3.

- A grid edge is bipolar if one endpoint is positive and one endpoint is negative.

- A grid square or cube or any face of a square or cube is active if it has at least one negative and one positive vertex.

Interior and boundary edges and cubes. We need to distinguish grid edges and cubes that are on the boundary of the grid from edges and cubes that are in the interior.

Definition 3.4.

- An interior grid edge is a grid edge that is contained in four grid cubes.

- A boundary grid edge is a grid edge that is contained in three or fewer grid cubes.

Definition 3.5.

- An interior grid cube is a grid cube whose edges are all interior grid edges.

- A boundary grid cube is a grid cube that contains at least one boundary grid edge.

Let Γ be a regular grid in $\mathbb{R}^3$. We use the notation Γ_{inner} to denote the regular subgrid of Γ consisting of the interior grid cubes of Γ. We let Γ_{outer} represent the st of boundary grid cubes of Γ.

Centroid. Let $P = \{p_1, p_2, \ldots, p_k\}$ be a set of k points in $\mathbb{R}^2$ or $\mathbb{R}^3$.

Definition 3.6. The centroid of P is

$$\frac{p_1 + p_2 + \ldots + p_k}{k}.$$

The centroid of p represents a point in the "center" of P.

3.2 Surface Nets

Gibson described the first dual contouring algorithm for regular grids, which she called SURFACE NETS [Gibson, 1998a, Gibson, 1998b]. We first describe a two-dimensional version of her algorithm.

3.2.1 2D Surface Nets

Let Γ be a two-dimensional regular grid with scalar values assigned to the vertices of Γ. Add an isocontour vertex $w_{\mathbf{c}}$ to each active grid square $\mathbf{c} \in \Gamma$. For each bipolar grid edge separating two squares, $\mathbf{c}$ and $\mathbf{c}'$, connect vertices $w_{\mathbf{c}}$ and $w_{\mathbf{c}'}$ by an isocontour edge (Figure 3.1(c)).

We would like to position the isocontour vertex within each grid square at the "center" of the isocontour patch in the grid square. To do so, we use linear interpolation as described in Section 1.7.2 to approximate the intersection of the isocontour and all the bipolar edges and then take the centroid of all the approximation points.

More specifically, let σ be the isovalue and s_p be the scalar value at a grid vertex p. For each bipolar edge $\mathbf{e} = [p, q]$, let $w_{\mathbf{e}}$ be the point $(1 - \alpha)p + \alpha q$ where $\alpha = (\sigma - s_p)/(s_q - s_p)$. If $w_{\mathbf{e}_1}, \ldots, w_{\mathbf{e}_k}$ are the interpolation points in the grid square $\mathbf{c}$, then locate vertex $w_{\mathbf{c}}$ at $(w_{\mathbf{e}_1} + \cdots + w_{\mathbf{e}_k})/k$, the centroid of $w_{\mathbf{e}_1}, \ldots, w_{\mathbf{e}_k}$.

The 2D SURFACE NETS algorithm is presented in Figure 3.2.

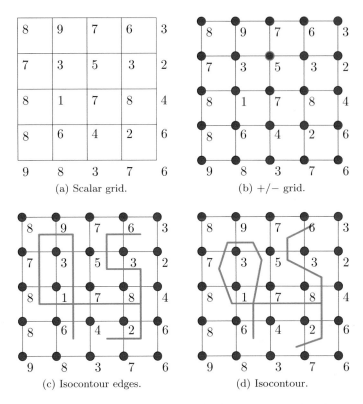

Figure 3.1. (a) 2D scalar grid. (b) Positive (red) or negative (blue) labels on grid vertices. Vertex v with scalar value s_v is positive (red) if $s_v \geq 5$ and negative (blue) if $s_v < 5$. (c) SURFACE NETS isocontour with vertices at square centers (before computing centroids). (d) SURFACE NETS isocontour with isovalue 5.

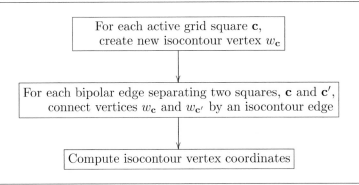

Figure 3.2. SURFACE NETS 2D.

Figure 3.1(d) contains an example of the isocontour produced by the algorithm. Note that an isocontour vertex may be incident on four isocontour edges. Thus the isocontour is not a simple closed curve or a 1-manifold.

3.2.2 3D Surface Nets

The three-dimensional SURFACE NETS algorithm is a direct generalization of the two-dimensional version. (See Figure 3.3.) Let Γ be a three-dimensional regular grid with scalar values assigned to the vertices of Γ. Add an isosurface vertex $w_{\mathbf{c}}$ to each grid cube $\mathbf{c}$ with at least one positive and at least one negative vertex. For each bipolar grid edge contained in four cubes, $\mathbf{c}_1$, $\mathbf{c}_2$, $\mathbf{c}_3$, and $\mathbf{c}_4$, add an isosurface quadrilateral with vertices $w_{\mathbf{c}_1}$, $w_{\mathbf{c}_2}$, $w_{\mathbf{c}_3}$, and $w_{\mathbf{c}_4}$. The cyclic order of the vertices around the quadrilateral should match the cyclic order of the cubes around the edge.

To position the isosurface vertex within each grid cube, use linear interpolation as described in Section 1.7.2 to approximate the intersection of the isosurface and all the bipolar edges. For each bipolar edge $\mathbf{e} = [p, q]$, let $w_{\mathbf{e}}$ be the point $(1-\alpha)p + \alpha q$ where $\alpha = (\sigma - s_p)/(s_q - s_p)$. Take the centroid, $(w_{\mathbf{e}_1} + \cdots + w_{\mathbf{e}_k})/k$, of all the approximation points as the location of the isosurface vertex.

Figure 3.4 contains examples of the isosurface produced by the algorithm. If every grid edge is a bipolar edge (see configuration 4F in Figure 3.5), then an isosurface vertex may be incident on twelve isosurface quadrilaterals. Thus the isosurface is not necessarily a 2-manifold with boundary.

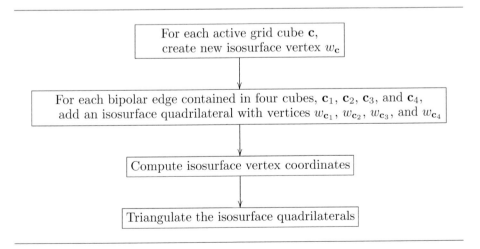

Figure 3.3. SURFACE NETS 3D.

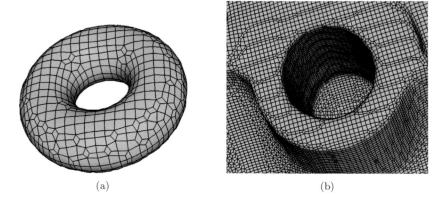

(a)   (b)

Figure 3.4. Examples of isosurfaces and quadrilateral meshes produced by SUR-FACENETS. (a) Isosurface (isovalue 3) forming a torus. Scalar data set is a $20 \times 20 \times 20$ regular grid with origin $(0, 0, 0)$ measuring the distance to a circle with radius 6 centered at $(9.5, 9.5, 9.5)$. (b) Isosurface (isovalue 80) of CT scan of an engine block (close-up). Data set created by General Electric.

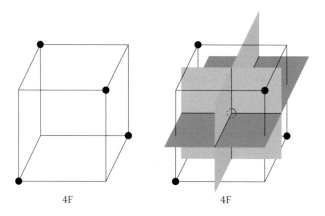

4F   4F

Figure 3.5. Configuration 4F. The isosurface vertex is incident on 12 isosurface quadrilaterals. The isosurface is not a manifold.

The algorithm just described uses the intersection of the isosurface and the bipolar edges to position the isosurface vertex within a grid cube. If the surface normals at the intersection points can also be computed, then these normals can be used in determining the location of the isosurface vertex. In particular, the normals can position isosurface vertices on sharp corners or edges in the isosurface, to accurately represent such corners or edges [Ju et al., 2002, Zhang et al., 2004].

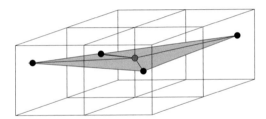

Figure 3.6. Triangle patch of four triangles whose boundary is a quadrilateral **q** around bipolar edge **e**. Each triangle is formed by joining $w_\mathbf{e}$ (magenta vertex) to one of the four edges of **q**.

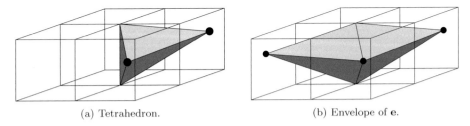

(a) Tetrahedron.   (b) Envelope of **e**.

Figure 3.7. (a) Convex hull of bipolar edge **e** and a quadrilateral edge. (b) Union of the four tetrahedra around **e** forming the envelope of **e**.

3.2.3 Triangulation

Dual contouring creates a quadrilateral mesh but these quadrilaterals are not necessarily planar, i.e., their four vertices do not necessarily lie on a single plane. In fact, unless isosurface vertices are placed at the center of each grid cube, it is highly unlikely that the four quadrilateral vertices will lie on a single plane. To create a piecewise linear surface, one needs to replace each quadrilateral with a patch of triangles whose boundary is the quadrilateral.

The simplest approach is to add a diagonal to each quadrilateral, splitting it into two triangles. Unfortunately, adding diagonals can create self-intersections in the isosurface [Ju and Udeshi, 2006]. The problem is that a diagonal from one quadrilateral can intersect the triangles of another quadrilateral.

Ju and Udeshi [Ju and Udeshi, 2006] solve this problem by breaking a quadrilateral into four, not two triangles. Consider the bipolar edge **e** contained in four cubes, $\mathbf{c}_1$, $\mathbf{c}_2$, $\mathbf{c}_3$, and $\mathbf{c}_4$, listed in order around **e**. Let $w_\mathbf{e}$ be the point on **e** constructed using linear interpolation as described in the previous section. The quadrilateral around **e** has four edges. Replace this quadrilateral with four triangles formed from joining $w_\mathbf{e}$ to each of the four edges (Figure 3.6).

Why does this new triangulation not have self-intersections? The convex hull of quadrilateral edge $(w_{\mathbf{c}_1}, w_{\mathbf{c}_2})$ and **e** is a tetrahedron (Figure 3.7(a)). Similarly,

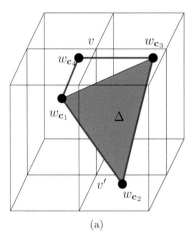 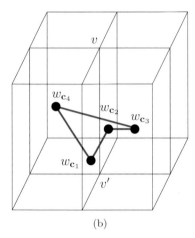

(a)   (b)

Figure 3.8. Isosurface triangles outside the envelope of (v, v'). Quadrilateral edges are colored red. (a) Plane containing triangle $\Delta = (w_{\mathbf{c}_1}, w_{\mathbf{c}_2}, w_{\mathbf{c}_3})$ does not strictly separate v from v'. (b) Plane containing v and quadrilateral diagonal $(w_{\mathbf{c}_1}, w_{\mathbf{c}_3})$ does not strictly separate $w_{\mathbf{c}_2}$ from $w_{\mathbf{c}_4}$.

the convex hull of the other quadrilateral edges and $\mathbf{e}$ are also tetrahedra. The union of these four tetrahedra is a polyhedron with eight triangular faces (Figure 3.7(b)), although this polyhedron is not necessarily convex. Ju and Udeshi call this polyhedron the **envelope** of $\mathbf{e}$. The four triangles sharing vertex $w_{\mathbf{e}}$ are contained in the envelope of $\mathbf{e}$.

The envelopes of two different grid edges intersect only on shared vertices, edges, or facets. Since the four triangles sharing vertex $w_{\mathbf{e}}$ are contained in the envelope for $\mathbf{e}$, they only intersect triangles from other quadrilaterals on shared vertices or edges. Thus, the resulting triangulated surface has no self-intersections. In contrast, the two triangles formed by adding a diagonal to the quadrilateral around $\mathbf{e}$ may not be contained in the envelope of $\mathbf{e}$ and may intersect the interior of triangles from other quadrilaterals.

Adding four triangles per quadrilateral instead of two increases the number of triangles representing the isosurface by a factor of two. If the two triangles formed by a diagonal lie inside the envelope, then there is no need to use four triangles. By the argument above, they will only intersect triangles from other quadrilaterals on shared edges or vertices.

Orientation tests can be used to determine if the two triangles formed by a diagonal lie inside the envelope of $\mathbf{e}$ [Wang, 2011]. Without loss of generality, assume the diagonal is $(w_{\mathbf{c}_1}, w_{\mathbf{c}_3})$. Let v and v' be the endpoints of $\mathbf{e}$. The two triangles $\Delta_1 = \Delta(w_{\mathbf{c}_1}, w_{\mathbf{c}_2}, w_{\mathbf{c}_3})$ and $\Delta_2 = \Delta(w_{\mathbf{c}_1}, w_{\mathbf{c}_3}, w_{\mathbf{c}_4})$ lie inside the envelope of (v, v') if and only if the planes through each triangle strictly separate v from v' and the plane through v and diagonal $(w_{\mathbf{c}_1}, w_{\mathbf{c}_3})$ strictly separates $w_{\mathbf{c}_2}$ from $w_{\mathbf{c}_4}$. Figure 3.8 illustrates violations of those two conditions.

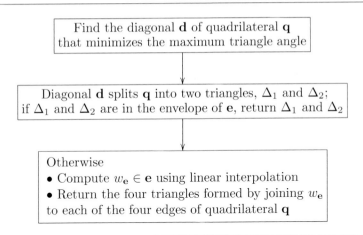

Figure 3.9. Triangulation of isosurface quadrilateral $\mathbf{q}$ around bipolar edge $\mathbf{e}$.

Joining v and v' to triangles Δ_1 and Δ_2 gives four oriented tetrahedra: $\mathbf{t}_1 = (w_{\mathbf{c}_1}, w_{\mathbf{c}_2}, w_{\mathbf{c}_3}, v)$, $\mathbf{t}_2 = (w_{\mathbf{c}_1}, w_{\mathbf{c}_2}, w_{\mathbf{c}_3}, v')$, $\mathbf{t}_3 = (w_{\mathbf{c}_1}, w_{\mathbf{c}_3}, w_{\mathbf{c}_4}, v)$, and $\mathbf{t}_4 = (w_{\mathbf{c}_1}, w_{\mathbf{c}_3}, w_{\mathbf{c}_4}, v')$. The plane through Δ_1 strictly separates v from v' if and only if the orientations of $\mathbf{t}_1$ and $\mathbf{t}_2$ are nonzero and not equal. The plane through Δ_2 strictly separates v from v' if and only if the orientations of $\mathbf{t}_3$ and $\mathbf{t}_4$ are nonzero and not equal. The plane through v and diagonal $(w_{\mathbf{c}_1}, w_{\mathbf{c}_3})$ strictly separates $w_{\mathbf{c}_2}$ from $w_{\mathbf{c}_4}$ if and only if $\mathbf{t}_1$ and $\mathbf{t}_3$ have the same orientation. Thus, we can determine if the triangles Δ_1 and Δ_2 lie inside the envelope of (v, v') by computing and comparing the orientations of tetrahedra $\mathbf{t}_1$, $\mathbf{t}_2$, $\mathbf{t}_3$, and $\mathbf{t}_4$. (See Appendix B.6 on computing the orientation of four points.)

Each quadrilateral has two possible choices for a diagonal. We use the diagonal that minimizes the maximum triangle angle. Using this diagonal reduces the number of long, skinny triangles.

We select diagonal $\mathbf{d}$, which minimizes the maximum triangle angle. Diagonal $\mathbf{d}$ splits quadrilateral $\mathbf{q}$ into two triangles, Δ_1 and Δ_2. If Δ_1 and Δ_2 are in the envelope of bipolar edge $\mathbf{e}$, we replace quadrilateral $\mathbf{q}$ with triangles Δ_1 and Δ_2. If they are not, then we calculate the interpolated point $w_{\mathbf{e}}$ and form the four triangles with vertex $w_{\mathbf{e}}$ as described above. (See Figure 3.9.)

Forming a patch of triangles is one way of determining the surface bounded by a quadrilateral. Bilinear interpolation, Hermite surfaces, Bézier patches, and B-Splines can also be used to create smoother surfaces. As long as those surfaces are contained within the envelope of the corresponding bipolar edge, the resulting isosurface will not self-intersect.

3.2.4   Isosurface Properties

The SURFACE NETS algorithm produces an isosurface whose vertices may and usually do lie in the interior of the grid cubes. Thus the boundary of that isosurface is not usually on the boundary of the grid. A path connecting positive and negative grid vertices can bypass the isosurface in boundary grid cubes. However, if we restrict the path to interior grid cubes, then the path must intersect the isosurface. Restricted to interior grid cubes, the isosurface separates the positive and negative grid vertices.

If none of the grid vertices have scalar value equal to the isovalue, then the isosurface is minimal in the sense that it intersects interior bipolar grid edges exactly once and does not intersect positive or negative grid edges. The isosurface is not necessarily a manifold, even if none of the grid vertices have scalar value equal to the isovalue.

There are numerous variations on SURFACE NETS depending on the placement of the vertices within each grid cube and the construction of the surface bounded by each quadrilateral. The properties listed below apply to SURFACE NETS as described in the previous section, where vertices are placed at the centroid of edge-isosurface intersection points and the quadrilaterals are triangulated into two or four triangles based on the envelope.

SURFACE NETS returns a finite set, Υ, of triangles. The isosurface is the union of these triangles. The vertices of the isosurface are the triangle vertices. As defined in Section 3.1, Γ_{inner} is the subgrid of interior grid cubes of Γ.

The following properties apply to all isosurfaces produced by the SURFACE NETS algorithm.

Property 1.   *The isosurface is piecewise linear.*

Property 2.   *Set Υ contains at most $4m$ triangles where m is the number of bipolar grid edges.*

Property 3.   *The isosurface does not contain any grid vertices whose scalar values do not equal the isovalue.*

Property 4.   *The isosurface intersects at exactly one point every interior bipolar grid edge whose positive endpoint is strictly positive.*

Property 5.   *The isosurface does not intersect any negative or strictly positive grid edges.*

Property 6.   *The isosurface strictly separates the strictly positive vertices of Γ_{inner} from the negative vertices of Γ_{inner}.*

Properties 3–5 imply that the isosurface intersects a minimum number of grid edges.

The following properties apply to the SURFACE NETS isosurfaces whose isovalues do not equal the scalar value of any grid vertex.

Property 7. *Set* Υ *does not contain any zero-area triangles or duplicate triangles and the triangles in* Υ *form a triangulation of the isosurface.*

3.2.5   Proof of Isosurface Properties 1–5

Property 1. *The isosurface is piecewise linear.*

Proof: The isosurface is a finite union of triangles and thus is piecewise linear.☐

Property 2. *Set* Υ *contains at most* 4m *triangles where* m *is the number of bipolar grid edges.*

Proof: SURFACE NETS constructs a quadrilateral for each bipolar edge $\mathbf{e}$ contained in four grid cubes and splits that quadrilateral into at most four triangles. Since the number of such grid edges is at most m, the number of triangles is at most $4m$. ☐

Property 3. *The isosurface does not contain any grid vertices whose scalar values do not equal the isovalue.*

Proof: Let v be a grid vertex whose scalar value does not equal the isovalue. If v were to lie on the isosurface, then it would have to be contained in some isosurface triangle formed from some bipolar edge $\mathbf{e}$. Edge $\mathbf{e}$ would have to be in a grid cube incident on v. We consider two cases depending upon whether $\mathbf{e}$ itself is incident on v.

Consider a bipolar edge $\mathbf{e}$ incident on v. Let P be the plane through v orthogonal to $\mathbf{e}$. SURFACE NETS uses linear interpolation to determine a point $w_{\mathbf{e}} \in \mathbf{e}$ that approximates the intersection of $\mathbf{e}$ and the isosurface. Since the scalar value of v does not equal the isovalue, this point is not v. For every grid cube $\mathbf{c}$ containing $\mathbf{e}$, the point $w_{\mathbf{c}}$ is the centroid of a set of points in $\mathbf{c}$ that includes point $w_{\mathbf{e}}$. Thus, $w_{\mathbf{e}}$ pulls $w_{\mathbf{c}}$ away from the plane P and none of the triangles formed from the quadrilateral around $\mathbf{e}$ contain v.

Now consider a bipolar edge $\mathbf{e}$ that is not incident on v. Let $\mathbf{c}$ be a grid cube containing $\mathbf{e}$ and v and let $\mathbf{f}$ be a facet of $\mathbf{c}$ that contains v but does not intersect $\mathbf{e}$. Let P be the plane containing $\mathbf{f}$. Let $w_{\mathbf{e}} \in \mathbf{e}$ be the linear interpolation point on $\mathbf{e}$. For every grid cube $\mathbf{c}'$ containing $\mathbf{e}$ and v, point $w_{\mathbf{e}}$ pulls $w_{\mathbf{c}'}$ away from plane P. Thus, none of the triangles formed from the quadrilateral around $\mathbf{e}$ contain v. ☐

As defined in Section 3.1, an interior grid edge $\mathbf{e}$ is a grid edge contained in four grid cubes. To prove Property 4, we first show that every interior bipolar grid edge intersects the isosurface.

Lemma 3.7. *If* $\mathbf{e}$ *is an interior bipolar grid edge, then* $\mathbf{e}$ *intersects the isosurface.*

Proof: Let $\mathbf{q}$ be the quadrilateral around $\mathbf{e}$. Let $\Sigma_{\mathbf{e}}$ be the isosurface patch generated by splitting $\mathbf{q}$ into triangles. By construction, $\Sigma_{\mathbf{e}}$ is contained in the envelope of $\mathbf{e}$ and its boundary is $\mathbf{q}$. Since $\mathbf{e}$ passes through $\mathbf{q}$ and has endpoints on the envelope of $\mathbf{e}$, edge $\mathbf{e}$ intersects $\Sigma_{\mathbf{e}}$. $\qquad\square$

As defined in Section 3.1, a grid cube or grid face is active if it contains at least one negative and at least one positive vertex. For each active grid cube $\mathbf{c}$, algorithm SURFACE NETS constructs a point $w_{\mathbf{c}} \in \mathbf{c}$ by approximating the intersections of the isosurface and the cube edges and taking the centroid of these intersections.

Lemma 3.8. *If v is a strictly positive vertex of an active cube $\mathbf{c}$ and $\mathbf{f}$ is a facet of $\mathbf{c}$ containing v, then $w_{\mathbf{c}}$ does not lie on $\mathbf{f}$.*

Proof: Let $\mathbf{f}'$ be the facet of $\mathbf{c}$ parallel to $\mathbf{f}$. We consider three cases based on whether $\mathbf{f}'$ is active, negative, or positive.

Case I: Facet $\mathbf{f}'$ is active.

Since $\mathbf{f}'$ is active, it has a bipolar edge $\mathbf{e}'$. Some point $r' \in \mathbf{e}'$ approximates the intersection of the isosurface and $\mathbf{e}'$. Point $w_{\mathbf{c}}$ is the centroid of a set of points in $\mathbf{c}$ that includes point r'. Since r' is not on $\mathbf{f}$, point r' pulls $w_{\mathbf{c}}$ away from $\mathbf{f}$ and $w_{\mathbf{c}}$ does not lie on $\mathbf{f}$.

Case II: Facet $\mathbf{f}'$ is negative.

Let $\mathbf{e} = (v, v')$ be the edge from v to a vertex v' in $\mathbf{f}'$. Since $\mathbf{f}'$ is negative, vertex v' is negative and edge $\mathbf{e}$ is bipolar. Some point $r \in \mathbf{e}$ approximates the intersection of the isosurface and $\mathbf{e}$. Since v is strictly positive, point r does not equal v and does not lie on $\mathbf{f}$. Thus r pulls $w_{\mathbf{c}}$ away from $\mathbf{f}$ and $w_{\mathbf{c}}$ does not lie on $\mathbf{f}$.

Case III: Facet $\mathbf{f}'$ is positive.

Since cube $\mathbf{c}$ is active, cube $\mathbf{c}$ must contain some negative vertex v''. Since $\mathbf{f}'$ is positive, negative vertex v'' must lie on $\mathbf{f}$. Let $\mathbf{e} = (v'', v')$ be the edge from v'' to a vertex v' in $\mathbf{f}'$. Since $\mathbf{f}'$ is positive, edge $\mathbf{e}$ is bipolar. Some point $r \in \mathbf{e}$ approximates the intersection of the isosurface and $\mathbf{e}$. Since v'' is negative, point r' does not equal v and does not lie on $\mathbf{f}$. Thus r' pulls $w_{\mathbf{c}}$ away from $\mathbf{f}$ and $w_{\mathbf{c}}$ does not lie on $\mathbf{f}$. $\qquad\square$

Property 4. *The isosurface intersects at exactly one point every interior bipolar grid edge whose positive endpoint is strictly positive.*

Proof: Let $\mathbf{e}$ be an interior edge with one endpoint scalar value greater than the isovalue and one endpoint scalar value less than the isovalue. Since neither scalar value equals the isovalue, the isosurface does not contain either endpoint (Property 3).

Let c_1, c_2, c_3, and c_4 be the four grid cubes containing e. By Lemma 3.7, the isosurface intersects e. By Lemma 3.8, none of the w_{c_i} lie on facets containing e. If the quadrilateral around e is split into four triangles, then each of these triangles intersect e at point w_e, proving the property. If the quadrilateral around e is split into two triangles by a diagonal d, then either both triangles intersect e at $d \cap e$ or one triangle intersects e at a single point and the other does not intersect e. □

Property 5. *The isosurface does not intersect any negative or strictly positive grid edges.*

Proof: Let e be a grid edge whose scalar values are both greater or both less than the isovalue. Since the scalar values at the endpoints of e do not equal the isovalue, the isosurface does not contain any endpoint (Property 3).

Since e is not a bipolar edge, edge e does not generate any isosurface triangles. Let e' be a bipolar grid edge. By construction, all the triangles generated by e' are contained in the envelope of e'. If e' does not intersect e, the envelope of e' does not intersect e and none of the isosurface triangles generated by e' intersect e. If e' shares an endpoint with e, then the envelope of e' intersects e only at this endpoint. Since the isosurface does not contain this endpoint, none of the triangles generated by e' intersect e. Thus e does not intersect the isosurface.□

3.2.6   Proof of Isosurface Property 6

To prove Property 6, we first show that a simplified version of the SURFACE NETS isosurface separates positive from negative vertices. We then apply a homotopy map from the simplified version to the SURFACE NETS isosurface, which preserves the separation.

For each cube c, let a_c be the center of c. For each edge e, let a_e be the midpoint of e. Remap the set of triangles Υ by mapping each isosurface vertex w_c to the cube center a_c and each isosurface vertex w_e to the edge midpoint a_e. Let $\widehat{\Upsilon}$ be the resulting set of triangles. Each interior bipolar edge corresponds to two or four triangles in $\widehat{\Upsilon}$ that together form a square that is orthogonal to the edge and passes through the edge midpoint. Set $|\widehat{\Upsilon}|$ is the union of all the triangles in $\widehat{\Upsilon}$. (See the example in Figure 3.10.)

Let a_0 be the center of the grid cube with the lowest x, y, and z coordinates and let a_1 be the center of the grid cube with the highest x, y, and z coordinates. If grid Γ is composed of unit cubes and has origin $(0,0,0)$ and vertex dimensions $n_x \times n_y \times n_z$, then a_0 is the point $(0.5, 0.5, 0.5)$ and a_1 is the point $(n_x - 0.5, n_y - 0.5, n_z - 0.5)$. Let $\mathbb{X}_\Gamma^*$ be the rectangular region from a_0 to a_1. (See Figure 3.11 for a 2D illustration.) The isosurface $|\widehat{\Upsilon}|$ is contained in $\mathbb{X}_\Gamma^*$. The boundary of this isosurface is contained in $\partial \mathbb{X}_\Gamma^*$.

We will show that a path that lies in $\mathbb{X}_\Gamma^*$ from a positive vertex of Γ_{inner} to a negative vertex of Γ_{inner} must intersect $|\widehat{\Upsilon}|$ (Lemma 3.13). We first consider a

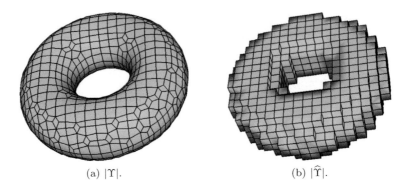

(a) $|\Upsilon|$.   (b) $|\widehat{\Upsilon}|$.

Figure 3.10. Dual contouring isosurfaces of torus. Only quadrilateral edges are displayed. (a) Isosurface $|\Upsilon|$ produced by SURFACE NETS. (b) Isosurface $|\widehat{\Upsilon}|$ produced by positioning vertices at cube centers and edge midpoints.

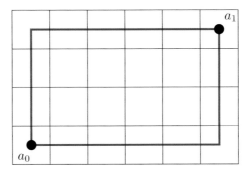

Figure 3.11. Two-dimensional illustration of rectangular region $\mathbb{X}_\Gamma^*$ (blue). Point a_0 is the center of the grid cube with the lowest x and y coordinates. Point a_1 is the center of the grid cube with the largest x and y coordinates.

path ζ in a single grid cube and show how it can be mapped to the edges and faces of the cube (Lemmas 3.9 and 3.10 and Corollary 3.11). We show that a path ζ from a negative vertex to a positive vertex of a grid cube will intersect $|\widehat{\Upsilon}|$ (Corollary 3.12). Finally, we divide grid cubes into negative and positive regions separated by $|\widehat{\Upsilon}|$ and prove that these regions agree on the faces between grid cubes. Any path in $\mathbb{X}_\Gamma^*$ from a negative vertex to a positive vertex of Γ_{inner} would have have to cross from a negative region to a positive one and would have to intersect $|\widehat{\Upsilon}|$ (Lemma 3.13).

We start by showing how paths in grid cube $\mathbf{c}$ can be mapped to faces and edges of $\mathbf{c}$. Define a mapping $\mu_\mathbf{c} : (\mathbf{c} - a_\mathbf{c}) \to \partial\mathbf{c}$ from $(\mathbf{c} - a_\mathbf{c})$ to the boundary of $\mathbf{c}$ as follows. For each point $x \in \mathbf{c} - a_\mathbf{c}$, there is a ray $\mathbf{r}_x$ from $a_\mathbf{c}$ through x that intersects $\partial\mathbf{c}$ at exactly one point. Let $\mu_\mathbf{c}(x)$ equal $\mathbf{r}_x \cap \partial\mathbf{c}$. Mapping $\mu_\mathbf{c}$ is called a radial projection of $(\mathbf{c} - a_\mathbf{c})$. (See Figure 3.12(a).)

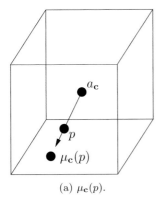
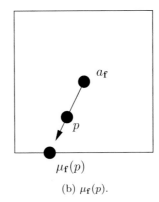

(a) $\mu_{\mathbf{c}}(p)$.   (b) $\mu_{\mathbf{f}}(p)$.

Figure 3.12. Radial projections. (a) Radial projection of $p \in \mathbf{c} - \{a_{\mathbf{c}}\}$ to $\mu_{\mathbf{c}}(p) \in \partial\mathbf{c}$. Point $a_{\mathbf{c}}$ is the center of cube $\mathbf{c}$. (b) Radial projection of $p \in \mathbf{f} - \{a_{\mathbf{f}}\}$ to $\mu_{\mathbf{f}}(p) \in \partial\mathbf{f}$. Point $a_{\mathbf{f}}$ is the center of face $\mathbf{f}$.

We define a similar mapping $\mu_{\mathbf{f}}$ for each grid facet $\mathbf{f}$. For each grid facet $\mathbf{f}$, let $a_{\mathbf{f}}$ be the center of $\mathbf{f}$. For each point $x \in \mathbf{f} - a_{\mathbf{f}}$, there is a ray $\mathbf{r}_x$ from $a_{\mathbf{f}}$ through x that intersects $\partial\mathbf{f}$ at exactly one point. Let $\mu_{\mathbf{f}}(x)$ equal $\mathbf{r}_x \cap \partial\mathbf{f}$. Mapping $\mu_{\mathbf{f}}$ is a radial projection of $(\mathbf{f} - a_{\mathbf{f}})$. (See Figure 3.12(b).)

Lemma 3.9.

1. *For all $x \in \mathbf{c} - a_{\mathbf{c}}$, $x \in |\widehat{\Upsilon}|$ if and only if $\mu_{\mathbf{c}}(x_{\mathbf{c}}) \in |\widehat{\Upsilon}|$.*

2. *For all $x \in \mathbf{f} - a_{\mathbf{f}}$, $x \in |\widehat{\Upsilon}|$ if and only if $\mu_{\mathbf{f}}(x_{\mathbf{f}}) \in |\widehat{\Upsilon}|$.*

Proof of 1: If x is in $|\widehat{\Upsilon}|$, then x lies in some square in $|\widehat{\Upsilon}|$ with vertex $a_{\mathbf{c}}$. Point $\mu_{\mathbf{c}}(x)$ also lies in this square so $\mu_{\mathbf{c}}(x)$ is in $|\widehat{\Upsilon}|$.

On the other hand, if $\mu_{\mathbf{c}}(x)$ is in $|\widehat{\Upsilon}|$, then $\mu_{\mathbf{c}}(x)$ lies in some square in $|\widehat{\Upsilon}|$ with vertex $a_{\mathbf{c}}$. Line segment $(a_{\mathbf{c}})$ is contained in this square and x lies on this lines segment, so x is in $|\widehat{\Upsilon}|$. □

The proof of Statement 2 in Lemma 3.9 is similar to the proof of Statement 1 and is omitted.

Let p be a point in $\mathbf{c} - a_{\mathbf{c}}$. Note that p is in $\mathbb{X}_{\Gamma}^{*}$ if and only if $\mu_{\mathbf{c}}(p)$ is in $\mathbb{X}_{\Gamma}^{*}$. Similarly, if p is a point in $\mathbf{f} - a_{\mathbf{f}}$, then p is in $\mathbb{X}_{\Gamma}^{*}$ if and only if $\mu_{\mathbf{f}}(p)$ is in $\mathbb{X}_{\Gamma}^{*}$.

The 1-skeleton of a cube is the union of all the vertices and edges of the cube.

Lemma 3.10. *Let $\mathbf{c}$ be a grid cube of Γ, let $\mathbf{e}$ be an edge of $\mathbf{c}$, and let ζ be a path in $\mathbf{c} \cap \mathbb{X}_{\Gamma}^{*}$ from a negative vertex of $\mathbf{c}$ to some point $p \in \mathbf{e} \cap \mathbb{X}_{\Gamma}^{*}$. If ζ does not intersect $|\widehat{\Upsilon}|$, then there is a negative endpoint v of $\mathbf{e}$ such that v is in $\mathbb{X}_{\Gamma}^{*}$ and the closed line segment $[v, p]$ does not intersect $|\widehat{\Upsilon}|$.*

Proof: If all the vertices in $\mathbf{c}$ are negative, then $|\widehat{\Upsilon}|$ does not intersect $\mathbf{c}$. The closed line segment from a vertex of $\mathbf{e}$ in $\mathbb{X}_\Gamma^*$ to p does not intersect $|\widehat{\Upsilon}|$, satisfying the conclusion.

Assume some vertex of $\mathbf{c}$ is positive. Since $\mathbf{c}$ contains at least one negative vertex, $|\widehat{\Upsilon}|$ intersects $\mathbf{c}$ and contains $a_\mathbf{c}$, the center of $\mathbf{c}$.

Let ζ be a path in $\mathbf{c} \cap \mathbb{X}_\Gamma^*$ that does not intersect $|\widehat{\Upsilon}|$ from a negative vertex v of $\mathbf{c}$ to point $p \in \mathbf{e} \cap \mathbb{X}_\Gamma^*$. By Lemma 3.9, there is a continuous map $\mu_\mathbf{c}$ that takes $\mathbf{c} - a_\mathbf{c}$ to $\partial\mathbf{c}$, maps points in $|\widehat{\Upsilon}|$ to points on $|\widehat{\Upsilon}|$ and points in $\mathbf{c} - |\widehat{\Upsilon}|$ to points in $\partial\mathbf{c} - |\widehat{\Upsilon}|$. Since path ζ does not intersect the $|\widehat{\Upsilon}|$, path ζ does not contain $a_\mathbf{c}$. Apply the mapping $\mu_\mathbf{c}$ to ζ to get a path ζ' lying in $\partial\mathbf{c}$ that connects v and p. Since ζ does not intersect $|\widehat{\Upsilon}|$, neither does ζ'. Since ζ is in $\mathbb{X}_\Gamma^*$, path ζ' is in $\mathbb{X}_\Gamma^*$.

Path ζ' may pass through the center $a_\mathbf{g}$ of some facets $\mathbf{g}$ of $\mathbf{c}$. If it does, then we can perturb it slightly to form a path $\zeta'' \subset \partial\mathbf{c}$ that does not pass through the center $a_\mathbf{g}$ of any facet $\mathbf{g}$ of $\mathbf{c}$ and remains in $\mathbb{X}_\Gamma^*$. If the perturbation is small enough, then ζ'' will not intersect $|\widehat{\Upsilon}|$.

By Lemma 3.9, there is a continuous map $\mu_\mathbf{g}$ that takes $\mathbf{g} - a_\mathbf{g}$ to $\partial\mathbf{g}$, maps points in $|\widehat{\Upsilon}|$ to points on $|\widehat{\Upsilon}|$, and points in $\mathbf{f} - |\widehat{\Upsilon}|$ to points in $\partial\mathbf{f} - |\widehat{\Upsilon}|$. Since ζ'' does not contain $a_\mathbf{c}$, applying the mapping $\mu_\mathbf{g}$ to ζ'' for each facet $\mathbf{g}$ gives a path ζ''' that intersects neither the interior of $\mathbf{c}$ nor that of any facet of $\mathbf{c}$. Thus, path ζ''' is contained in the 1-skeleton of the cube. Since ζ'' is in $\mathbb{X}_\Gamma^*$, path ζ''' is also in $\mathbb{X}_\Gamma^*$. Path ζ''' connects a negative vertex of $\mathbf{c}$ to p.

Every bipolar edge of $\mathbf{c}$ intersects $|\widehat{\Upsilon}|$ at its midpoint. Since path ζ''' does not intersect $|\widehat{\Upsilon}|$, it does not contain any bipolar edge of $\mathbf{c}$. To reach p, path ζ''' must pass through a sequence of grid vertices $(v_1, v_2, \ldots, v_k)$ of $\mathbf{c}$. Since ζ''' does not contain any bipolar edge of $\mathbf{c}$ and v_1 is negative, each of the vertices v_i must be negative. In particular, vertex v_k must be negative. Since v_k is the last grid vertex contained in ζ''', it must be an endpoint of edge $\mathbf{e}$. Since ζ''' is in $\mathbb{X}_\Gamma^*$, vertex v_k is in $\mathbb{X}_\Gamma^*$.

Path ζ''' contains a subpath $\zeta_\mathbf{e} \subset \mathbf{e}$ that connects v_k to p. Since $\zeta_\mathbf{e}$ does not intersect $|\widehat{\Upsilon}|$, the closed line segment $[v_k, p]$ does not intersect $|\widehat{\Upsilon}|$, proving the lemma.   □

Corollary 3.11. *Let $\mathbf{c}$ be a grid cube of grid Γ, let $\mathbf{f}$ be a face of $\mathbf{c}$, and let ζ be a path in $\mathbf{c} \cap \mathbb{X}_\Gamma^*$ from a negative vertex of $\mathbf{c}$ to some point $p \in \mathbf{f} \cap \mathbb{X}_\Gamma^*$. If ζ does not intersect $|\widehat{\Upsilon}|$, then there is a path in $\mathbf{f} \cap \mathbb{X}_\Gamma^*$ that does not intersect $|\widehat{\Upsilon}|$ from a negative vertex of $\mathbf{f}$ to p.*

Proof: We consider two cases depending upon whether p is at the center of $\mathbf{f}$.

Case I: Point p is not at the center of $\mathbf{f}$.

Since p is not at the center of $\mathbf{f}$, the map μ can be applied to p. Point $\mu_\mathbf{f}(p)$ is on an edge $\mathbf{e}$ of $\mathbf{f}$. Since p is in $\mathbb{X}_\Gamma^*$, so is $\mu_\mathbf{f}(p)$. By Lemma 3.10, edge $\mathbf{e}$ has a negative endpoint v such that v is in $\mathbb{X}_\Gamma^*$ and the closed line segment $[v, \mu_{\mathbf{f}}(p)]$ does not intersect $|\widehat{\Upsilon}|$. Since p is not in $|\widehat{\Upsilon}|$, line segment $[\mu_{\mathbf{f}}(p), p]$ does not intersect $|\widehat{\Upsilon}|$. Since p and $\mu_{\mathbf{f}}(p)$ are in $\mathbb{X}_\Gamma^*$, line segment $[\mu_{\mathbf{f}}(p), p]$ is in $\mathbb{X}_\Gamma^*$. Thus the path $(v, \mu_{\mathbf{f}}(p), p)$ is in $\mathbf{f} \cap \mathbb{X}_\Gamma^*$ and does not intersect $|\widehat{\Upsilon}|$.

Case II: Point p is at the center of $\mathbf{f}$.

Perturb p slightly to a point p' such that closed line segment $[p, p']$ does not intersect $|\widehat{\Upsilon}|$ and p' is in $\mathbb{X}_\Gamma^*$. By Case I, there is a path $\zeta_{\mathbf{f}}$ in $\mathbf{f} \cap \mathbb{X}_\Gamma^*$ from some negative vertex v of $\mathbf{f}$ to p'. Adding $[p', p]$ to this path gives a path ζ in $\mathbf{f} \cap \mathbb{X}_\Gamma^*$ from v to p that does not intersect $|\widehat{\Upsilon}|$. $\square$

Corollary 3.12. *Let $\mathbf{c}$ be a grid cube of Γ. If ζ is a path in $\mathbf{c} \cap \mathbb{X}_\Gamma^*$ from a negative vertex to a positive vertex of $\mathbf{c}$, then ζ intersects $|\widehat{\Upsilon}|$.*

Proof: Assume ζ was a path in $\mathbf{c} \cap \mathbb{X}_\Gamma^*$ from a negative vertex of $\mathbf{c}$ to a positive vertex v of $\mathbf{c}$ that did not intersect $|\widehat{\Upsilon}|$. Let $\mathbf{e}$ be an edge of $\mathbf{c}$ containing v. By Lemma 3.10, edge $\mathbf{e}$ has a negative endpoint v' such that v' is in $\mathbb{X}_\Gamma^*$ and the closed line segment $[v, v']$ does not intersect $|\widehat{\Upsilon}|$. Since v is positive and v' is negative, vertices v and v' must be distinct and (v, v') must equal $\mathbf{e}$. Since $\mathbf{e}$ has a positive endpoint v and a negative endpoint v', edge $\mathbf{e} = (v, v')$ intersects $|\widehat{\Upsilon}|$ at the midpoint of $\mathbf{e}$, a contradiction. Thus ζ intersects $|\widehat{\Upsilon}|$. $\square$

Note that Corollary 3.12 requires that ζ be contained in $\mathbb{X}_\Gamma^*$. If ζ is not in $\mathbb{X}_\Gamma^*$, then there may be a path that does not intersect $|\widehat{\Upsilon}|$ from a negative vertex to a positive vertex in $\mathbf{c}$.

Lemma 3.13. *Every path ζ in $\mathbb{X}_\Gamma^*$ from a positive vertex of Γ_{inner} to a negative vertex of Γ_{inner} intersects $|\widehat{\Upsilon}|$.*

Proof: For each grid cube $\mathbf{c}$ of grid Γ, let the negative region for $\mathbf{c} \cap \mathbb{X}_\Gamma^*$ be the set of points that can be reached by a path in $\mathbb{X}_\Gamma^*$ from a negative vertex such that the path does not intersect $|\widehat{\Upsilon}|$. Let the positive region for $\mathbf{c} \cap \mathbb{X}_\Gamma^*$ be all the other points in $\mathbf{c} \cap \mathbb{X}_\Gamma^*$. The positive region contains $|\widehat{\Upsilon}| \cap \mathbf{c}$ and is closed. Any point within the positive region that does not lie on $|\widehat{\Upsilon}|$ has a neighborhood in $\mathbf{c} \cap \mathbb{X}_\Gamma^*$ contained in the positive region.

By construction of $\widehat{\Upsilon}$, no grid vertex is contained in $|\widehat{\Upsilon}|$. Since a zero-length path connects a negative vertex to itself, every negative vertex of $\mathbf{c}$ in $\mathbf{c} \cap \mathbb{X}_\Gamma^*$ is contained in the negative region. On the other hand, by Corollary 3.12 any path in $\mathbf{c} \cap \mathbb{X}_\Gamma^*$ from a negative vertex of $\mathbf{c}$ to a positive vertex $\mathbf{c}$ must intersect $|\widehat{\Upsilon}|$.

We claim that positive and negative regions agree on shared boundaries of grid cubes. Let R_1^- and R_2^- be the negative regions of $\mathbf{c}_1 \cap \mathbb{X}_\Gamma^*$ and $\mathbf{c}_2 \cap \mathbb{X}_\Gamma^*$ for two adjacent grid cubes, $\mathbf{c}_1$ and $\mathbf{c}_2$. Let $\mathbf{f}$ be the intersection of the two grid cubes. Note that $\mathbf{f}$ may be a grid vertex, a grid edge, or a square grid facet.

By Corollary 3.11, if point p lies in $R_1^- \cap \mathbf{f}$, then some path in $\mathbf{f} \cap \mathbb{X}_\Gamma^*$ does not intersect $|\widehat{\Upsilon}|$ and connects a negative vertex to p. Since $\mathbf{f}$ is a face of $\mathbf{c}_2$, this path also lies in $\mathbf{c}_2$ and so p is in R_2^-. A symmetric argument shows that if p is in R_2^-, then p is in R_1^-. Thus the positive and negative regions agree on shared boundaries of grid cubes.

Let R^+ be the union of the positive regions of $\mathbf{c} \cap \mathbb{X}_\Gamma^*$ over all the grid cubes $\mathbf{c} \in \Gamma$. Consider a path ζ in $\mathbb{X}_\Gamma^*$ from a positive grid vertex to a negative one. The positive grid vertex lies in R^+ while the negative one does not. Thus ζ must intersect some point p on the boundary of R^+ where it crosses out of R^+. Every neighborhood of p must contain points that are not in R^+.

Since R^+ is closed, point p lies in R^+. Since positive regions agree on the grid cube boundaries, point p lies in the positive region of each grid cube containing p. Assume p is not in $|\widehat{\Upsilon}|$. Within each grid cube containing p, some neighborhood of p is contained in the positive region for that grid cube. The union of those neighborhoods is a neighborhood of p within the grid and is contained in R^+. Thus ζ does not cross out of R^+ at p. We conclude that p must lie on $|\widehat{\Upsilon}|$ and that ζ intersects $|\widehat{\Upsilon}|$. $\qquad\square$

Set $|\Upsilon|$ is the union of all the triangles in Υ. Therefore, set $|\Upsilon|$ is the set of points on the isosurface created by SURFACE NETS.

Property 6.   *The isosurface strictly separates the strictly positive vertices of* Γ_{inner} *from the negative vertices of* Γ_{inner}.

Proof: Each vertex of a triangle in $\widehat{\Upsilon}$ is the midpoint $a_\mathbf{e}$ of some bipolar edge $\mathbf{e}$ or the center $a_\mathbf{c}$ of some active cube $\mathbf{c}$. For bipolar edges $\mathbf{e}$ and active cubes $\mathbf{c}$, define the homotopy

$$\eta(a_\mathbf{e}, \alpha) = (1 - \alpha)a_\mathbf{e} + \alpha w_\mathbf{e},$$

$$\eta(a_\mathbf{c}, \alpha) = (1 - \alpha)a_\mathbf{c} + \alpha w_\mathbf{c}.$$

Linearly extend the homotopy η to all the triangles in $\widehat{\Upsilon}$ giving a homotopy $\eta : |\widehat{\Upsilon}| \to |\Upsilon|$.

By Lemma 3.13, every path in $\mathbb{X}_\Gamma^*$ from a positive vertex of Γ_{inner} to a negative vertex of Γ_{inner} intersects set $|\widehat{\Upsilon}|$. Set $|\widehat{\Upsilon}|$ does not contain any vertices of Γ_{inner}. Map η is a homotopy map that never passes through any negative or strictly positive vertices in Γ_{inner}. Since $w_\mathbf{c}$ is contained in cube $\mathbf{c}$, set $\{\eta(x, \alpha) : x \in |\widehat{\Upsilon}| \cap \partial \mathbb{X}_\Gamma^*\}$ is a subset of $\mathbb{R}^3 - |\Gamma_{\mathrm{inner}}|$. By Lemma B.19 in Appendix B, every path in $|\Gamma_{\mathrm{inner}}|$ from a negative to a strictly positive vertex of Γ_{inner} intersects $|\Upsilon|$. Thus $|\Upsilon|$ separates the negative vertices of Γ_{inner} from the strictly positive vertices of Γ_{inner}. By Property 3, isosurface $|\Upsilon|$ does not contain any negative or strictly positive vertices of Γ_{inner}. Thus $|\Upsilon|$ strictly separates the negative vertices of Γ_{inner} from the strictly positive vertices of Γ_{inner}. $\qquad\square$

3.2.7   Proof of Isosurface Property 7

Property 7. *Set* Υ *does not contain any zero-area triangles or duplicate triangles and the triangles in* Υ *form a triangulation of the isosurface.*

Proof: Let $\mathbf{q}$ be a quadrilateral around edge $\mathbf{e}$. Since the isovalue does not equal the scalar value of any grid vertex, the points $w_\mathbf{c}$ are in the interior of their respective cubes $\mathbf{c}$. Thus no two vertices of $\mathbf{e}$ coincide and no four vertices of $\mathbf{e}$ are colinear. Three vertices can be colinear, but the degenerate triangle containing the three vertices has a 180-degree angle and so the diagonal splitting this triangle will be selected. Thus, if $\mathbf{q}$ is split into two triangles, neither triangle will have zero area.

Let $(w_{\mathbf{c}_1}, w_{\mathbf{c}_2})$ be an edge of $\mathbf{q}$. Since no point $w_\mathbf{c}$ lies on a grid facet, points $w_{\mathbf{c}_1}$, $w_{\mathbf{c}_2}$, and $w_\mathbf{e}$ are not colinear. Thus triangle $\Delta(w_{\mathbf{c}_1}, w_{\mathbf{c}_2}, w_\mathbf{e})$ does not have zero area.

Each quadrilateral around an edge $\mathbf{e}$ is replaced by two or four triangles. By construction, this triangular patch is contained in the envelope of bipolar edge $\mathbf{e}$. Since the envelopes of two different bipolar edges intersect only at shared vertices, edges, or facets, the triangles in different triangular patches intersect only at shared vertices or edges. Thus Υ forms a triangulation of the isosurface.   □

3.2.8   Non-Manifold Isosurface Patches

Let $\mathbf{c}$ be some active grid cube. Since all the isosurface quadrilaterals created by bipolar edges of $\mathbf{c}$ are incident on $w_\mathbf{c}$, there are no topological choices about the construction of the isosurface patch in $\mathbf{c}$. (There is a choice of the triangulation, but this choice affects the geometry, not the topology, of the isosurface.) Thus, the ambiguous MARCHING CUBES configurations listed in Section 2.3.5 do not create any ambiguity for the SURFACE NETS algorithm.

However, the ambiguous MARCHING CUBES configurations do create another problem for SURFACE NETS. SURFACE NETS applied to any of these configurations creates a non-manifold in the neighborhood of $w_\mathbf{c}$. The non-manifold around $w_\mathbf{c}$ may be viewed as the price to pay for avoiding any ambiguity.

The non-manifold configurations and their isosurface patches are illustrated in Figure 3.13. Note that even the isosurface patches for configurations 2C and 6C are not manifolds, even though none of the facets of these configurations are ambiguous. Note also that the isosurface quadrilaterals are artificially truncated. They actually extend further into adjacent cubes.

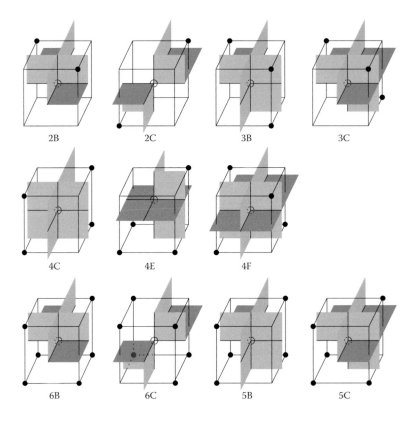

Figure 3.13. Non-manifold isosurface patches. Black vertices are positive.

3.2.9   Convex Polyhedral Meshes

Algorithm SURFACE NETS extends easily to convex polyhedral meshes. A mesh element is active if it contains at least one negative and at least one positive vertex. For each active mesh element $\mathbf{c}$, construct an isosurface vertex $w_{\mathbf{c}}$. For each bipolar mesh edge contained in mesh elements $\mathbf{c}_1, \mathbf{c}_2, \ldots, \mathbf{c}_k$, add an isosurface polygon with vertices $w_{\mathbf{c}_1}, w_{\mathbf{c}_2}, \ldots, w_{\mathbf{c}_k}$. Note that this polygon is not necessarily a quadrilateral.

To position the isosurface vertices within each grid cube, use linear interpolation as described in Section 1.7.2 to approximate the intersection of the isosurface and all the bipolar edges. Take the centroid of all the approximation points as the location of the isosurface vertex.

The isosurface polygon $(w_{\mathbf{c}_1}, w_{\mathbf{c}_2}, \ldots, w_{\mathbf{c}_k})$ is not necessarily planar. To create a piecewise linear surface, triangulate the isosurface polygons.

3.3 Dual Marching Cubes

The SURFACE NETS algorithm places a single isosurface vertex in each active grid cube. This single vertex causes the surface to be a non-manifold for some configurations. The DUAL MARCHING CUBES algorithm by Nielson [Nielson, 2004] attempts to remedy this problem by sometimes adding more than one vertex per cube.

For each bipolar grid edge, DUAL MARCHING CUBES creates an isosurface quadrilateral connecting four isosurface vertices in the four incident grid cubes. If the grid cube contains more than one isosurface vertex, the appropriate vertex is chosen by using a lookup table. The isosurface quadrilaterals are then split into two triangles.

As will be discussed in Section 3.3.3, there is a correspondence between the DUAL MARCHING CUBES lookup table and the MARCHING CUBES lookup table (Figure 2.16). The name DUAL MARCHING CUBES comes from this correspondence between the two lookup tables.

For most configurations in Figure 3.13, DUAL MARCHING CUBES creates two, three or four isosurface vertices in the grid cube. However, for two configurations, 2B and 3B, DUAL MARCHING CUBES still adds only a single isosurface vertex. As discussed in Section 3.3.4, these configurations can create non-manifold surfaces.

A small modification of DUAL MARCHING CUBES changes the isosurface patches for some instances of configurations 2B and 3B, avoiding non-manifold edges created by those configurations. The modified algorithm, called MANIFOLD DUAL MARCHING CUBES, produces an isosurface which is guaranteed to be a manifold if the isovalue does not equal the scalar value of any grid vertex. The modified algorithm is described in Section 3.3.5.

We first describe a two dimensional version of the algorithm, which we call DUAL MARCHING SQUARES.

3.3.1 Dual Marching Squares

Input to the DUAL MARCHING SQUARES algorithm is an isovalue and a set of scalar values at the vertices of a two-dimensional regular grid. The algorithm has three steps. (See Figure 3.14.) Read in the isocontour lookup table from a preconstructed data file. For each interior bipolar grid edge **e**, retrieve from the lookup table the two isocontour vertices associated with **e** and add an isocontour edge between those vertices. Assign geometric locations to the isocontour vertices based on the scalar values at the square edge endpoints. We explain the last two steps below.

Since each square vertex is either positive or negative and a square has four vertices, there are $2^4 = 16$ different configurations of square vertex labels. These configurations are listed in Figure 2.2 in Chapter 2.

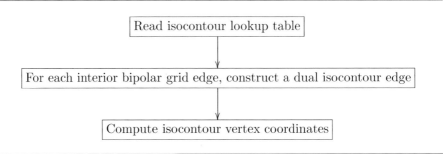

Figure 3.14. DUAL MARCHING SQUARES.

The number and connectivity of the isocontour vertices within each square $\mathbf{c}$ is determined from the configuration of the square's vertex labels. If the configuration has two bipolar edges, then the configuration has a single vertex, $w_{\mathbf{c}}^1$, associated with both bipolar edges. This vertex will be an endpoint of the isocontour edges dual to each bipolar edge. If the configuration has four bipolar edges, then the configuration has two isocontour vertices, $w_{\mathbf{c}}^1$ and $w_{\mathbf{c}}^2$, one for each positive vertex. Associate the isocontour vertex corresponding to negative vertex v with the two square edges incident on v. Similarly, associate the isocontour vertex corresponding to negative vertex v' with the two square edges incident on v'. Each vertex will be an endpoint of two isocontour edges dual to the associated square edges. The isocontour generated for each configuration is illustrated in Figure 3.18.

The isocontour lookup table stores the association between isocontour vertices and square edges. An entry in the table is referenced as $\mathsf{Table}[\kappa, e]$, where κ is a configuration of square vertex labels and $\mathbf{e}$ is a bipolar edge of κ. Each entry is 1 or 2 corresponding to isocontour vertex w^1 or vertex w^2, respectively. If a configuration κ has two bipolar edges, $\mathbf{e}_1$ and $\mathbf{e}_2$, then $\mathsf{Table}[\kappa, \mathbf{e}_1] = \mathsf{Table}[\kappa, \mathbf{e}_2] = 1$. representing the single isocontour vertex, w^1, for configuration κ. If a configuration κ has four bipolar edges, $\mathbf{e}_1$, $\mathbf{e}_2$, $\mathbf{e}_3$, and $\mathbf{e}_4$, and $\mathbf{e}_1$ and $\mathbf{e}_2$ share a negative vertex, and $\mathbf{e}_3$ and $\mathbf{e}_4$ share a negative vertex, then $\mathsf{Table}[\kappa, \mathbf{e}_1] = \mathsf{Table}[\kappa, \mathbf{e}_2] = 1$, representing vertex w^1, and $\mathsf{Table}[\kappa, \mathbf{e}_3] = \mathsf{Table}[\kappa, \mathbf{e}_4] = 2$, representing w^2. Vertices w^1 and w^2 are the two isocontour vertices for configuration κ.

The algorithm for positioning the isocontour vertices, $w_{\mathbf{c}}^m$, is similar to the one in SURFACE NETS, but uses only the bipolar edges associated with $w_{\mathbf{c}}^m$. For each bipolar edge $\mathbf{e}$, approximate the intersection $w_{\mathbf{e}}$ of $\mathbf{e}$ and the isocontour using linear interpolation as described in Section 1.7.2. Point $w_{\mathbf{e}}$ equals $(1-\alpha)p + \alpha q$ where $\alpha = (\sigma - s_p)/(s_q - s_p)$. Locate isocontour vertex $w_{\mathbf{c}}^m$ at the midpoint of line segment $(w_{\mathbf{e}}, w_{\mathbf{e}'})$, where $\mathbf{e}$ and $\mathbf{e}'$ are the edges associated with $w_{\mathbf{c}}^m$.

The DUAL MARCHING SQUARES algorithm is presented in Figures 3.14–3.17. Function `LinearInterpolation`, called by this algorithm, is defined in Algorithm 1.1 in Section 1.7.2.

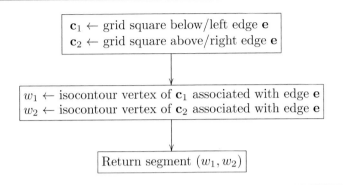

Figure 3.15. Construct isocontour edge dual to grid edge $\mathbf{e}$.

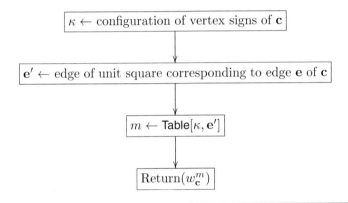

Figure 3.16. Return isocontour vertex of grid square $\mathbf{c}$ associated with edge $\mathbf{e}$.

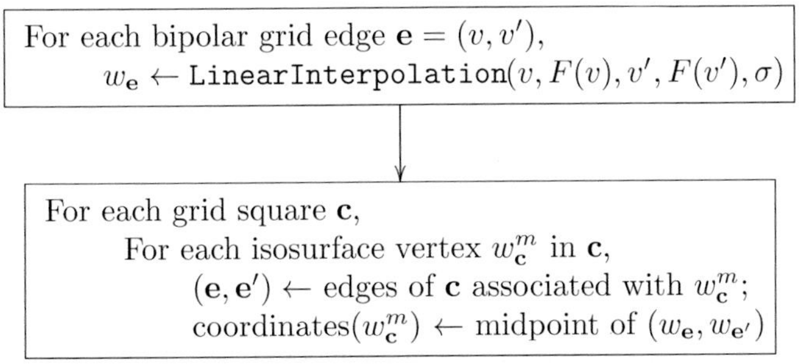

For each bipolar grid edge $\mathbf{e} = (v, v')$,
    $w_{\mathbf{e}} \leftarrow$ LinearInterpolation$(v, F(v), v', F(v'), \sigma)$

For each grid square $\mathbf{c}$,
    For each isosurface vertex $w_{\mathbf{c}}^m$ in $\mathbf{c}$,
        $(\mathbf{e}, \mathbf{e}') \leftarrow$ edges of $\mathbf{c}$ associated with $w_{\mathbf{c}}^m$;
        coordinates$(w_{\mathbf{c}}^m) \leftarrow$ midpoint of $(w_{\mathbf{e}}, w_{\mathbf{e}'})$

Figure 3.17. Compute vertex coordinates. $F(v)$ is the scalar value of v and σ is the isovalue.

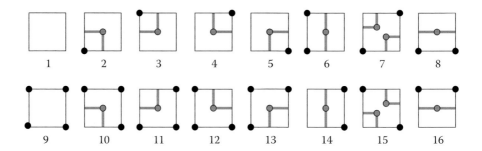

Figure 3.18. DUAL MARCHING SQUARES isocontours.

3.3.2   3D Dual Marching Cubes

Input to the DUAL MARCHING CUBES algorithm is an isovalue and a set of scalar values at the vertices of a three-dimensional regular grid. The algorithm has four steps. (See Figure 3.19.) Read in the isocontour lookup table from a preconstructed data file. For each interior bipolar grid edge **e**, retrieve from the lookup table the four isosurface vertices associated with **e** and add an isosurface quadrilateral between those vertices. Assign geometric locations to the isocontour vertices based on the scalar values at the square edge endpoints. Triangulate the isosurface quadrilaterals. We explain the last three steps next.

Since each cube vertex is either positive or negative and a cube has eight vertices, there are $2^8 = 256$ different configurations of cube vertex labels. These configurations are listed in Figure 2.15 in Chapter 2.

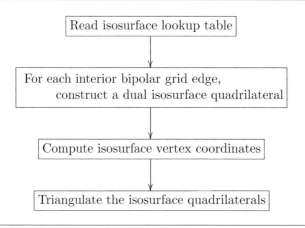

Figure 3.19. DUAL MARCHING CUBES.

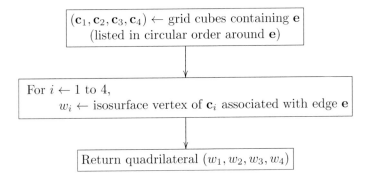

Figure 3.20. Construct isosurface quadrilateral dual to grid edge $\mathbf{e}$.

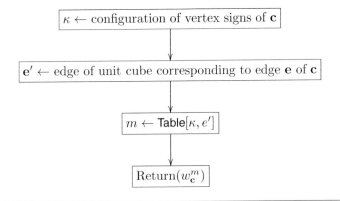

Figure 3.21. Return isosurface vertex of grid cube $\mathbf{c}$ associated with edge $\mathbf{e}$.

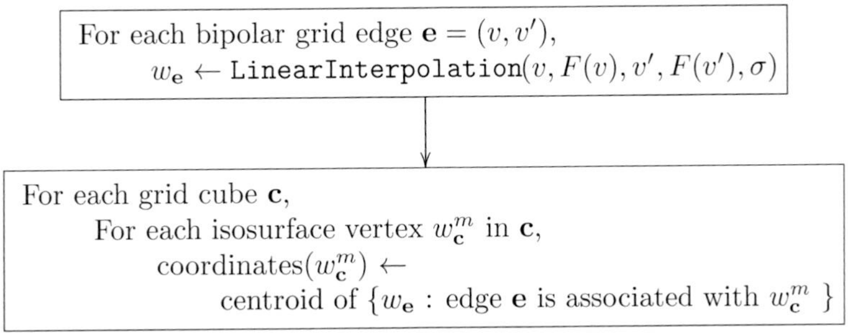

> For each bipolar grid edge $\mathbf{e} = (v, v')$,
> $w_\mathbf{e} \leftarrow$ LinearInterpolation$(v, F(v), v', F(v'), \sigma)$

> For each grid cube $\mathbf{c}$,
>   For each isosurface vertex $w_\mathbf{c}^m$ in $\mathbf{c}$,
>   coordinates$(w_\mathbf{c}^m) \leftarrow$
>   centroid of $\{w_\mathbf{e} :$ edge $\mathbf{e}$ is associated with $w_\mathbf{c}^m \}$

Figure 3.22. Compute vertex coordinates. $F(v)$ is the scalar value of v. σ is the isovalue.

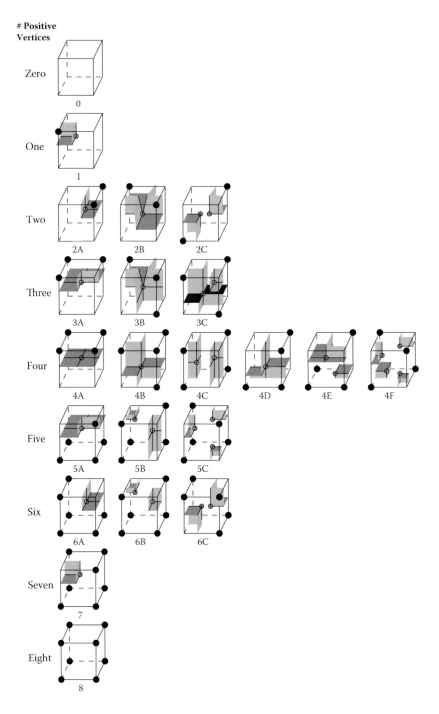

Figure 3.23. DUAL MARCHING CUBES isosurface patches.

The number and connectivity of the isosurface vertices within each cube is determined from the configuration of the cube's vertex labels. There are 22 distinct cube configurations after rotation and/or reflection. The isosurface vertices and edges for the twenty-two distinct configurations are represented in Figure 3.23.

Configuration 1 with a single positive vertex has a single isosurface vertex associated with its three bipolar edges. This vertex will be an endpoint of the isosurface edges dual to each bipolar edge. Configuration 2A with two adjacent positive vertices also has a single isosurface vertex associated with its four bipolar edges. Configuration 2C with two nonadjacent positive vertices, v and v', has two isosurface vertices. Isosurface quadrilaterals incident to one isosurface vertex intersect the three cube edges incident on v. Isosurface quadrilaterals incident to the other isosurface vertex intersect the three cube edges incident on v'. Configuration 4F generates four isosurface vertices, the most for any configuration.

The isosurface lookup table, Table stores the association between isosurface vertices and cube edges. An entry in the table is referenced as $\mathsf{Table}[\kappa, e]$, where κ is a configuration of cube vertex labels and $\mathbf{e}$ is a bipolar edge of κ. Each entry is 1, 2, 3, or 4 corresponding to isosurface vertex w_1, w_2, w_3, or w_4. An additional table, NumVertices, stores the number of isosurface vertices for configuration κ.

For instance, when κ is configuration 1 with three bipolar edges, $\mathbf{e}_1, \mathbf{e}_2$, and $\mathbf{e}_3$, then $\mathsf{Table}[\kappa, \mathbf{e}_1] = \mathsf{Table}[\kappa, \mathbf{e}_2] = \mathsf{Table}[\kappa, \mathbf{e}_3] = 1$ and $\mathsf{NumVertices}[\kappa] = 1$. When κ is configuration 2C with two positive grid vertices, v and v', with grid edges $\mathbf{e}_1, \mathbf{e}_2$, and $\mathbf{e}_3$ incident on v and grid edges $\mathbf{e}'_1, \mathbf{e}'_2$, and $\mathbf{e}'_3$ incident on v', then $\mathsf{Table}[\kappa, \mathbf{e}_1] = \mathsf{Table}[\kappa, \mathbf{e}_2] = \mathsf{Table}[\kappa, \mathbf{e}_3] = 1$ and $\mathsf{Table}[\kappa, \mathbf{e}'_1] = \mathsf{Table}[\kappa, \mathbf{e}'_2] = \mathsf{Table}[\kappa, \mathbf{e}'_3] = 2$ and $\mathsf{NumVertices}[\kappa] = 2$. (See Figure 3.24.)

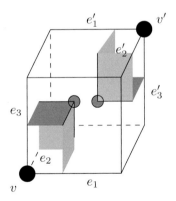

Figure 3.24. Configuration 2C: Grid edges intersected by the DUAL MARCHING CUBES isosurface.

Input   : F is a 3D array of scalar values.
            Coord is a 3D array of (x, y, z) coordinates.
            σ is an isovalue.
Result   : Set Quad of isosurface quadrilaterals and array VertexCoord of
            vertex coordinates.

DualMarchingCubes(F,Coord, σ, Quad)

1 Read Dual Marching Cubes lookup table into Table;
   /* Assign "+" or "−" signs to each vertex   */
2 foreach grid vertex (i, j, k) do
3 | if $F[i, j, k] < \sigma$ then $\mathsf{Sign}[i, j, k] \leftarrow$ "−";
4 | else $\mathsf{Sign}[i, j, k] \leftarrow$ "+";   /* $F[i, j, k] \geq \sigma$ */
5 end
6 ConstructQuadrilaterals (Sign, Quad);
7 ComputeVertexCoordinates (F, Coord, σ, Sign, VertexCoord);

Algorithm 3.1. DUAL MARCHING CUBES.

The algorithm for positioning the isosurface vertices is similar to the one in SURFACE NETS but uses only the bipolar edges associated with $w_{\mathbf{c}}^m$. For each bipolar edge $\mathbf{e}$, approximate the intersection $w_{\mathbf{e}}$ of $\mathbf{e}$ and the isosurface using linear interpolation as described in Section 1.7.2. Point $w_{\mathbf{e}}$ equals $(1 - \alpha)p + \alpha q$ where $\alpha = (\sigma - s_p)/(s_q - s_p)$. Let $\mathbf{e}_1, \mathbf{e}_2, \ldots, \mathbf{e}_k$ be the bipolar grid edges associated with isosurface vertex $w_{\mathbf{c}}^m$. Locate isosurface vertex $w_{\mathbf{c}}^m$ at the centroid, $(w_{\mathbf{e}_1} + w_{\mathbf{e}_2} + \cdots + w_{\mathbf{e}_k})/k$, of the points $w_{\mathbf{e}_1}, w_{\mathbf{e}_2}, \ldots, w_{\mathbf{e}_k}$, where $\mathbf{e}_1, \mathbf{e}_2, \ldots, \mathbf{e}_k$ are the bipolar grid edges associated with isosurface vertex $w_{\mathbf{c}}^m$.

As with SURFACE NETS, we replace each quadrilateral around bipolar edge $\mathbf{e}$ by two or four triangles, depending on the envelope of $\mathbf{e}$. To construct the isosurface patch with two triangles, we use the diagonal that minimizes the maximum triangle angle.

The DUAL MARCHING CUBES algorithm is presented in Algorithm 3.1. Function LinearInterpolation, called by this algorithm, is defined in Algorithm 1.1 in Section 1.7.2.

3.3.3 Lookup Table Duality

The DUAL MARCHING CUBES lookup table in Figure 3.23 is dual to the MARCHING CUBES lookup table in Figure 2.16. Let κ be a configuration of positive and negative cube vertices.

Each connected component λ_{dual} of the isosurface patch for κ in the DUAL MARCHING CUBES lookup table corresponds to a connected component λ of the isosurface patch for κ in the MARCHING CUBES lookup table. Each isosurface

---

ConstructQuadrilaterals(Sign, Quad)

1 Quad ← ∅;
2 foreach interior grid bipolar grid edge e do
   /* construct isosurface quadrilateral dual to e */
3 for $l = 1$ to 4 do
4   (i, j, k) ← index of l'th grid cube (in circular order) containing e;
5   κ ← (Sign$[i, j, k]$, Sign$[i, j, k{+}1]$, …, Sign$[i{+}1, j{+}1, k{+}1]$);
6   m ← GetVertexIndex(i, j, k, e, κ);
7   QuadVert$[l]$ ← (i, j, k, m);
8 end
9 Quad ← Quad ∪ {(QuadVert$[1]$, QuadVert$[2]$, QuadVert$[3]$, QuadVert$[4]$)};
10 end

ComputeVertexCoordinates(F, Coord, σ, Sign, VertexCoord)
  /* Approximate intersection of isocontour and bipolar edges */
1 foreach bipolar grid edge e with endpoints (i_1, j_1, k_1) and (i_2, j_2, k_2) do
2 w_e ← LinearInterpolation
3   (Coord$[i_1, j_1, k_1]$, F$[i_1, j_1, k_1]$, Coord$[i_2, j_2, k_2]$, F$[i_2, j_2, k_2]$, σ);
4 end
  /* Compute vertex coordinates */
5 foreach grid cube (i, j, k) do
6 κ ← (Sign$[i, j, k]$, Sign$[i, j, k{+}1]$, …, Sign$[i{+}1, j{+}1, k{+}1]$);
7 for m ← 1 to NumVertices$[\kappa]$ do
8   VertexCoord$[i, j, m]$ ← $(0, 0, 0)$;
9   NumIncident$[m]$ ← 0;
10 end
11 foreach bipolar edge e of grid cube (i, j, k) do
12   m ← GetVertexIndex(i, j, k, e, κ);
13   VertexCoord$[i, j, k, m]$ ← VertexCoord$[i, j, k, m]$ + w_e;
14   NumIncident$[m]$ ← NumIncident$[m]$ + 1;
15 end
16 for m ← 1 to NumVertices$[\kappa]$ do
17   VertexCoord$[i, j, m]$ ← VertexCoord$[i, j, m]$/NumIncident$[m]$;
18 end
19 end

Function GetVertexIndex(i, j, k, e, κ)

1 e' ← edge of unit cube corresponding to edge e in cube (i, j, k);
2 m ← Table$[\kappa, e']$;
3 return (m);

Algorithm 3.2. Subroutines for DUAL MARCHING CUBES.

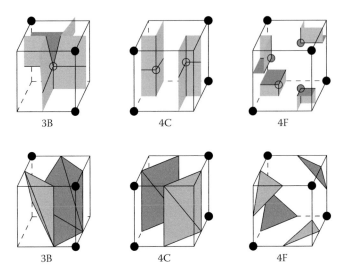

Figure 3.25. Examples of isosurface patches from the DUAL MARCHING CUBES lookup table for configurations 3B, 4C, and 4F and the corresponding patches for MARCHING CUBES.

quadrilateral in λ_{dual} intersecting bipolar edge **e** corresponds to an isosurface vertex in λ lying on bipolar edge **e**.

On the other hand, λ_{dual} contains one and only one isosurface vertex. Thus, each isosurface vertex of the isosurface patch for κ in the DUAL MARCHING CUBES lookup table corresponds to a connected component of the isosurface patch for κ in the MARCHING CUBES lookup table.

Figure 3.25 contains three examples of isosurface patches for the DUAL MARCHING CUBES lookup table and the corresponding isosurface patches for MARCHING CUBES.

3.3.4 Non-Manifold Isosurface Patches

By adding more than one isosurface vertex per grid cube, DUAL MARCHING CUBES avoids many of the SURFACE NETS non-manifold constructions discussed in Section 3.2.8. Unfortunately, DUAL MARCHING CUBES still has some non-manifold constructions. As shown in Figure 3.26, if cubes with configurations 2B and 3B are adjacent, then four isosurface quadrilaterals can meet in a single isosurface edge. The problem is that the two near-vertical edges in configurations 2B and 3B in Figure 3.23 have the same two endpoints, and so merge into a single isosurface edge.

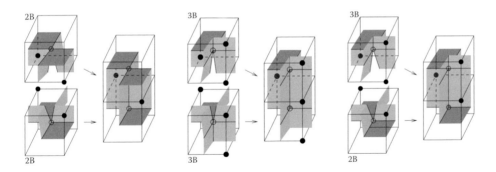

Figure 3.26. Dual Marching Cubes non-manifold constructions.

3.3.5 Manifold Dual Marching Cubes

The non-manifold problem described in the previous section could be solved by adding more complex isosurface patches for configurations 2B and 3B. However, there is a simple solution that avoids adding complexity to the isosurface patches.

Note that configurations 2B and 3B contain exactly one ambiguous facet. They are the only configurations in Figure 3.23 containing exactly one ambiguous facet. The only case in which four isosurface quadrilaterals share an edge is when a grid cube with configuration 2B or 3B shares its ambiguous facet with another grid cube with configuration 2B or 3B.

As discussed in Section 2.3.5 and illustrated in Figure 2.22, every ambiguous configuration has two "natural" Marching Cubes isosurface patches, one separating negative vertices and one separating positive vertices. Similarly, every ambiguous configuration has two "natural" Dual Marching Cubes isosurface patches. The two Dual Marching Cubes isosurface patches for configurations 2B and 3B are illustrated in Figure 3.27. For each configuration, one isosurface patch has a single connected component and the other isosurface patch has two connected components.

Manifold Dual Marching Cubes avoids creating non-manifold edges by using isosurface patches 2B-II and 3B-II instead of 2B-I and 3B-I whenever two cubes with configurations 2B and 3B share their ambiguous facets. Note that the isosurface patches for both cubes must be replaced, otherwise the patches will not align along their boundaries as discussed in Section 2.3.5. Moreover, if a cube **c** with configuration 2B or 3B shares an ambiguous facet with a cube **c**′ with configuration other than 2B or 3B, isosurface patch 2B-I or 3B-I should be used. Cube **c**′ will have two isosurface vertices connected by two distinct isosurface edges to the isosurface vertex in **c**, and each isosurface edge will have only two incident quadrilaterals. If isosurface patch 2B-II or 3B-II were used for cube **c**, then there would be misalignment between the isosurface patches in **c** and **c**′.

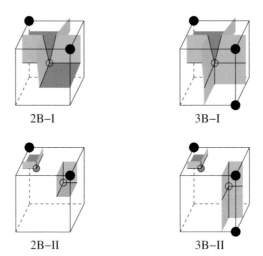

2B–I   3B–I

2B–II   3B–II

Figure 3.27. Isosurface patches 2B-I, 2B-II, 3B-I, and 3B-II for MANIFOLD DUAL MARCHING CUBES.

MANIFOLD DUAL MARCHING CUBES uses a modification of the algorithm in Figure 3.21 for retrieving the isosurface vertex of grid cube c associated with edge e. Let Table1 be the DUAL MARCHING CUBES isosurface lookup table as described in Section 3.3.2. Let Table2 be the same as Table1, except that configurations 2B and 3B use isosurface patches 2B-II and 3B-II. If cube c with configuration type 2B or 3B shares its ambiguous facet with a cube with configuration type 2B or 3B, then return Table2$[\kappa, e']$ instead of Table1$[\kappa, e']$ (Figure 3.28). All other routines are exactly the same.

3.3.6 Isosurface Properties

Properties 1–6 of the MANIFOLD DUAL MARCHING CUBES isosurface are the same as Properties 1–6 of the SURFACE NETS isosurface. The isosurface produced by MANIFOLD DUAL MARCHING CUBES depends on the placement of the vertices within each grid cube and the construction of the surface bounded by each quadrilateral. The properties listed below apply to MANIFOLD DUAL MARCHING CUBES where vertices are placed at the centroid of edge-isosurface intersection points and each quadrilateral is triangulated into two or four triangles. A quadrilateral around grid edge e is triangulated into two triangles, Δ_1 and Δ_2, using the diagonal that minimizes the maximum triangle angle. A quadrilateral is triangulated into four triangles if Δ_1 or Δ_2 are not in the envelope of e.

MANIFOLD DUAL MARCHING CUBES returns a finite set, Υ, of triangles. The isosurface is the union of these triangles. The vertices of the isosurface are

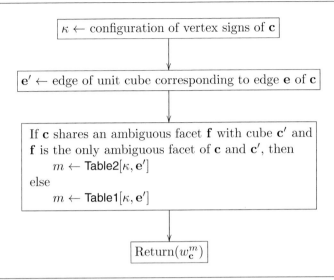

Figure 3.28. MANIFOLD DUAL MARCHING CUBES: Return isosurface vertex of grid cube **c** associated with edge **e**.

the triangle vertices. As defined in Section 3.1, Γ_{inner} is the subgrid of interior grid cubes of Γ.

The following properties apply to all isosurfaces produced by MANIFOLD DUAL MARCHING CUBES.

Property 1. *The isosurface is piecewise linear.*

Property 2. *Set Υ contains at most $4m$ triangles where m is the number of bipolar grid edges.*

Property 3. *The isosurface does not contain any grid vertices whose scalar values do not equal the isovalue.*

Property 4. *The isosurface intersects at exactly one point every interior bipolar grid edge whose positive endpoint is strictly positive.*

Property 5. *The isosurface does not intersect any negative or strictly positive grid edges.*

Property 6. *The isosurface strictly separates the strictly positive grid vertices of Γ_{inner} from the negative vertices of Γ_{inner}.*

Properties 3–5 imply that the isosurface intersects a minimum number of grid edges.

As defined in Section 3.1, the Γ_{outer} is the set of boundary grid cubes of Γ. The following properties apply to a MANIFOLD DUAL MARCHING CUBES isosurface whose isovalues do not equal the scalar value of any grid vertex.

Property 7. *The isosurface is a piecewise linear, orientable 2-manifold with boundary.*

Property 8. *The boundary of the isosurface is contained in Γ_{outer}.*

Property 9. *Set Υ does not contain any zero-area triangles or duplicate triangles and the triangles in Υ form a triangulation of the isosurface.*

3.3.7   Proof of Isosurface Properties

The proof of Properties 1–6 are similar to the proofs in Sections 3.2.5–3.2.7 for SURFACE NETS and are omitted.

We first prove that Υ is a triangulation of the isosurface (Property 9). To prove this property, we need some lemmas about the separation of isosurface vertices and about the separation of envelopes around bipolar edges.

Lemma 3.14. *Let $\mathbf{e}$ and $\mathbf{e}'$ be distinct bipolar edges of cube $\mathbf{c}$. Let w and w' be distinct isosurface vertices in cube $\mathbf{c}$ where w is associated with edge $\mathbf{e}$ and w' is associated with edge $\mathbf{e}'$. If no vertex of $\mathbf{c}$ has scalar value equal to the isovalue, then some plane separates $w \cup \mathbf{e}$ from $w' \cup \mathbf{e}'$ and does not contain w or w'.*

Proof: Cube $\mathbf{c}$ must have a configuration that contains at least two distinct isosurface vertices. For configurations 2B, 2C, 3B, 3C, 4E, 4F, 5B, 5C, 6B, and 6C, either w or w' are incident on exactly three isosurface quadrilaterals. (Configurations 2B and 3B have two isosurface vertices when we use isosurface patches 2B-II and 3B-II.) Without loss of generality, assume that w is incident on exactly three isosurface quadrilaterals.

Isosurface vertex w is the centroid of exactly three points on three cube edges, and these three cube edges are incident on a single grid vertex v. The convex hull of v and its three incident cube edges are a tetrahedron. Let $\mathbf{f}$ be the facet of this tetrahedron that does not contain v, and let $\mathbf{h}$ be the plane containing $\mathbf{f}$. Plane $\mathbf{h}$ separates $w \cup \mathbf{e}$ from $w' \cup \mathbf{e}'$. Because no vertex of $\mathbf{c}$ has scalar value equal to the isovalue, point w is the centroid of three points that lie on one side of $\mathbf{h}$, so w is not on $\mathbf{h}$. Similarly, point w' is the centroid of points that lie on the other side of $\mathbf{h}$, so w' is not on $\mathbf{h}$.

For configurations 4C, let $\mathbf{h}$ be the plane through the two positive grid edges. Plane $\mathbf{h}$ separates $w \cup \mathbf{e}$ from $w' \cup \mathbf{e}'$. Since no vertex of $\mathbf{c}$ has scalar value equal to the isovalue, point w is the centroid of four points that lie on one side of $\mathbf{h}$, and point w' is the centroid of four points that lie on the other side. Thus, neither w nor w' lie on $\mathbf{h}$.   □

Each interior bipolar edge $\mathbf{e}$ is associated with four isosurface vertices, one in each grid cube containing $\mathbf{e}$. These four isosurface vertices determine an envelope around $\mathbf{e}$ as described in Section 3.2.3.

Lemma 3.15. *The envelopes of two different interior bipolar grid edges intersect only at shared vertices, edges, or facets.*

Proof: Let $\mathbf{e}$ and $\mathbf{e}'$ be two different interior bipolar grid edges. Let $\mathbf{c}$ be a grid cube containing both $\mathbf{e}$ and $\mathbf{e}'$. Let w be the isosurface vertex in cube $\mathbf{c}$ associated with $\mathbf{e}$. Let w' be the isosurface vertex in cube $\mathbf{c}'$ associated with $\mathbf{e}'$. By Lemma 3.14, if w does not equal w', then there is some plane that separates $w \cup \mathbf{e}$ from $w' \cup \mathbf{e}'$. This plane separates the two envelopes. The two envelopes may only intersect if they share isosurface vertices and/or a vertex at $\mathbf{e} \cap \mathbf{e}'$. If they share such vertices, then their intersection is a vertex, edge, or facet determined by such shared vertices. □

Property 9. *Set Υ does not contain any zero-area triangles or duplicate triangles and the triangles in Υ form a triangulation of the isosurface.*

The proof that set Υ does not contain any zero-area triangles or duplicate triangles is exactly the same as the proof in Section 3.2.7 for SURFACE NETS, and is omitted.

Proof that set Υ is a triangulation: Each envelope around an interior bipolar edge contains two or four isosurface triangles. The nonempty intersection of any two triangles within an envelope is a vertex or edge of each. By Lemma 3.15, envelopes of different bipolar edges intersect only at shared vertices, edges, or facets. Thus, the triangles in different envelopes intersect only at shared edges or vertices. □

Finally, we prove that the isosurface is a 2-manifold.

Property 7. *The isosurface is a piecewise linear, orientable 2-manifold with boundary.*

Property 8. *The boundary of the isosurface is contained in Γ_{outer}.*

Proof: Since Υ is a triangulation of the isosurface, we need only check that every isosurface edge is contained in at most two isosurface triangles and that the neighborhood of every isosurface vertex is a disk. Inspection of the isosurface patches in Figure 3.23 shows that this is true for all configurations except for configurations 2B and 3B. As discussed in Section 3.3.4, when a cube $\mathbf{c}$ with configuration 2B or 3B shares its ambiguous facet with another cube $\mathbf{c}'$ with configuration 2B or 3B, an isosurface edge is shared by four isosurface triangles. However, in this case, the MANIFOLD DUAL MARCHING CUBES algorithm replaces isosurface patches 2B-1 or 3B-I with 2B-II or 3B-II, respectively, in both cubes. Each edge in these isosurface patches lies in only two triangles, and the neighborhood of each vertex is a disk. Because $\mathbf{c}$ and $\mathbf{c}'$ have only one ambiguous facet, the new isosurface patches in $\mathbf{c}$ and $\mathbf{c}'$ still align with all of their neighbors.

Let w be an isosurface vertex in grid cube $\mathbf{c} \in \Gamma_{\text{inner}}$. The triangles containing w form a topological disk, and w is in the interior of this disk. Every isosurface edge incident on w is contained in exactly two triangles. Thus, neither w nor any isosurface edge incident on w is on the boundary of the isosurface. The only vertices on the boundary of the isosurface are isosurface vertices contained in Γ_{outer}. The only edges on the boundary of the isosurface are edges connecting vertices in Γ_{outer} and are also contained in Γ_{outer}. Thus, the boundary of the isosurface is contained in Γ_{outer}. $\qquad\square$

3.4   Comparison of Algorithms

Table 3.1 compares isosurfaces constructed by MARCHING CUBES, SURFACE NETS, and MANIFOLD DUAL MARCHING CUBES on engine, carp, and tooth data sets. The vertex and triangle counts for Nielson's original DUAL MARCHING CUBES are almost exactly the same as for MANIFOLD DUAL MARCHING CUBES and are omitted. As can be seen, the number of isosurface vertices and triangles is about the same in each. The only significant difference is Tooth with isovalue 386. This is a particularly noisy isosurface with many small components of noise.

Figure 3.29 displays isosurfaces constructed by MARCHING CUBES, SURFACE NETS, DUAL MARCHING CUBES, and MANIFOLD DUAL MARCHING CUBES.

Data set	Isovalue	Algorithm	# Vertices	# Triangles
Engine	80.5	MARCHING CUBES	298,000	594,000
Engine	80.5	SURFACE NETS	297,000	592,000
Engine	80.5	MANIFOLD DUAL MC	297,000	592,000
Carp	1150.5	MARCHING CUBES	670,000	1,342,000
Carp	1150.5	SURFACE NETS	662,000	1,340,000
Carp	1150.5	MANIFOLD DUAL MC	672,000	1,340,000
Tooth	386.5	MARCHING CUBES	3,374,000	6,898,000
Tooth	386.5	SURFACE NETS	2,904,000	6,690,000
Tooth	386.5	MANIFOLD DUAL MC	3,345,000	6,690,000
Tooth	640.5	MARCHING CUBES	64,000	128,000
Tooth	640.5	SURFACE NETS	64,000	128,000
Tooth	640.5	MANIFOLD DUAL MC	64,000	128,000

Table 3.1. Comparison of isosurfaces produced by MARCHING CUBES, SURFACE NETS, and MANIFOLD DUAL MARCHING CUBES, including the number of vertices and number of triangles in each isosurface.

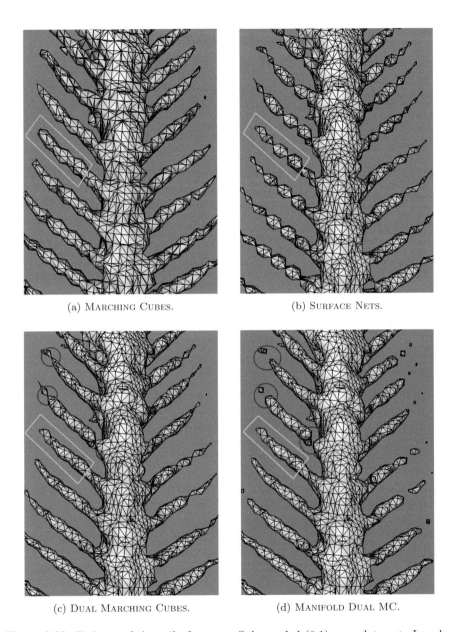

(a) MARCHING CUBES.

(b) SURFACE NETS.

(c) DUAL MARCHING CUBES.

(d) MANIFOLD DUAL MC.

Figure 3.29. Skeleton of the tail of a carp. Subsampled (8:1) carp data set. Isovalue 1150.5. (a) MARCHING CUBES isosurface. (b) SURFACE NETS triangulated isosurface. (c) DUAL MARCHING CUBES triangulated isosurface. (d) MANIFOLD DUAL MARCHING CUBES triangulated isosurface. (Data set was created by Michael Scheuring, Computer Graphics Group, University of Erlangen, Germany.)

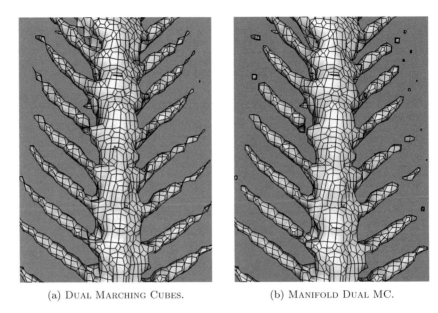

(a) DUAL MARCHING CUBES.   (b) MANIFOLD DUAL MC.

Figure 3.30. Isosurface of quadrilaterals from the carp data set, isovalue 1150.5. (a) DUAL MARCHING CUBES isosurface. (b) MANIFOLD DUAL MARCHING CUBES isosurface.

The isosurface is the skeleton of a carp. The isovalue, 1150.5, does not equal the scalar value of any grid vertex, so the MARCHING CUBES and MANIFOLD DUAL MARCHING CUBES isosurfaces are manifolds. (The scalar values of the grid vertices are all integers.) The data set was subsampled (8:1) to better show the mesh edges and to illustrate the differences among the algorithms. The isosurface is particularly challenging for isosurface reconstruction algorithms because of the narrow bone structures. (The subsampling adds to the difficulty.) Figure 3.30 displays the quadrilaterals in the DUAL MARCHING CUBES and MANIFOLD DUAL MARCHING CUBES isosurfaces.

Visual comparison of the MARCHING CUBES isosurface (Figure 3.29(a)) with the other isosurfaces (Figures 3.29(b), (c), and (d)) shows many more thin, small angle triangles in the MARCHING CUBES isosurface. The small, thin triangles create distracting visual artifacts in the isosurface. They also make the mesh unsuitable for finite element methods [Shewchuk, 2002].

The MARCHING CUBES isosurface is a manifold, whereas the SURFACE NETS and DUAL MARCHING CUBES isosurfaces are not. Many of the long bones are pinched in the SURFACE NETS isosurface. (Compare the rectangular green regions in Figures 3.29(a), (b), and (c).) The DUAL MARCHING CUBES isosurface does a much better job of modeling the fine bone structure than the SURFACE NETS algorithm does. However, even the DUAL MARCHING CUBES has some pinching to a line segment. (See regions circled in red in Figure 3.29(c).) This pinching is caused by the non-manifold cases discussed in Section 3.3.4 and illustrated in Figure 3.26.

The MANIFOLD DUAL MARCHING CUBES isosurface is a manifold. The non-manifold pinched regions have been eliminated by changes in the isosurface patches around those regions. However, those changes have also changed the isosurface topology, disconnecting some of the bones. (See regions circled in red in Figure 3.29(d).)

We should note that the pinching in Figure 3.29 is a result of the subsampling of the data set. As shown in Figures 3.31(b) and (c), the full-resolution SURFACE NETS and DUAL MARCHING CUBES isosurfaces have no such pinching and look similar to the full-resolution MARCHING CUBES isosurface (Figure 3.29(a)). Similarly, the full-resolution MANIFOLD DUAL MARCHING CUBES isosurface in Figure 3.31(d) does not have disconnected bones and also looks similar to the other three isosurfaces.

The pinching in Figure 3.29 is also a result of the long, narrow bone structure being modeled by the isosurface. For many isosurfaces, the subsampled SURFACE NETS isosurface has no such problems.

Even though the full-resolution SURFACE NETS isosurface in Figure 3.31(b) looks much better than its subsampled version, the isosurface still has about 8,000 non-manifold edges and 1,100 non-manifold vertices. Even the full-resolution DUAL MARCHING CUBES isosurface has 875 non-manifold edges. The MANIFOLD DUAL MARCHING CUBES algorithm changes the isosurface patches in 1,750 ($= 2 \times 875$) cubes from 2B-I or 3B-I to 2B-II or 3B-II to avoid such non-manifold edges.

3.5   Notes and Comments

Gibson, in [Gibson, 1998a, Gibson, 1998b], introduced the SURFACE NETS algorithm. The term *dual contouring* was used in [Ju et al., 2002] to describe isosurface construction algorithms similar to SURFACE NETS, which place vertices inside grid cubes.

Surface normals at the intersection points of grid edges and the isosurface can be used to locate isosurface vertices on sharp corners or edges in the isosurface. Using these surface normals with dual contouring, Ju et al. [Ju et al., 2002] constructed isosurfaces with sharp features. Further details of their algorithm are included in [Schaefer and Warren, 2002]. To allow finer control over small features, Varadhan et al. [Varadhan et al., 2003] modified the dual contouring algorithm to model two intersections of the isosurface and each grid edge. Zhang, Hong, and Kaufman [Zhang et al., 2004] modified the algorithm further to model even more complicated structures within a grid cube. These papers assume that the input includes a continuous scalar function, either explicitly defined or

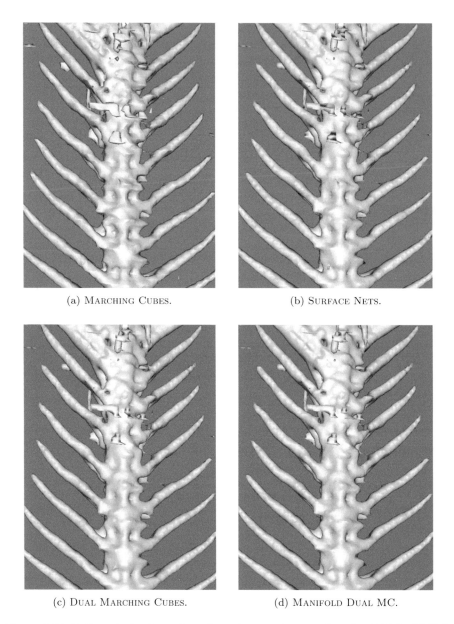

(a) MARCHING CUBES.   (b) SURFACE NETS.

(c) DUAL MARCHING CUBES.   (d) MANIFOLD DUAL MC.

Figure 3.31. Full-resolution isosurfaces from the carp data set, isovalue 1150.5. (a) Full-resolution MARCHING CUBES isosurface. (b) Full-resolution SURFACE NETS isosurface. (c) Full-resolution DUAL MARCHING CUBES isosurface. (d) Full-resolution MANIFOLD DUAL MARCHING CUBES isosurface.

defined as the signed distance to a polygonal mesh, and they compute the surface normals using that function.

Dual contouring algorithms are easily adaptable to multiresolution isosurface construction. Constructing multiresolution isosurfaces using dual contouring is discussed in Section 10.2.

DUAL MARCHING CUBES by Nielson [Nielson, 2004] reads a "dual contouring" lookup table to place and connect isosurface vertices in grid cubes. The DUAL MARCHING CUBES lookup table is "dual" to the MARCHING CUBES lookup table. Schaefer, Ju, and Warren [Schaefer et al., 2007] generalize Nielson's algorithm to multiresolution isosurfaces.

Ashida and Badler [Ashida and Badler, 2003] and Greß and Klein [Greß and Klein, 2004] give dual contouring algorithms that produce isosurfaces similar to the DUAL MARCHING CUBES isosurface with multiple isosurface vertices in a grid cube. Their algorithms determine connectivity between isosurface vertices directly from the grid without using a lookup table.

Ju and Udeshi [Ju and Udeshi, 2006] discuss the problem of self-intersections in the dual contouring isosurface caused by triangulating the isosurface quadrilateral. They present a solution by adding a vertex on the bipolar edge surrounded by the quadrilateral and adding four triangles incident on this vertex. Wang [Wang, 2011] describes orientation tests for determining when four triangles are necessary.

CHAPTER 4

MULTILINEAR INTERPOLATION

The MARCHING CUBES and dual contouring algorithms construct isosurfaces as approximations to the level set $\phi^{-1}(\sigma)$. However, the scalar function ϕ is not given; only function values at the vertices of the regular grid are known. In what sense does an isosurface approximate a level set $\phi^{-1}(\sigma)$ when ϕ is unknown?

One way to answer this question is to build a continuous scalar function, $\widehat{\phi}$, from the regular scalar grid. The level set $\widehat{\phi}^{-1}(\sigma)$ is well-defined and the isosurface can be measured against $\widehat{\phi}$. Natural scalar functions on the 2D square and the 3D cube are the bilinear and trilinear interpolants. Applying the bilinear interpolant to each 2D grid square and combining the results gives a function on the entire grid. Similarly, combining the trilinear interpolants applied to each 3D grid cube gives a function on a 3D grid.

As discussed in Sections 2.2.5 and 2.3.5, MARCHING SQUARES and MARCHING CUBES have ambiguous configurations where there are multiple ways of connecting the isosurface vertices into an isosurface patch. In Chapter 2, the ambiguous configurations were resolved by choosing a single isosurface patch for each configuration. In 3D, it is necessary to make sure that these choices are compatible, i.e., the isosurface patches in any two adjacent cubes match on their shared facet.

The bilinear and trilinear interpolants give an alternate method for resolving ambiguous configurations. Instead of always using the same isosurface patch for a configuration, use the isosurface patch that matches the topology of the level set of the bilinear or trilinear interpolants. By doing so, one builds an isosurface that is homeomorphic to the level set $\widehat{\phi}^{-1}(\sigma)$ where $\widehat{\phi}$ is the function determined by the bilinear or trilinear interpolant.

In Section 4.1, we discuss bilinear interpolation and its use in resolving the ambiguous cases in MARCHING SQUARES. The ASYMPTOTIC DECIDER algorithm by Nielson and Hamann [Nielson and Hamann, 1991] uses bilinear interpolation on grid facets to resolve ambiguities on the grid facets. In Section 4.2, we present a variation of the algorithm from [Nielson and Hamann, 1991]. Finally, in Section 4.3, we briefly discuss trilinear interpolation and its use in isosurface construction.

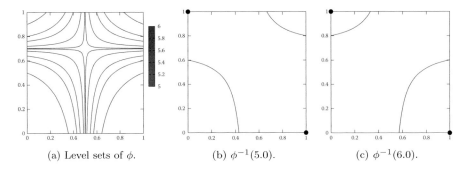

(a) Level sets of ϕ.   (b) $\phi^{-1}(5.0)$.   (c) $\phi^{-1}(6.0)$.

Figure 4.1. (a) Level sets of the bilinear function $\phi(x, y) = 2(1 - x)(1 - y) + 7x(1 - y)$ $+ 9(1 - x)y + 4xy$. Function ϕ is the bilinear interpolant of scalar values $s_{0,0} = 2$ and $s_{0,1} = 9$ and $s_{1,0} = 7$ and $s_{1,1} = 4$. (b) Level set $\phi^{-1}(5.0)$. Vertices $(1, 0)$ and $(0, 1)$ have scalar value greater than 5.0. (c) Level set $\phi^{-1}(6.0)$. Vertices $(1, 0)$ and $(0, 1)$ have scalar value greater than 6.0.

4.1   Bilinear Interpolation: 2D

4.1.1   The Bilinear Interpolant

Consider the unit square $\mathbf{c}_U$ with vertices $(0, 0), (0, 1), (1, 0)$, and $(1, 1)$ and with scalar values $s_{0,0}, s_{0,1}, s_{1,0}$, and $s_{1,1}$, respectively. Points in the unit square are (x, y) where $0 \leq x \leq 1$ and $0 \leq y \leq 1$. The function $\phi : \mathbb{R}^2 \to \mathbb{R}$ where

$$\phi(x, y) = (1 - x, x) \begin{pmatrix} s_{0,0} & s_{0,1} \\ s_{1,0} & s_{1,1} \end{pmatrix} \begin{pmatrix} 1 - y \\ y \end{pmatrix} \tag{4.1}$$

is the bilinear interpolant of the scalar values at the vertices. Equivalently,

$$\phi(x, y) = (1 - x)(1 - y)s_{0,0} + x(1 - y)s_{1,0} + (1 - x)ys_{0,1} + xys_{1,1}.$$

The bilinear interpolant can be constructed on grid squares other than $\mathbf{c}_U$ by translating them to $\mathbf{c}_U$, and then applying the bilinear interpolant on $\mathbf{c}_U$.

For almost all $\sigma \in \mathbb{R}$, the level set $\{(x, y) : \phi(x, y) = \sigma\}$ consists of a hyperbola with two branches. Figure 4.1(a) contains an example of the level sets of a bilinear function. The function is the bilinear interpolant of the scalar values 2, 7, 9, and 4, at the square vertices $(0, 0), (1, 0), (0, 1)$, and $(1, 1)$, respectively. Figure 4.1(b) contains an example of the level set $\phi^{-1}(5.0)$. Figure 4.1(b) contains an example of the level set $\phi^{-1}(6.0)$.

The two branches of the hyperbola defined by $\phi^{-1}(\sigma)$ may not both intersect the unit square. In Figures 4.2(b) and 4.2(c), only one hyperbola branch

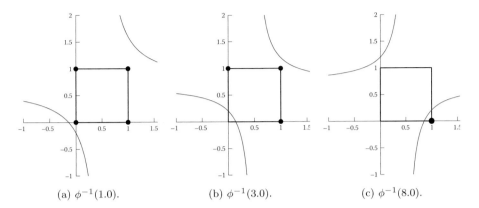

(a) $\phi^{-1}(1.0)$.   (b) $\phi^{-1}(3.0)$.   (c) $\phi^{-1}(8.0)$.

Figure 4.2. (a) Level set $\phi^{-1}(1.0)$. Neither branch of the hyperbola intersects the unit square. All four unit square vertices have scalar value greater than 1.0. (b) Level set $\phi^{-1}(3.0)$. One hyperbola branch intersects the unit square. Vertices $(1,0)$, $(0,1)$, and $(1,1)$ have scalar value greater than 3.0. (c) Level set $\phi^{-1}(8.0)$. One hyperbola branch intersects the unit square. Vertex $(1,0)$ has scalar value greater than 8.0.

intersects the unit square. In Figures 4.2(a), neither branch of the hyperbola representing $\phi^{-1}(1)$ intersect the unit square.

As in Chapter 2, label a vertex as positive, "+", if its scalar value is greater than or equal to the isovalue σ, and negative, "−", if its scalar value is below the isovalue σ. Grid edges whose two endpoints are positive are called positive edges. Grid edges whose two endpoints are negative are called negative edges.

When neither branch of the hyperbola intersects the square (Figure 4.2(a)), either all the square vertices are positive or all the square vertices are negative. When one hyperbola branch intersects the square (Figures 4.2(b) and 4.2(c)), either exactly one vertex is positive or one vertex is negative or one edge is positive and one edge is negative. None of these configurations are ambiguous. The level set, $\phi^{-1}(\sigma)$, intersects the square boundary at exactly two points. The Marching Squares algorithm connects these two points by a single line segment.

If both branches of the hyperbola defined by $\phi^{-1}(\sigma)$ intersect the unit square, then two diagonally opposite square vertices are positive and the other two vertices are negative. This is the 2D ambiguous configuration. (See Figure 4.3.) Conversely, if two diagonally opposite square vertices are positive and the other two vertices are negative, then both hyperbola branches defined by $\phi^{-1}(\sigma)$ must intersect the unit square.

There are two possible ways that the two hyperbola branches can intersect the unit square. Either the two hyperbola branches separate the two negative square vertices (Figure 4.1(b)) or they separate the two positive vertices (Figure 4.1(c)).

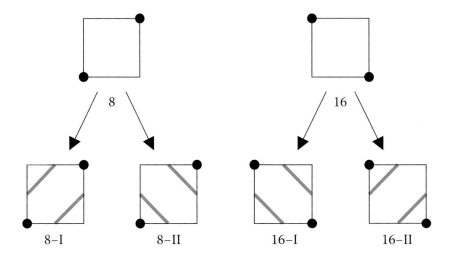

Figure 4.3. Ambiguous square configurations.

Similarly, there are two possible ways to resolve a 2D ambiguous configuration with a pair of line segments. Either the two isocontour line segments separate the two negative square vertices (8-I or 16-I in Figure 4.3) or the two isocontour line segments separate the two positive vertices (8-II or 16-II in Figure 4.3).

4.1.2   The 2D Asymptotic Decider

The 2D ASYMPTOTIC DECIDER uses the bilinear interpolant to resolve 2D ambiguous configurations. For each ambiguous grid square $\mathbf{c}$, let $\phi_{\mathbf{c}}$ be the bilinear interpolant determined by the scalar values at the vertices of $\mathbf{c}$. The level set, $\phi_{\mathbf{c}}^{-1}(\sigma)$, of this bilinear interpolant is two hyperbola branches that both intersect the grid square. If the hyperbola branches separate the negative vertices, then the ASYMPTOTIC DECIDER adds two isocontour line segments separating the negative vertices (8-I or 16-I). Otherwise, the hyperbola branches separate the positive vertices and the ASYMPTOTIC DECIDER adds two isocontour line segments separating the positive vertices (8-II or 16-II).

There is no need to actually construct the hyperbola branches to determine how it intersects the unit square. Instead, one can determine the contour value α such that $\phi^{-1}(\alpha)$ is the intersecting horizontal and vertical asymptotes. If the isovalue σ is less than α, then the hyperbola branches separate the negative vertices as in Figure 4.1(b). Otherwise, the hyperbola branches separate the positive vertices as in Figure 4.1(c).

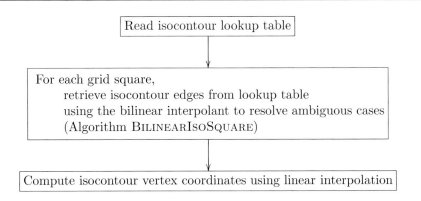

Figure 4.4. ASYMPTOTIC DECIDER: 2D.

Equation 4.1 expands out to

$$\phi(x,y) \;=\; s_{0,0} + x(s_{1,0} - s_{0,0}) + y(s_{0,1} - s_{0,0})$$
$$+ \; xy(s_{0,0} - s_{0,1} - s_{1,0} + s_{1,1}).$$

The asymptotes are given by either choosing x so that the coefficient of y becomes zero or choosing y so that the coefficient of x become zero:

$$\left\{ (x,y) : x = \frac{s_{0,0} - s_{0,1}}{s_{0,0} - s_{0,1} - s_{1,0} + s_{1,1}} \right\},$$

$$\left\{ (x,y) : y = \frac{s_{0,0} - s_{1,0}}{s_{0,0} - s_{0,1} - s_{1,0} + s_{1,1}} \right\}.$$

Plugging these values back into $\phi(x,y)$ gives

$$\alpha = \frac{s_{0,0}s_{1,1} - s_{1,0}s_{0,1}}{s_{0,0} - s_{0,1} - s_{1,0} + s_{1,1}}.$$

The ambiguous configuration is resolved by comparing the isovalue to this α. If the isovalue is less than α, then two line segments are added that separate the negative vertices as in 8-I and 16-I of Figure 4.3. If the isovalue is greater or equal to α, then two line segments are added that separate the positive vertices as in 8-II and 16-II of the figure.

Figure 4.5 contains Algorithm BILINEARISOSQUARE for resolving ambiguities using the bilinear interpolant. The 2D ASYMPTOTIC DECIDER algorithm is in Figure 4.4.

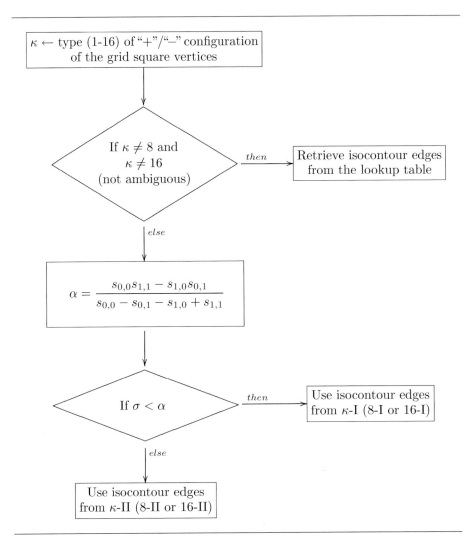

Figure 4.5. BILINEARISOSQUARE: Retrieve isocontour edges (isovalue σ) from the lookup table for a grid square using the bilinear interpolant to resolve ambiguous cases; $s_{x,y}$ is the scalar value at vertex (x, y).

4.2   The Asymptotic Decider: 3D

The MARCHING CUBES algorithm presented in Chapter 2 resolves ambiguous configurations arbitrarily. Nielson and Hamann [Nielson and Hamann, 1991] proposed using the bilinear interpolant to determine isosurface patches for the

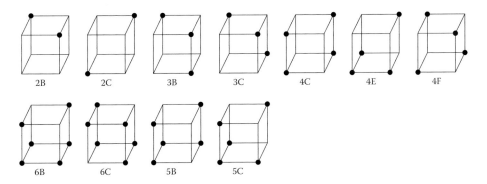

Figure 4.6. Ambiguous cube configurations.

ambiguous 3D configurations. We give an overview of their algorithm (Section 4.2.1), and then present a variation based on pyramidal decomposition (Section 4.2.2), which is easy to implement.

4.2.1   Resolving Ambiguous Facets

The intersection of an isosurface and a grid facet is an isocontour on that facet. Instead of creating the isosurface patch within each cube, first create the isocontour on each grid facet. Each grid facet has a configuration of positive and negative vertices given in Figure 2.2. Configurations 8 and 16 are the ambiguous 2D configurations. For all the nonambiguous configurations, use the isocontour table in Figure 2.3 to determine the line segment forming the isocontour. For the two ambiguous configurations, use the algorithm BILINEARISOSQUARE in Figure 4.5 to choose between isocontours 8-I and 8-II or 16-I and 16-II given in Figure 4.3.

Each edge $\mathbf{e}$ of a grid cube $\mathbf{c}$ lies on two facets, $\mathbf{f}$ and $\mathbf{f}'$, of $\mathbf{c}$. If edge $\mathbf{e}$ has one positive and one negative endpoint, then both $\mathbf{f}$ and $\mathbf{f}'$ contain an isocontour edge with endpoint on $\mathbf{e}$. The isocontour edges on all six facets of $\mathbf{c}$ join together to form one or more cycles forming an isocontour on the boundary of $\mathbf{c}$. The remaining problem is to add an isosurface patch inside the cube whose boundary edges exactly match the computed isocontour edges.

Ambiguous configurations 2B, 3B, 5B, and 6B all have a single facet with an ambiguous configuration. Each of these configurations has two possible isocontours on its boundary. These isocontours are easy to connect with isosurface patches. Figure 4.7 contains two isosurface patches for each such configuration, corresponding to the two possible isocontours. The choice of the 2D isocontour in the single ambiguous facet determines the choice of isosurface patch in 3D. The 2D isocontour separates the positive vertices in the facet if and only if the 3D isosurface separates the positive vertices in the cube.

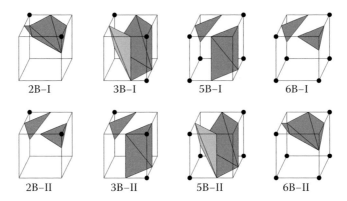

Figure 4.7. Isosurface patches for ambiguous cube configurations with one ambiguous facet: 2B, 3B, 5B, and 6B.

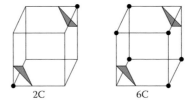

Figure 4.8. Isosurface patches for ambiguous cube configurations with zero ambiguous facets: 2C and 6C.

Configurations 2C and 6C have no ambiguous facets. They are ambiguous because there are two possible isosurface patches for these configurations, one separating two opposing vertices with two triangles and one connecting them by a tunnel. (See Figure 2.22.) As in the MARCHING CUBES algorithm, the ASYMPTOTIC DECIDER always uses the isosurface patches formed by two triangles (Figure 4.8).

All other ambiguous 3D configurations have two or more facets with ambiguous 2D configurations. Unfortunately, constructing the isosurface patch in these cases is more difficult.

Consider configuration 4C where the top and bottom facets have ambiguous configurations. If the isocontours in both the top and bottom facets separate the negative vertices, then the isosurface patch in 4C-I has the desired isocontour boundary. If the isocontours in both the top and bottom facets separate the positive vertices, then the isosurface patch in 4C-II has the desired isocontour boundary. However, what if the isocontour on the top boundary separates the positive vertices while the isocontour on the bottom separates the negative vertices? (See configuration 4C-III in Figure 4.9.)

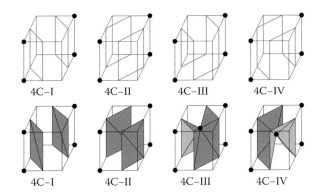

Figure 4.9. Four isocontour configurations for ambiguous configuration 4C and their isosurface patches. Isosurface patches in 4C-III and 4C-IV require an additional point.

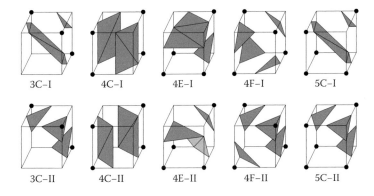

Figure 4.10. Isosurface patches for cubes with two or more ambiguous facets.

It would be preferable to construct an isosurface patch for 4C-III that uses only the cube vertices. Unfortunately, it is impossible to do so.[1] Figure 4.9 contains isosurface patches for isocontour configurations 4C-III and 4C-IV, which use additional points as isosurface vertices.

The remaining configurations are 3C, 4E, 4F, and 5C. Each of these configurations has two or more ambiguous facets. For some isocontours on the cube boundary, there are natural isosurface patches. Figure 4.10 contains two isosurface patches for each configuration. Let κ be the configuration type, i.e., κ is 3C, 4E, 4F, and 5C. If in each ambiguous facet the isocontour edges separate the negative vertices (cases 8-I and 16-I from Figure 4.3), then isosurface patches κ-I (3C-I, 4E-I, 4F-I, or 5C-I) have the appropriate isocontours. If in each ambigu-

---

[1] See [Nielson and Hamann, 1991] for a simple proof.

ous facet the isocontour edges separate the positive vertices (cases 8-II and 16-II from Figure 4.3), then isosurface patches κ-I (3C-I, 4E-I, 4F-I, or 5C-I) have the appropriate isocontours. As with configuration 4C, the problem arises when in some facets the isocontour edges separate the negative vertices while in other facets the isocontour edges separate the positive vertices.

It is possible to exhaustively analyze the possible isocontours for each configuration and construct isosurface patches as done for configuration 4C. Some of these isosurface patches require adding a new point in the interior of the cube as an isosurface vertex.

The exhaustive analysis of all the isocontour configurations and their isosurface patches is complicated. Writing a program to handle all the configurations is difficult. Instead of the exhaustive analysis, we show how isosurface patches can be constructed for the isocontours by decomposing the cube into pyramids.

4.2.2   Isosurface Lookup Table for Pyramids

By adding a point p to the center of a cube, a cube can be partitioned into six pyramids (Figure 4.11). Each of the pyramids contains the point p as its apex and one of six facets as its base. Each of pyramids has four triangles formed by p and the four edges of a facet.

Partitioning a cube into six pyramids does not partition any of the cube facets. The intersection of every pyramid with the boundary of the cube is a face of the cube. Thus a regular grid that has some cubes partitioned into pyramids is still a convex polyhedral mesh, i.e., the intersection of every pair of mesh elements is a vertex, edge, or facet of each. (See Appendix B.5 for the definition of convex polyhedral meshes.) The partitioned grid is a convex polyhedral mesh no matter which subset of cubes is partitioned into pyramids.

A pyramid with a square base has five vertices and $2^5 = 32$ configurations of "+" and "−" vertices. Twelve of these configurations are distinct. The distinct configurations and their isosurface patches are shown in Figure 4.12.

Configurations 2C and 3C are the only ambiguous pyramid configurations. Configuration 3C is the complement of 2C, with the positive vertices exchange for negative ones. Both configuration 2C and 3C have one ambiguous facet. There are two possible isosurface patches for each configuration, one separating the positive vertices and one separating the negative vertices. Figure 4.13 shows the isosurface patches for each configuration.

Algorithm BILINEARISOSQUARE can be used to choose isosurface patches for the ambiguous pyramid configurations. Consider a pyramid whose "+"/"−" vertex labels form configuration 2C. Apply BILINEARISOSQUARE to the square pyramid base **f** to determine the isocontour on **f**. If the isocontour separates the negative vertices on **f**, then use isosurface patch 2C-I. Otherwise, the isocontour separates the positive vertices on **f**. Use isosurface patch 2C-II. Similarly, use BILINEARISOSQUARE to choose the isosurface patch for configuration 3C.

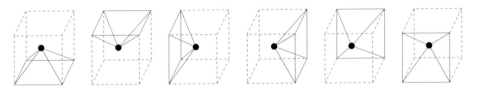

Figure 4.11. Six pyramids partitioning the cube. Apex is at cube center.

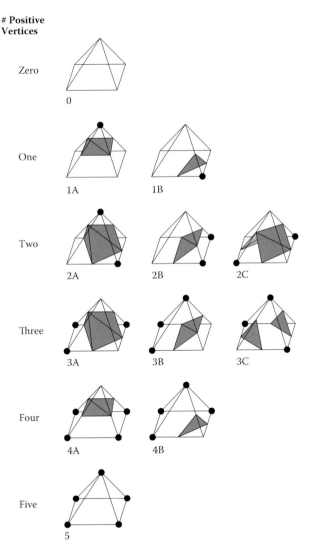

Figure 4.12. Isosurfaces for the twelve pyramid configurations.

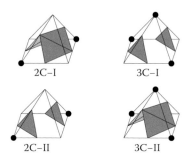

Figure 4.13. Ambiguous pyramid configurations 2C and 3C.

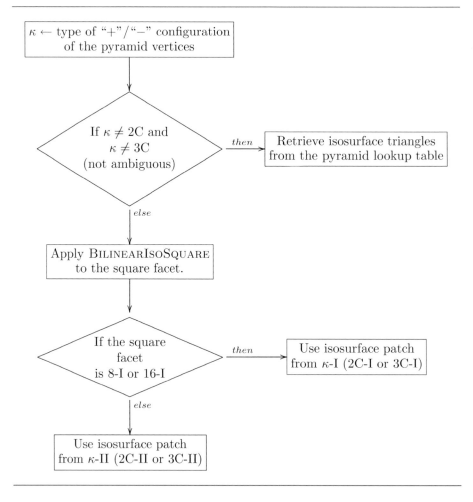

Figure 4.14. BILINEARISOPYRAMID: Retrieve isosurface patch from the lookup table for a grid pyramid using the bilinear interpolant to resolve ambiguous cases.

Algorithm BILINEARISOPYRAMID for retrieving the isosurface patch from a pyramid is shown in Figure 4.14.

4.2.3 Algorithm

Input to the 3D ASYMPTOTIC DECIDER is an isovalue and a set of scalar values at the vertices of a three-dimensional regular grid. The algorithm has four steps. (See Figure 4.15.) Read the MARCHING CUBES isosurface lookup table (Figure 2.16) from a preconstructed data file. Read the pyramid isosurface lookup table (Figure 4.12) from a preconstructed data file. For each cube, construct a set of isosurface triangles representing the combinatorial structure of the isosurface using a subdivision into pyramids and BILINEARISOSQUARE to resolve ambiguous cases. Assign geometric locations to the isosurface vertices based on the scalar values at the cube edge endpoints. We explain the third step, constructing the isosurface patches, next.

Let **c** be the grid cube under consideration. If no facet of **c** is ambiguous, then retrieve the isosurface triangles from the cube isosurface lookup table. Note that if **c** has the ambiguous configuration 2C or 3C, the isosurface patch is still retrieved directly from the lookup table.

Let κ be the type of "+"/"−" configuration of the cube vertices as given by Figure 2.15, i.e., $\kappa = 0$, 1, 2A, 2B, 2C, etc. Apply the algorithm BILINEARISOSQUARE to each of the ambiguous facets of **c** to construct the isocontours in these facets. If every ambiguous facet is of type 8-I or 16-I, then use isosurface

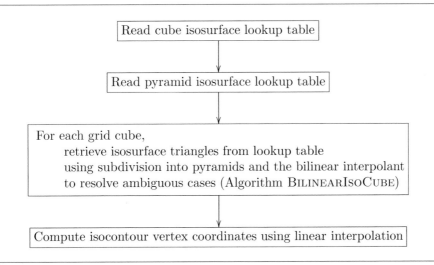

Figure 4.15. ASYMPTOTIC DECIDER: 3D.

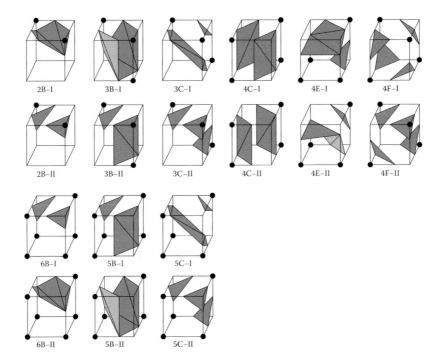

Figure 4.16. Isosurface patches for the 3D configurations with ambiguous facets.

patch κ-I from Figure 4.16. If every ambiguous facet is of type 8-II or 16-II, then use isosurface patch κ-2 from the figure. Otherwise, partition the cube into six pyramids sharing an apex at the center of the cube. Apply Algorithm BILINEARISOPYRAMID to construct an isosurface patch in each pyramid. Algorithm BILINEARISOCUBE is presented in Figure 4.17.

4.3   Trilinear Interpolation

Let $\mathbf{c}_U$ be the unit cube $\{(x, y, z) : x, y, z \in [0, 1]\}$. Let $s_{x,y,z}$ be the scalar value of the vertex with coordinates (x, y, z). The trilinear interpolant of the scalar values $s_{x,y,z}$ is the function

$$\phi(x, y, z) = (1 - x)(1 - y)(1 - z)s_{0,0,0} + x(1 - y)(1 - z)s_{1,0,0}$$
$$+ (1 - x)y(1 - z)s_{0,1,0} + (1 - x)(1 - y)zs_{0,0,1}$$
$$+ xy(1 - z)s_{1,1,0} + x(1 - y)zs_{1,0,1} + (1 - x)yzs_{0,1,1} + xyzs_{1,1,1}.$$

The trilinear interpolant on a grid cube $\mathbf{c}$ other than $\mathbf{c}_U$ is by translating $\mathbf{c}$ to $\mathbf{c}_U$ and then applying the trilinear interpolant to $\mathbf{c}_U$.

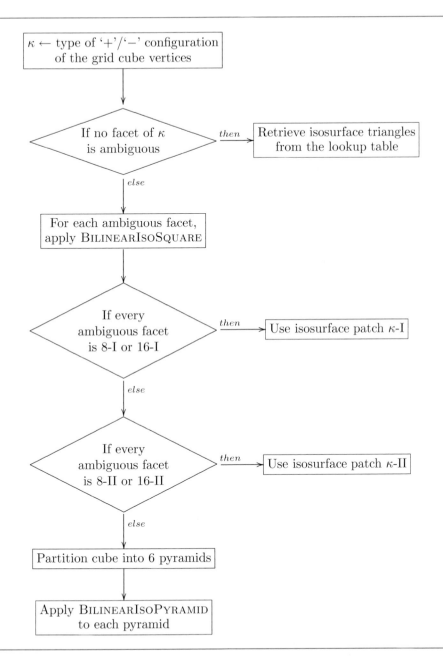

Figure 4.17. BILINEARISOCUBE: Retrieve isocontour triangles (isovalue σ) from the lookup table for a grid cube using subdivision into pyramids and the bilinear interpolant to resolve ambiguous cases.

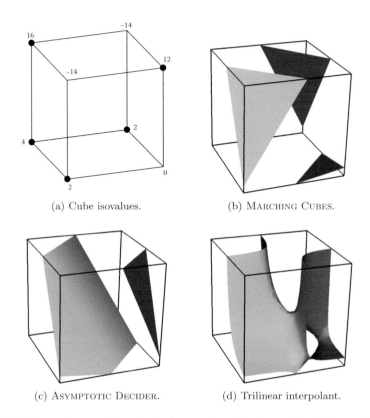

(a) Cube isovalues.   (b) MARCHING CUBES.

(c) ASYMPTOTIC DECIDER.   (d) Trilinear interpolant.

Figure 4.18. Isosurfaces and level sets for a cube. Isovalue $= 1$. Cube configuration 5C (from Figure 2.16). Front faces are colored green. Back faces are colored magenta. (a) Scalar values at cube vertices. Black vertices have scalar value greater than 1. (b) MARCHING CUBES isosurface. (c) ASYMPTOTIC DECIDER isosurface. (d) Level set of the trilinear interpolant.

The trilinear interpolant is a generalization of the bilinear interpolant to three dimensions. The restriction of the trilinear interpolant to a grid facet is the bilinear interpolant on that facet.

The 3D ASYMPTOTIC DECIDER uses the bilinear interpolant to determine how the isosurface intersects grid facets. However, this is not the same as using the trilinear interpolant to determine how the isosurface should intersect the grid cubes. Figure 4.18 compares the MARCHING CUBES isosurface, Σ_{MC}, the ASYMPTOTIC DECIDER isosurface, Σ_{AD}, and the level set, Σ_{TI}, of the trilinear interpolant of a cube. Restricted to the boundary of the cube, Σ_{AD} and Σ_{TI} are topologically the same and intersect the grid edges in the same order. However, Σ_{AD} has two connected components, both disks, while Σ_{TI} is a cylinder.

Lookup tables for the trilinear interpolant can be found in [Carr and Max, 2010, Chernyaev, 1995, Lopes and Brodlie, 2003, Nielson, 2003b]. The lookup table has about fifty distinct cases. Configurations are based on the relative value of the isovalue and the scalar values at the grid vertices, located at critical points on the cube facets and at critical points inside the cube. Isosurfaces built using the trilinear interpolant lookup table have the same topological structure as the level set of the trilinear function.

4.4 Notes and Comments

Nielson and Hamann [Nielson and Hamann, 1991] proposed the ASYMPTOTIC DECIDER algorithm, a modification of MARCHING CUBES using the bilinear interpolant on grid cube faces. Matveyev [Matveyev, 1994] suggested an implementation of the ASYMPTOTIC DECIDER that does not require explicitly computing the face saddle.

Matveyev [Matveyev, 1994], Natarajan [Natarajan, 1994], and Chernyaev [Chernyaev, 1995] presented approaches to constructing an isosurface that matches the topology of the level set of the trilinear interpolant. An excellent review of these papers is in the introduction to [Lopes and Brodlie, 2003].

Chernyaev [Chernyaev, 1995] gave a modified isosurface patch lookup table for constructing such isosurfaces. For certain configurations of "+" or "−" labels, the lookup table contains multiple possible isosurface patches. Computations of the facet and body saddle points of the trilinear interpolant of the grid cube determine which patch is chosen. Lin and Ching [Lin and Ching, 1997] suggested a more efficient method for computing body saddles.

Later descriptions of algorithms based on the trilinear interpolant are in [Cignoni et al., 2000, Lewiner et al., 2003, Lopes and Brodlie, 2003, Nielson, 2003b, Carr and Snoeyink, 2009, Carr and Max, 2010]. These papers give more explicit algorithm descriptions, better lookup tables, or better explanations of the topological cases determined by the trilinear interpolant.

Zhang and Qian [Zhang and Qian, 2012] give a dual contouring algorithm to produce an isosurface matching the topology of the level set of the trilinear interpolant.

Etiene et al., in [Etiene et al., 2012], describe a technique for verifying that an isosurface matches the topology of the level set of the trilinear interpolant.

CHAPTER 5

ISOSURFACE PATCH
CONSTRUCTION

The MARCHING CUBES algorithm can be generalized to other convex mesh elements such as pyramids or bipyramids. Each such mesh element has a different isosurface lookup table. The table entries correspond to different configurations of positive and negative vertices. Each entry contains a list of edge triples, $(\mathbf{e}_1, \mathbf{e}_2, \mathbf{e}_3)$, representing triangles with vertices on edges $\mathbf{e}_1$, $\mathbf{e}_2$, and $\mathbf{e}_3$.

MARCHING CUBES can also be generalized to a mixture of convex mesh elements, such as cubes, pyramids, and tetrahedra. Care must be taken to ensure that the isosurfaces align properly on the ambiguous facets, i.e., that the ambiguous configurations are resolved consistently across facets of adjacent mesh elements.

One could generate a set of tables for each of the various mesh elements by hand and check for consistency. Instead of doing so, we describe an algorithm for automatically generating isosurface patches within convex mesh elements. The idea behind this algorithm was independently proposed by Lachaud and Montanvert [Lachaud and Montanvert, 2000] and Bhaniramka, Wenger, and Crawfis [Bhaniramka et al., 2000, Bhaniramka et al., 2004a]. As discussed in Chapters 6 and 7, the algorithm can be used to generate isosurface patches in dimensions higher than three and to generate interval volumes.

We first present the algorithm for constructing the entries in the isosurface lookup table and then present the algorithm for isosurface construction. The isosurface construction algorithm and the properties of the resulting isosurface parallel exactly the MARCHING SQUARES, MARCHING CUBES, and MARCHING TETRAHEDRA algorithms in Chapter 2. Before discussing constructing the isosurface lookup table, we define some terminology and notation that will be used in this chapter.

5.1   Definitions and Notation

Vertex and edge properties. The definitions of positive, strictly positive, and negative for convex polyhedral meshes are similar to the definitions in Section 2.1.

Definition 5.1.

- A mesh vertex is positive, "+", if its scalar value is greater than or equal to the isovalue σ.

- A mesh vertex is negative, "−", if its scalar value is less than σ.

- A positive vertex is strictly positive if its scalar value does not equal σ.

Let $\mathbf{c}$ be a convex polyhedron and Γ a convex polyhedral mesh where every vertex of $\mathbf{c}$ and Γ is labeled negative, positive, or strictly positive. We characterize the edges of $\mathbf{c}$ and Γ by the labels at their endpoints.

Definition 5.2.

- An edge of $\mathbf{c}$ or Γ is positive if both its endpoints are positive.

- An edge of $\mathbf{c}$ or Γ is negative if both its endpoints are negative.

- A positive edge of $\mathbf{c}$ or Γ is strictly positive if both its endpoints are strictly positive.

- An edge of $\mathbf{c}$ or Γ is bipolar if one endpoint is positive and one endpoint is negative.

Denote by $V_{\mathbf{c}}^{+}$ the set of positive vertices of polyhedron $\mathbf{c}$ and by $V_{\mathbf{c}}^{-}$ the set of negative vertices of $\mathbf{c}$. Denote by $M_{\mathbf{c}}$ the set of midpoints of bipolar edges of $\mathbf{c}$. The algorithm for isosurface patch construction uses the convex hull of a set of points. Convex hulls are discussed in Appendix A.2. The convex hull of a set of points P is denoted $\mathrm{conv}(P)$.

Isosurface vertex sets. In Section 5.2, we will generate isosurface patches from the midpoints, $M_{\mathbf{c}}$, of bipolar edges of $\mathbf{c}$. Later in this chapter, we will generate isosurface patches from other points along the bipolar edges of $\mathbf{c}$. In Chapter 7, we will generate interval volumes using points that are $1/3$ of the distance from some edge endpoint. Thus, it is useful to have a general characterization of the set of points used in isosurface patch construction.

Definition 5.3.

- An isosurface vertex set $U_{\mathbf{c}}$ of convex polyhedron $\mathbf{c}$ is a finite set of points lying on the edges of $\mathbf{c}$ such that every bipolar edge of $\mathbf{c}$ contains one and only one point of $U_{\mathbf{c}}$ and no point of $U_{\mathbf{c}}$ lies on a strictly positive vertex or negative vertex or negative edge of $\mathbf{c}$.

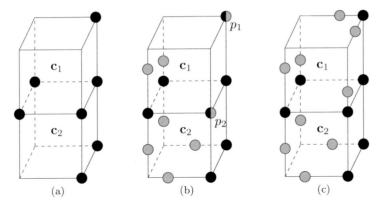

Figure 5.1. (a) Grid Γ of two cubes. Black vertices are positive. (b) An isosurface vertex set U (green vertices) for grid Γ. Points p_1 and p_2 are both positive grid vertices and elements of U. Thus isosurface vertex set U is not proper. (c) A *proper* isosurface vertex set U' (green vertices) for grid Γ.

- An isosurface vertex set U of mesh Γ is a finite set of points lying on the edges of Γ such that every bipolar edge of Γ contains one and only one point of U and no point of U lies on a strictly positive vertex or negative vertex or negative edge of Γ.

Figure 5.1(b) contains an example of an isosurface vertex set.

A grid vertex whose scalar value equals the isovalue is positive. As a result, the MARCHING CUBES isosurface can contain a positive grid vertex but never a negative grid vertex. Similarly, the isosurface vertex set can contain a positive mesh vertex but not a negative one.

Points of an isosurface vertex set may lie on positive edges of the mesh or polyhedron. For instance, in Figure 5.1(b) point p_2 lies on three edges in cube c_2, and all three of those edges are positive. Points of an isosurface vertex set may also lie on positive vertices that are not strictly positive. Thus a point of the isosurface vertex set could be an endpoint of a bipolar edge. For instance, in Figure 5.1(b) point $p_1 \in U$ is the endpoint of two bipolar edges in c_1 and point $p_2 \in U$ is the endpoint of one bipolar edge in c_1.

Definition 5.4.

- An isosurface vertex set U_c of convex polyhedron c is proper if every point in U_c lies in the interior of a bipolar edge of c.

- An isosurface vertex set U of mesh Γ is proper if every point in U lies in the interior of a bipolar edge of Γ.

The set M_c of midpoints of bipolar edges of c is an example of a proper isosurface vertex set of c. Figure 5.1(c) contains an example of a proper isosurface vertex set that is not the set of midpoints of bipolar edges.

The algorithm for isosurface patch construction uses the convex hull of a set of points. Convex hulls are discussed in Appendix A.2. The convex hull of a set of points is denoted $\mathrm{conv}(P)$.

5.2 Isosurface Patch Construction

Let $\mathbf{c}$ be a convex polyhedron in $\mathbb{R}^3$ where each vertex of $\mathbf{c}$ has a positive or a negative label. We show how to construct an isosurface patch in $\mathbf{c}$ whose vertices are $M_{\mathbf{c}}$, the midpoints of bipolar edges of $\mathbf{c}$.

Compute the set $M_{\mathbf{c}}$ of midpoints of the bipolar edges of $\mathbf{c}$. Construct the convex hull of $M_{\mathbf{c}} \cup V_{\mathbf{c}}^{+}$, the midpoints of bipolar edges and the positive vertices. The boundary of the convex hull[1] is composed of convex polygons. Triangulate the boundary of the convex hull by triangulating each convex polygon without adding any additional vertices. Orient the triangles so that the induced normals point toward the convex hull. Let $T_{\mathbf{c}}^{+}$ be the oriented triangles in this triangulation that do not lie on the boundary of $\mathbf{c}$. We shall prove (Lemma 5.5) that the vertices of $T_{\mathbf{c}}^{+}$ are all from $M_{\mathbf{c}}$. The triangles in $T_{\mathbf{c}}^{+}$ form the triangulated isosurface patch whose vertices are $M_{\mathbf{c}}$. This isosurface patch separates the positive vertices from the negative vertices of $\mathbf{c}$. We call the algorithm IsoMid3D. Algorithm IsoMid3D is presented in Figure 5.2 and illustrated in Figure 5.3.

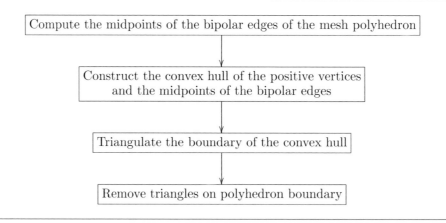

Figure 5.2. Algorithm IsoMid3D for construction of 3D isosurface patches from midpoints of bipolar edges.

---

[1]Data structures representing the convex hull of a set of points usually represent the boundary of the convex hull. Thus computer scientists often fail to distinguish between the convex hull of a set of points and its boundary. We will try to keep that distinction clear and refer explicitly to the boundary wherever appropriate.

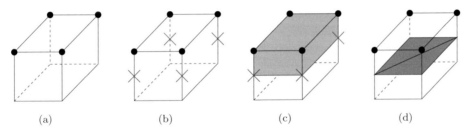

Figure 5.3. Isosurface patch generation. (a) Cube configuration. (b) Midpoints of bipolar edges. (c) The convex hull of $M_{\mathbf{c}} \cup V_{\mathbf{c}}^{+}$, the set of midpoints, and positive vertices. (d) Removal of triangles on cube boundary, leaving isosurface.

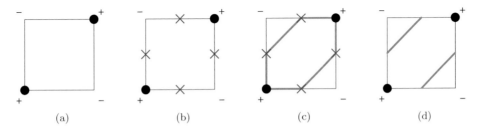

Figure 5.4. Isocontour generation. (a) Square configuration. (b) Midpoints of bipolar edges. (c) The convex hull of $M_{\mathbf{c}} \cup V_{\mathbf{c}}^{+}$, the set of midpoints, and positive vertices. (d) Removal of line segments on square boundary, leaving isocontour.

The combinatorial structure of the isosurface within $\mathbf{c}$ is determined from the configuration of the polyhedron's vertex labels. The isosurface does not intersect any edge whose endpoints have the same label.

As will be discussed in Chapter 6, the algorithm for isosurface patch construction can be generalized to any dimension. For illustrative purposes, it is useful to present here the two-dimensional version. Let $\mathbf{c}$ be a square in $\mathbb{R}^2$ whose vertices have positive or negative labels. Define $M_{\mathbf{c}}$ as the midpoints of bipolar edges. Construct the convex hull of $M_{\mathbf{c}} \cup V_{\mathbf{c}}^{+}$ and let $L_{\mathbf{c}}^{+}$ be the line segments on the boundary of the convex hull that do not lie on the boundary of $\mathbf{c}$. The line segments in $L_{\mathbf{c}}^{+}$ form the isocontour in $\mathbf{c}$.

Figure 5.4 gives an example applied to square configuration 7 from Figure 2.2 (Chapter 2). The construction produces exactly the isocontour given in Figure 2.3. More generally, the construction applied to any of the configurations in Figure 2.2 produces the corresponding isocontours in Figure 2.3.

Figure 5.6 presents examples of the 3D algorithm applied to the cube configurations 1, 2A, 2B, and 4C from Figure 2.15. For each configuration in Figure 2.15, the construction produces the isosurface patches given by Figure 2.16. The only

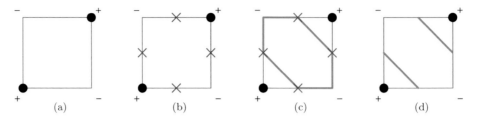

Figure 5.5. Isocontour generation using negative vertices. (a) Square configuration. (b) Midpoints of bipolar edges. (c) The convex hull of $M_{\mathbf{c}} \cup V_{\mathbf{c}}^{-}$, the set of midpoints, and negative vertices. (d) Removal of line segments on square boundary, leaving isocontour.

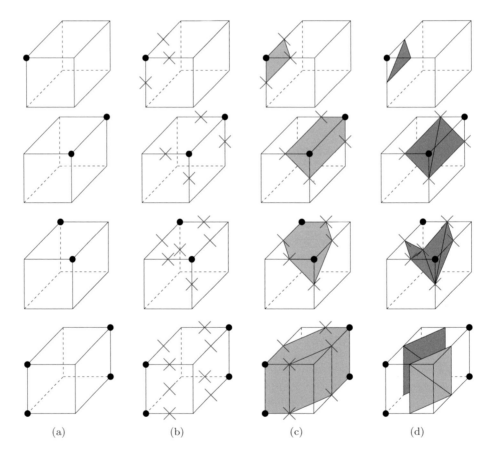

Figure 5.6. Isosurface patch generation. (a) Cube configuration. (b) Midpoints of bipolar edges. (c) The convex hull of $M_{\mathbf{c}} \cup V_{\mathbf{c}}^{+}$, the set of midpoints, and positive vertices. (d) Removal of triangles on cube boundary, leaving isosurface.

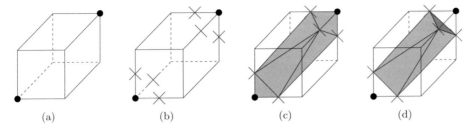

Figure 5.7. Isosurface patch generation. (a) Cube configuration 2C. (b) Midpoints of bipolar edges. (c) The convex hull of $M_\mathbf{c} \cup V_\mathbf{c}^+$, the set of midpoints, and positive vertices. (d) Removal of triangles on cube boundary, leaving isosurface.

exception is configuration 2C. The isosurface patch construction for configuration 2C is illustrated in Figure 5.7. The isosurface patch is not the same as the one given in Figure 2.16 for configuration 2C. Instead, it is the alternate isosurface 2C-I from the isosurface patches for ambiguous configurations given in Figure 2.22.

The 3D algorithm applied to the tetrahedra configurations in Figure 2.25 gives the isosurface patches in Figure 2.27.

Algorithm ISOMID3D could be modified in two simple ways. First, the set of midpoints, $M_\mathbf{c}$, could be replaced by any set of points $U_\mathbf{c}$ from the interiors of bipolar edges. The choice of points does not change the topology of the isosurface patch, although it could change its triangulation.

Second, instead of using the positive vertices, $V_\mathbf{c}^+$, one could construct the convex hull of $M_\mathbf{c} \cup V_\mathbf{c}^-$, triangulate its boundary, and let $T_\mathbf{c}^-$ be the triangles in this triangulation that do not lie on the boundary of $\mathbf{c}$. For each of the ambiguous configurations in Figure 2.22, set $T_\mathbf{c}^-$ is the alternate isosurface patch given in that figure.

In two dimensions, $L_\mathbf{c}^-$ is the line segments on the boundary of $\mathrm{conv}(M_\mathbf{c} \cup V_\mathbf{c}^-)$ that do not lie on the boundary of $\mathbf{c}$. Figure 5.5 gives a 2D example of isocontour construction using the negative vertices, $V_\mathbf{c}^-$, on configuration 7 from Figure 2.2. This generates the isocontour 7-II from Figure 2.12. For each of the ambiguous configurations in Figure 2.12, set $L_\mathbf{c}^-$ is the alternate isocontour given in that figure.

Any triangle in the triangulation of $\mathrm{conv}(M_\mathbf{c} \cup V_\mathbf{c}^+)$ that contains a vertex of $V_\mathbf{c}^+$ is on the boundary of $\mathbf{c}$ and thus not a triangle of $T_\mathbf{c}^+$ (Corollary 5.7). However, the converse is not true. There may be triangles that lie on some facet of $\mathbf{c}$ but whose vertices all come from $M_\mathbf{c}$. (See Figure 5.8.) Since these triangles lie on a facet of $\mathbf{c}$, they are not included in $T_\mathbf{c}^+$.

Isosurface patches in adjacent mesh elements should properly align on their boundaries. As was discussed in Section 2.3.5, if some isosurface patches separate positive vertices while others separate negative vertices, proper alignment may not occur. The isosurface patches separating positive vertices are defined by

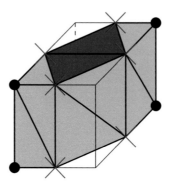

Figure 5.8. Triangulation of the boundary of $\mathrm{conv}(M_\mathbf{c} \cup V_\mathbf{c}^+)$ from the fourth row of Figure 5.6. Blue triangles lie on a facet of $\mathbf{c}$ but their vertices are all from $M_\mathbf{c}$.

$T_\mathbf{c}^+$ while the isosurface patches separating negative vertices are defined by $T_\mathbf{c}^-$. Thus, while either $T_\mathbf{c}^+$ or $T_\mathbf{c}^-$ can be used to generate the isosurface lookup tables, only one or the other should be used for all table entries. This applies not only within a single table but between the different lookup tables for different convex mesh elements. Because the isosurface patches must align on the common boundary between two such mesh elements, only $T_\mathbf{c}^+$ or only $T_\mathbf{c}^-$ should be used for all the isosurface lookup tables.

As discussed in Section 2.3.5, configuration 2C is the only configuration where the isosurface lookup table in Figure 2.16 separates the positive vertices, not the negative vertices. Because no facet of configuration 2C has an ambiguous configuration, isosurface patches 2C-I and 2C-II from Figure 2.22 have the same boundary. Thus, either patch can be used. If the isosurface table is constructed from the positive vertices (i.e., the isosurface triangulation is $T_\mathbf{c}^+$,) then the tube connecting the two positive vertices in 2C-I will be the isosurface patch. This is the only difference between the automatically constructed lookup table for cubes and the table in Figure 2.16.

5.3   Isosurface Table Construction

As in the previous section, let $\mathbf{c}$ be a convex polyhedron in $\mathbb{R}^3$. Let n be the number of vertices of $\mathbf{c}$. In the isosurface extraction algorithm, the vertices of $\mathbf{c}$ receive positive and negative labels representing their relationship to the isovalue. A vertex that has scalar value below the isovalue will receive a negative label, "$-$". A vertex with scalar value above or equal to the isovalue receives a positive label, "$+$".

Since each vertex is either positive or negative, there are 2^n configurations of positive and negative vertex labels. The isosurface lookup table contains 2^n

entries, one for each configuration κ. Each entry is a list of triples of bipolar edges of $\mathbf{c}$. Each triple $(\mathbf{e}_1, \mathbf{e}_2, \mathbf{e}_3)$ represents an oriented triangle whose vertices lie on $\mathbf{e}_1$ and $\mathbf{e}_2$ and $\mathbf{e}_3$. The list of triples define the combinatorial structure of the isosurface patch for configuration κ. This isosurface patch is constructed using algorithm IsoMid3D in the previous section. The triangles in this isosurface patch are the set $T_\mathbf{c}^+$. The isosurface patch intersects every bipolar edge of $\mathbf{c}$ exactly once and does not intersect any positive or negative polyhedron edges.

5.4 Marching Polyhedra Algorithm

We assume that the mesh is composed of a fixed, predefined set of polyhedra classes. Elements within each class are affine transformations of one another.[2] For instance, all cubes and rectangular regions are affine transformations of the unit cube and are in the same class. Similarly, all tetrahedra are affine transformations of each other and all pyramids with rectangular bases are affine transformations of each other. One "generic" convex polyhedron is chosen from each class and an isosurface lookup table is built on that polyhedron with the algorithm described in the previous section.

We will call the variation of MARCHING CUBES applied to convex polyhedra, the MARCHING POLYHEDRA algorithm. (See Figure 5.9.) The MARCHING POLYHEDRA algorithm follows the steps of the MARCHING CUBES and MARCHING TETRAHEDRA algorithms. Input to the MARCHING POLYHEDRA algorithm is an isovalue, a three-dimensional convex polyhedral mesh, a set of scalar values at the vertices of the mesh and a set of polyhedra classes.

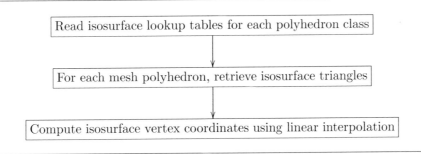

Figure 5.9. MARCHING POLYHEDRA.

---

[2]A more general approach is to divide mesh elements into classes of combinatorially equivalent polyhedra.

The algorithm has three steps. The first step is reading the isosurface lookup tables for each class from preconstructed data files. The next two steps of the algorithm follow the MARCHING CUBES and MARCHING TETRAHEDRA algorithms. For each mesh element, retrieve from the appropriate lookup table a set of isosurface triangles representing the combinatorial structure of the isosurface. The vertices of these triangles form the isosurface vertices. Assign geometric locations to the isosurface vertices based on the scalar values at the mesh edge endpoints.

To map each isosurface triangle to a geometric triangle, we use linear interpolation to position the isosurface vertices as described in Section 1.7.2. Each isosurface vertex v lies on a grid edge $[p, q]$. If s_p and s_q are the scalar values at p and q and σ is the isovalue, then map v to $(1 - \alpha)p + \alpha q$ where $\alpha = (\sigma - s_p)/(s_q - s_p)$.

5.4.1   Isosurface Properties

We claim the following properties for the isosurface produced by the MARCHING POLYHEDRA algorithm. These properties are the same as those for the isosurface produced by MARCHING CUBES. As with MARCHING CUBES, the isosurface is a 2-manifold with boundary under appropriate conditions.

MARCHING POLYHEDRA returns a finite set, Υ, of oriented triangles. The isosurface is the union of these triangles. The vertices of the isosurface are the triangle vertices.

The following properties apply to all isosurfaces produced by the MARCHING POLYHEDRA algorithm.

Property 1.   *The isosurface is piecewise linear.*

Property 2.   *The vertices of the isosurface lie on mesh edges.*

Property 3.   *The isosurface intersects every bipolar mesh edge at exactly one point.*

Property 4.   *The isosurface does not intersect any negative or strictly positive mesh edges.*

Property 5.   *The isosurface separates positive mesh vertices from negative ones and strictly separates strictly positive mesh vertices from negative mesh vertices.*

Properties 3 and 4 imply that the isosurface intersects a minimum number of mesh edges. As with MARCHING CUBES, if both endpoints of a mesh edge have scalar value equal to the isovalue, then the isosurface may intersect the mesh edge zero, one, or two times or may contain the mesh edge.

By Property 3, the isosurface intersects every bipolar mesh edge. However, the bipolar mesh edge may be intersected by zero-area isosurface triangles.

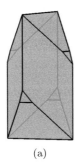
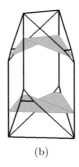

(a)   (b)   (c)

Figure 5.10. (a) Convex mesh polyhedron. (b) Isosurface patch. (c) Self intersecting interpolated isosurface patch.

Under appropriate conditions, the isosurface produced by MARCHING POLY-HEDRA is a 2-manifold with boundary. As with the MARCHING TETRAHEDRA algorithm, the isovalue should not equal the scalar value of any vertex and the underlying space must be a manifold. However, a third condition is necessary. Consider the convex polyhedron in Figure 5.10. The algorithm in Section 5.2 will construct a proper isosurface patch, but linear interpolation on the vertices may cause that isosurface patch to self-intersect.[3] To avoid this pathological case, we use a restricted set of mesh elements. While this set can certainly be enlarged, we tried to include the most typical ones.

Consider a convex polyhedral mesh and isovalue that has the following three conditions:

- The isovalue does not equal the scalar value of any mesh vertex.

- The mesh is a partition of a 3-manifold with boundary.

- The mesh elements are affine transformations of cubes, tetrahedra, square pyramids, triangular bipyramids, and octahedra.

Under these conditions, the isosurface produced by MARCHING POLYHEDRA has the following three properties:

Property 6. *The isosurface is a piecewise linear, oriented 2-manifold with boundary.*

Property 7. *The boundary of the isosurface lies on the boundary of the mesh.*

Property 8. *Set Υ does not contain any zero-area triangles or duplicate triangles, and the triangles in Υ form a triangulation of the isosurface.*

---

[3]Applying first linear interpolation, computing the convex hull of the interpolated vertices and the positive vertices, and then constructing the isosurface patch from this convex hull avoids this problem of self-intersection. The ISOHULL algorithm in Section 5.5 does exactly this. The drawback is the necessity of constructing a convex hull for each mesh element intersected by the isosurface.

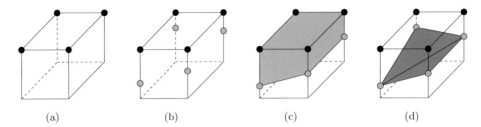

Figure 5.11. (a) Set $V_\mathbf{c}^+$ (black vertices). (b) Isosurface vertex set $U_\mathbf{c}$ (green vertices). (c) Set $R_\mathbf{c}^+(U_\mathbf{c})$, the convex hull of $U_\mathbf{c} \cup V_\mathbf{c}^+$. (d) Set $T_\mathbf{c}^+(U_\mathbf{c})$, the triangles in the triangulation of $\partial R_\mathbf{c}^+(U_\mathbf{c})$ that do not lie on the cube boundary.

5.4.2   Proofs of Isosurface Properties 1–4

Isosurface vertex sets and proper isosurface vertex sets were defined in Section 5.1. The set of midpoints of bipolar edges is an example of a proper isosurface vertex set.

The MARCHING POLYHEDRA algorithm uses isosurface patches constructed from the set of midpoints of bipolar edges. However, the properties of these isosurface patches hold for any proper isosurface vertex set. Many of the properties also hold for isosurface vertex sets that are not proper. We will need properties of isosurface vertex sets, both proper and not proper, later in this chapter. Therefore, we state and prove the lemmas in this section and the following two sections in the more general setting of isosurface vertex sets instead of restricting ourselves to sets of midpoints.

Let $U_\mathbf{c}$ be an isosurface vertex set of a convex polyhedron $\mathbf{c}$. Let $R_\mathbf{c}^+(U_\mathbf{c})$ equal $\mathrm{conv}(U_\mathbf{c} \cup V_\mathbf{c}^+)$, the convex hull of $U_\mathbf{c}$ and the positive vertices of $\mathbf{c}$ (Figure 5.11(c)). Triangulate the boundary of $R_\mathbf{c}^+(U_\mathbf{c})$ by triangulating each convex polygon in the boundary without adding any additional vertices. Orient the triangles so that the induced normals point toward $R_\mathbf{c}^+(U_\mathbf{c})$. Let $T_\mathbf{c}^+(U_\mathbf{c})$ be the oriented triangles in the triangulation of $\partial R_\mathbf{c}^+(U_\mathbf{c})$ that do not lie on the boundary of $\mathbf{c}$ (Figure 5.11(d)). Set $T_\mathbf{c}^+(U_\mathbf{c})$ is itself a triangulation. We will often abbreviate $R_\mathbf{c}^+(U_\mathbf{c})$ and $T_\mathbf{c}^+(U_\mathbf{c})$, as $R_\mathbf{c}^+$ and $T_\mathbf{c}^+$, respectively.

The set $M_\mathbf{c}$ of midpoints of bipolar edges of $\mathbf{c}$ is a proper isosurface vertex set of $\mathbf{c}$. Abbreviate $R_\mathbf{c}^+(M_\mathbf{c})$ as $R_\mathbf{c}^{M+}$ and $T_\mathbf{c}^+(M_\mathbf{c})$ as $T_\mathbf{c}^{M+}$.

Proofs of Properties 1 and 2 are the same as in Section 2.3.4 for MARCHING CUBES and are omitted.

To prove Property 3, we need to show that $U_\mathbf{c}$ is the set of vertices of $T_\mathbf{c}^+(U_\mathbf{c})$. We first show that the set of vertices of $T_\mathbf{c}^+(U_\mathbf{c})$ is a subset of $U_\mathbf{c}$.

Lemma 5.5. *Let $\mathbf{c}$ be a convex polyhedron in $\mathbb{R}^3$ where each vertex of $\mathbf{c}$ has a positive or a negative label. If $U_\mathbf{c}$ is an isosurface vertex set for $\mathbf{c}$, then the set of vertices of $T_\mathbf{c}^+(U_\mathbf{c})$ is a subset of $U_\mathbf{c}$.*

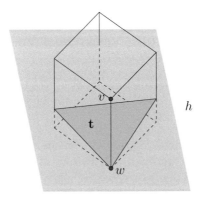

Figure 5.12. Isosurface triangle **t** with a positive vertex w that is also a cube vertex. The interior of bipolar edge (w, v) is separated from the isosurface patch by the plane h containing **t**.

Proof: Let w be a vertex of $T_{\mathbf{c}}^+(U_{\mathbf{c}})$ (abbreviated as $T_{\mathbf{c}}^+$). The vertices of $R_{\mathbf{c}}^+ = \operatorname{conv}(U_{\mathbf{c}} \cup V_{\mathbf{c}}^+)$ are $U_{\mathbf{c}} \cup V_{\mathbf{c}}^+$. Since $T_{\mathbf{c}}^+$ is a subset of a triangulation of $\partial R_{\mathbf{c}}^+$ that does not add any vertices, vertex w is in $U_{\mathbf{c}} \cup V_{\mathbf{c}}^+$.

If w is not in $V_{\mathbf{c}}^+$, then w must be in $U_{\mathbf{c}}$, establishing the claim. Assume that vertex w is in $V_{\mathbf{c}}^+$.

Since w is in $T_{\mathbf{c}}^+$, it must be incident on some triangle **t** that is not on a facet of **c**. (See Figure 5.12.) Let h be the plane containing **t**. Since h intersects the interior of **c** and h contains w, there is some edge (w, v) of **c** that is separated from $R_{\mathbf{c}}^+$ by h and does not lie on h. Thus,

$$(w, v) \cap R_{\mathbf{c}}^+ = w.$$

Since vertex v is not in $R_{\mathbf{c}}^+$, vertex v must have a negative label. Thus, (w, v) is a bipolar edge. By definition, the isosurface vertex set $U_{\mathbf{c}}$ contains some point p on (w, v). Since p is in $R_{\mathbf{c}}^+$ and $(w, v) \cap R_{\mathbf{c}}^+$ equals w, point p must equal w. Thus w is in $U_{\mathbf{c}}$, proving the claim. $\square$

Lemma 5.5 applies to any isosurface vertex set. The corollaries apply to proper isosurface vertex sets.

Corollary 5.6. *Let* **c** *be a convex polyhedron in* $\mathbb{R}^3$ *where each vertex of* **c** *has a positive or a negative label. If* $U_{\mathbf{c}}$ *is a proper isosurface vertex set for* **c**, *then the set of vertices of* $T_{\mathbf{c}}^+(U_{\mathbf{c}})$ *equals* $U_{\mathbf{c}}$.

Proof: By Lemma 5.5, if w is a vertex of $T_{\mathbf{c}}^+$, then w is in $U_{\mathbf{c}}$. On the other hand, if w is in $U_{\mathbf{c}}$, then some triangle **t** in the triangulation of $\partial\operatorname{conv}(U_{\mathbf{c}} \cup V_{\mathbf{c}}^+)$ contains w and is not on a facet of **c**. This triangle is in $T_{\mathbf{c}}^+$. Thus w is a vertex of $T_{\mathbf{c}}^+$. $\square$

Corollary 5.7. *Let* **c** *be a convex polyhedron in* $\mathbb{R}^3$ *where each vertex of* **c** *has a positive or a negative label. If* $U_{\mathbf{c}}$ *is a proper isosurface vertex set for* **c**, *then no vertex of* **c** *is in the set of vertices of* $T_{\mathbf{c}}^+(U_{\mathbf{c}})$.

Proof: By Lemma 5.5, the set of vertices of $T_{\mathbf{c}}^+(U_{\mathbf{c}})$ is a subset of $U_{\mathbf{c}}$. Since isosurface vertex set $U_{\mathbf{c}}$ is proper, no vertex of **c** is in $U_{\mathbf{c}}$. Thus, no vertex of **c** is in the set of vertices of $T_{\mathbf{c}}^+(U_{\mathbf{c}})$. □

Let μ be the affine transformation that maps each mesh element **c** to the corresponding predefined "generic" polyhedron. If the vertices of **c** are labeled "+" or "−", then vertices of $\mu(\mathbf{c})$ receive "+" or "−" labels from the corresponding vertices of **c**. Thus, $\mu(v)$ is positive if and only if v is positive and $\mu(v)$ is negative if and only if v is negative.

Property 3.   *The isosurface intersects every bipolar mesh edge at exactly one point.*

Property 4.   *The isosurface does not intersect any negative or strictly positive mesh edges.*

Proof of Properties 3 & 4: The isosurface is constructed by retrieving triangles from a lookup table and then determining their vertex positions using linear interpolation. Let **c** be a mesh element containing a mesh edge **e**. Vertices of **c** are labeled "+" or "−" based on the relationship of their scalar value to the isovalue. The isosurface patch for $\mu(\mathbf{c})$ is a subset of $\mathrm{conv}(M_{\mu(\mathbf{c})} \cup V_{\mu(\mathbf{c})}^+)$, the convex hull of the midpoints of bipolar edges and positive vertices of $\mu(\mathbf{c})$.

Since **c** is convex, only isosurface triangles with vertices on **e** intersect **e**. By Corollary 5.6, the vertices of $T_{\mu(\mathbf{c})}^+(M_{\mu(\mathbf{c})})$ (abbreviated $T_{\mu(\mathbf{c})}^{M+}$) are midpoints of edges of $\mu(\mathbf{c})$. Thus, the isosurface patch in $\mu(\mathbf{c})$ intersects $\mu(\mathbf{e})$ at most once.

If **e** is bipolar, then so is $\mu(\mathbf{e})$. The midpoint of $\mu(\mathbf{e})$ is in $M_{\mu(\mathbf{c})}$. Set $T_{\mu(\mathbf{c})}^{M+}$ contains some triangle containing this midpoint. Thus, the isosurface patch in $\mu(\mathbf{c})$ intersects $\mu(\mathbf{e})$ at exactly one point p. Point $p \in \mu(\mathbf{e})$ maps to a single point on **e** determined by linear interpolation. Thus the isosurface intersects **e** at exactly one point.

If **e** is negative or strictly positive, then no isosurface vertex lies on **e**. Thus the isosurface does not intersect negative or strictly positive grid edges. □

5.4.3   Proof of Isosurface Property 5

The MARCHING POLYHEDRA algorithm retrieves isosurface patches for a mesh element **c** from the isosurface lookup table for $\mu(\mathbf{c})$, where $\mu(\mathbf{c})$ is a predefined "generic" polyhedron and μ is an affine transformation. The algorithm then embeds those isosurface patches using linear interpolation on the isosurface

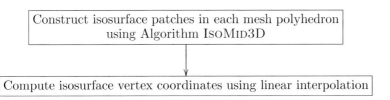

Figure 5.13. Variation of MARCHING POLYHEDRA without a lookup table.

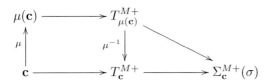

Figure 5.14. The MARCHING POLYHEDRA algorithm constructs the isosurface patch $\Sigma_{\mathbf{c}}^{M+}(\sigma)$ by mapping polyhedron $\mathbf{c}$ to $\mu(\mathbf{c})$, retrieving $T_{\mu(\mathbf{c})}^{M+}$ from the isosurface lookup table and mapping $T_{\mu(\mathbf{c})}^{M+}$ to $\Sigma_{\mathbf{c}}^{M+}(\sigma)$ using linear interpolation. The isosurface patch $\Sigma_{\mathbf{c}}^{M+}(\sigma)$ could be constructed directly from $\mathbf{c}$ by computing $T_{\mathbf{c}}^{M+}$ directly on $\mathbf{c}$ and then mapping $T_{\mathbf{c}}^{M+}$ to $\Sigma_{\mathbf{c}}^{M+}(\sigma)$.

vertices. Isosurface patches in the lookup table are constructed by applying Algorithm IsoMid3D to $\mu(\mathbf{c})$. In place of using the lookup table, we could have constructed isosurface patches by applying IsoMid3D directly to each mesh element $\mathbf{c}$. (See Figure 5.13.) We argue that doing so would give the exact same isosurface.

If $\mu : \mathbb{R}^3 \to \mathbb{R}^3$ is some affine transformation of $\mathbb{R}^3$, then $\mu(\mathrm{conv}(P))$ equals $\mathrm{conv}(\mu(P))$ for any point set $P \subseteq \mathbb{R}^3$. Thus, $\mu(R_{\mathbf{c}}^{M+})$ equals $R_{\mu(\mathbf{c})}^{M+}$ and $\mu(T_{\mathbf{c}}^{M+})$ equals $T_{\mu(\mathbf{c})}^{M+}$. Applying μ^{-1} to $T_{\mu(\mathbf{c})}^{M+}$ gives a triangulation $T_{\mathbf{c}}^{M+}$, which is isomorphic to $T_{\mu(\mathbf{c})}^{M+}$. (See Figure 5.14.)

MARCHING POLYHEDRA embeds $T_{\mu(\mathbf{c})}^{M+}$ in $\mathbf{c}$ using linear interpolation on the isosurface vertices. The location of the isosurface vertices depends upon the isovalue σ. Let $\Sigma_{\mathbf{c}}^{M+}(\sigma)$ (abbreviated $\Sigma_{\mathbf{c}}^{M+}$) represent the resulting isosurface patch for $\mathbf{c}$.

Because $T_{\mu(\mathbf{c})}^{M+}$ and $T_{\mathbf{c}}^{M+}$ are isomorphic, the algorithm could have constructed $\Sigma_{\mathbf{c}}^{M+}$ from $T_{\mathbf{c}}^{M+}$ by reembedding $T_{\mathbf{c}}^{M+}$ in $\mathbf{c}$ using linear interpolation on its vertices. Each vertex v_i of $T_{\mathbf{c}}^{M+}$ is the midpoint m_i of some edge $\mathbf{e}_i$. Vertex v_i is remapped to the interpolated point $w_i \in \mathbf{e}_i$. Map each vertex v_i of $T_{\mathbf{c}}^{M+}$ to $(1 - \alpha)v_i + \alpha w_i$. Linearly extend the mapping to each triangle in $T_{\mathbf{c}}^{M+}$.

The algorithm uses $\mu(\mathbf{c})$ to construct $\Sigma_{\mathbf{c}}^{M+}$ because it is faster to retrieve the precomputed values of $T_{\mu(\mathbf{c})}^{M+}$ from the lookup table than to compute $T_{\mathbf{c}}^{M+}$. However, for the proofs of Properties 5–8 we will view $\Sigma_{\mathbf{c}}^{M+}$ as a homotopy of $T_{\mathbf{c}}^{M+}$.

To prove Property 5, it is necessary to show that there are no cracks between isosurface patches in adjacent polyhedra. This was done by case analysis for MARCHING CUBES and MARCHING TETRAHEDRA. Case analysis will not work for the MARCHING POLYHEDRA algorithm and we need a more general proof.

We first show that for each polyhedron $\mathbf{c}$, set $|T_{\mathbf{c}}^{M+}|$ separates the positive vertices of $\mathbf{c}$ from the negative ones (Corollary 5.12). ($|T_{\mathbf{c}}^{M+}|$ represents the union of all the triangles of $T_{\mathbf{c}}^{M+}$.) We next show that for two adjacent polyhedra, $\mathbf{c}_1$ and $\mathbf{c}_2$, region $R_{\mathbf{c}_1}^{M+}$ and $R_{\mathbf{c}_2}^{M+}$ agree on their adjacent boundaries (Corollary 5.15). This implies that there are no "cracks" between $|T_{\mathbf{c}_1}^{M+}|$ and $|T_{\mathbf{c}_2}^{M+}|$ in adjacent polyhedra and that $\bigcup_{\mathbf{c}\in\Gamma} |T_{\mathbf{c}}^{M+}|$ separates the positive vertices of Γ from the negative ones. We also show that the triangulations $T_{\mathbf{c}_1}^{M+}$ and $T_{\mathbf{c}_2}^{M+}$ agree on their adjacent boundaries, i.e., $\mathbf{t}_1 \cap \mathbf{t}_2$ is a face of $\mathbf{t}_1$ and a face of $\mathbf{t}_2$ for every $\mathbf{t}_1 \in T_{\mathbf{c}_1}^{M+}$ and $\mathbf{t}_2 \in T_{\mathbf{c}_2}^{M+}$ (Corollary 5.17). Thus, $\bigcup_{\mathbf{c}\in\Gamma} T_{\mathbf{c}}^{M+}$ is a triangulation (Lemma 5.21). We define a homotopy map η that maps this triangulation to the isosurface. Since this homotopy map does not cross any vertices, the resulting isosurface separates the positive vertices of Γ from the negative ones.

As previously noted, we present the lemmas in this section in the more general setting of isosurface vertex sets instead of restricting the lemmas to the midpoints of bipolar edges. Some of the lemmas apply only to isosurface vertex sets that are proper, while others have no such restriction.

We start by showing that if $U_{\mathbf{c}}$ is a proper isosurface vertex set, then $R_{\mathbf{c}}^+(U_{\mathbf{c}})$ is three-dimensional.

Lemma 5.8. *Let $\mathbf{c}$ be a convex polyhedron in $\mathbb{R}^3$ where each vertex of $\mathbf{c}$ has a positive or a negative label. If $U_{\mathbf{c}}$ is a proper isosurface vertex set of $\mathbf{c}$ and at least one vertex of $\mathbf{c}$ is positive, then $R_{\mathbf{c}}^+(U_{\mathbf{c}})$ is a three-dimensional convex set.*

Proof: Let v be a positive vertex of $\mathbf{c}$. Since v is positive, some interior point of every edge incident on v is in $R_{\mathbf{c}}^+$. Since there are at least three edges incident on v and these edges are not coplanar, $R_{\mathbf{c}}^+$ is three-dimensional. Since $R_{\mathbf{c}}^+$ is the convex hull of the set $U_{\mathbf{c}} \cup V_{\mathbf{c}}^+$, set $R_{\mathbf{c}}^+$ is convex.   □

Corollary 5.9. *Let $\mathbf{c}$ be a convex polyhedron in $\mathbb{R}^3$ where each vertex of $\mathbf{c}$ has a positive or a negative label. If $U_{\mathbf{c}}$ is a proper isosurface vertex set of $\mathbf{c}$, then $R_{\mathbf{c}}^+(U_{\mathbf{c}})$ is either a three-dimensional convex set or the empty set.*

Proof: By definition, points in $U_{\mathbf{c}}$ do not lie on negative edges of $\mathbf{c}$. Thus, if $V_{\mathbf{c}}^+$ is empty, then $U_{\mathbf{c}}$ must also be empty. By Lemma 5.8, if $V_{\mathbf{c}}^+$ is not empty, then $R_{\mathbf{c}}^+(U_{\mathbf{c}})$ is not empty.   □

Figure 5.15. Let $\mathbb{Y}_2$ be the red region and $\mathbb{Y}_1$ be the rectangle containing $\mathbb{Y}_2$. The yellow curve, $\mathbb{Y}_2 \cap \mathrm{cl}(\mathbb{Y}_1 - \mathbb{Y}_2)$, separates the red region, $\mathbb{Y}_1$, from the blue region, $\mathrm{cl}(\mathbb{Y}_1 - \mathbb{Y}_2)$.

Lemma 5.8 and its corollary require $U_{\mathbf{c}}$ to be proper. When $U_{\mathbf{c}}$ is an isosurface vertex set that is not proper, $R_{\mathbf{c}}^+(U_{\mathbf{c}})$ can have any dimension from zero to three.

Let $\mathbb{Y}_2 \subset \mathbb{Y}_1$ be a d-dimensional region in $\mathbb{R}^d$. Set $\mathrm{cl}(\mathbb{Y}_1 - \mathbb{Y}_2)$ is the closure of $\mathbb{Y}_1 - \mathbb{Y}_2$. The set $\mathbb{Y}_2 \cap \mathrm{cl}(\mathbb{Y}_1 - \mathbb{Y}_2)$ separates $\mathbb{Y}_1$ from $\mathbb{Y}_1 - \mathbb{Y}_2$. (See Appendix B, Lemma B.8.) For instance, in Figure 5.15 the red region is $\mathbb{Y}_2$, the rectangle is $\mathbb{Y}_1$ and the blue region is $\mathrm{cl}(\mathbb{Y}_1 - \mathbb{Y}_2)$. The green curve, $\mathbb{Y}_2 \cap \mathrm{cl}(\mathbb{Y}_1 - \mathbb{Y}_2)$, separates the red region, $\mathbb{Y}_2$, from the blue region, $\mathrm{cl}(\mathbb{Y}_1 - \mathbb{Y}_2)$.

Let $\mathbf{c}$ be a convex polyhedron in $\mathbb{R}^3$ where each vertex of $\mathbf{c}$ has a positive or a negative label and let $U_{\mathbf{c}}$ be an isosurface vertex set (not necessarily proper) of $\mathbf{c}$. We define $S_{\mathbf{c}}^+(U_{\mathbf{c}})$ (abbreviated $S_{\mathbf{c}}^+$) to be the set that separates $R_{\mathbf{c}}^+(U_{\mathbf{c}})$ (abbreviated $R_{\mathbf{c}}^+$) from $\mathrm{cl}(\mathbf{c} - R_{\mathbf{c}}^+(U_{\mathbf{c}}))$, as follows:

$$S_{\mathbf{c}}^+(U_{\mathbf{c}}) = R_{\mathbf{c}}^+(U_{\mathbf{c}}) \cap \mathrm{cl}(\mathbf{c} - R_{\mathbf{c}}^+(U_{\mathbf{c}})).$$

In abbreviated form, $S_{\mathbf{c}}^+$ is defined as

$$S_{\mathbf{c}}^+ = R_{\mathbf{c}}^+ \cap \mathrm{cl}(\mathbf{c} - R_{\mathbf{c}}^+).$$

Algorithm IsoMid3D in Section 5.2 returns a set of triangles $T_{\mathbf{c}}^{M+}$ forming an isosurface patch. The corollary to the following lemma shows that this isosurface patch is exactly $S_{\mathbf{c}}^+(M_{\mathbf{c}})$.

Lemma 5.10. *Let $\mathbf{c}$ be a convex polyhedron in $\mathbb{R}^3$ where each vertex of $\mathbf{c}$ has a positive or a negative label and let $U_{\mathbf{c}}$ be an isosurface vertex set of $\mathbf{c}$. If $R_{\mathbf{c}}^+(U_{\mathbf{c}})$ has dimension three, then $T_{\mathbf{c}}^+(U_{\mathbf{c}})$ is a triangulation of $S_{\mathbf{c}}^+(U_{\mathbf{c}})$.*

Proof: We show that $\bigcup_{\mathbf{t} \in T_{\mathbf{c}}^+} \mathbf{t}$ equals $S_{\mathbf{c}}^+$. By definition, $S_{\mathbf{c}}^+$ equals $R_{\mathbf{c}}^+ \cap \mathrm{cl}(\mathbf{c} - R_{\mathbf{c}}^+)$. Every triangle $\mathbf{t} \in T_{\mathbf{c}}^+$ is in both $R_{\mathbf{c}}^+$ and $\mathrm{cl}(\mathbf{c} - R_{\mathbf{c}}^+)$. Thus, $\bigcup_{\mathbf{t} \in T_{\mathbf{c}}^+} \mathbf{t} \subseteq S_{\mathbf{c}}^+$.

Assume p is a point in $S_{\mathbf{c}}^+ = R_{\mathbf{c}}^+ \cap \mathrm{cl}(\mathbf{c} - R_{\mathbf{c}}^+)$. Since $R_{\mathbf{c}}^+$ is not empty, some vertex v of $\mathbf{c}$ is positive. Since $p \in \mathrm{cl}(\mathbf{c} - R_{\mathbf{c}}^+)$, point p is not in the interior of $R_{\mathbf{c}}^+$ and so is on $\partial R_{\mathbf{c}}^+$. Moreover, some triangle $\mathbf{t}$ in the triangulation of $\partial R_{\mathbf{c}}^+$ containing p is in $\mathrm{cl}(\mathbf{c} - R_{\mathbf{c}}^+)$. Since $R_{\mathbf{c}}^+$ is three-dimensional and $\mathbf{t}$ is in $\mathrm{cl}(\mathbf{c} - R_{\mathbf{c}}^+)$, triangle $\mathbf{t}$ cannot lie on $\partial \mathbf{c}$. Thus, $\mathbf{t}$ intersects $\mathrm{int}(\mathbf{c})$ and so is in $T_{\mathbf{c}}^+$. Thus, $S_{\mathbf{c}}^+$

is a subset of $\bigcup_{\mathbf{t} \in T_{\mathbf{c}}^+} \mathbf{t}$. We already showed that $\bigcup_{\mathbf{t} \in T_{\mathbf{c}}^+} \mathbf{t}$ is a subset of $S_{\mathbf{c}}^+$ so $S_{\mathbf{c}}^+$ equals $\bigcup_{\mathbf{t} \in T_{\mathbf{c}}^+} \mathbf{t}$.

Since $T_{\mathbf{c}}^+$ is a subset of a triangulation, it is also a triangulation. Thus, $T_{\mathbf{c}}^+$ is a triangulation of $S_{\mathbf{c}}^+$. $\qquad\square$

Corollary 5.11. *Let $\mathbf{c}$ be a convex polyhedron in $\mathbb{R}^3$ where each vertex of $\mathbf{c}$ has a positive or a negative label. If $U_{\mathbf{c}}$ is a proper isosurface vertex set, then $T_{\mathbf{c}}^+(U_{\mathbf{c}})$ is a triangulation of $S_{\mathbf{c}}^+(U_{\mathbf{c}})$.*

Proof: By Lemma 5.8, $R_{\mathbf{c}}^+(U_{\mathbf{c}})$ is three-dimensional. By Lemma 5.10, $T_{\mathbf{c}}^+(U_{\mathbf{c}})$ is a triangulation of $S_{\mathbf{c}}^+(U_{\mathbf{c}})$. $\qquad\square$

The following corollary is not used in the proof of Property 5. (It is superseded by Corollary 5.20.) However, it illustrates how $|T_{\mathbf{c}}^+|$ behaves within a single polyhedron $\mathbf{c}$.

Corollary 5.12. *Let $\mathbf{c}$ be a convex polyhedron in $\mathbb{R}^3$ where each vertex of $\mathbf{c}$ has a positive or a negative label. If $U_{\mathbf{c}}$ is a proper isosurface vertex set, then $|T_{\mathbf{c}}^+(U_{\mathbf{c}})|$ separates the positive vertices of $\mathbf{c}$ from the negative ones.*

Proof: By Corollary 5.11, $|T_{\mathbf{c}}^+|$ equals $S_{\mathbf{c}}^+$. The definition of $S_{\mathbf{c}}^+$ is $R_{\mathbf{c}}^+ \cap \mathrm{cl}(\mathbf{c} - R_{\mathbf{c}}^+)$. By Lemma B.8 in Appendix B, $S_{\mathbf{c}}^+$ separates $R_{\mathbf{c}}^+$ from $(\mathbf{c} - R_{\mathbf{c}}^+)$. Region $R_{\mathbf{c}}^+$ contains all the positive vertices of $\mathbf{c}$ while $(\mathbf{c} - R_{\mathbf{c}}^+)$ contains all the negative ones. Thus $|T_{\mathbf{c}}^+| = S_{\mathbf{c}}^+$ separates the positive vertices of $\mathbf{c}$ from the negative ones. $\qquad\square$

We next show that if $\mathbf{c}_1$ and $\mathbf{c}_2$ are adjacent polyhedra, then $R_{\mathbf{c}_1}^+$ and $R_{\mathbf{c}_2}^+$ agree on shared faces of $\mathbf{c}_1$ and $\mathbf{c}_2$. It will follow that there are no "cracks" between adjacent isosurface patches.

We first extend our definitions of $R_{\mathbf{c}}^+$ from $\mathbf{c}$ to the faces of $\mathbf{c}$. Let $\mathbf{c}$ be a convex polyhedron in $\mathbb{R}^3$ where each vertex of $\mathbf{c}$ has a positive or a negative label. For each face $\mathbf{f}$ of $\mathbf{c}$, let $V_{\mathbf{f}}^+$ be the set of positive vertices of $\mathbf{f}$. Let $U_{\mathbf{f}}$ be the points of $U_{\mathbf{c}}$ that lie on $\mathbf{f}$. Let $R_{\mathbf{f}}^+(U_{\mathbf{f}})$ (abbreviated as $R_{\mathbf{f}}^+$) equal $\mathrm{conv}(U_{\mathbf{f}} \cup V_{\mathbf{f}}^+)$. In these definitions, face $\mathbf{f}$ can have dimension 0, 1, 2, or 3.

Lemma 5.8 states that if $U_{\mathbf{c}}$ is a proper isosurface vertex set, then $R_{\mathbf{c}}^+(U_{\mathbf{c}})$ is three-dimensional. The following lemma generalizes Lemma 5.8 to $R_{\mathbf{f}}^+(U_{\mathbf{f}})$.

Lemma 5.13. *Let $\mathbf{c}$ be a convex polyhedron in $\mathbb{R}^3$ where each vertex of $\mathbf{c}$ has a positive or a negative label and let $\mathbf{f}$ be a k-dimensional face of $\mathbf{c}$. If $U_{\mathbf{c}}$ is a proper isosurface vertex set of $\mathbf{c}$ and at least one vertex of $\mathbf{f}$ is positive, then $R_{\mathbf{f}}^+(U_{\mathbf{f}})$ is a k-dimensional convex set.*

Proof: Let v be a positive vertex of $\mathbf{f}$. Since v is positive, some interior point of every edge of $\mathbf{f}$ incident on v is in $R_{\mathbf{f}}^+$. Since there are at least k edges incident on v and these edges are not contained in any $(k-1)$-dimensional subspace, $R_{\mathbf{f}}^+$ is k-dimensional. Since $R_{\mathbf{f}}^+$ is the convex hull of the set $U_{\mathbf{c}} \cup V_{\mathbf{c}}^+$, set $R_{\mathbf{f}}^+$ is convex. $\qquad\square$

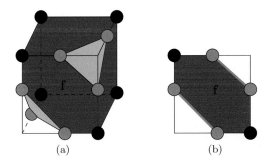

Figure 5.16. (a) Cube $\mathbf{c}$, green isosurface vertex set $U_{\mathbf{c}}$, and red region $R_{\mathbf{c}}^+$. Black cube vertices are positive. (b) Front facet $\mathbf{f}$ of cube $\mathbf{c}$, set $U_{\mathbf{f}} = U_{\mathbf{c}} \cap \mathbf{f}$, and red region $R_{\mathbf{f}}^+$. Region $R_{\mathbf{c}}^+ \cap \mathbf{f}$ equals $R_{\mathbf{f}}^+$.

We show that if $\mathbf{f}$ is a face of $\mathbf{c}$, then $R_{\mathbf{c}}^+ \cap \mathbf{f}$ equals $R_{\mathbf{f}}^+$. For example, Figure 5.16(a) displays a cube $\mathbf{c}$, its positive vertices, isosurface vertex set $U_{\mathbf{c}}$, and region $R_{\mathbf{c}}^+$. Figure 5.16(b) displays the front facet $\mathbf{f}$ of cube $\mathbf{c}$, the set $U_{\mathbf{c}} \cap \mathbf{f}$, and region $R_{\mathbf{f}}^+$. As claimed, $R_{\mathbf{c}}^+ \cap \mathbf{f}$ equals $R_{\mathbf{f}}^+$.

Lemma 5.14. *Let $\mathbf{c}$ be a convex polyhedron in $\mathbb{R}^3$ where each vertex of $\mathbf{c}$ has a positive or a negative label and let $U_{\mathbf{c}}$ be an isosurface vertex set of $\mathbf{c}$. For every face $\mathbf{f}$ of polyhedron $\mathbf{c}$, set $R_{\mathbf{c}}^+(U_{\mathbf{c}}) \cap \mathbf{f}$ equals $R_{\mathbf{f}}^+(U_{\mathbf{c}} \cap \mathbf{f})$.*

Proof: Since $\mathbf{f}$ is a face of $\mathbf{c}$,

$$R_{\mathbf{c}}^+ \cap \mathbf{f} = \mathrm{conv}(U_{\mathbf{c}} \cup V_{\mathbf{c}}^+) \cap \mathbf{f} = \mathrm{conv}((U_{\mathbf{c}} \cup V_{\mathbf{c}}^+) \cap \mathbf{f}).$$

Since $(U_{\mathbf{c}} \cup V_{\mathbf{c}}^+) \cap \mathbf{f}$ equals $U_{\mathbf{f}} \cup V_{\mathbf{f}}^+$,

$$\mathrm{conv}((U_{\mathbf{c}} \cup V_{\mathbf{c}}^+) \cap \mathbf{f}) = \mathrm{conv}(U_{\mathbf{f}} \cup V_f^+) = R_{\mathbf{f}}^+.$$

Thus,

$$R_{\mathbf{c}}^+ \cap \mathbf{f} = R_{\mathbf{f}}^+. \qquad \square$$

As an immediate corollary, $R_{\mathbf{c}_1}^+$ and $R_{\mathbf{c}_2}^+$ agree on shared faces of $\mathbf{c}_1$ and $\mathbf{c}_2$.

Corollary 5.15. *Let $\mathbf{c}_1$ and $\mathbf{c}_2$ be adjacent convex polyhedra in $\mathbb{R}^3$ where each vertex of $\mathbf{c}_1$ and $\mathbf{c}_2$ has a positive or a negative label. Let $U_{\mathbf{c}_1}$ and $U_{\mathbf{c}_2}$ be isosurface vertex sets of $\mathbf{c}_1$ and $\mathbf{c}_2$, respectively. If $\mathbf{f}$ is a face of both $\mathbf{c}_1$ and $\mathbf{c}_2$, then $R_{\mathbf{c}_1}^+(U_{\mathbf{c}_1}) \cap \mathbf{f}$ equals $R_{\mathbf{c}_2}^+(U_{\mathbf{c}_2}) \cap \mathbf{f}$.*

Proof: Applying Lemma 5.14,

$$R_{\mathbf{c}_1}^+ \cap \mathbf{f} = R_f^+ = R_{\mathbf{c}_2}^+ \cap \mathbf{f}.$$

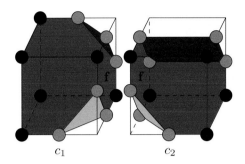

Figure 5.17. Adjacent cubes $\mathbf{c}_1$ and $\mathbf{c}_2$, green isosurface set $U_{\mathbf{c}}$, and red regions $R_{\mathbf{c}_1}^+$ and $R_{\mathbf{c}_2}^+$. Black cube vertices are positive. Shared facet $\mathbf{f}$ lies between the two cubes. Region $R_{\mathbf{c}_1}^+ \cap \mathbf{f}$ equals $R_{\mathbf{c}_2}^+ \cap \mathbf{f}$.

Figure 5.17 illustrates Corollary 5.15. In the statement of Lemma 5.14 and its corollary, face $\mathbf{f}$ can have dimensions 0, 1, 2, or 3, although the three-dimensional case is trivial.

Corollary 5.15 indicates that $R_{\mathbf{c}_1}$ and $R_{\mathbf{c}_2}$ agree on adjacent polyhedra $\mathbf{c}_1$ and $\mathbf{c}_2$. We need also to show that triangulations $T_{\mathbf{c}_1}$ and $T_{\mathbf{c}_2}$ agree on their adjacent boundaries. We first show that the nonempty intersection of a triangle in $T_{\mathbf{c}}^+$ and a face $\mathbf{f}$ of $\mathbf{c}$ is a vertex or edge of $R_{\mathbf{f}}^+$.

If $\mathbf{f}$ is two-dimensional and $U_{\mathbf{c}}$ is proper, then $R_{\mathbf{f}}^+(U_{\mathbf{f}})$ is a convex polygon. The edges and vertices of $R_{\mathbf{f}}^+(U_{\mathbf{f}})$ are the edges and vertices of its boundary. If $\mathbf{f}$ is zero- or one-dimensional or $U_{\mathbf{c}}$ is not proper, then $R_{\mathbf{f}}^+(U_{\mathbf{f}})$ may be a line segment or a point. In this case, $R_{\mathbf{f}}^+(U_{\mathbf{f}})$ is either a single vertex or an edge and two vertices.

Lemma 5.16. *Let $\mathbf{c}$ be a convex polyhedron in $\mathbb{R}^3$ where each vertex of $\mathbf{c}$ has a positive or a negative label, let $U_{\mathbf{c}}$ be an isosurface vertex set of $\mathbf{c}$, and let $\mathbf{f}$ be a proper face of $\mathbf{c}$. If $\mathbf{t}$ is a triangle in $T_{\mathbf{c}}(U_{\mathbf{c}})$ that intersects $\mathbf{f}$, then $\mathbf{t} \cap \mathbf{f}$ is a vertex or edge of $\mathbf{t}$ and a vertex or edge of $R_{\mathbf{f}}^+(U_{\mathbf{c}} \cap \mathbf{f})$.*

Proof: If $R_{\mathbf{c}}^+$ has dimension two or less, then $T_{\mathbf{c}}^+$ is the empty set and the lemma is trivially true. Assume that $R_{\mathbf{c}}^+$ is three-dimensional.

Let $\mathbf{g}$ equal $\mathbf{t} \cap \mathbf{f}$. Since $\mathbf{f}$ is on the boundary of $\mathbf{c}$ and $\mathbf{t}$ is contained in $\mathbf{c}$, region $\mathbf{g} = \mathbf{t} \cap \mathbf{f}$ is a face of $\mathbf{t}$. Since $T_{\mathbf{c}}^+$ does not contain any triangles on the boundary of $\mathbf{c}$, triangle $\mathbf{t}$ is not contained in $\mathbf{f}$. Thus $\mathbf{g}$ is a vertex or edge of $\mathbf{t}$.

By Lemma 5.14, the intersection $R_{\mathbf{c}}^+ \cap \mathbf{f}$ equals $R_{\mathbf{f}}^+(U_{\mathbf{c}} \cap \mathbf{f})$ (abbreviated $R_{\mathbf{f}}^+$) and so $R_{\mathbf{f}}^+$ is a face of $R_{\mathbf{c}}^+$. Thus the triangulation of $\partial R_{\mathbf{c}}^+$ contains the vertices and edges of $R_{\mathbf{f}}^+$. Since $\mathbf{t}$ is in the triangulation of $\partial R_{\mathbf{c}}^+$, its face $\mathbf{g}$ is a vertex or edge of $R_{\mathbf{f}}^+$. $\qquad\square$

As a corollary to Lemma 5.16, triangulations $T_{\mathbf{c}_1}$ and $T_{\mathbf{c}_2}$ agree on their adjacent boundaries.

Corollary 5.17. *Let $\mathbf{c}_1$ and $\mathbf{c}_2$ be adjacent convex polyhedra in $\mathbb{R}^3$ where each vertex of $\mathbf{c}_1$ and $\mathbf{c}_2$ has a positive or a negative label. Let $U_{\mathbf{c}_1}$ and $U_{\mathbf{c}_2}$ be isosurface vertex sets of $\mathbf{c}_1$ and $\mathbf{c}_2$, respectively, If $\mathbf{t}_1$ is a triangle in $T_{\mathbf{c}_1}^+(U_{\mathbf{c}_1})$ and $\mathbf{t}_2$ is a triangle in $T_{\mathbf{c}_1}^+(U_{\mathbf{c}_2})$, then $\mathbf{t}_1 \cap \mathbf{t}_2$ is a face of $\mathbf{t}_1$ and a face of $\mathbf{t}_2$.*

Proof: If $\mathbf{t}_1$ does not intersect $\mathbf{t}_2$, then the corollary is trivially true. Assume $\mathbf{t}_1$ intersects $\mathbf{t}_2$. The intersection $\mathbf{t}_1 \cap \mathbf{t}_2$ lies on some face $\mathbf{f}$ of both $\mathbf{c}_1$ and $\mathbf{c}_2$.

Let $\mathbf{g}_1$ equal $\mathbf{t}_1 \cap \mathbf{f}$ and $\mathbf{g}_2$ equal $\mathbf{t}_2 \cap \mathbf{f}$. Since $\mathbf{t}_1 \cap \mathbf{t}_2$ is a subset of $\mathbf{f}$, the intersection $\mathbf{t}_1 \cap \mathbf{t}_2$ equals $\mathbf{g}_1 \cap \mathbf{g}_2$.

By Lemma 5.16, $\mathbf{g}_1$ and $\mathbf{g}_2$ are vertices or edges of $R_{\mathbf{f}}^+ (= R_{\mathbf{f}}^+(U_{\mathbf{c}} \cap \mathbf{f}).)$ Since the edges of $R_{\mathbf{f}}^+$ intersect only at the vertices of $R_{\mathbf{f}}^+$, the intersection $\mathbf{g}_1 \cap \mathbf{g}_2$ is a face of $\mathbf{g}_1$. By Lemma 5.16, $\mathbf{g}_1$ is a face of $\mathbf{t}_1$, so the intersection $\mathbf{g}_1 \cap \mathbf{g}_2$ is a face of $\mathbf{t}_1$. Since $\mathbf{t}_1 \cap \mathbf{t}_2$ equals $\mathbf{g}_1 \cap \mathbf{g}_2$, the intersection $\mathbf{t}_1 \cap \mathbf{t}_2$ is a face of $\mathbf{t}_1$. Similarly, the intersection $\mathbf{t}_1 \cap \mathbf{t}_2$ is a face of $\mathbf{t}_2$. $\qquad\qquad\square$

In the statement of Lemma 5.16 and its corollary, $\mathbf{f}$ is a proper face of $\mathbf{c}$ and can have dimensions 0, 1, or 2. Note that Lemmas 5.14 and 5.16 and their corollaries do not require that isosurface vertex sets $U_{\mathbf{c}}$ or $U_{\mathbf{c}_1}$ or $U_{\mathbf{c}_2}$ be proper.

For a convex polyhedral mesh Γ, let U be an isosurface vertex set for Γ, and let $U_{\mathbf{c}}$ equals $\mathbf{c} \cap U$ for each polyhedron $\mathbf{c} \in \Gamma$. Note that $U_{\mathbf{c}}$ is an isosurface vertex set for $\mathbf{c}$ and that if U is proper, then each $U_{\mathbf{c}}$ is proper.

Define M_Γ, $\Sigma_\Gamma^{M+}(\sigma)$, $T_\Gamma^+(U)$, $R_\Gamma^+(U)$, and $S_\Gamma^+(U)$ as

$$M_\Gamma = \bigcup_{\mathbf{c} \in \Gamma} M_{\mathbf{c}},$$

$$\Sigma_\Gamma^{M+}(\sigma) = \bigcup_{\mathbf{c} \in \Gamma} \Sigma_{\mathbf{c}}^{M+}(\sigma),$$

$$T_\Gamma^+(U) = \bigcup_{\mathbf{c} \in \Gamma} T_{\mathbf{c}}^+(U_{\mathbf{c}}),$$

$$R_\Gamma^+(U) = \bigcup_{\mathbf{c} \in \Gamma} R_{\mathbf{c}}^+(U_{\mathbf{c}}),$$

$$S_\Gamma^+(U) = R_\Gamma^+(U) \cap \mathrm{cl}(|\Gamma| - R_\Gamma^+(U)).$$

$|\Gamma|$ is the union of all the elements of mesh Γ. Abbreviate $\Sigma_\Gamma^{M+}(\sigma)$, $T_\Gamma^+(U)$, $R_\Gamma^+(U)$, and $S_\Gamma^+(U)$ as Σ_Γ^{M+}, T_Γ^+, R_Γ^+, and S_Γ^+, respectively. Abbreviate $T_\Gamma^+(M_\Gamma)$ and $S_\Gamma^+(M_\Gamma)$ as T_Γ^{M+} and S_Γ^{M+}. Note that S_Γ^+ separates R_Γ^+ from $|\Gamma| - R_\Gamma^+$. (See Lemma B.8 in Appendix B.)

We show that $S_\Gamma^+(U)$ equals $\bigcup_{\mathbf{c} \in \Gamma} S_{\mathbf{c}}^+(U \cap \mathbf{c})$ if U is a proper isosurface vertex set.

Lemma 5.18. *Let Γ be a convex polyhedral mesh in $\mathbb{R}^3$ where each vertex of Γ has a positive or negative label. If U is a proper isosurface vertex set of Γ, then set $S_\Gamma^+(U)$ equals $\bigcup_{\mathbf{c} \in \Gamma} S_{\mathbf{c}}^+(U \cap \mathbf{c})$.*

Proof: Abbreviate $R_{\mathbf{c}}^+(U \cap \mathbf{c})$ and $S_{\mathbf{c}}^+(U \cap \mathbf{c})$ as $R_{\mathbf{c}}^+$ and $S_{\mathbf{c}}^+$, respectively. We first prove that $|\Gamma| - R_{\Gamma}^+$ equals $\bigcup_{\mathbf{c} \in \Gamma}(\mathbf{c} - R_{\mathbf{c}}^+)$.

Let p be a point in $|\Gamma| - R_{\Gamma}^+$. Point p lies in some polyhedron $\mathbf{c}_0 \in \Gamma$. Since p is not in R_{Γ}^+, point p is not in $R_{\mathbf{c}_0}^+$. Thus, p is in $\mathbf{c}_0 - R_{\mathbf{c}_0}^+$ and in $\bigcup_{\mathbf{c} \in \Gamma}(\mathbf{c} - R_{\mathbf{c}}^+)$. Thus, $|\Gamma| - R_{\Gamma}^+$ is a subset of $\bigcup_{\mathbf{c} \in \Gamma}(\mathbf{c} - R_{\mathbf{c}}^+)$.

Now assume that p is a point in $\bigcup_{\mathbf{c} \in \Gamma}(\mathbf{c} - R_{\mathbf{c}}^+)$. Point p lies in $\mathbf{c}_0 - R_{\mathbf{c}_0}^+$ for some polyhedron $\mathbf{c}_0 \in \Gamma$. Thus, point p is not in $R_{\mathbf{c}_0}^+$. By Corollary 5.15, point p does not lie in $R_{\mathbf{c}}^+$ for any polyhedron $\mathbf{c} \in \Gamma$. Thus, point p is in $|\Gamma| - R_{\Gamma}^+$ and $\bigcup_{\mathbf{c} \in \Gamma}(\mathbf{c} - R_{\mathbf{c}}^+)$ is a subset of $|\Gamma| - R_{\Gamma}^+$. Since $|\Gamma| - R_{\Gamma}^+$ is a subset of $\bigcup_{\mathbf{c} \in \Gamma}(\mathbf{c} - R_{\mathbf{c}}^+)$ and $\bigcup_{\mathbf{c} \in \Gamma}(\mathbf{c} - R_{\mathbf{c}}^+)$ is a subset of $|\Gamma| - R_{\Gamma}^+$, the two are equal.

We proceed to the proof of the lemma. Let p be a point in $\bigcup_{\mathbf{c} \in \Gamma} S_{\mathbf{c}}^+$. Point p is in $S_{\mathbf{c}_0}^+$ for some polyhedron $\mathbf{c}_0 \in \Gamma$. By definition, set $S_{\mathbf{c}_0}^+$ equals $R_{\mathbf{c}_0}^+ \cap \mathrm{cl}(\mathbf{c} - R_{\mathbf{c}_0}^+)$. Since point p is in $R_{\mathbf{c}_0}^+$, point p is in $R_{\Gamma}^+ = \bigcup_{\mathbf{c} \in \Gamma} R_{\mathbf{c}}^+$. Since point p is in $\mathrm{cl}(\mathbf{c}_0 - R_{\mathbf{c}_0}^+)$, point p is in

$$\bigcup_{\mathbf{c} \in \Gamma} \mathrm{cl}(\mathbf{c} - R_{\mathbf{c}}^+) = \mathrm{cl}\left(\bigcup_{\mathbf{c} \in \Gamma}(\mathbf{c} - R_{\mathbf{c}}^+)\right) = \mathrm{cl}(|\Gamma| - R_{\Gamma}^+).$$

Thus, $\bigcup_{\mathbf{c} \in \Gamma} S_{\mathbf{c}}^+$ is a subset of $R_{\Gamma}^+ \cap \mathrm{cl}(|\Gamma| - R_{\Gamma}^+)$. By definition, S_{Γ}^+ equals $R_{\Gamma}^+ \cap \mathrm{cl}(|\Gamma| - R_{\Gamma}^+)$. Thus, $\bigcup_{\mathbf{c} \in \Gamma} S_{\mathbf{c}}^+$ is a subset of S_{Γ}^+.

Now let p be a point in S_{Γ}^+. As previously noted, S_{Γ}^+ equals $R_{\Gamma}^+ \cap \mathrm{cl}(|\Gamma| - R_{\Gamma}^+)$. Since R_{Γ}^+ equals $\bigcup_{\mathbf{c} \in \Gamma} R_{\mathbf{c}}^+$, point p is in $R_{\mathbf{c}_1}^+$ for some $\mathbf{c}_1 \in \Gamma$. Since $\mathrm{cl}(|\Gamma| - R_{\Gamma}^+)$ equals $\bigcup_{\mathbf{c} \in \Gamma} \mathrm{cl}(\mathbf{c} - R_{\mathbf{c}}^+)$, point p is in $\mathrm{cl}(\mathbf{c}_2 - R_{\mathbf{c}_2}^+)$ for some $\mathbf{c}_2 \in \Gamma$. If $\mathbf{c}_1$ equals $\mathbf{c}_2$, then p is in $S_{\mathbf{c}_1}^+$ and therefore in $\bigcup_{\mathbf{c} \in \Gamma} S_{\mathbf{c}}^+$. Assume $\mathbf{c}_1$ does not equal $\mathbf{c}_2$.

The intersection of $\mathbf{c}_1$ and $\mathbf{c}_2$ is a face $\mathbf{f}$ of both polyhedra, and point p is in $\mathbf{f}$. By Lemma 5.14, point p is in $R_{\mathbf{c}_2}^+$. Thus, p is in $R_{\mathbf{c}_2}^+ \cap \mathrm{cl}(\mathbf{c}_2 - R_{\mathbf{c}_2}^+)$. By definition, $S_{\mathbf{c}_2}^+$ equals $R_{\mathbf{c}_2}^+ \cap \mathrm{cl}(\mathbf{c}_2 - R_{\mathbf{c}_2}^+)$. Thus, p is in $\bigcup_{\mathbf{c} \in \Gamma} S_{\mathbf{c}}^+$ and S_{Γ}^+ is a subset of $\bigcup_{\mathbf{c} \in \Gamma} S_{\mathbf{c}}^+$.

Since $\bigcup_{\mathbf{c} \in \Gamma} S_{\mathbf{c}}^+$ is a subset of S_{Γ}^+ and S_{Γ}^+ is a subset of $\bigcup_{\mathbf{c} \in \Gamma} S_{\mathbf{c}}^+$, set S_{Γ}^+ equals $\bigcup_{\mathbf{c} \in \Gamma} S_{\mathbf{c}}^+$. $\qquad\square$

Corollary 5.19. *Let Γ be a convex polyhedral mesh in $\mathbb{R}^3$ where each vertex of Γ has a positive or negative label. If U is a proper isosurface vertex set of Γ, then $|T_{\Gamma}^+(U)|$ equals $S_{\Gamma}^+(U)$.*

Proof: By Lemma 5.18, S_{Γ}^+ equals $\bigcup_{\mathbf{c} \in \Gamma} S_{\mathbf{c}}^+(U_{\mathbf{c}})$. By Corollary 5.11, $|T_{\mathbf{c}}^+(U_{\mathbf{c}})|$ equals $S_{\mathbf{c}}^+(U_{\mathbf{c}})$ for each $\mathbf{c}$. Thus,

$$|T_{\Gamma}^+(U)| = \bigcup_{\mathbf{c} \in \Gamma} |T_{\mathbf{c}}^+(U_{\mathbf{c}})| = \bigcup_{\mathbf{c} \in \Gamma} S_{\mathbf{c}}^+(U_{\mathbf{c}}) = S_{\Gamma}^+(U). \qquad\square$$

As another corollary, $|T_{\Gamma}^+(U)|$ separates the positive vertices of Γ from the negative ones. ($|T_{\Gamma}^+(U)|$ is the union of all the triangles in $T_{\Gamma}^+(U)$.)

Corollary 5.20. *Let Γ be a convex polyhedral mesh in $\mathbb{R}^3$ where each vertex of Γ has a positive or negative label. If U is a proper isosurface vertex set of Γ, then $|T_\Gamma^+(U)|$ strictly separates the positive vertices of Γ from the negative ones.*

Proof: By Corollary 5.19, $|T_\Gamma^+|$ equals S_Γ^+. By definition, S_Γ^+ equals $R_\Gamma^+ \cap \mathrm{cl}(|\Gamma| - R_\Gamma^+)$. By Lemma B.8 in Appendix B, S_Γ^+ separates R_Γ^+ from $(|\Gamma| - R_\Gamma^+)$. Region R_Γ^+ contains all the positive vertices of Γ while $(|\Gamma| - R_\Gamma^+)$ contains all the negative ones. Thus $|T_\Gamma^+|$ separates the positive vertices of Γ from the negative ones. By Corollary 5.7, no vertex of Γ is a vertex of T_Γ^+. Thus $|T_\Gamma^+|$ does not contain any vertices of Γ and $|T_\Gamma^+|$ strictly separates the positive vertices of Γ from the negative ones. □

$T_\Gamma^+(U)$ is defined as $\bigcup_{\mathbf{c} \in \Gamma} T_\mathbf{c}^+(U_\mathbf{c})$, the union of sets of triangles. Each $T_\Gamma^+(U_\mathbf{c})$ is a subset of a triangulation of the boundary of $R_\mathbf{c}^+(U_\mathbf{c})$ so each $T_\mathbf{c}^+(U_\mathbf{c})$ is itself a triangulation. However, we have yet to establish that $T_\Gamma^+(U)$ is a triangulation. We need this fact to define a homotopy on $|T_\Gamma^+(U)|$.

Lemma 5.21. *Let Γ be a convex polyhedral mesh in $\mathbb{R}^3$ where each vertex of Γ has a positive or negative label. If U is a proper isosurface vertex set of Γ, then $T_\Gamma^+(U)$ is a triangulation of $S_\Gamma^+(U)$.*

Proof: Consider two triangles $\mathbf{t}_1, \mathbf{t}_2 \in T_\Gamma^+(U)$. By definition, $\mathbf{t}_1$ is in $T_{\mathbf{c}_1}(U \cap \mathbf{c}_1)$ for some polyhedron $\mathbf{c}_1$ and $\mathbf{t}_2$ is in $T_{\mathbf{c}_2}(U \cap \mathbf{c}_2)$ for some polyhedron $\mathbf{c}_2$.

By construction, $T_{\mathbf{c}_1}^+(U \cap \mathbf{c}_1)$ is itself a triangulation since it is a subset of a triangulation of $\partial R_{\mathbf{c}_1}(U_\mathbf{c} \cap \mathbf{c}_1)$. Thus, if $\mathbf{c}_1$ equals $\mathbf{c}_2$, then $\mathbf{t}_1 \cap \mathbf{t}_2$ is a face of $\mathbf{t}_1$ and a face of $\mathbf{t}_2$. On the other hand, if $\mathbf{c}_1$ does not equal $\mathbf{c}_2$, then, by Corollary 5.17, $\mathbf{t}_1 \cap \mathbf{t}_2$ is a face of $\mathbf{t}_1$ and a face of $\mathbf{t}_2$. This proves that $T_\Gamma^+(U)$ is a triangulation.

By Corollary 5.19, $|T_\Gamma^+(U)|$ equals $S_\Gamma^+(U)$. Thus $T_\Gamma^+(U)$ is a triangulation of $S_\Gamma^+(U)$. □

M_Γ is the set of all midpoints of bipolar edges of Γ. Since $T_\Gamma^+(M_\Gamma)$ (abbreviated T_Γ^{M+}) is a triangulation of $S_\Gamma^+(M_\Gamma)$ (abbreviated S_Γ^{M+}), the homotopy η can be extended to all of S_Γ^{M+}. Each vertex v_i of T_Γ^{M+} is the midpoint m_i of some edge $\mathbf{e}_i$. Vertex v_i is remapped to the interpolated point $w_i \in \mathbf{e}_i$. For each vertex v_i of T_Γ^{M+}, define $\eta(v_i, \alpha)$ as $(1 - \alpha)v_i + \alpha w_i$. Extend η to all of S_Γ^{M+} by linearly extending η on each triangle in T_Γ^{M+}. Because T_Γ^{M+} is a triangulation of S_Γ^{M+}, homotopy η is well-defined on every point of S_Γ^{M+}. Thus $\eta : S_\Gamma^{M+} \times [0, 1]$ is a homotopy from S_Γ^{M+} to $\Sigma_\Gamma^{M+}(\sigma)$ where $\eta(S_\Gamma^{M+}, 0)$ equals S_Γ^{M+} and $\eta(S_\Gamma^{M+}, 1)$ equals $\Sigma_\Gamma^{M+}(\sigma)$.

Our last lemma shows that Σ_Γ^{M+} does not contain mesh vertices whose scalar values do not equal the isovalue.

Lemma 5.22. *Let Γ be a convex polyhedral mesh in $\mathbb{R}^3$ where each vertex of Γ has a scalar value. If v is a vertex of Γ whose scalar value does not equal $\sigma \in \mathbb{R}$, then $\Sigma_\Gamma^{M+}(\sigma)$ does not contain v.*

Proof: Let v be a mesh vertex whose scalar value, s_v, does not equal σ. By Property 3, every edge $\mathbf{e}$ incident on v intersects $\Sigma_\Gamma^{M+}(\sigma)$ (abbreviated Σ_Γ^{M+}) at most once. Assume $\mathbf{e}$ intersects Σ_Γ^{M+}. Since $\mathbf{e} \cap \Sigma_\Gamma^{M+}$ is determined by linear interpolation and s_v does not equal σ, point $\mathbf{e} \cap \Sigma_\Gamma^{M+}$ is not equal to v. Thus, v does not lie on Σ_Γ^{M+}. $\qquad\square$

Applying Corollary 5.20, the homotopy map η and Lemma 5.22, gives Property 5.

Property 5. *The isosurface separates positive mesh vertices from negative ones and strictly separates strictly positive mesh vertices from negative mesh vertices.*

Proof: By Corollary 5.20, set $|T_\Gamma^{M+}|$ strictly separates the positive vertices of Γ from the negative ones. Map η is a homotopy map from S_Γ^{M+} to $\Sigma_\Gamma^{M+}(\sigma)$, which never passes through any vertices of Γ. Vertices of S_Γ^{M+} that lie on the boundary of $|\Gamma|$ remain on the boundary of $|\Gamma|$ throughout the homotopy η. Thus $\eta(S_\Gamma^{M+} \cap \partial|\Gamma|, \alpha)$ is a subset of $\partial|\Gamma|$ for all $\alpha \in [0,1]$. By Lemma B.18, $\Sigma_\Gamma^{M+}(\sigma)$ separates the positive vertices of Γ from the negative ones.

By Lemma 5.22, $\Sigma_\Gamma^{M+}(\sigma)$ does not contain any mesh vertices whose scalar values equal the isovalue σ. Therefore, $\Sigma_\Gamma^{M+}(\sigma)$ strictly separates vertices with scalar values greater than σ from vertices with scalar value less than σ. $\qquad\square$

5.4.4   Proof of Isosurface Properties 6, 7, and 8

In Section 5.4.3, we proved that if $\mathbf{c}_1$ and $\mathbf{c}_2$ are adjacent polyhedra, then $R_{\mathbf{c}_1}^+(U_{\mathbf{c}_1})$ and $R_{\mathbf{c}_2}^+(U_{\mathbf{c}_2})$ agree on shared faces of $\mathbf{c}_1$ and $\mathbf{c}_2$ (Corollary 5.15). To prove Property 6, we need to show that if $U_{\mathbf{c}_1}$ and $U_{\mathbf{c}_2}$ are proper, then $S_{\mathbf{c}_1}^+(U_{\mathbf{c}_1})$ and $S_{\mathbf{c}_2}^+(U_{\mathbf{c}_2})$ agree on shared faces of $\mathbf{c}_1$ and $\mathbf{c}_2$. We first extend the definition of $S_{\mathbf{c}}$ from $\mathbf{c}$ to faces of $\mathbf{c}$.

Let $\mathbf{f}$ be a face of $\mathbf{c}$. As in Section 5.4.3, set $V_{\mathbf{f}}^+$ is the set of positive vertices of $\mathbf{f}$, set $U_{\mathbf{f}}$ is the points of $U_{\mathbf{c}}$ that lie on $\mathbf{f}$, and $R_{\mathbf{f}}^+(U_{\mathbf{f}})$ (abbreviated $R_{\mathbf{f}}^+$) equals $\mathrm{conv}(U_{\mathbf{f}} \cup V_{\mathbf{f}}^+)$. Define $S_{\mathbf{f}}^+(U_{\mathbf{f}})$ (abbreviated $S_{\mathbf{f}}^+$) as

$$S_{\mathbf{f}}^+(U_{\mathbf{f}}) = R_{\mathbf{f}}^+(U_{\mathbf{f}}) \cap \mathrm{cl}(\mathbf{f} - R_{\mathbf{f}}^+(U_{\mathbf{f}})).$$

As in Section 5.4.3, face $\mathbf{f}$ need not be two-dimensional.

Lemma 5.14 shows that $R_{\mathbf{c}}^+ \cap \mathbf{c}$ equals $R_{\mathbf{f}}^+$. To prove that $S_{\mathbf{c}_1}^+ \cap \mathbf{f}$ equals $S_{\mathbf{c}_2}^+ \cap \mathbf{f}$, we need a similar lemma for the complements of $R_{\mathbf{c}}^+$ and $R_{\mathbf{f}}^+$.

Lemma 5.23. *Let $\mathbf{c}$ be a convex polyhedron in $\mathbb{R}^3$ where each vertex of $\mathbf{c}$ has a positive or a negative label, and let $U_{\mathbf{c}}$ be a proper isosurface vertex set of $\mathbf{c}$. For every face $\mathbf{f}$ of polyhedron $\mathbf{c}$, $\mathrm{cl}(\mathbf{c} - R_{\mathbf{c}}^+(U_{\mathbf{c}})) \cap \mathbf{f}$ equals $\mathrm{cl}(\mathbf{f} - R_{\mathbf{f}}^+(U_{\mathbf{c}} \cap \mathbf{f}))$.*

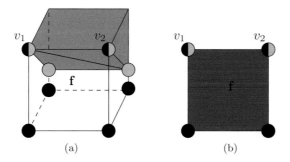

Figure 5.18. (a) Cube $\mathbf{c}$, green isosurface vertex set $U_{\mathbf{c}}$, and blue region $\mathbf{c} - R_{\mathbf{c}}^+$. Black cube vertices are positive. Vertices v_1 and v_2 are both cube vertices and elements of $U_{\mathbf{c}}$. (b) Front facet $\mathbf{f}$ of cube $\mathbf{c}$, set $U_{\mathbf{f}} = U_{\mathbf{c}} \cap \mathbf{f}$, and red region $R_{\mathbf{f}}^+$. Region $\mathrm{cl}(\mathbf{f} - R_{\mathbf{f}}^+)$ is the empty set while $\mathrm{cl}(\mathbf{c} - R_{\mathbf{c}}^+) \cap \mathbf{f}$ is line segment (v_1, v_2).

Proof: By Lemma 5.14, set $R_{\mathbf{f}}$ equals $R_{\mathbf{c}}^+ \cap \mathbf{f}$. Thus, $\mathbf{f} - R_{\mathbf{f}}$ equals $(\mathbf{c} - R_{\mathbf{c}}^+) \cap \mathbf{f}$ and $\mathrm{cl}(\mathbf{f} - R_{\mathbf{f}}^+)$ is a subset of $\mathrm{cl}(\mathbf{c} - R_{\mathbf{c}}^+) \cap \mathbf{f}$.

Let p be a point in $\mathrm{cl}(\mathbf{c} - R_{\mathbf{c}}^+) \cap \mathbf{f}$. If p is in $(\mathbf{c} - R_{\mathbf{c}}^+) \cap \mathbf{f}$, then p is in $\mathbf{f} - R_{\mathbf{f}}$, proving the lemma.

Assume that p is a point in $\mathrm{cl}(\mathbf{c} - R_{\mathbf{c}}^+) \cap \mathbf{f}$ but p is not in $(\mathbf{c} - R_{\mathbf{c}}^+) \cap \mathbf{f}$. Since p is in $\mathbf{f}$, point p is not in $\mathbf{c} - R_{\mathbf{c}}^+$. Thus p must be in $R_{\mathbf{c}}^+$.

Since p is in $\mathrm{cl}(\mathbf{c} - R_{\mathbf{c}}^+)$ and p is in $R_{\mathbf{c}}^+$, point p is in $S_{\mathbf{c}}^+ = R_{\mathbf{c}}^+ \cap \mathrm{cl}(\mathbf{c} - R_{\mathbf{c}}^+)$. By Lemma 5.10, point p lies on some triangle $\mathbf{t} \in T_{\mathbf{c}}^+$. Let h be the plane through triangle $\mathbf{t}$. By Corollary 5.6, the vertices of triangle $\mathbf{t}$ are in $U_{\mathbf{c}}$. Since $U_{\mathbf{c}}$ is proper, the vertices of $\mathbf{t}$ are in the interior of edges of $\mathbf{c}$. Thus, plane h intersects the interior of $\mathbf{f}$ and, for every neighborhood $\mathbb{N}_p \subseteq \mathbb{R}^3$ of p, set $\mathbb{N}_p \cap \mathbf{f}$ contains points in $\mathbf{f} - R_{\mathbf{f}}^+$. Thus, p is in $\mathrm{cl}(\mathbf{f} - R_{\mathbf{f}}^+)$ and $\mathrm{cl}(\mathbf{c} - R_{\mathbf{c}}^+) \cap f$ is a subset of $\mathrm{cl}(\mathbf{f} - R_{\mathbf{f}}^+)$.

Since $\mathrm{cl}(\mathbf{f} - R_{\mathbf{f}}^+)$ is a subset of $\mathrm{cl}(\mathbf{c} - R_{\mathbf{c}}^+) \cap \mathbf{f}$ and $\mathrm{cl}(\mathbf{c} - R_{\mathbf{c}}^+) \cap f$ is a subset of $\mathrm{cl}(\mathbf{f} - R_{\mathbf{f}}^+)$, set $\mathrm{cl}(\mathbf{f} - R_{\mathbf{f}}^+)$ equals $\mathrm{cl}(\mathbf{c} - R_{\mathbf{c}}^+) \cap \mathbf{f}$. $\qquad\square$

Lemma 5.23 requires that the isosurface vertex set be proper. Figure 5.18 contains an example of an isosurface set that is not proper and where $\mathrm{cl}(\mathbf{f} - R_{\mathbf{f}}^+)$ is empty while $\mathrm{cl}(\mathbf{c} - R_{\mathbf{c}}^+) \cap \mathbf{f}$ is a line segment in $\mathbf{f}$. Lemmas 5.14 and 5.16 and their corollaries do not require the isosurface vertex sets to be proper.

We apply Lemma 5.23 to show that $S_{\mathbf{f}}^+$ equals $S_{\mathbf{c}}^+ \cap \mathbf{f}$.

Lemma 5.24. *Let $\mathbf{c}$ be a convex polyhedron in $\mathbb{R}^3$ where each vertex of $\mathbf{c}$ has a positive or a negative label and let $U_{\mathbf{c}}$ be a proper isosurface vertex set of $\mathbf{c}$. For every face $\mathbf{f}$ of polyhedron $\mathbf{c}$, $S_{\mathbf{c}}^+(U_{\mathbf{c}}) \cap \mathbf{f}$ equals $S_{\mathbf{f}}^+(U_{\mathbf{c}} \cap \mathbf{f})$.*

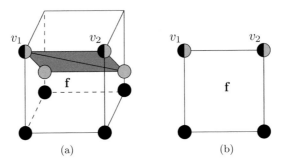

Figure 5.19. (a) Cube $\mathbf{c}$, green isosurface vertex set $U_{\mathbf{c}}$, and isosurface patch $S_{\mathbf{c}}^{+}$. Black cube vertices are positive. Vertices v_1 and v_2 are both cube vertices and elements of $U_{\mathbf{c}}$. (b) Front facet $\mathbf{f}$ of cube $\mathbf{c}$ and set $U_{\mathbf{f}} = U_{\mathbf{c}} \cap \mathbf{f}$. Set $S_{\mathbf{f}}^{+}$ is the empty set.

Proof: Set $S_{\mathbf{c}}^{+}$ equals $R_{\mathbf{c}}^{+} \cap \mathrm{cl}(\mathbf{c} - R_{\mathbf{c}}^{+})$. By Lemma 5.14, set $R_{\mathbf{c}}^{+} \cap \mathbf{f}$ equals $R_{\mathbf{f}}^{+}$. By Lemma 5.23, set $\mathrm{cl}(\mathbf{c} - R_{\mathbf{c}}^{+}) \cap \mathbf{f}$ equals $\mathrm{cl}(\mathbf{f} - R_{\mathbf{f}}^{+})$. Thus,

$$
\begin{aligned}
S_{\mathbf{c}}^{+} \cap \mathbf{f} &= R_{\mathbf{c}}^{+} \cap \mathrm{cl}(\mathbf{c} - R_{\mathbf{c}}^{+}) \cap \mathbf{f} \\
&= (R_{\mathbf{c}}^{+} \cap \mathbf{f}) \cap (\mathrm{cl}(\mathbf{c} - R_{\mathbf{c}}^{+}) \cap \mathbf{f}) \\
&= R_{\mathbf{f}}^{+} \cap \mathrm{cl}(\mathbf{f} - R_{\mathbf{f}}^{+}) \\
&= S_{\mathbf{f}}^{+}.
\end{aligned}
$$

$\square$

As a corollary to Lemma 5.24, we show that that $S_{\mathbf{c}_1}^{+}$ and $S_{\mathbf{c}_2}^{+}$ agree on a shared face of $\mathbf{c}_1$ and $\mathbf{c}_2$.

Corollary 5.25. *Let $\mathbf{c}_1$ and $\mathbf{c}_2$ be adjacent convex polyhedra in $\mathbb{R}^3$ where each vertex of $\mathbf{c}_1$ and $\mathbf{c}_2$ has a positive or a negative label. Let $U_{\mathbf{c}_1}$ and $U_{\mathbf{c}_2}$ be isosurface vertex sets for $\mathbf{c}_1$ and $\mathbf{c}_2$, respectively, where $U_{\mathbf{c}_1} \cap \mathbf{f}$ equals $U_{\mathbf{c}_2} \cap \mathbf{f}$. If $U_{\mathbf{c}_1}$ and $U_{\mathbf{c}_2}$ are proper isosurface vertex sets, then $S_{\mathbf{c}_1}^{+}(U_{\mathbf{c}_1}) \cap \mathbf{f}$ equals $S_{\mathbf{c}_2}^{+}(U_{\mathbf{c}_2}) \cap \mathbf{f}$.*

Proof: Applying Lemma 5.24,

$$
S_{\mathbf{c}_1}^{+} \cap \mathbf{f} = S_{\mathbf{f}} = S_{\mathbf{c}_2}^{+} \cap \mathbf{f}.
$$

Lemma 5.24 and its corollary require that the isosurface vertex sets be proper. Figure 5.19 contains an example of an isosurface set that is not proper and where $S_{\mathbf{f}}^{+}$ is the empty set while $S_{\mathbf{c}}^{+} \cap \mathbf{f}$ is a line segment in $\mathbf{f}$.

To prove Properties 6 and 7, we prove that if isosurface vertex set U is proper, then $S_{\Gamma}^{+}(U)$ is a piecewise linear, orientable 2-manifold whose boundary lies on $\partial |\Gamma|$.

Lemma 5.26. *Let Γ be a convex polyhedral mesh where each vertex of Γ has a positive or negative label. Let U be a proper isosurface vertex set for Γ. If the underlying points set, $|\Gamma|$, of Γ is a 3-manifold with boundary, then set $S_{\Gamma}^{+}(U)$ is a piecewise linear, orientable 2-manifold whose boundary lies on $\partial |\Gamma|$.*

Proof: For any mesh element $\mathbf{c} \in \Gamma$, region $R_{\mathbf{c}}^+$ is either empty or a three-dimensional convex set (Corollary 5.9). Since $R_{\mathbf{c}}^+$ is a three-dimensional convex set, the boundary of $R_{\mathbf{c}}^+$ is a two-dimensional manifold.

Let p be a point in S_{Γ}^+.

Case I: Point p is in the interior of some mesh element $\mathbf{c}$.

By Corollary 5.11, $T_{\mathbf{c}}^+$ is a triangulation of $S_{\mathbf{c}}^+$. Thus, $S_{\mathbf{c}}^+ \cap \text{int}(\mathbf{c})$ equals $\partial R_{\mathbf{c}}^+ \cap \text{int}(\mathbf{c})$. Since $\partial R_{\mathbf{c}}^+$ is a two-dimensional manifold, $\partial R_{\mathbf{c}}^+ \cap \text{int}(\mathbf{c})$ is a two-dimensional manifold. Thus some neighborhood of p in $S_{\mathbf{c}}^+$ is homeomorphic to $\mathbb{R}^2$.

Case II: Point p is in the interior of some two-dimensional facet $\mathbf{f} \in \Gamma$.

By Lemma 5.21, T_{Γ}^+ is a triangulation of S_{Γ}^+. Thus, point p is in some triangle $\mathbf{t}_1 \in T_{\Gamma}^+$. Triangle $\mathbf{t}_1$ is contained in $T_{\mathbf{c}_1}^+$ for some polyhedron $\mathbf{c}_1$. Since p lies on $\mathbf{t}_1 \cap \text{int}(\mathbf{f})$ and the vertices of $\mathbf{t}_1$ lie on edges of $\mathbf{c}$ (Lemma 5.5), point p lies in the interior of some edge $\hat{\mathbf{e}}$ of $\mathbf{t}_1$.

Since $\partial R_{\mathbf{c}_1}^+$ is two dimensional, there is another triangle $\mathbf{t}_1' \neq \mathbf{t}_1$ that is in the triangulation of $\partial R_{\mathbf{c}_1}^+$ and contains $\hat{\mathbf{e}}$. Since $R_{\mathbf{c}_1}^+ \cap \mathbf{f}$ is not empty, facet $\mathbf{f}$ must contain some positive vertex $v \in \mathbf{c}$. By Lemma 5.13, $R_{\mathbf{f}}^+$ is two-dimensional. By Lemma 5.14, $R_{\mathbf{c}_1}^+ \cap \mathbf{f}$ equals $R_{\mathbf{f}}^+$. Since $R_{\mathbf{c}_1}^+ \cap \mathbf{f}$ is two-dimensional, triangle $\mathbf{t}_1'$ lies in $\mathbf{f}$. Since triangle $\mathbf{t}_1'$ lies in $\mathbf{f}$, it is not in $T_{\mathbf{c}_1}^+$. Thus, triangle $\mathbf{t}_1$ is the only triangle of $T_{\mathbf{c}_1}^+$ containing $\hat{\mathbf{e}}$ and point p.

If p is in the interior of $|\Gamma|$, then $\mathbf{f}$ lies on some other polyhedron $\mathbf{c}_2 \in \Gamma$. By Corollary 5.24, $S_{\mathbf{c}_1}^+ \cap \mathbf{f}$ equals $S_{\mathbf{c}_2}^+ \cap \mathbf{f}$. By definition, $\mathbf{t}_1$ is not contained in $\mathbf{f}$ and so is not in $\mathbf{c}_2$. By Corollary 5.11, $T_{\mathbf{c}_2}^+$ is a triangulation of $S_{\mathbf{c}_2}^+$. Thus some triangle $\mathbf{t}_2 \in T_{\mathbf{c}_2}^+$ contains $\hat{\mathbf{e}}$. As argued for $T_{\mathbf{c}_1}^+$, triangle $\mathbf{t}_2$ is the only triangle in $T_{\mathbf{c}_2}^+$ that contains $\hat{\mathbf{e}}$. Since point p lies on an edge $\hat{\mathbf{e}}$ between two triangles, a neighborhood of p in S_{Γ}^+ is homeomorphic to $\mathbb{R}^2$.

If p is on the boundary of $|\Gamma|$, then p is not contained in any polyhedron of $|\Gamma|$ and so is not in any triangle other than $\mathbf{t}_1$. Since p lies on the boundary of $\mathbf{t}_1$, some neighborhood of p is homeomorphic to a half-space in $\mathbb{R}^2$.

Case III: Point p lies on some mesh edge $\mathbf{e}$ of Γ.

Consider any triangle $\mathbf{t} \in T_{\Gamma}^+$ containing p. Triangle $\mathbf{t}$ lies in some mesh element $\mathbf{c} \in \Gamma$. By Lemma 5.5, the vertices of $\mathbf{t}$ are a subset of $U \cap \mathbf{c}$. Since only one point of U lies on $\mathbf{e}$, triangle $\mathbf{t}$ intersects $\mathbf{e}$ at most once. Thus point p is a vertex of $\mathbf{t}$. Since U is proper, point p lies in the interior of $\mathbf{e}$.

Let $\mathbf{c}$ be a polyhedron containing $\mathbf{e}$. Let $\tau_{\mathbf{c}}$ be the set of triangles that contain p in the triangulation of $\partial R_{\mathbf{c}}^+$. Since $\partial R_{\mathbf{c}}^+$ is a two-dimensional manifold, the triangles of $\tau_{\mathbf{c}}$ form a disk. Point p lies in the interior of this disk. Since p is in $S_{\mathbf{c}}^+$, at least one of these triangles lies in $T_{\mathbf{c}}^+$.

Let $\mathbf{f}$ and $\mathbf{f}'$ be the two facets of $\mathbf{c}$ containing $\mathbf{e}$. Since $R_{\mathbf{c}}^+$ intersects $\mathbf{e}$, an endpoint v of $\mathbf{e}$ must be positive. Since v is positive, $R_{\mathbf{f}}^+$ and $R_{\mathbf{f}'}^+$ are two-dimensional. Thus, $\mathbf{f}$ contains at least one triangle of $\tau_{\mathbf{c}}$ and so does $\mathbf{f}'$.

The triangles of $\tau_{\mathbf{c}}$ that lie in $\mathbf{f}$ or $\mathbf{f}'$ form a disk with p on its boundary. Removing these triangles from $\tau_{\mathbf{c}}$ leaves a set of triangles forming another disk with p on its boundary. Thus, the triangles in $T_{\mathbf{c}}^+$ that contain p form a disk with p on the boundary.

If p is in the interior of $|\Gamma|$, then the polyhedron containing p form a cycle around edge $\mathbf{e}$. Within each polyhedron $\mathbf{c}$ containing p, some subset of $S_{\mathbf{c}}^+$ is a disk containing p on its boundary. By Corollary 5.25, these disks agree on their boundaries. Together they form a disk surrounding p. Thus, some neighborhood of p in S_{Γ}^+ is homeomorphic to $\mathbb{R}^2$.

If p is on the boundary of $|\Gamma|$, then the polyhedron containing p form a chain around edge $\mathbf{e}$. Within each polyhedron $\mathbf{c}$ containing p, some subset of $S_{\mathbf{c}}^+$ is a disk containing p on its boundary. By Corollary 5.25, these disks agree on their boundaries. Together they form a disk with p on the boundary. Thus, some neighborhood of p in S_{Γ}^+ is homeomorphic to a half-space in $\mathbb{R}^2$.

Since every point $p \in S_{\Gamma}^+$ has a neighborhood in S_{Γ}^+ homeomorphic to $\mathbb{R}^2$ or a half-space in $\mathbb{R}^2$, set S_{Γ}^+ is a 2-manifold with boundary. Since the neighborhood of p in S_{Γ}^+ is homeomorphic to $\mathbb{R}^2$ if p is in the interior of $|\Gamma|$ and homeomorphic to a half-space in $\mathbb{R}^2$ if p is on the boundary of $|\Gamma|$, the boundary of S_{Γ}^+ equals $|\Gamma| \cap S_{\Gamma}^+$.

By Lemma 5.21, set T_{Γ}^+ is a triangulation of S_{Γ}^+. Each triangles in T_{Γ}^+ is oriented so that the induced normal points to the positive region. Thus the triangles have consistent orientations and the manifold is orientable.   □

The isosurface $\Sigma_{\Gamma}^{M+}(\sigma)$ is an embedding of $S_{\Gamma}^+(M_{\Gamma})$ into $|\Gamma|$. If $S_{\Gamma}^+(M_{\Gamma})$ is a 2-manifold with boundary and this embedding is one-to-one, then $\Sigma_{\Gamma}^{M+}(\sigma)$ is also a 2-manifold with boundary. However, as shown in Figure 5.10, the mapping from $S_{\mathbf{c}}^+(M_{\mathbf{c}})$ to $\Sigma_{\mathbf{c}}^{M+}(\sigma)$ may not be one-to-one, even for a single mesh element $\mathbf{c}$. We claim that if $\mathbf{c}$ is a cube, tetrahedron, square pyramid, octahedron, or bipyramid, then the embedding is one-to-one.

Lemma 5.27. *Let σ be a scalar value. Let $\mathbf{c}$ be a cube, tetrahedron, square pyramid, octahedron, or bipyramid where each vertex v of $\mathbf{c}$ has a scalar value, s_v, and each vertex v has a negative label if $s_v < \sigma$ or a positive label if $s_v \geq \sigma$. If no vertex of $\mathbf{c}$ has scalar value equal to σ, then the mapping from $S_{\mathbf{c}}^+(M_{\mathbf{c}})$ to $\Sigma_{\mathbf{c}}^{M+}(\sigma)$ is one-to-one.*

Unfortunately, the only proof we have for Lemma 5.27 is a tedious case analysis for each polyhedron. Thus, we omit the proof of this lemma.

Property 6. *The isosurface is a piecewise linear, oriented 2-manifold with boundary.*

Property 7. *The boundary of the isosurface lies on the boundary of the mesh.*

Proof of Properties 6 & 7: By Lemma 5.26, the set $S_\Gamma^+(M_\Gamma)$ (abbreviated S_Γ^+) is a piecewise linear, orientable 2-manifold with boundary whose boundary lies on $\partial|\Gamma|$. Set $\Sigma_\Gamma^{M+}(\sigma)$ (abbreviated Σ_Γ^{M+}) is an embedding of S_Γ^+ into $|\Gamma|$. We claim that this embedding is one-to-one.

Let p be some point in Σ_Γ^{M+}. Since no mesh vertex value equals the isovalue, set Σ_Γ^{M+} does not contain any mesh vertices (Lemma 5.22). Moreover, any point of Σ_Γ^{M+} on a mesh edge $\mathbf{e}$ is the image of $\mathbf{e} \cap S_\Gamma^+$, which can only be the midpoint of $\mathbf{e}$. Thus, if p lies on a mesh edge $\mathbf{e}$, then it is the image of only one point in S_Γ^+.

Assume p is on the interior of a mesh facet $\mathbf{f}$. Point p lies on a line segment in $\Sigma_\Gamma^{M+} \cap \mathbf{f}$ connecting two mesh edges $\mathbf{e}$ and $\mathbf{e}'$. This line segment is the image of the edge $\hat{\mathbf{e}}$ in S_Γ^+ connecting the midpoints of $\mathbf{e}$ and $\mathbf{e}'$. Since the mapping of $\hat{\mathbf{e}}$ to Σ_Γ^{M+} is linear, this mapping is one-to-one. Thus p is again the image of only one point in S_Γ^+.

Finally, assume p lies in the interior of some mesh element $\mathbf{c}$. By Lemma 5.27, the mapping from $S_\mathbf{c}^+(M_\mathbf{c})$ to $\Sigma_\mathbf{c}^{M+}(\sigma)$ is one-to-one. Since $\Sigma_\Gamma^{M+} \cap \mathrm{int}(\mathbf{c})$ equals $\Sigma_\mathbf{c}^{M+}(\sigma) \cap \mathrm{int}(\mathbf{c})$, point p is not the image of any points in $S_\Gamma^+ - S_\mathbf{c}^+(M_\mathbf{c})$. Thus p is the image of only one point in S_Γ^+.

Since the mapping from S_Γ^+ to Σ_Γ^{M+} is one-to-one and S_Γ^+ is an orientable 2-manifold with boundary, set Σ_Γ^{M+} is an orientable 2-manifold with boundary. Since ∂S_Γ^+ lies on $\partial|\Gamma|$ and all points on $\partial S_\Gamma^+ \cap \partial|\Gamma|$ are mapped to $\partial|\Gamma|$, the boundary of Σ_Γ^{M+} also lies on $\partial|\Gamma|$.   □

The last property is that Υ does not contain any zero-area or duplicate triangles and forms a triangulation of the isosurface.

Property 8. *Set Υ does not contain any zero-area triangles or duplicate triangles, and the triangles in Υ form a triangulation of the isosurface.*

Proof: Since no mesh vertex has scalar value equal to the isovalue, no isosurface vertex lies on a mesh vertex. By Property 3, each bipolar mesh edge contains only one isosurface vertex. Thus, the linear interpolation on isosurface vertices does not create any zero-area or duplicate isosurface triangles.

Set Υ is the image of T_Γ^+ under the embedding which takes S_Γ^+ to Σ_Γ^{M+}. Since this embedding is one-to-one and T_Γ^+ defines a triangulation of S_Γ^+ (Lemma 5.21), set Υ defines a triangulation of Σ_Γ^{M+}.   □

5.5   Isohull

Instead of using the isosurface patch construction algorithm in Section 5.2 to construct an isosurface lookup table, one could use the algorithm directly on the mesh elements. The disadvantage is that this will typically be slower than using a lookup table. However, it does have some advantages. Firstly, one can avoid the problem of self-intersection (Figure 5.10) by constructing the isosurface patches directly on the isosurface vertices. Thus, Properties 6 and 7 in Section 5.4.1 are no longer restricted to meshes composed of cubes, tetrahedra, square pyramids, etc., but apply to any convex polyhedral mesh. Secondly and perhaps more importantly, one can avoid the construction of degenerate triangles when isosurface vertices lie on mesh vertices. As discussed in Section 5.5.5, one can also have a mixed algorithm, which uses table lookup for well-behaved mesh elements and constructs the isosurface patch directly in all other cases.

5.5.1 Isosurface Patch Construction

The isosurface patch construction algorithm in Section 5.5.1 uses edge midpoints to construct isosurface patches. We generalize that algorithm to use points other than midpoints, including edge endpoints.

Let $\mathbf{c}$ be a convex polyhedron in $\mathbb{R}^3$ where each vertex of $\mathbf{c}$ is labeled positive or negative and positive vertices may be labeled strictly positive. Edges of $\mathbf{c}$ are positive, negative, or bipolar as defined in Section 5.1. Positive edges may be strictly positive.

Isosurface vertex sets and proper isosurface vertex sets are defined in Section 5.1. Let $U_{\mathbf{c}}$ be an isosurface vertex set of $\mathbf{c}$, although not necessarily a proper one. Let $V_{\mathbf{c}}^+$ be the vertices of $\mathbf{c}$ with positive labels, let $V_{\mathbf{c}}^-$ be the vertices with negative labels, and let $R_{\mathbf{c}}^+(U_{\mathbf{c}})$ equal $\mathrm{conv}(U_{\mathbf{c}} \cup V_{\mathbf{c}}^+)$, the convex hull of $U_{\mathbf{c}} \cup V_{\mathbf{c}}^+$. Construct $R_{\mathbf{c}}^+(U_{\mathbf{c}})$.

We distinguish three cases, based on the dimension of $R_{\mathbf{c}}^+(U_{\mathbf{c}})$. If $R_{\mathbf{c}}^+(U_{\mathbf{c}})$ has dimension three, then $\partial R_{\mathbf{c}}^+(U_{\mathbf{c}})$, the boundary of $R_{\mathbf{c}}^+(U_{\mathbf{c}})$, is composed of two-dimensional convex polygons. Triangulate $\partial R_{\mathbf{c}}^+(U_{\mathbf{c}})$ by triangulating each convex polygon without adding any additional vertices. Orient the triangles so that the induced normals point to $R_{\mathbf{c}}^+(U_{\mathbf{c}})$. Let $\widetilde{T}_{\mathbf{c}}^+(U_{\mathbf{c}})$ be the set of oriented triangles in the triangulation of $\partial R_{\mathbf{c}}^+(U_{\mathbf{c}})$ that do not lie on facets of $\mathbf{c}$. If $R_{\mathbf{c}}^+(U_{\mathbf{c}})$ has dimension two and is a subset of some facet of $\mathbf{c}$, then let $\widetilde{T}_{\mathbf{c}}^+(U_{\mathbf{c}})$ be the set of all triangles in the triangulation of $R_{\mathbf{c}}^+(U_{\mathbf{c}})$. Orient the triangles so that the induced normal points away from $\mathbf{c}$. Otherwise, let $\widetilde{T}_{\mathbf{c}}^+(U_{\mathbf{c}})$ be the empty set.

The triangles in $\widetilde{T}_{\mathbf{c}}^+(U_{\mathbf{c}})$ form the triangulated isosurface patch whose vertices are $U_{\mathbf{c}}$. This isosurface patch separates $V_{\mathbf{c}}^-$ from $V_{\mathbf{c}}^+$. The algorithm is presented in Figure 5.20 and illustrated in Figure 5.21.

Note the difference between $T_{\mathbf{c}}^+(U_{\mathbf{c}})$, defined in Section 5.4.2, and $\widetilde{T}_{\mathbf{c}}^+(U_{\mathbf{c}})$. Set $T_{\mathbf{c}}^+(U_{\mathbf{c}})$ is a triangulation of $\mathrm{cl}(\partial R_{\mathbf{c}}^+(U_{\mathbf{c}}) - \partial \mathbf{c})$. If $R_{\mathbf{c}}^+(U_{\mathbf{c}})$ has dimension three, then $\widetilde{T}_{\mathbf{c}}^+(U_{\mathbf{c}})$ equals $T_{\mathbf{c}}^+(U_{\mathbf{c}})$. If $R_{\mathbf{c}}^+(U_{\mathbf{c}})$ has dimension two and is a subset of some facet of $\mathbf{c}$, then $\widetilde{T}_{\mathbf{c}}^+(U_{\mathbf{c}})$ is the triangulation of $R_{\mathbf{c}}^+(U_{\mathbf{c}})$ while $T_{\mathbf{c}}^+(U_{\mathbf{c}})$ is empty.

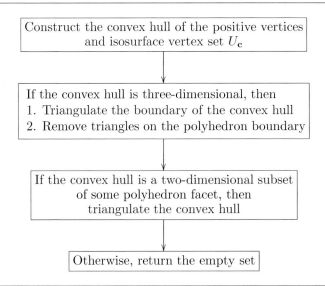

Figure 5.20. Algorithm IsoPatch3D for 3D isosurface patch construction from isosurface vertex set $U_\mathbf{c}$.

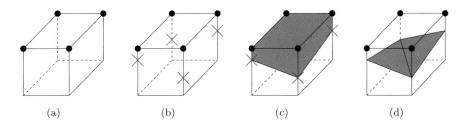

| (a) | (b) | (c) | (d) |

Figure 5.21. Isosurface patch generation. (a) Cube configuration. (b) Isosurface vertex set $U_\mathbf{c}$. (c) Set $R_\mathbf{c}^+(U_\mathbf{c})$, the convex hull of $U_\mathbf{c}$ and positive vertices. (d) Isosurface composed of triangles in $R_\mathbf{c}^+(U_\mathbf{c}) \cap \mathrm{cl}(\mathbf{c} - R_\mathbf{c}^+(U_\mathbf{c}))$.

Algorithm IsoPatch3D is almost the same as the midpoint-based algorithm IsoMid3D in Section 5.2. The only difference is the case where $R_\mathbf{c}^+(U_\mathbf{c})$ is two-dimensional. If we let $U_\mathbf{c}$ equal $M_\mathbf{c}$, the set of midpoints of the bipolar edges of $\mathbf{c}$, then algorithm IsoPatch3D constructs an isosurface patch on the midpoints of the bipolar edges of $\mathbf{c}$. Since the set $R_\mathbf{c}^+(M_\mathbf{c}) = R_\mathbf{c}^+$ is always three-dimensional (Corollary 5.9), this isosurface patch is the same as the one constructed by Iso-Mid3D. Thus the examples in Figure 5.6 also illustrate algorithm IsoPatch3D.

In the case where $R_\mathbf{c}^+(U_\mathbf{c})$ is two-dimensional and a subset of $\partial \mathbf{c}$, algorithm IsoPatch3D returns a set of triangles that lie on the boundary of $\mathbf{c}$. As illustrated in Figure 5.22, omitting such triangles might leave a hole in the isosurface.

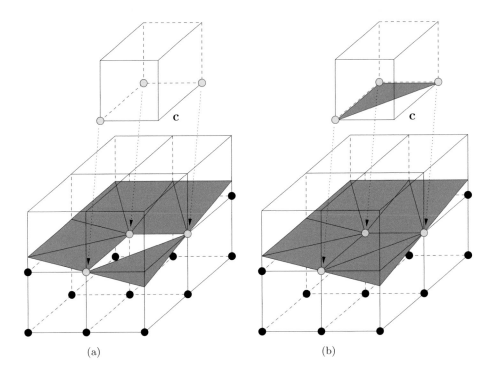

Figure 5.22. Example where $R_{\mathbf{c}}^+(U_{\mathbf{c}})$ is a two-dimensional subset of a facet of $\mathbf{c}$. Gold vertices have scalar value equal to the isovalue. Black vertices have scalar value greater than the isovalue. Set $R_{\mathbf{c}}^+(U_{\mathbf{c}})$ is a single triangle lying on the facet of $\mathbf{c}$ containing the three gold vertices. (a) Removing this triangle creates a hole in the isosurface. (b) Algorithm IsoPatch3D retains this triangle.

Algorithm IsoPatch3D could use the negative vertices $V_{\mathbf{c}}^-$ in place of $V_{\mathbf{c}}^+$. Let $R_{\mathbf{c}}^-(U_{\mathbf{c}})$ equal $\operatorname{conv}(U_{\mathbf{c}} \cup V_{\mathbf{c}}^+)$ and triangulate $\partial R_{\mathbf{c}}^+(U_{\mathbf{c}})$. If $R_{\mathbf{c}}^-(U_{\mathbf{c}})$ is three-dimensional, let $\widetilde{T}_{\mathbf{c}}^-(U_{\mathbf{c}})$ be the set of triangles in the triangulation of $\partial R_{\mathbf{c}}^+(U_{\mathbf{c}})$ that do not lie on the boundary of $\mathbf{c}$. If $R_{\mathbf{c}}^-(U_{\mathbf{c}})$ is two-dimensional and is a subset of some facet of $\mathbf{c}$, let $\widetilde{T}_{\mathbf{c}}^-(U_{\mathbf{c}})$ be the set of all triangles in the triangulation of $R_{\mathbf{c}}^-(U_{\mathbf{c}})$. Otherwise, let $\widetilde{T}_{\mathbf{c}}^-(U_{\mathbf{c}})$ be the empty set.

In order for $\widetilde{T}_{\mathbf{c}}^-(U_{\mathbf{c}})$ to separate $V_{\mathbf{c}}^-$ from $V_{\mathbf{c}}^+$, the restrictions on $U_{\mathbf{c}}$ must also change. Instead of points in $U_{\mathbf{c}}$ not lying on negative vertices or negative edges of $\mathbf{c}$, points in $U_{\mathbf{c}}$ must not lie on positive vertices or positive edges of $\mathbf{c}$. Without this restriction, $\widetilde{T}_{\mathbf{c}}^-(U_{\mathbf{c}})$ may not separate $V_{\mathbf{c}}^-$ from $V_{\mathbf{c}}^+$. More significantly, isosurfaces constructed from isosurface patches without this restriction may have holes. (See Figure 5.23.) As previously discussed, one should not mix isosurface patches of the form $\widetilde{T}_{\mathbf{c}}^+(U_{\mathbf{c}})$ with isosurface patches of the form $\widetilde{T}_{\mathbf{c}}^-(U_{\mathbf{c}})$.

Isosurface patches constructed from $U_{\mathbf{c}}$ can be different than isosurface patches constructed from edge midpoints that are then remapped to $U_{\mathbf{c}}$. Figure 5.24

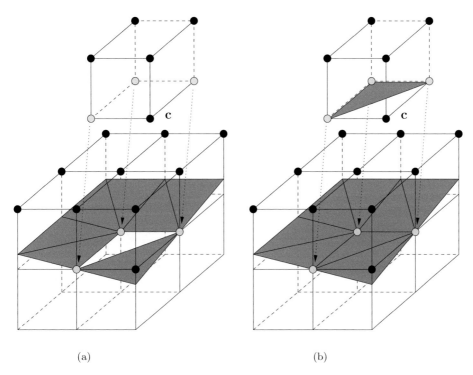

(a)   (b)

Figure 5.23. Example of isosurface construction using $R_{\mathbf{c}}^-(U_{\mathbf{c}})$. Gold vertices have scalar value equal to the isovalue. Black vertices have scalar value greater than the isovalue. (a) If vertices with scalar value equal to the isovalue are positive (i.e., in set $V_{\mathbf{c}}^+$), then $V_{\mathbf{c}}^-$ and $R_{\mathbf{c}}^-(U_{\mathbf{c}})$ are empty, creating a hole in the isosurface. (b) If vertices with scalar value equal to the isovalue are negative (i.e., in set $V_{\mathbf{c}}^-$), then $V_{\mathbf{c}}^-$ contains the three gold vertices and $R_{\mathbf{c}}^-(U_{\mathbf{c}})$ contains a triangle.

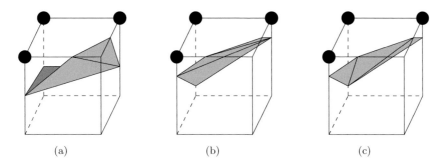

(a)   (b)   (c)

Figure 5.24. Example of different triangulations produced by algorithms IsoMid3D and IsoPatch3D. (a) Triangulation produced by IsoMid3D with vertices at midpoints. (b) Triangulation produced by mapping vertices of the IsoMid3D triangulation to points U. (c) Triangulation produced by IsoPatch3D directly on isosurface vertex set U.

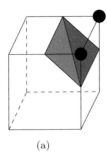 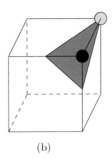

(a)   (b)

Figure 5.25. Example of degenerate triangulation produced by algorithm IsoMid3D but not by IsoPatch3D. (a) Triangulation produced by IsoMid3D with vertices at midpoints. Gray triangle becomes degenerate when vertices are mapped to points U in (b). (b) Triangulation produced by IsoPatch3D directly on isosurface vertex set U. Note that the triangulation produced by mapping vertices of the IsoMid3D triangulation to points U would look the same, but would contain an invisible degenerate triangle corresponding to the gray triangle in (a).

shows such differences. Because the triangulation of the patch constructed from $U_{\mathbf{c}}$ is different from the triangulation of the patch constructed from edge midpoints, the isosurface is different. Figure 5.25 shows the difference when a point of $U_{\mathbf{c}}$ lies on a vertex. The midpoint construction creates degenerate triangles while the construction from $U_{\mathbf{c}}$ does not.

For simplicity, we often abbreviate $R_{\mathbf{c}}^+(U_{\mathbf{c}})$ and $\widetilde{T}_{\mathbf{c}}^+(U_{\mathbf{c}})$ as $R_{\mathbf{c}}^+$ and $\widetilde{T}_{\mathbf{c}}^+$, respectively, whenever set $U_{\mathbf{c}}$ is clear from context.

5.5.2   Algorithm

We use algorithm IsoPatch3D from the previous section to construct isosurfaces. We call the algorithm IsoHull3D. Algorithm IsoHull3D is similar to the Marching Polyhedra algorithm, but instead of first constructing isosurface patches and then interpolating their vertices, it first computes the vertices and then construct isosurface patches directly on those vertices. Input to IsoHull3D is an isovalue, a three-dimensional convex polyhedral mesh and a set of scalar values at the vertices of the mesh.

The algorithm has two steps. The first step is computing the location of the isosurface vertices. For each mesh edge $[p, q]$ with one positive and one negative endpoint, apply linear interpolation as in Section 1.7.2 to compute a point on $[p, q]$ representing the intersection of the isosurface with $[p, q]$. If s_p and s_q are the scalar values at p and q and σ is the isovalue, then map v to $(1 - \alpha)p + \alpha q$ where $\alpha = (\sigma - s_p)/(s_q - s_p)$. Let U be the set of all such interpolated points over all mesh edges with one positive and one negative endpoint.

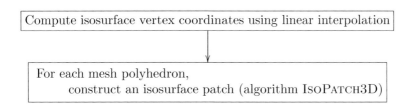

Figure 5.26. Algorithm IsoHull3D.

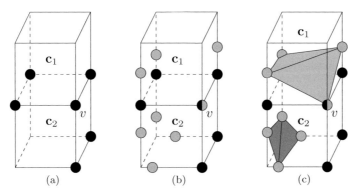

(a)   (b)   (c)

Figure 5.27. (a) Grid Γ of two cubes. Black vertices are positive. Note that vertex v lies on a bipolar edge in cube $\mathbf{c}_1$ but not in $\mathbf{c}_2$. (b) Green vertices forming an isosurface vertex set U for grid Γ. Vertex v is both a positive grid vertex and an element of U. Thus isosurface vertex set U is not proper. Set $U_{\mathbf{c}_2} = U \cap \mathbf{c}_2$ is an isosurface vertex set for $\mathbf{c}_2$ even though $v \in U_{\mathbf{c}_2}$ does not lie on any bipolar edge in $\mathbf{c}_2$. (c) Isosurface formed from the isosurface vertex set U.

If the positive endpoint of $[p, q]$ is strictly positive, then $0 < \alpha < 1$ and $\alpha p + (1 - \alpha)q$ is in the interior of $[p, q]$. Thus no point of U lies on a strictly positive vertex of $\mathbf{c}$.

Linear interpolation on multiple mesh edges that share the same endpoint could return that endpoint for each such edge. Since U is a set, it contains only a single copy of that endpoint.

The second step is constructing isosurface patches on those vertices. For each mesh polyhedron $\mathbf{c}$, let $U_{\mathbf{c}}$ be the points of U that lie in $\mathbf{c}$. Each bipolar edge of $\mathbf{c}$ contains one and only one point of $U_{\mathbf{c}}$, and no point of $U_{\mathbf{c}}$ lies on a strictly positive vertex or a negative vertex or a negative edge. Apply algorithm IsoPatch3D (Section 5.5.1) using the points $U_{\mathbf{c}}$. Return the union of all the triangle sets $\widetilde{T}_{\mathbf{c}}^{+}(U_{\mathbf{c}})$ over all mesh polyhedron $\mathbf{c}$.

Note that not every point of $U_{\mathbf{c}}$ must lie on a bipolar edge of $\mathbf{c}$. (See Figure 5.27.) Since there is no restriction on points of $U_{\mathbf{c}}$ lying on positive edges, algorithm IsoPatch3D can still be applied.

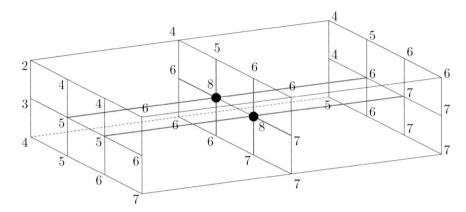

Figure 5.28. With isovalue 8, ten grid edges (colored magenta) incident on the two interior vertices are bipolar. For each grid cube **c**, set $R_\mathbf{c}^+$ is either a point or a line segment so $\tilde{S}_\mathbf{c}^+$ is the empty set. None of the bipolar edges intersect the IsoHull3D isosurface. (IsoHull3D returns the empty set as the isosurface.)

5.5.3   Isosurface Properties

IsoHull3D returns a finite set, Υ, of oriented triangles. The isosurface is the union of these triangles. The vertices of the isosurface are the triangle vertices.

We claim the following properties for the isosurface and triangulation produced by IsoHull3D. All properties other than Properties 3 and 5 are the same as the Marching Cubes isosurface properties. As with Marching Cubes, the isosurface is a 2-manifold with boundary under appropriate conditions.

Property 3 is similar to Property 3 of Marching Cubes and Marching Polyhedra isosurfaces, but no longer guarantees that bipolar edges are intersected by the isosurface. If the scalar value of one endpoint of the bipolar edge equals the isovalue, the isosurface may not intersect the bipolar edge. For instance, there are ten bipolar edges in the grid in Figure 5.28 but none of them intersect the IsoHull3D isosurface.

Property 5 is also a property (Property 8) of Marching Cubes and Marching Polyhedra isosurfaces, but only under the restriction that isovalue does not equal the scalar value of any mesh vertex. There is no such restriction on Property 5 for IsoHull3D isosurfaces. Because IsoHull3D does not position isosurface vertices after constructing the triangulation, it does not create zero-area triangles.

The following properties apply to all isosurfaces produced by the IsoHull3D algorithm.

Property 1.   *The isosurface is piecewise linear.*

Property 2.   *The vertices of the isosurface lie on mesh edges.*

Property 3. *The isosurface intersects every bipolar mesh edge at most once and intersects every bipolar edge with a strictly positive endpoint at exactly one point.*

Property 4. *The isosurface does not intersect any negative or strictly positive mesh edges.*

Property 5. *Set Υ does not contain any zero-area triangles.*

To claim that the isosurface separates strictly positive mesh vertices from strictly negative ones, we need the following condition on the mesh:

- The mesh is a partition of a 3-manifold with boundary.

Under this condition we have the following separation property:

Property 6. *The isosurface strictly separates strictly positive mesh vertices from negative mesh vertices.*

Note that Property 6 no longer claims anything about mesh vertices with scalar value equal to the isovalue. These vertices may be on either side of the isosurface.

As with MARCHING CUBES, if both endpoints of a mesh edge have scalar value equal to the isovalue, then the isosurface may intersect the mesh edge zero, one, or two times or may contain the mesh edge.

Under appropriate conditions, the isosurface produced by ISOHULL3D is a 2-manifold with boundary. As with the MARCHING POLYHEDRA algorithm, the isovalue should not equal the scalar value of any vertex and the underlying space must be a manifold. However, unlike the MARCHING POLYHEDRA algorithm, there is no restriction on the mesh elements. ISOHULL3D will not create a self-intersecting surface, even for the pathological case in Figure 5.10.

Consider a convex polyhedral mesh and isovalue that has the following two conditions:

- The isovalue does not equal the scalar value of any mesh vertex.

- The mesh is a partition of a 3-manifold with boundary.

Under these conditions, the isosurface produced by ISOHULL3D has the following three properties:

Property 7. *The isosurface is a piecewise linear, orientable 2-manifold with boundary.*

Property 8. *The boundary of the isosurface lies on the boundary of the mesh.*

Property 9. *Set Υ does not contain any zero-area triangles or duplicate triangles and the triangles in Υ form a triangulation of the isosurface.*

The fact that Υ does not contain any zero-area triangles is already stated in Property 5 and is restated in Property 9 for completeness.

5.5.4 Proof of Isosurface Properties

Proofs of Properties 1 and 2 are the same as in Section 2.3.4 for MARCHING CUBES and are omitted. Many of the other proofs are similar to those for MARCHING POLYHEDRA in Sections 5.4.2, 5.4.3, and 5.4.4.

As defined in Section 5.5.1, let $R_{\mathbf{c}}^+(U_{\mathbf{c}})$ (abbreviated $R_{\mathbf{c}}^+$) be the convex hull of $U_{\mathbf{c}} \cup V_{\mathbf{c}}^+$. If $R_{\mathbf{c}}^+$ is three-dimensional, then let $\widetilde{T}_{\mathbf{c}}^+(U_{\mathbf{c}})$ (abbreviated $\widetilde{T}_{\mathbf{c}}^+$) be the triangles in a triangulation of $\partial R_{\mathbf{c}}^+$ that do not lie on the boundary of $\mathbf{c}$. Orient the triangles so that the induced normals point toward $R_{\mathbf{c}}^+$. If $R_{\mathbf{c}}^+$ is two-dimensional and $R_{\mathbf{c}}^+$ is a subset of some facet of $\mathbf{c}$, then let $\widetilde{T}_{\mathbf{c}}^+$ be the triangles in a triangulation of $R_{\mathbf{c}}^+$. Orient the triangles so that the induced normals point away from $\mathbf{c}$. Otherwise, let $\widetilde{T}_{\mathbf{c}}^+$ be the empty set. Set $\bigcup_{\mathbf{c} \in \Gamma} \widetilde{T}_{\mathbf{c}}^+$ is the set Υ returned by ISOHULL3D.

To prove Property 3, we need the following lemma, which shows that the vertices of $\widetilde{T}_{\mathbf{c}}^+(U_{\mathbf{c}})$ are all from the set $U_{\mathbf{c}}$. This lemma is similar to Lemma 5.5 from Section 5.4.2 and its proof relies upon that lemma.

Lemma 5.28. *Let $\mathbf{c}$ be a convex polyhedron in $\mathbb{R}^3$ where each vertex of $\mathbf{c}$ has a positive or a negative label. If $U_{\mathbf{c}}$ is an isosurface vertex set for $\mathbf{c}$, then the set of vertices of $\widetilde{T}_{\mathbf{c}}^+(U_{\mathbf{c}})$ is a subset of $U_{\mathbf{c}}$.*

Proof: As defined in Section 5.4.2, $T_{\mathbf{c}}^+(U_{\mathbf{c}})$ (abbreviated $T_{\mathbf{c}}^+$) are the triangles in a triangulation of $\partial R_{\mathbf{c}}^+$ that do not lie on the boundary of $\mathbf{c}$. If $R_{\mathbf{c}}^+(U_{\mathbf{c}})$ (abbreviated $R_{\mathbf{c}}^+$) is three-dimensional, then $\widetilde{T}_{\mathbf{c}}^+(U_{\mathbf{c}})$ (abbreviated $\widetilde{T}_{\mathbf{c}}^+$) equals $T_{\mathbf{c}}^+$. By Lemma 5.5, the vertices of $T_{\mathbf{c}}^+$ are a subset of $U_{\mathbf{c}}$, proving the claim.

If $R_{\mathbf{c}}^+$ is a two-dimensional subset of some facet of $\mathbf{c}$, then $\widetilde{T}_{\mathbf{c}}^+$ is a triangulation of $R_{\mathbf{c}}^+$. Let w be a vertex of $\widetilde{T}_{\mathbf{c}}^+$. Since $R_{\mathbf{c}}^+$ equals $\mathrm{conv}(U_{\mathbf{c}} \cup V_{\mathbf{c}}^+)$, vertex w is in $U_{\mathbf{c}} \cup V_{\mathbf{c}}^+$.

If w is not in $V_{\mathbf{c}}^+$, then w must be in $U_{\mathbf{c}}$, establishing the claim. Assume that vertex w is in $V_{\mathbf{c}}^+$.

Let $\mathbf{f}$ be the facet of $\mathbf{c}$ containing $R_{\mathbf{c}}^+$. Let (w, v) be the edge of $\mathbf{c}$ incident on w that is not contained in $\mathbf{f}$. Since $R_{\mathbf{c}}^+$ is contained in $\mathbf{f}$,

$$(w, v) \cap R_{\mathbf{c}}^+ = w.$$

Since vertex v is not in $R_{\mathbf{c}}^+$, vertex v must have a negative label. Thus, (w, v) is a bipolar edge. By definition, the isosurface vertex set $U_{\mathbf{c}}$ contains some point p on (w, v). Since p is in $R_{\mathbf{c}}^+$ and $(w, v) \cap R_{\mathbf{c}}^+$ equals w, point p must equal w. Thus w is in $U_{\mathbf{c}}$, proving the claim. □

The set $|\widetilde{T}_{\mathbf{c}}^+(U_{\mathbf{c}})|$ is the union of the triangles in $\widetilde{T}_{\mathbf{c}}^+$.

Corollary 5.29. *Let $\mathbf{c}$ be a convex polyhedron in $\mathbb{R}^3$ where each vertex of $\mathbf{c}$ has a positive or a negative label, and let $U_{\mathbf{c}}$ be an isosurface vertex set for $\mathbf{c}$. If $\mathbf{e}$ is a bipolar edge of $\mathbf{c}$, then*

$$\mathbf{e} \cap |\widetilde{T}_{\mathbf{c}}^+(U_{\mathbf{c}})| \subseteq \mathbf{e} \cap U_{\mathbf{c}}.$$

Proof: Since $\mathbf{c}$ is convex, only isosurface triangles with vertices on $\mathbf{e}$ intersect $\mathbf{e}$. By Lemma 5.28, every vertex of $\widetilde{T}^+_{\mathbf{c}}(U_{\mathbf{c}})$ (abbreviated $\widetilde{T}^+_{\mathbf{c}}$) must be in $U_{\mathbf{c}}$. Set $U_{\mathbf{c}}$ intersects $\mathbf{e}$ in a single point $\mathbf{e} \cap U_{\mathbf{c}}$. Thus $\mathbf{e} \cap U_{\mathbf{c}}$ is the only point of $\mathbf{e}$ that could lie on $|\widetilde{T}^+_{\mathbf{c}}|$ and so $\mathbf{e} \cap |\widetilde{T}^+_{\mathbf{c}}| \subseteq \mathbf{e} \cap U_{\mathbf{c}}$.   □

If $R^+_{\mathbf{c}}$ is zero- or one-dimensional, then $\widetilde{T}^+_{\mathbf{c}}$ is the empty set. Thus, it is possible that $\mathbf{e} \cap |\widetilde{T}^+_{\mathbf{c}}|$ is empty, while $\mathbf{e} \cap U_{\mathbf{c}}$ is not.

Property 3.   *The isosurface intersects every bipolar mesh edge at most once and intersects every bipolar edge with a strictly positive endpoint at exactly one point.*

Property 4.   *The isosurface does not intersect any negative or strictly positive mesh edges.*

Proof of Properties 3 & 4: Let $\mathbf{e}$ be a mesh edge. Since mesh elements are convex, only isosurface triangles with vertices on $\mathbf{e}$ intersect $\mathbf{e}$.

Assume $\mathbf{e}$ is bipolar and $\mathbf{e}$ intersects $|\widetilde{T}^+_{\mathbf{c}}|$. By Corollary 5.29, for every mesh element $\mathbf{c}$ containing $\mathbf{e}$, set $|\widetilde{T}^+_{\mathbf{c}}| \cap \mathbf{e}$ is a subset of $\mathbf{e} \cap U_{\mathbf{c}}$. Since $U_{\mathbf{c}}$ equals $U \cap \mathbf{c}$, for every mesh element $\mathbf{c}$ containing $\mathbf{e}$, set $|\widetilde{T}^+_{\mathbf{c}}| \cap \mathbf{e}$ is a subset of $\mathbf{e} \cap U$. Since $\mathbf{e} \cap U$ is a single point, the isosurface intersects $\mathbf{e}$ at most once.

Assume the positive endpoint of $\mathbf{e}$ is strictly positive. By definition, set $U_{\mathbf{c}}$ does not contain either the strictly positive or the negative endpoint of $\mathbf{e}$. Thus set $U_{\mathbf{c}}$ must contain some point p in the interior of $\mathbf{e}$. Point p is a vertex of some triangle $\mathbf{t} \in \widetilde{T}_{\mathbf{c}}(U_{\mathbf{c}})$. Thus p is on the isosurface and the isosurface intersects $\mathbf{e}$ at exactly one point.

If $\mathbf{e}$ is negative or strictly positive, then no isosurface vertex lies on $\mathbf{e}$. Thus the isosurface does not intersect negative or strictly positive grid edges.   □

Because IsoHull3D does not position isosurface vertices after constructing the triangulation, it does not create zero-area triangles.

Property 5.   *Set Υ does not contain any zero-area triangles.*

Proof: Set Υ equals $\bigcup_{\mathbf{c} \in \Gamma} \widetilde{T}^+_{\mathbf{c}}$. For each mesh element $\mathbf{c}$, set $\widetilde{T}^+_{\mathbf{c}}(U_{\mathbf{c}})$ is the subset of the triangles in a triangulation of $\partial R^+_{\mathbf{c}}(U_{\mathbf{c}})$. Since none of the triangles in the triangulation of $\partial R^+_{\mathbf{c}}(U_{\mathbf{c}})$ have zero area, none of the triangles in $\widetilde{T}^+_{\mathbf{c}}(U_{\mathbf{c}})$ have zero area. Thus, no triangle returned by IsoHull3D has zero area.   □

In Section 5.4.3, we defined $S^+_{\mathbf{c}}$ as $R^+_{\mathbf{c}} \cap \mathrm{cl}(\mathbf{c} - R^+_{\mathbf{c}})$. When $R^+_{\mathbf{c}}$ is three-dimensional or is a two-dimensional subset of $\partial \mathbf{c}$, set $\widetilde{T}^+_{\mathbf{c}}$ is a triangulation of $S^+_{\mathbf{c}}$. (See the proof of 5.30 below.) However, if $R^+_{\mathbf{c}}$ has dimension zero or one or has dimension two but is not a subset of $\partial \mathbf{c}$, then $S_{\mathbf{c}}$ will be nonempty while $\widetilde{T}^+_{\mathbf{c}}$ will be the empty set. We need to modify $S^+_{\mathbf{c}}$ so that it corresponds to the region triangulated by $\widetilde{T}^+_{\mathbf{c}}$.

Define $\widetilde{S}_{\mathbf{c}}^+(U_{\mathbf{c}})$ as

$$\widetilde{S}_{\mathbf{c}}^+(U_{\mathbf{c}}) = \begin{cases} R_{\mathbf{c}}^+(U_{\mathbf{c}}) \cap \mathrm{cl}(R_{\mathbf{c}}^+(U_{\mathbf{c}})) & \text{if } \dim(R_{\mathbf{c}}^+(U_{\mathbf{c}})) = 3, \\ R_{\mathbf{c}}^+(U_{\mathbf{c}}) & \text{if } \dim(R_{\mathbf{c}}^+(U_{\mathbf{c}})) = 2 \text{ and } R_{\mathbf{c}}^+(U_{\mathbf{c}}) \subset \partial\mathbf{c}, \\ \emptyset & \text{otherwise,} \end{cases}$$

where $\dim(R_{\mathbf{c}}^+(U_{\mathbf{c}}))$ is the dimension of $R_{\mathbf{c}}^+(U_{\mathbf{c}})$. Abbreviate $\widetilde{S}_{\mathbf{c}}^+(U_{\mathbf{c}})$ as $\widetilde{S}_{\mathbf{c}}^+$. Set $\widetilde{S}_{\mathbf{c}}$ equals $S_{\mathbf{c}}$ when $R_{\mathbf{c}}^+$ has dimension three and $\mathrm{int}(R_{\mathbf{c}}^+)$, the interior of $R_{\mathbf{c}}^+$, is the empty set when $R_{\mathbf{c}}^+$ has a dimension less than three. Thus $\widetilde{S}_{\mathbf{c}}$ strictly separates $\mathrm{int}(R_{\mathbf{c}}^+)$ from $(\mathbf{c} - R_{\mathbf{c}}^+)$.

We claim that $\widetilde{T}_{\mathbf{c}}^+$ is a triangulation of $\widetilde{S}_{\mathbf{c}}^+$ and thus $|\widetilde{T}_{\mathbf{c}}^+|$ equals $\widetilde{S}_{\mathbf{c}}^+$.

Lemma 5.30. *Let $\mathbf{c}$ be a convex polyhedron in $\mathbb{R}^3$ where each vertex of $\mathbf{c}$ has a positive or a negative label. If $U_{\mathbf{c}}$ is an isosurface vertex set of $\mathbf{c}$, then $\widetilde{T}_{\mathbf{c}}^+(U_{\mathbf{c}})$ is a triangulation of $\widetilde{S}_{\mathbf{c}}^+(U_{\mathbf{c}})$.*

Proof: As defined in Section 5.4.2, $T_{\mathbf{c}}^+$ are the triangles in a triangulation of $\partial R_{\mathbf{c}}^+$ that do not lie on the boundary of $\mathbf{c}$ and $S_{\mathbf{c}}^+$ equals $R_{\mathbf{c}}^+ \cap \mathrm{cl}(\mathbf{c} - R_{\mathbf{c}}^+)$. If $R_{\mathbf{c}}^+$ is three-dimensional, then $\widetilde{T}_{\mathbf{c}}^+$ equals $T_{\mathbf{c}}^+$ and $\widetilde{S}_{\mathbf{c}}^+$ equals $S_{\mathbf{c}}^+$. By Lemma 5.10, $T_{\mathbf{c}}^+$ is a triangulation of $S_{\mathbf{c}}^+$, proving the lemma.

Since $R_{\mathbf{c}}^+$ is convex, $R_{\mathbf{c}}^+$ is a subset of $\mathbf{c}$ if and only if it is a subset of some facet of $\mathbf{c}$. If $R_{\mathbf{c}}^+$ is two-dimensional and a subset of $\mathbf{c}$, then $\widetilde{S}_{\mathbf{c}}^+$ equals $R_{\mathbf{c}}^+$ and $\widetilde{T}_{\mathbf{c}}^+$ is a triangulation of $R_{\mathbf{c}}^+$. Thus, $\widetilde{T}_{\mathbf{c}}^+$ is a triangulation of $\widetilde{S}_{\mathbf{c}}^+$.

If $R_{\mathbf{c}}^+$ has dimension zero or one, then both $\widetilde{T}_{\mathbf{c}}^+$ and $\widetilde{S}_{\mathbf{c}}^+$ are empty, again proving the lemma. $\qquad\square$

In Section 5.4.2, we defined $R_{\Gamma}^+(U)$ as the union of $R_{\mathbf{c}}^+(U_{\mathbf{c}})$ over all the polyhedra $\mathbf{c} \in \Gamma$. When U is a proper isosurface vertex set, each region $R_{\mathbf{c}}^+(U_{\mathbf{c}})$ is three-dimensional and $R_{\Gamma}^+(U)$ is three-dimensional. However, if U is not proper, then some regions $R_{\mathbf{c}}^+(U_{\mathbf{c}})$ can have dimensions zero, one, or two and $R_{\Gamma}^+(U)$ can have subregions with dimensions zero, one, or two. To strip these subregions from $R_{\Gamma}^+(U)$, we take the closure of the interior of $R_{\Gamma}^+(U)$, denoted $\mathrm{cl}(\mathrm{int}(R_{\Gamma}^+(U)))$.

Let Γ be a convex polyhedral mesh where every vertex of Γ has a positive or a negative label. Let $|\Gamma|$ be the union of all the polyhedra in Γ, and let U be an isosurface vertex set of Γ. For each polyhedron $\mathbf{c} \in \Gamma$, let $U_{\mathbf{c}}$ equal $U \cap \mathbf{c}$. Define $\widetilde{T}_{\Gamma}^+(U)$, $R_{\Gamma}^+(U)$, $\widetilde{R}_{\Gamma}^+(U)$, and $\widetilde{S}_{\Gamma}^+(U)$ (abbreviated $\widetilde{T}_{\Gamma}^+$, R_{Γ}^+, $\widetilde{R}_{\Gamma}^+$, and $\widetilde{S}_{\Gamma}^+$, respectively) as

$$\widetilde{T}_{\Gamma}^+(U) = \bigcup_{\mathbf{c} \in \Gamma} \widetilde{T}_{\mathbf{c}}^+(U_{\mathbf{c}}),$$

$$R_{\Gamma}^+(U) = \bigcup_{\mathbf{c} \in \Gamma} R_{\mathbf{c}}^+(U_{\mathbf{c}}),$$

$$\widetilde{R}_{\Gamma}^+(U) = \mathrm{cl}(\mathrm{int}(R_{\Gamma}^+(U))),$$

$$\widetilde{S}_{\Gamma}^+(U) = \widetilde{R}_{\Gamma}^+(U) \cap \mathrm{cl}(|\Gamma| - \widetilde{R}_{\Gamma}^+(U)).$$

(The definition of R_Γ^+ was given in Section 5.4.3 and is repeated here for completeness.) Set $\widetilde{T}_\Gamma(U)$ is the set Υ returned by IsoHull3D. By Lemma B.8 in Appendix B, $\widetilde{S}_\Gamma^+$ separates $\widetilde{R}_\Gamma^+$ from $|\Gamma| - \widetilde{R}_\Gamma^+$.

$\widetilde{S}_\Gamma$ and $\widetilde{S}_c$ have substantially different definitions. If Γ is composed of a single convex polyhedron, then $\widetilde{S}_\Gamma$ and $\widetilde{S}_c$ may not be equal. Thus, $\widetilde{S}_\Gamma^+(U)$ does not necessarily equal $\bigcup_{c \in \Gamma} \widetilde{S}_c^+(U_c)$. Nevertheless, we claim that $\widetilde{S}_\Gamma^+(U)$ is a subset of $\bigcup_{c \in \Gamma} \widetilde{S}_c^+(U_c)$.

Lemma 5.31. *Let Γ be a convex polyhedral mesh in $\mathbb{R}^3$ where each vertex of Γ has a positive or negative label. If U is an isosurface vertex set of Γ and U_c equals $U \cap c$ for each polyhedron $c \in \Gamma$, then $\widetilde{S}_\Gamma^+(U)$ is a subset of $\bigcup_{c \in \Gamma} \widetilde{S}_c^+(U_c)$.*

Proof: Let p be a point in $\widetilde{S}_\Gamma^+$. Triangulate $\widetilde{S}_\Gamma^+$ so that every triangle is contained in some grid cube. Since p has local dimension two in $\widetilde{S}_\Gamma^+$, point p must lie on some triangle $\mathbf{t} \subseteq \widetilde{S}_\Gamma^+$. Since $\widetilde{R}_\Gamma^+$ is a subset of R_Γ^+ and R_Γ^+ equals $\bigcup_{c \in \Gamma} R_c^+$, triangle $\mathbf{t}$ is in $R_{c_1}^+$ for some $c_1 \in \Gamma$.

By Lemma B.2 in Appendix B, $\mathrm{cl}(|\Gamma| - \widetilde{R}_\Gamma^+)$ equals $\mathrm{cl}(|\Gamma| - R_\Gamma^+)$. Thus, $\mathbf{t}$ is in $\mathrm{cl}(|\Gamma| - R_\Gamma^+)$. Since $\mathrm{cl}(|\Gamma| - R_\Gamma^+)$ equals $\bigcup_{c \in \Gamma} \mathrm{cl}(c - R_c^+)$, triangle $\mathbf{t}$ is in $\mathrm{cl}(c_2 - R_{c_2}^+)$ for some $c_2 \in \Gamma$. If c_1 equals c_2, then $\mathbf{t}$ is in $\widetilde{S}_{c_2}^+$ and therefore in $\bigcup_{c \in \Gamma} \widetilde{S}_c^+$. Assume c_1 does not equal c_2.

The intersection of c_1 and c_2 is a facet $\mathbf{f}$ of both polyhedra, and triangle $\mathbf{t}$ is in $\mathbf{f}$. By Lemma 5.14, triangle $\mathbf{t}$ is in $R_{c_2}^+$. Since triangle $\mathbf{t}$ is in $\mathrm{cl}(c_2 - R_{c_2}^+)$, region $R_{c_2}^+$ must be contained in $\mathbf{f}$. By definition, $\widetilde{S}_{c_2}^+$ equals $R_{c_2}^+$ and thus contains $\mathbf{t}$. Thus p is $\widetilde{S}_{c_2}^+$ and is therefore in $\bigcup_{c \in \Gamma} \widetilde{S}_c^+$. $\square$

Property 6 follows from Lemma 5.31.

Property 6. *The isosurface strictly separates strictly positive mesh vertices from negative mesh vertices.*

Proof: Let v_1 and v_2 be vertices of Γ with scalar values above and below the isovalue, respectively. Since v_1 is strictly positive and v_2 is negative, neither vertex is in set U. By Lemma 5.28, neither v_1 nor v_2 is a vertex of $\widetilde{T}_c^+$ for any c and thus neither lies on $|\widetilde{T}_c^+|$ for any c.

By Lemmas 5.30 and 5.31, $\widetilde{S}_\Gamma^+$ is a subset of $|\widetilde{T}_\Gamma^+|$. Since $|\widetilde{T}_\Gamma^+|$ equals $\bigcup_{c \in \Gamma} |\widetilde{T}_c^+|$, neither v_1 nor v_2 lie on $|\widetilde{T}_\Gamma^+|$. The strictly positive vertex v_1 lies in the interior of $\widetilde{R}_\Gamma^+$ while the negative vertex v_2 lies in $|\Gamma| - \widetilde{R}_\Gamma^+$. By Lemma B.8, $\widetilde{S}_\Gamma^+$ strictly separates $\mathrm{int}(\widetilde{R}_\Gamma^+)$ from $|\Gamma| - \widetilde{R}_\Gamma^+$. Thus $|\widetilde{T}_\Gamma^+|$ strictly separates v_1 from v_2. $\square$

The proof of Properties 7 and 8 for IsoHull3D is similar to the proof of Properties 6 and 7 for the Marching Polyhedra algorithm. Because IsoHull3D does not move isosurface vertices after the creation of isosurface patches, there is

no problem of self-intersection within a mesh element as for MARCHING POLY-
HEDRA (Figure 5.10). Thus there is no restriction other than convexity on the
type of mesh elements.

Property 7. *The isosurface is a piecewise linear, orientable 2-manifold with
boundary.*

Property 8. *The boundary of the isosurface lies on the boundary of the mesh.*

Proof of Properties 7 & 8: Since no vertex of Γ has scalar value equal to the isovalue, all points in U lie in the interior of bipolar edges. Thus U is a proper
isosurface vertex set for Γ. By Corollary 5.9, R_c^+ is three-dimensional or empty
for each polyhedron $\mathbf{c}$. Thus set $\widetilde{R}_\Gamma^+(U)$ equals $R_\Gamma^+(U)$, set $\widetilde{T}_\Gamma^+(U)$ equals $T_\Gamma^+(U)$,
and set $\widetilde{S}_\Gamma^+(U)$ equals $S_\Gamma^+(U)$.

Applying Lemma 5.26 from Section 5.4.4, $\widetilde{S}_\Gamma^+(U) = S_\Gamma^+(U)$ is a piecewise
linear, orientable 2-manifold whose boundary lies on $\partial|\Gamma|$. By Lemma 5.21,
$\widetilde{T}_\Gamma^+(U) = T_\Gamma^+(U)$ is a triangulation of $\widetilde{S}_\Gamma^+(U) = S_\Gamma^+(U)$. Thus $\Upsilon = |\widetilde{T}_\Gamma^+(U)|$ is
a piecewise linear, orientable 2-manifold with boundary whose boundary lies on
the boundary of Γ. $\square$

The last property is that Υ does not contain any zero-area or duplicate triangles and forms a triangulation of the isosurface.

Property 9. *Set Υ does not contain any zero-area triangles or duplicate triangles
and the triangles in Υ form a triangulation of the isosurface.*

Proof: By Property 5, Υ does not contain any zero-area triangles.

Since no mesh vertex has scalar value equal to the isovalue, all points in
U lie in the interior of bipolar edges. and U is a proper isosurface vertex set.
Thus every isosurface triangle is contained in one and only one mesh element.
Since the isosurface triangles within a mesh element are unique, Υ contains no
duplicate triangles.

Since U is a proper isosurface vertex set, set $\widetilde{T}_\Gamma^+(U)$ equals T_Γ^+ and $\widetilde{S}_\Gamma^+(U)$
equals $S_\Gamma^+(U)$. By Lemma 5.21, $\widetilde{T}_\Gamma^+(U) = T_\Gamma^+(U)$ is a triangulation of $\widetilde{S}_\Gamma^+(U) =
S_\Gamma^+(U)$. Thus $\Upsilon = \widetilde{T}_\Gamma^+(U)$ is a triangulation of the isosurface. $\square$

5.5.5 IsoHull3D with Table Lookup

The ISOHULL3D algorithm can be combined with table lookup. A table can
be constructed and used for the most common polyhedral mesh elements such
as cubes, tetrahedra, and pyramids, which satisfy Lemma 5.27. For all other
mesh elements or for any mesh element containing a vertex whose scalar value
equals the isovalue, the direct isosurface patch construction can be performed as
in Section 5.5.1.

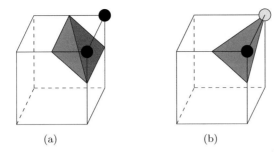

(a)   (b)

Figure 5.29. Example of degenerate triangulation produced by algorithm IsoMid3D. (a) Triangulation produced by IsoMid3D with vertices at midpoints. Gray triangle becomes degenerate when vertices are mapped to a gold vertex whose scalar value equals the isovalue. (b) Resulting isosurface after mapping vertices to a gold vertex whose scalar value equals the isovalue. The isosurface contains an invisible degenerate triangle corresponding to the gray triangle in (a).

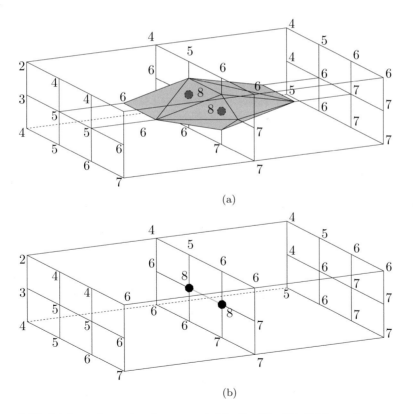

Figure 5.30. (a) Isosurface triangles generated by IsoMid3D. (b) Isosurface composed of degenerate triangles forming a line segment.

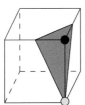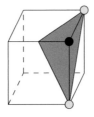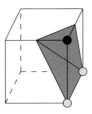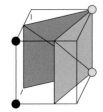

Figure 5.31. Examples of isosurfaces produced with "+" and "−" and "=" labels. Black vertices are labelled positive and gold vertices are labelled equals.

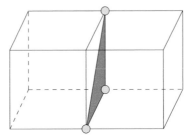

Figure 5.32. Isosurface consisting of two duplicate triangles that do not enclose any three-dimensional regions. The three gold vertices have scalar value equal to the isovalue. All other vertices have scalar value less than the isovalue. A single isosurface triangle is generated from the isosurface lookup table for each cube. Since the two triangles lie in the common facet shared by the two cubes, the two triangles are identical.

5.5.6 Expanded Isosurface Tables

The MARCHING CUBES, MARCHING TETRAHEDRA, and MARCHING POLYHEDRA algorithms construct isosurfaces by treating points with scalar value equal to the isovalue as points whose scalar value is above the isovalue. The subsequent remapping of isosurface vertices creates degenerate triangles as in Figure 5.29. It also creates one- or zero-dimensional isosurface components as in Figure 5.30.

The isosurface patch construction algorithm in Section 5.5.1 can be used to properly handle points whose scalar value equals the isovalue. As before, label points with scalar value greater than the isovalue with a "+" and label points with scalar value below the isovalue with a "−". However, label points whose value equals the isovalue with an "=". Each vertex now has three possible labellings. A cube $\mathbf{c}$ has eight vertices and so there are 3^8 or 6561 possible configurations. Let $U_{\mathbf{c}}$ be the set of all midpoints of edges with one "+" and one "−" endpoint and the set of all "=" vertices. Let $R_{\mathbf{c}}^+(U_{\mathbf{c}})$ equal $\mathrm{conv}(U_{\mathbf{c}} \cup V_{\mathbf{c}}^+)$. Construct the isosurface patches as in Section 5.5.1. Examples of some of the components are in Figure 5.31.

Since no isosurface vertex is mapped to a mesh vertex unless it was already there, the remapping of isosurface vertices does not create any degenerate triangles. It also does not create any zero- or one-dimensional isosurface components. It can still create some two-dimensional components that do not enclose any three-dimensional regions. (See Figure 5.32.)

5.6   Notes and Comments

Using the convex hull to generate isosurface patches was independently proposed by Lachaud and Montanvert [Lachaud and Montanvert, 2000] and Bhaniramka et al. [Bhaniramka et al., 2000, Bhaniramka et al., 2004a]. Generating isosurface lookup tables based on three vertex labels, "−", "+", and "=" is described by Raman and Wenger in [Raman and Wenger, 2008].

CHAPTER 6

ISOSURFACE GENERATION IN 4D

Time varying scalar data is inherently four-dimensional, with the fourth dimension being time. Tremendous advancements in memory size and processing speed have greatly the reduced the cost of gathering or generating such data sets.

Typically, time varying data is visualized by constructing and rendering isosurfaces at each three-dimensional time slice. Rendered images are composed to form an animation of the isosurface in time. In this chapter we discuss a different approach of creating a single continuous isosurface from the four-dimensional data. This isosurface can be sliced along the time but it can also be sliced along different axes to give different representations of the data. It can also be decimated to create a compact representation of the isosurface and smoothed to create an isosurface that varies smoothly with time.

Parametric studies are another source of four-dimensional scalar data sets. Multiple three-dimensional scalar data sets are generated by varying a single parameter in a simulation or experiment. Adjoining the three-dimensional data sets gives a four-dimensional scalar data set with the fourth dimension representing the varying parameter.

Perhaps the most useful application of isosurface generation in $\mathbb{R}^4$ is the automatic construction of three-dimensional interval volumes. An interval volume is a region bounded by isosurfaces. In Chapter 7 we will define and discuss interval volumes and their construction by lifting to $\mathbb{R}^4$.

In $\mathbb{R}^3$, generic level sets[1] and their isosurface approximations are two-dimensional, either two-dimensional manifolds or the union of such manifolds. In $\mathbb{R}^4$, generic level sets and isosurfaces are three-dimensional. Piecewise linear isosurfaces in $\mathbb{R}^4$ are composed of tetrahedra.

---

[1] As noted in Chapter 1, a level set can be a the entire space or a single point. In fact, it can be any subset of the space. However, if $f : \mathbb{R}^3 \to \mathbb{R}$ and all its derivatives are differentiable, then $f^{-1}(c)$ is a 2-manifold for "almost all" values of c. Similarly, if $f : \mathbb{R}^4 \to \mathbb{R}$ and all its derivatives are differentiable, then $f^{-1}(c)$ is a 3-manifold for "almost all" values of c.

Both variations of MARCHING CUBES and the dual contouring algorithms extend to four dimensions. In this chapter, we describe four-dimensional variations of the algorithms MARCHING CUBES, ISoHULL3D, and SURFACE NETS.

6.1   Definitions and Notation

Visibility. In this chapter, we define two types of triangulations, lexicographic triangulations and incremental triangulations. In analyzing these triangulations, we need the following definition.

Definition 6.1. Let $\{p_1, \ldots, p_m\}$ be a set of points, let τ be some triangulation of $\mathrm{conv}(p_1, \ldots, p_m)$, and let q be a point that is not in $\mathrm{conv}(p_1, \ldots, p_m)$. A simplex **t** is visible from q if, for each $p \in \mathbf{t}$, the open line segment (p, q) does not intersect $\mathrm{conv}(p_1, \ldots, p_m)$.

Figure 6.1 illustrates edges in 2D that are visible from a point q_1 and triangles in 3D that are visible from a point q_2.

Vertex and edge properties. The definitions of vertex and edge properties and isosurface vertex sets given here are the same as the 3D definitions in Section 5.1. For review, we summarize them here.

A mesh vertex is positive if its scalar value is greater than or equal to σ and negative if its scalar value is less than σ. A positive vertex is strictly positive if its scalar value does not equal the isovalue.

Let **c** be a polytope in $\mathbb{R}^4$ where each vertex of **c** has a positive or a negative label. Set $V_{\mathbf{c}}^+$ is the set of positive vertices of **c**, set $V_{\mathbf{c}}^-$ is the set of negative vertices of **c**, and set $M_{\mathbf{c}}$ is the set of midpoints of bipolar edges of **c**.

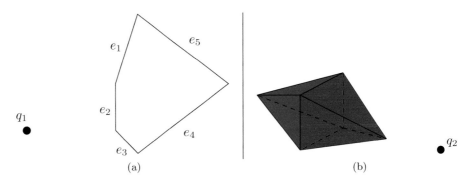

Figure 6.1. (a) Edges e_1, e_2, and e_3 are visible from q_1. Edges e_4 and e_5 are not visible from q_1. (b) Red triangles are visible from q_2. Blue triangles are not visible from q_2.

An edge of $\mathbf{c}$ is positive if both its endpoints are positive. An edge of $\mathbf{c}$ is negative if both its endpoints are negative. A positive edge is strictly positive if both its endpoints are strictly positive. An edge of $\mathbf{c}$ is bipolar if one endpoint is positive and one endpoint is negative.

Isosurface vertex sets. Let Γ be a convex polytopal mesh in $\mathbb{R}^4$ and $\mathbf{c}$ a convex polytope in $\mathbb{R}^4$ where every vertex of Γ and $\mathbf{c}$ is labeled negative, positive, or strictly positive. An isosurface vertex set U of Γ (respectively $U_\mathbf{c}$ of $\mathbf{c}$) is a finite set of points lying on the edges of Γ (respectively $\mathbf{c}$) such that every bipolar edge of Γ (respectively $\mathbf{c}$) contains one and only one point of U (respectively $U_\mathbf{c}$), and no point of U (respectively $U_\mathbf{c}$) lies on a strictly positive vertex or negative vertex or negative edge of Γ (respectively $\mathbf{c}$.) A proper isosurface vertex set is an isosurface vertex set U of Γ (respectively $U_\mathbf{c}$ of $\mathbf{c}$) such that every point in U (respectively $U_\mathbf{c}$) lies in the interior of a bipolar edge of Γ (respectively $\mathbf{c}$.) The set $M_\mathbf{c}$ of midpoints of bipolar edges of $\mathbf{c}$ is an example of a proper isosurface vertex set of $\mathbf{c}$.

Interior and boundary edges and cubes. The definitions of interior and boundary edges and cubes for a 4D grid parallel the 3D definitions from Section 3.1.

Definition 6.2. Let Γ be a regular grid in $\mathbb{R}^4$.

- An interior grid edge in $\mathbb{R}^4$ is a grid edge that is contained in eight grid cubes.

- An boundary grid edge in $\mathbb{R}^4$ is a grid edge that is contained in seven or fewer grid cubes.

Definition 6.3.

- An interior grid cube is a grid cube whose edges are all interior grid edges.

- A boundary grid cube is a grid cube that contains at least one boundary grid edge.

Let Γ be a regular grid in $\mathbb{R}^4$. We use the notation Γ_{inner} to denote the regular subgrid of Γ composed of the interior grid cubes of Γ.

Centroid. The centroid of a set of points $P = \{p_1, p_2, \ldots, p_k\}$ is

$$\frac{p_1 + p_2 + \ldots + p_k}{k}.$$

6.2   Isosurface Table Generation in 4D

The algorithm for isosurface table generation in 4D is a direct generalization of the 3D algorithm described in Chapter 5. The only difference is that the triangulation step must ensure that triangulations of adjacent isosurface patches are compatible. As in 3D, we present an algorithm that constructs isosurface patches in any (bounded) four-dimensional convex polytope, not just a hypercube.

6.2.1   Lexicographic Triangulations

As in three dimensions, the boundary of isosurface patches in 4D should align with the boundary of adjacent isosurface patches. In 3D, the boundary of an isosurface patch consists of line segments, so the only issue is matching these line segments. In 4D, the boundary consists of triangulated convex polygons. Not only should the convex polygons in adjacent patches match but the triangulation of these convex polygons should also match. To ensure this matching, we use a specific triangulation that we call a lexicographic triangulation.

Definition 6.4. Point $p = (x_1, \ldots, x_d)$ precedes point $q = (y_1, \ldots, y_d)$ in lexicographic order if, for some $k \leq d$,

- $x_i = y_i$ for all $i < k$,

- $x_k < y_k$.

For instance, point $(5.0, 2.0, 3.0, 6.0)$ precedes point $(5.0, 2.0, 6.0, 2.0)$ in lexicographic order. Lexicographic order is similar to the dictionary ordering of equal length words. Denote p precedes q by $p \prec q$. A sequence of distinct points, $(p_1, \ldots, p_m)$ is listed in lexicographic order if $p_i \prec p_j$ whenever $i < j$.

For a set of simplices τ, let $F^*(\tau)$ be the set τ and all faces of elements of τ. If τ is a triangulation and p is a vertex of some simplex in τ, then $\tau - \{p\}$ is the set of simplices in $F^*(\tau)$ that do not contain vertex p. Let $\text{conv}(p_1, \ldots, p_m)$ denote the convex hull of point set $\{p_1, \ldots, p_m\}$.

The definition of a lexicographic triangulation is recursive.

Definition 6.5. Let $(p_1, \ldots, p_m)$ be a sequence of points in $\mathbb{R}^d$ listed in lexicographic order. A triangulation τ of $\text{conv}(p_1, \ldots, p_m)$ is a lexicographic triangulation if

- τ is a single vertex, or

- $\tau - p_m$ is a lexicographic triangulation of $\text{conv}(p_1, \ldots, p_{m-1})$.

If one includes all the lower-dimensional faces, the lexicographic triangulation of the convex hull of a set of points is unique. More precisely, if τ and τ'

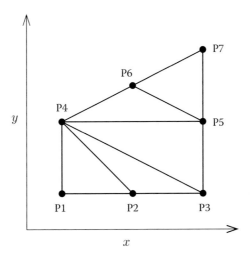

Figure 6.2. A lexicographic triangulation of a point set. The lexicographic order of the vertices is $(p_1, p_2, p_3, p_4, p_5, p_6, p_7)$.

are lexicographic triangulations of $\mathrm{conv}(p_1, \ldots, p_m)$, then $F^*(\tau)$ equals $F^*(\tau')$ (Lemma 6.9).

The dimension of $\mathrm{conv}(p_1, \ldots, p_m)$ may be $k < d$, in which case the largest simplices in τ are k-simplices. Note that the dimension of $\mathrm{conv}(p_1, \ldots, p_{m-1})$ may be less than the dimension of $\mathrm{conv}(p_1, \ldots, p_m)$. Thus, even if τ contains k-simplices, the largest simplices in $\tau - p_m$ could be $(k-1)$-simplices. Figure 6.2 contains an example of a set of points and their lexicographic triangulation.

The definition of lexicographic triangulation extends to the boundary of a convex hull. The boundary of a k-dimensional convex polytope is composed of $(k-1)$ dimensional convex facets. The lexicographic triangulation of $\partial \mathrm{conv}(p_1, \ldots, p_m)$, which is the boundary of $\mathrm{conv}(p_1, \ldots, p_m)$, is the union of the lexicographic triangulations of all the facets. Equivalently, let τ be the lexicographic triangulation of $\mathrm{conv}(p_1, \ldots, p_m)$. The lexicographic triangulation of $\partial \mathrm{conv}(p_1, \ldots, p_m)$ is the triangulation of $\partial \mathrm{conv}(p_1, \ldots, p_m)$ determined by the $(k-1)$-dimensional simplices of τ that lie on $\partial \mathrm{conv}(p_1, \ldots, p_m)$ (Lemma 6.10). It follows from the second definition that the intersection of any two simplices in the lexicographic triangulation of $\partial \mathrm{conv}(p_1, \ldots, p_m)$ is a face of each simplex.

The recursive definition of a lexicographic triangulation leads directly to an incremental algorithm for constructing one. Initially, let τ represent a triangulation consisting of a single point. At the start of the ith iteration, the triangulation forms a lexicographic triangulation of $\mathrm{conv}(p_1, \ldots, p_{i-1})$. With the addition of point p_i, we also add simplices connecting p_i to those simplices on the boundary of $\mathrm{conv}(p_1, \ldots, p_{i-1})$ that are visible from p_i. The enlarged set of simplices forms a triangulation of $\mathrm{conv}(p_1, \ldots, p_i)$.

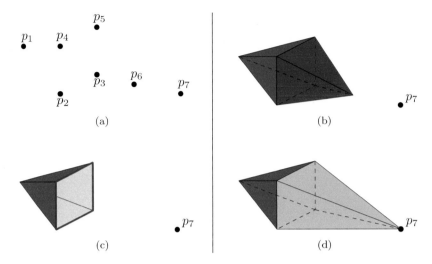

Figure 6.3. Illustration of the INCREMENTAL CONVEX HULL algorithm. (a) Points $(p_1, \ldots, p_7)$ in $\mathbb{R}^3$. (b) Convex hull of $\{p_1, \ldots, p_6\}$. Red simplices are in $\operatorname{conv}(p_1, \ldots, p_6)$ but not in $\operatorname{conv}(p_1, \ldots, p_7)$. (c) Convex hull of $\{p_1, \ldots, p_6\}$, minus deleted red simplices. Deleting red simplices creates a "hole" whose boundary is colored red. (d) Convex hull of $\{p_1, \ldots, p_7\}$. New simplices are colored green.

6.2.2   The Incremental Convex Hull Algorithm

The INCREMENTAL CONVEX HULL algorithm is an algorithm for constructing the convex hull of a set $\{p_1, \ldots, p_m\}$ of points in $\mathbb{R}^d$. More precisely, the algorithm constructs a triangulation of $\partial\operatorname{conv}(p_1, \ldots, p_m)$, the boundary of the convex hull of $\{p_1, \ldots, p_m\}$.

The algorithm starts with the set $\{p_1\}$, which is its own convex hull. It adds points one at a time, constructing a triangulation of $\partial\operatorname{conv}(p_1, \ldots, p_{i+1})$ from the triangulation of $\partial\operatorname{conv}(p_1, \ldots, p_i)$. The algorithm identifies and removes those simplices of $\partial\operatorname{conv}(p_1, \ldots, p_i)$ that are not in $\partial\operatorname{conv}(p_1, \ldots, p_{i+1})$ (Figure 6.3(c)). Removal of the simplices creates a "hole" whose boundary is a set of $(d-2)$-dimensional simplices. The algorithm adds p_{i+1} to each of the boundary simplices to form a set of $(d-1)$-dimensional simplices that fill the hole (Figure 6.3(d)). A detailed 3D version of the algorithm is given in [O'Rourke, 1998] and a higher-dimensional version in [Clarkson et al., 1993].

As described previously, the INCREMENTAL CONVEX HULL algorithm returns a triangulation of the boundary of the convex hull. A variation of the algorithm [Preparata and Shamos, 1985, Edelsbrunner, 1987] returns the facets of the convex hull instead of a triangulation. The variation is called the BENEATH-BEYOND METHOD in [Preparata and Shamos, 1985, Edelsbrunner, 1987]. Returning a triangulation of $\partial\operatorname{conv}(p_1, \ldots, p_m)$ is actually simpler than returning the facets, since the simplices in the triangulation have a simple, uniform description. Because we want a triangulation of isosurface patches, we use the algorithm [O'Rourke, 1998, Clarkson et al., 1993], which returns a triangulation.

The triangulation constructed by the INCREMENTAL CONVEX HULL algorithm is determined by the order in which the points p_i are processed. If points are added in lexicographic order, the algorithm constructs a lexicographic triangulation of the boundary of $\mathrm{conv}(p_1, p_2, \ldots, p_m)$.

6.2.3 Isosurface Patch Construction

Let $\mathbf{c}$ be a polytope in $\mathbb{R}^4$ where each vertex of $\mathbf{c}$ has a positive or a negative label. Let $M_{\mathbf{c}}$ be the midpoints of the bipolar edges of $\mathbf{c}$. We show how to construct an isosurface patch in $\mathbf{c}$ using set $M_{\mathbf{c}}$ as its vertices.

Compute the set $M_{\mathbf{c}}$ of midpoints of the bipolar edges of $\mathbf{c}$. As defined in Section 6.1, set $V_{\mathbf{c}}^+$ is the set of positive vertices of $\mathbf{c}$. Construct the convex hull of $M_{\mathbf{c}} \cup V_{\mathbf{c}}^+$ and the lexicographic triangulation of its boundary. Orient the tetrahedra in this triangulation so that the induced normals point toward the convex hull. Let $T_{\mathbf{c}}^+$ be the set of oriented tetrahedra that do not lie on facets of $\mathbf{c}$. We shall prove (Corollary 6.7) that the vertices of $T_{\mathbf{c}}^+$ are all from $M_{\mathbf{c}}$. The tetrahedra in $T_{\mathbf{c}}^+$ form the triangulated isosurface patch whose vertices are $M_{\mathbf{c}}$. This isosurface patch separates the positive vertices from the negative vertices of $\mathbf{c}$.

We call the algorithm ISOMID4D. Algorithm ISOMID4D is presented in Figure 6.4.

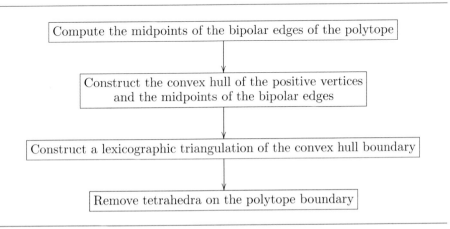

Figure 6.4. Algorithm ISOMID4D for construction of isosurface patches from midpoints of bipolar edges of a 4D polytope.

The combinatorial structure of the isosurface within $\mathbf{c}$ is determined from the configuration of the polytope's vertex labels. The isosurface does not intersect any edge whose endpoints have the same label.

As in three dimensions, instead of using the positive vertices, $V_\mathbf{c}^+$, one could use the negative vertices, $V_\mathbf{c}^-$. Construct the convex hull of $M_\mathbf{c} \cup V_\mathbf{c}^-$ and the lexicographic triangulation of its boundary. Let $T_\mathbf{c}^-$ be the tetrahedra in this triangulation that do not lie on the boundary of $\mathbf{c}$. The tetrahedra in $T_\mathbf{c}^-$ form an alternate isosurface patch for $\mathbf{c}$.

Any tetrahedron in the triangulation of $\mathrm{conv}(M_\mathbf{c} \cup V_\mathbf{c}^+)$ that contains a vertex of $V_\mathbf{c}^+$ is on the boundary of $\mathbf{c}$ and thus not a tetrahedron of $T_\mathbf{c}^+$ (Corollary 6.8). However, the converse is not true. There may be tetrahedra that lie on some facet of $\mathbf{c}$ but whose vertices all come from $M_\mathbf{c}$. Since these tetrahedra lie on a facet of $\mathbf{c}$, they are not included in $T_\mathbf{c}^+$.

A primary requirement is that isosurface patches in adjacent grid elements properly align on their boundaries. As was discussed in Section 2.3.5, if some isosurface patches separate positive vertices while others separate negative vertices, proper alignment may not occur. The isosurface patches separating positive vertices are defined by $T_\mathbf{c}^+$ while the isosurface patches separating negative vertices are defined by $T_\mathbf{c}^-$. Thus, while either $T_\mathbf{c}^+$ or $T_\mathbf{c}^-$ can be used to generate the isosurface lookup tables, only one or the other should be used for all table entries.

Other triangulations such as the one in [Max, 2001] can be used in place of the lexicographic triangulation of the convex hull boundary. The key requirement is that two adjacent isosurface patches have the same triangulation on their common border. In Section 6.3.5, we will show that lexicographic triangulations satisfy this requirement (Lemma 6.18). We prefer lexicographic triangulations because such triangulations are naturally created by the INCREMENTAL CONVEX HULL algorithm when vertices are added in lexicographic order.

6.2.4   Isosurface Table Construction

Let $\mathbf{c}$ be the unit hypercube in $\mathbb{R}^4$ whose vertices are $(0,0,0,0)$, $(1,0,0,0)$, ..., $(1,1,1,1)$. The combinatorial structure of a four-dimensional hypercube is depicted in Figure 6.5. In the isosurface extraction algorithm, the sixteen vertices of $\mathbf{c}$ receive positive and negative labels representing their relationship to the isovalue. A vertex that has scalar value below the isovalue will receive a negative label, "$-$". A vertex with scalar value above or equal to the isovalue receives a positive label, "$+$".

Since each vertex is either positive or negative, there are $2^{16} = 65,536$ configurations of positive and negative vertex labels.[2] The isosurface lookup table contains 2^{16} entries, one for each configuration κ. Each entry is a list of quadru-

---

[2]Under rotational symmetry and reflections, the number of distinct configurations is 222 [Banks et al., 2004].

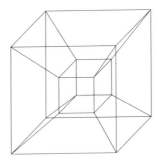

Figure 6.5. Four-dimensional hypercube with 16 vertices, 32 edges, 24 two-dimensional faces, and 8 three-dimensional facets.

ples of bipolar edges of **c**. Each quadruple (e_1, e_2, e_3, e_4) represents an oriented tetrahedron whose vertices lie on e_1 and e_2 and e_3 and e_4. The list of quadruples define the combinatorial structure of the isosurface patch for configuration κ. This isosurface patch is constructed using algorithm IsoMID4D in the previous section. The tetrahedra in this isosurface patch are the set $T_{\mathbf{c}}^+$. The isosurface patch intersects every bipolar edge of **c** exactly once and does not intersect any other hypercube edges.

Two configurations can be adjacent to one another if the face they share in common has the same set of positive and negative vertices. The boundaries of the isosurface patches for each configuration should align on the common face. Note that a 4D hypercube has eight three-dimensional cubes as faces as well as lower-dimensional squares, edges, and points. As in $\mathbb{R}^3$, two reasonable isosurface patches for adjacent configurations can have boundaries that do not align on the common face and so are incompatible. Isosurfaces constructed using such incompatible isosurface patches for adjacent configurations may have "holes" and may not be a 3-manifold.

Compatibility of isosurface patches in $\mathbb{R}^4$ is more subtle than in $\mathbb{R}^3$. The isosurface lookup table stores a set of tetrahedra forming a triangulation (tetrahedralization) of the isosurface patches. If two hypercubes share a three-dimensional cube as a facet, then the intersection between their two isosurface patches is a two-dimensional region (or the empty set.) The triangulation of each isosurface patch determines a triangulation of this two-dimensional region. The two triangulations determined by the two isosurface patches should be the same. Thus triangulations of isosurface patches should be constructed so that the triangulations align on a common face. Triangulation alignment between isosurface patches was not a problem in 3D since isosurface patches in adjacent cubes in 3D intersect in a single line segment.

If triangulations of isosurface patches do not align on a common face, then isosurface tetrahedra will intersect in subsets of faces. Thus, the isosurface tetrahedra will not form a triangulation of the isosurface. More importantly, the

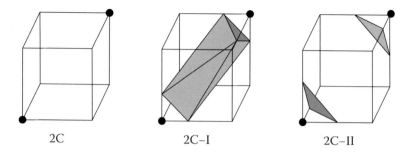

Figure 6.6. Ambiguous 3D configuration 2C. Isosurface patches 2C-I and 2C-II.

geometric position of each isosurface tetrahedron is determined by the location of the isosurface vertices. Depending upon the location of isosurface vertices, "holes" or "pockets" could form in the isosurface or the isosurface could intersect itself. Algorithm IsoMid4D constructs an isosurface lookup table where isosurfaces in any two adjacent configurations are compatible. The isosurface patch boundaries and their triangulations align on the common face of any two adjacent configurations.

6.2.5 Opposite Positive Vertices

Configuration 2C in 3D has exactly two positive vertices on opposite corners of the cube (Figure 6.6). As discussed in Section 5.2, this configuration is special. This configuration has two "natural" isosurface patches, one consisting of a cylinder connecting the two positive vertices (2C-I in Figure 6.6) and one consisting of two triangles splitting each of the positive vertices from the rest of the cube (2C-II in Figure 6.6). Because the intersection of each of these isosurface patches and the cube boundary is exactly the same, either isosurface patch can be used without creating any inconsistencies with adjacent configurations. Algorithm IsoMid3D generates a cylindrical isosurface patch around those two vertices but this cylindrical patch can be replaced by two triangles.

A similar property holds in 4D when exactly two positive vertices are on opposite corners of the hypercube in 4D. (See Figure 6.7(a).) Algorithm IsoMid4D generates a "cylindrical"[3] isosurface patch around those two vertices. This cylindrical patch can be replaced by two tetrahedra splitting the two positive vertices from the fourteen negative ones. Since no hypercube facet contains these two vertices, the replacement by two tetrahedra does not change the boundary of the isosurface patch and so is compatible with the isosurface patches in all the other configurations. There are eight configurations with exactly two positive vertices on opposite corners of the hypercube.

---

[3]The isosurface patch will be homeomorphic to $\mathbb{S}^2 \times [0, 1]$.

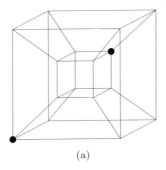 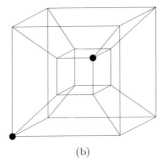

(a)   (b)

Figure 6.7. (a) Positive vertices (black dots) at opposite corners of the hypercube. (b) Positive vertices (black dots) at opposite corners of a 3D facet of the hypercube.

Note that the two positive vertices must lie on opposite corners of the 4D hypercube, not of some 3D facet of the hypercube. (See Figure 6.7(b).) Switching the isosurface for two positive tetrahedra splitting off opposite corners of a 3D facet can create inconsistencies with adjacent configurations.

6.3   Marching Hypercubes

6.3.1   Algorithm

The basic steps of the MARCHING HYPERCUBES algorithm are the same as the steps of the 3D MARCHING CUBES algorithm. Just as a three-dimensional regular grid is composed of three-dimensional cubes, a four-dimensional regular grid is composed of four-dimensional hypercubes. The algorithm iterates over the hypercubes of the four-dimensional grid.

Input to the MARCHING HYPERCUBES algorithm is an isovalue and a set of scalar values at the vertices of a four-dimensional regular grid. The algorithm has three steps. (See Figure 6.8.) Read the isosurface lookup table from a preconstructed data file. For each grid hypercube, retrieve from the lookup table a set of oriented isosurface tetrahedra representing the combinatorial structure of the isosurface. The vertices of these tetrahedra form the isosurface vertices. Assign geometric locations to the isosurface vertices based on the scalar values at the hypercube edge endpoints. Construction of the MARCHING HYPERCUBES lookup table was described in Section 6.2.4. We explain the last two steps of the algorithm below.

The isosurface lookup table is constructed on the unit hypercube with vertices $(0,0,0,0), (1,0,0,0), (0,1,0,0), \ldots, (0,1,1,1), (1,1,1,1)$. To construct isosurface

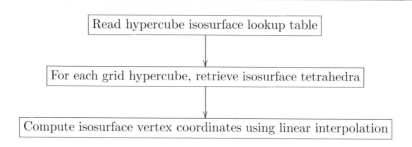

Figure 6.8. MARCHING HYPERCUBES.

edges in grid hypercube (i_x, i_y, i_z, i_w), we have to map unit hypercube edges to edges of hypercube (i_x, i_y, i_z, i_w). Each vertex $v = (v_x, v_y, v_z, v_w)$ of the unit hypercube maps to

$$v + (i_x, i_y, i_z, i_w) = (v_x, v_y, v_z, v_w) + (i_x, i_y, i_z, i_w)$$
$$= (v_x + i_x, v_y + i_y, v_z + i_z, v_w + i_w).$$

Each edge $\mathbf{e}$ of the unit square with endpoints (v, v') maps to edge

$$\mathbf{e} + (i_x, i_y, i_z, i_w) = (v + (i_x, i_y, i_z, i_w), v' + (i_x, i_y, i_z, i_w)).$$

Finally, each edge quadruple $(\mathbf{e}_1, \mathbf{e}_2, \mathbf{e}_3, \mathbf{e}_4)$ maps to

$$(\mathbf{e}_1 + (i_x, i_y, i_z, i_w), \mathbf{e}_2 + (i_x, i_y, i_z, i_w), \mathbf{e}_3 + (i_x, i_y, i_z, i_w), \mathbf{e}_4 + (i_x, i_y, i_z, i_w)).$$

The isosurface lookup table determines only the edge containing each isosurface vertex. The geometric location of the isosurface vertex along that edge is not defined by the lookup table.

The vertices of the isosurface tetrahedra are the isosurface vertices. To map each isosurface tetrahedra to a geometric tetrahedron, we use linear interpolation to position the isosurface vertices as described in Section 1.7.2. Each isosurface vertex v lies on a grid edge $[p, q]$. If s_p and s_q are the scalar values at p and q and σ is the isovalue, then map v to $(1 - \alpha)p + \alpha q$ where $\alpha = (\sigma - s_p)/(s_q - s_p)$.

The MARCHING HYPERCUBES algorithm is presented in Algorithm 6.1. Function `LinearInterpolation`, called by this algorithm, is defined in Algorithm 1.1 in Section 1.7.2.

6.3.2 Isosurface Properties

Properties for the MARCHING HYPERCUBES isosurface are similar to those for the MARCHING CUBES isosurface. Under appropriate conditions the isosurface is a 3-manifold with boundary.

Input : F is a 4D array of scalar values.
Coord is a 4D array of (x, y, z, w) coordinates.
σ is an isovalue.
Result : A set Υ of isosurface tetrahedra.

MarchingHypercubes(F, Coord, σ, Υ)

1 Read Marching Hypercubes lookup table into Table;
/* Assign "+" or "−" signs to each vertex */
2 foreach grid vertex (i_x, i_y, i_z, i_w) do
3 | if F$[i_x, i_y, i_z, i_w] < \sigma$ then Sign$[i_x, i_y, i_z, i_w] \leftarrow$ "−";
4 | else Sign$[i_x, i_y, i_z, i_w] \leftarrow$ "+"; /* F$[i_x, i_y, i_z, i_w] \geq \sigma$ */
5 end
6 T $\leftarrow \emptyset$;
/* For each grid hypercube, retrieve isosurface tetrahedra */
7 foreach grid hypercube (i_x, i_y, i_z, i_w) do
8 | $I \leftarrow (i_x, i_y, i_z, i_w)$;
/* Cube vertices are $I, I + (1,0,0,0), \ldots, I + (1,1,1,1)$ */
9 | $\kappa \leftarrow$ (Sign$[I]$, Sign$[I + (1,0,0,0)], \ldots,$ Sign$[I + (1,1,1,1)])$;
10 | foreach edge quadruple $(e_1, e_2, e_3, e_4) \in$ Table$[\kappa]$ do
11 | | Insert edge quadruple $(e_1 + I, e_2 + I, e_3 + I, e_4 + I)$ into T;
12 | end
13 end
/* Compute isosurface vertex coordinates using linear interpolation */
14 foreach bipolar grid edge e with endpoints $I = (i_x, i_y, i_z, i_w)$ and $J = (j_x, j_y, j_z, j_w)$ do
15 | $w_e \leftarrow$ LinearInterpolation (Coord$[I]$, F$[I]$, Coord$[J]$, F$[J]$, σ);
16 end
/* Convert T to set of tetrahedra */
17 $\Upsilon \leftarrow \emptyset$;
18 foreach quadruple of edges $(e_1, e_2, e_3, e_4) \in$ T do
19 | $\Upsilon \leftarrow \Upsilon \cup \{(w_{e_1}, w_{e_2}, w_{e_3}, w_{e_4})\}$;
20 end

Algorithm 6.1. MARCHING HYPERCUBES.

MARCHING HYPERCUBES returns a finite set, Υ, of tetrahedra. The isosurface is the union of these tetrahedra. The vertices of the isosurface are the tetrahedra vertices.

The following properties apply to all isosurfaces produced by the MARCHING HYPERCUBES algorithm.

Property 1. *The isosurface is piecewise linear.*

Property 2. *The vertices of the isosurface lie on grid edges.*

Property 3.   *The isosurface intersects every bipolar grid edge at exactly one point.*

Property 4.   *The isosurface does not intersect any negative or strictly positive grid edges.*

Property 5.   *The isosurface separates positive grid vertices from negative ones and strictly separates strictly positive grid vertices from negative grid vertices.*

Properties 3 and 4 imply that the isosurface intersects a minimum number of grid edges. As with MARCHING CUBES, if both endpoints of a grid edge have scalar value equal to the isovalue, then the isosurface may intersect the grid edge zero, one, or two times or may contain the grid edge.

By Property 3, the isosurface intersects every bipolar grid edge. However, the bipolar grid edge may be intersected by zero-volume isosurface tetrahedra.

The following properties apply to the MARCHING HYPERCUBES isosurfaces whose isovalues do not equal the scalar value of any grid vertex.

Property 6.   *The isosurface is a piecewise linear, orientable 3-manifold with boundary.*

Property 7.   *The boundary of the isosurface lies on the boundary of the grid.*

Property 8.   *Set Υ does not contain any zero-volume tetrahedra or duplicate tetrahedra, and the tetrahedra in Υ form a triangulation of the isosurface.*

6.3.3   Proof of Isosurface Properties 1–4

Most of the proofs of the MARCHING HYPERCUBES isosurface properties are simply 4D versions of the 3D proofs of the MARCHING POLYHEDRA isosurface properties presented in Sections 5.4.2–5.4.4. We highly recommend that the reader review the proofs in that section before looking at the 4D versions here. For lemmas with proofs that are exactly the same, we have omitted the proofs and refer back to the proofs in Sections 5.4.2–5.4.4.

Isosurface vertex sets and proper isosurface vertex sets were defined in Section 6.1. As with the proofs for MARCHING POLYHEDRA, we state the lemmas in this section and the following three sections in the general setting of isosurface vertex sets instead of restricting ourselves to the set of midpoints used in ISOMID4D.

Let $\mathbf{c}$ be a convex polytope in $\mathbb{R}^4$. Let $U_{\mathbf{c}}$ be an isosurface vertex set of $\mathbf{c}$. Let $R_{\mathbf{c}}^+(U_{\mathbf{c}})$ equal $\text{conv}(U_{\mathbf{c}} \cup V_{\mathbf{c}}^+)$, the convex hull of $U_{\mathbf{c}}$ and the positive vertices of $\mathbf{c}$. Construct the lexicographic triangulation of $\partial R_{\mathbf{c}}^+(U_{\mathbf{c}})$, which is the boundary of $R_{\mathbf{c}}^+(U_{\mathbf{c}})$. Let $T_{\mathbf{c}}^+(U_{\mathbf{c}})$ be the tetrahedra in the lexicographic triangulation of $\partial R_{\mathbf{c}}^+(U_{\mathbf{c}})$ that do not lie on the boundary of $\mathbf{c}$. Since $T_{\mathbf{c}}^+(U_{\mathbf{c}})$ is the subset of a triangulation, set $T_{\mathbf{c}}^{+}(U_{\mathbf{c}})$ is itself a triangulation. We will often abbreviate $R_{\mathbf{c}}^{+}(U_{\mathbf{c}})$ and $T_{\mathbf{c}}^{+}(U_{\mathbf{c}})$, as $R_{\mathbf{c}}^{+}$ and $T_{\mathbf{c}}^{+}$, respectively.

The set $M_{\mathbf{c}}$ of midpoints of bipolar edges of $\mathbf{c}$ is a proper isosurface vertex set of $\mathbf{c}$. Abbreviate $R_{\mathbf{c}}^{+}(M_{\mathbf{c}})$ as $R_{\mathbf{c}}^{M+}$ and $T_{\mathbf{c}}^{+}(M_{\mathbf{c}})$ as $T_{\mathbf{c}}^{M+}$.

Property 1. *The isosurface is piecewise linear.*

Property 2. *The vertices of the isosurface lie on grid edges.*

Proof of Properties 1 & 2: The MARCHING HYPERCUBES isosurface is the union of a finite set of tetrahedra, so it is piecewise linear. By construction, the vertices of these tetrahedra lie on the grid edges. $\qquad\square$

To prove Property 3, we need to show that $U_{\mathbf{c}}$ is the set of vertices of $T_{\mathbf{c}}^{+}(U_{\mathbf{c}})$. We first show that the set of vertices of $T_{\mathbf{c}}^{+}(U_{\mathbf{c}})$ is a subset of $U_{\mathbf{c}}$. The proof of the following lemma is almost the same as the proof of Lemma 5.5 from Section 5.4.2, with the separating plane replaced by a separating hyperplane.

Lemma 6.6. *Let $\mathbf{c}$ be a convex polytope in $\mathbb{R}^4$ where each vertex of $\mathbf{c}$ has a positive or a negative label. If $U_{\mathbf{c}}$ is an isosurface vertex set for $\mathbf{c}$, then the set of vertices of $T_{\mathbf{c}}^{+}(U_{\mathbf{c}})$ is a subset of $U_{\mathbf{c}}$.*

Proof: Let w be a vertex of $T_{\mathbf{c}}^{+} = T_{\mathbf{c}}^{+}(U_{\mathbf{c}})$. The vertices of $R_{\mathbf{c}}^{+} = \operatorname{conv}(U_{\mathbf{c}} \cup V_{\mathbf{c}}^{+})$ are $U_{\mathbf{c}} \cup V_{\mathbf{c}}^{+}$. Since $T_{\mathbf{c}}^{+}(U_{\mathbf{c}})$ is a subset of a triangulation of $\partial R_{\mathbf{c}}^{+}$ that does not add any vertices, vertex w is in $U_{\mathbf{c}} \cup V_{\mathbf{c}}^{+}$.

If w is not in $V_{\mathbf{c}}^{+}$, then w must be in $U_{\mathbf{c}}$, establishing the claim. Assume that vertex w is in $V_{\mathbf{c}}^{+}$.

Since w is in $T_{\mathbf{c}}^{+}$, it must be incident on some tetrahedron $\mathbf{t}$ that is not on a facet of $\mathbf{c}$. Let h be the hyperplane containing $\mathbf{t}$. Since h intersects the interior of $\mathbf{c}$ and h contains w, there is some edge (w, v) of $\mathbf{c}$ that is separated from $R_{\mathbf{c}}^{+}$ by h and does not lie on h. Thus,

$$(w, v) \cap R_{\mathbf{c}}^{+} = w.$$

Since vertex v is not in $R_{\mathbf{c}}^{+}$, vertex v must have a negative label. Thus, (w, v) is a bipolar edge. By definition, the isosurface vertex set $U_{\mathbf{c}}$ contains some point p on (w, v). Since p is in $R_{\mathbf{c}}^{+}$ and $(w, v) \cap R_{\mathbf{c}}^{+}$ equals w, point p must equal w. Thus w is in $U_{\mathbf{c}}$, proving the claim. $\qquad\square$

Proofs of the following corollaries to Lemma 6.6 are the same as the proofs to the corollaries of Lemma 5.5 in Section 5.4.2 and are omitted.

Corollary 6.7. *Let $\mathbf{c}$ be a convex polytope in $\mathbb{R}^4$ where each vertex of $\mathbf{c}$ has a positive or a negative label. If $U_{\mathbf{c}}$ is a proper isosurface vertex set for $\mathbf{c}$, then the set of vertices of $T_{\mathbf{c}}^{+}(U_{\mathbf{c}})$ equals $U_{\mathbf{c}}$.*

Corollary 6.8. *Let* **c** *be a convex polytope in* $\mathbb{R}^4$ *where each vertex of* **c** *has a positive or a negative label. If* $U_{\mathbf{c}}$ *is a proper isosurface vertex set for* **c**, *then no vertex of* **c** *is in the set of vertices of* $T_{\mathbf{c}}^+(U_{\mathbf{c}})$.

Let μ be the translation that maps each grid hypercube **c** to the unit hypercube used in constructing the isosurface lookup table. If the vertices of **c** are labeled "+" or "−", then vertices of $\mu(\mathbf{c})$ receive "+" or "−" labels from the corresponding vertices of **c**. Thus, $\mu(v)$ is positive if and only if v is positive and $\mu(v)$ is negative if and only if v is negative.

Property 3. *The isosurface intersects every bipolar grid edge at exactly one point.*

Property 4. *The isosurface does not intersect any negative or strictly positive grid edges.*

Proof of Properties 3 & 4: The isosurface is constructed by retrieving tetrahedra from a lookup table and then determining their vertex positions using linear interpolation. Let **c** be a grid hypercube containing a grid edge **e**. Vertices of **c** are labeled "+" or "−" based on the relationship of their scalar value to the isovalue. The isosurface patch for $\mu(\mathbf{c})$ is a subset of $\text{conv}(M_{\mu(\mathbf{c})} \cup V_{\mu(\mathbf{c})}^+)$, the convex hull of the midpoints of bipolar edges and positive vertices of $\mu(\mathbf{c})$.

Since **c** is convex, only isosurface tetrahedra with vertices on **e** intersect **e**. By Corollary 6.8, the vertices of $T_{\mu(\mathbf{c})}^+(M_{\mu(\mathbf{c})})$ (abbreviated $T_{\mu(\mathbf{c})}^{M+}$) are midpoints of edges of $\mu(\mathbf{c})$. Thus, the isosurface patch in $\mu(\mathbf{c})$ intersects $\mu(\mathbf{e})$ at most once.

If **e** is bipolar, then so is $\mu(\mathbf{e})$. The midpoint of $\mu(\mathbf{e})$ is in $M_{\mu(\mathbf{c})}$. Set $T_{\mu(\mathbf{c})}^{M+}$ contains some tetrahedron containing this midpoint. Thus, the isosurface patch in $\mu(\mathbf{c})$ intersects $\mu(\mathbf{e})$ at exactly one point p. Point $p \in \mu(\mathbf{e})$ maps to a single point on **e** determined by linear interpolation. Thus the isosurface intersects **e** at exactly one point.

If **e** is negative or strictly positive, then no isosurface vertex lies on **e**. Thus the isosurface does not intersect negative or strictly positive grid edges. □

6.3.4 Lexicographic Triangulation Properties

Before proving the rest of the isosurface properties, we need to state and prove two lemmas about lexicographic triangulations.

Let P be a finite set of points in $\mathbb{R}^d$. We claim that the convex hull of a finite set of points P has only one lexicographic triangulation when one includes all lower-dimensional faces in the triangulation. As previously defined, for a set τ of simplices, $F^*(\tau)$ is the set τ and all the faces of elements of τ.

Lemma 6.9. *If P is a finite set of points and τ and τ' are lexicographic triangulations of* $\text{conv}(P)$, *then $F^*(\tau)$ equals $F^*(\tau')$.*

Proof: The proof is by induction on the number of points in P.

If P has exactly one point, then $F^*(\tau)$ and $F^*(\tau')$ consist of the zero-dimensional simplex p_1 and the lemma is trivially true.

Assume that $F^*(\tau_m)$ equals $F^*(\tau'_m)$ for any two lexicographic triangulations τ_m and τ'_m of the convex hull of m points. We wish to show that $F^*(\tau_{m+1})$ equals $F^*(\tau'_{m+1})$ for any two lexicographic triangulations τ_{m+1} and τ'_{m+1} of the convex hull of $m + 1$ points.

Let $\{p_1, p_2, \ldots, p_{m+1}\}$ be a set of $m + 1$ points listed in lexicographic order. Let τ_{m+1} and τ'_{m+1} be lexicographic triangulations of $\text{conv}(p_1, \ldots, p_{m+1})$. Let τ_m and τ'_m equal $\tau_{m+1} - \{p_{m+1}\}$ and $\tau'_{m+1} - \{p_{m+1}\}$, respectively. By definition, τ_m and τ'_m are lexicographic triangulations of $\text{conv}(p_1, p_2, \ldots, p_m)$. By the inductive assumption, $F^*(\tau_m)$ equals $F^*(\tau'_m)$.

If p_{m+1} is in $\text{conv}(p_1, p_2, \ldots, p_m)$, then $F^*(\tau_{m+1})$ equals $F^*(\tau_m)$ and $F^*(\tau'_{m+1})$ equals $F^*(\tau'_m)$. Thus, $F^*(\tau_{m+1})$ equals $F^*(\tau'_{m+1})$.

Assume p_{m+1} is not in $\text{conv}(p_1, p_2, \ldots, p_m)$. A simplex $\mathbf{t}_1$ is in $F^*(\tau_{m+1})$ if and only if $\mathbf{t}_1$ is in $F^*(\tau_m)$ or $\mathbf{t}_1$ equals $\text{conv}(p_{m+1} \cup \mathbf{t}_0)$ where $\mathbf{t}_0 \in F^*(\tau_m)$ and $\mathbf{t}_0$ is visible from p_{m+1}. Since $F^*(\tau_m)$ equals $F^*(\tau'_m)$, any simplex in $F^*(\tau_m)$ that is visible from p_{m+1} is also a simplex in $F^*(\tau'_m)$ that is visible from p_{m+1} and vice versa. Thus $F^*(\tau_{m+1})$ equals $F^*(\tau'_{m+1})$, proving the claim. $\qquad\square$

Let P be a finite set of points in $\mathbb{R}^d$. If $\mathbf{f}$ is a face of $\text{conv}(P)$ and τ is a triangulation of $\text{conv}(P)$, let $\tau \cap \mathbf{f}$ denote the set $\{\mathbf{t} \cap \mathbf{f} : \mathbf{t} \in \tau \text{ and } \mathbf{t} \cap \mathbf{f} \neq \emptyset\}$.

Lemma 6.10. *Let P be a finite set of points in $\mathbb{R}^d$, and let $\mathbf{f}$ be a face of $\text{conv}(P)$. If τ is a lexicographic triangulation of $\text{conv}(P)$ and τ' is a lexicographic triangulation of $\text{conv}(P \cap \mathbf{f})$, then $F^*(\tau) \cap \mathbf{f}$ equals $F^*(\tau')$.*

Proof: The proof is by induction on the number of points in P.

If P has exactly one point, then τ consists of the zero-dimensional simplex p_1 whose only face is p_1 and the lemma is trivially true.

Assume the lemma is true when P has m points. We wish to show that the the lemma is true when P has $m + 1$ points.

Let $P_{m+1} = \{p_1, p_2, \ldots, p_{m+1}\}$ be a set of $m+1$ points listed in lexicographic order. Let $P_m = \{p_1, p_2, \ldots, p_m\}$ be the first m points in P_{m+1}. Let τ_m and τ_{m+1} be lexicographic triangulations of $\text{conv}(P_m)$ and $\text{conv}(P_{m+1})$, respectively. Let $\mathbf{f}$ be a face of $\text{conv}(P_{m+1})$. Let τ'_m and τ'_{m+1} be lexicographic triangulations of $\text{conv}(P_m \cap \mathbf{f})$ and $\text{conv}(P_{m+1} \cap \mathbf{f})$, respectively.

If p_{m+1} is not in $\mathbf{f}$, then $\mathbf{f}$ is also a face of $\text{conv}(P_m)$. By Lemma 6.9, set $F^*(\tau'_{m+1})$ equals $F^*(\tau'_m)$ and $F^*(\tau_{m+1}) \cap \mathbf{f}$ equals $F^*(\tau_m) \cap \mathbf{f}$. By the inductive assumption, $F^*(\tau'_m)$ equals $F^*(\tau_m) \cap \mathbf{f}$. Thus, $F^*(\tau'_{m+1})$ equals $F^*(\tau_{m+1}) \cap \mathbf{f}$.

Assume p_{m+1} is in $\mathbf{f}$. Let $\mathbf{g}$ equal $\mathbf{f} \cap \text{conv}(P_m)$. If $\mathbf{g}$ is the empty set, then $\text{conv}(P_{m+1} \cap \mathbf{f})$ is just the point p_{m+1} and $\mathbf{f}$ is the point p_{m+1}. Thus both $\tau_{m+1} \cap \mathbf{f}$ and τ'_{m+1} equal p_{m+1} and so $F^*(\tau_{m+1}) \cap \mathbf{f}$ equals $F^*(\tau'_{m+1})$.

Assume $\mathbf{g}$ is not the empty set. Let $\mathbf{t}_1$ be a simplex in $F^*(\tau_{m+1})$ containing p_{m+1}. Let $\mathbf{t}_0$ be the facet of $\mathbf{t}_1$ that does not contain p_{m+1}. By Lemma 6.9, facet

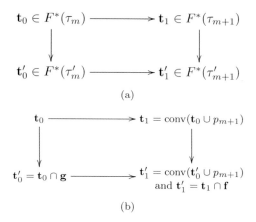

(a)

(b)

Figure 6.9. (a) Triangulations τ_m and τ_{m+1} are lexicographic triangulations of conv(P_m) and conv(P_{m+1}), respectively. Triangulations τ'_m and τ'_{m+1} are lexicographic triangulations of conv$(P_m \cap \mathbf{g})$ and conv$(P_{m+1} \cap \mathbf{f})$, respectively. (b) Simplices $\mathbf{t}_1$ and $\mathbf{t}'_1$ equal conv$(\mathbf{t}_0 \cup p_{m+1})$ and conv$(\mathbf{t}'_0 \cup p_{m+1})$, respectively. Simplices $\mathbf{t}'_0$ and $\mathbf{t}'_1$ equal $\mathbf{t}_0 \cap \mathbf{g}$ and $\mathbf{t}_1 \cap \mathbf{f}$, respectively.

$\mathbf{t}_0$ is a simplex in $F^*(\tau_m)$. Let $\mathbf{t}'_0$ equal $\mathbf{t}_0 \cap \mathbf{g}$. By the inductive assumption, $\mathbf{t}'_0$ is in $F^*(\tau'_m)$. By Lemma 6.9, $\mathbf{t}'_1 = \text{conv}(\mathbf{t}'_0 \cup p_{m+1})$ is in $F^*(\tau'_{m+1})$. Since $\mathbf{t}_1 \cap \mathbf{f}$ equals conv$(\mathbf{t}'_0 \cup \{p_{m+1}\})$, the simplex $\mathbf{t}_1 \cap \mathbf{f}$ equals $\mathbf{t}'_1$. (See Figure 6.9.) This proves that $F^*(\tau_{m+1}) \cap \mathbf{f}$ is a subset of $F^*(\tau'_{m+1})$.

Conversely, let $\mathbf{t}'_1$ be a simplex in $F^*(\tau'_{m+1})$ containing p_{m+1}. Let $\mathbf{t}'_0$ be the facet of $\mathbf{t}'_1$ which does not contain p_{m+1}. By Lemma 6.9, facet $\mathbf{t}'_0$ is a simplex in $F^*(\tau'_m)$. By the inductive assumption, simplex $\mathbf{t}'_0$ equals $\mathbf{t}_0 \cap \mathbf{g}$ for some $\mathbf{t}_0 \in F^*(\tau_m)$. By Lemma 6.9, $\mathbf{t}_1 = \text{conv}(\mathbf{t}_0 \cup p_{m+1})$ is a simplex of $F^*(\tau_{m+1})$. Since $\mathbf{t}_1 \cap \mathbf{f}$ equals conv$(\mathbf{t}'_0 \cup \{p_{m+1}\})$, the simplex $\mathbf{t}_1 \cap \mathbf{f}$ equals $\mathbf{t}'_1$. Thus $F^*(\tau'_{m+1})$ is a subset of $F^*(\tau_{m+1}) \cap \mathbf{f}$.

Since $F^*(\tau_{m+1}) \cap \mathbf{f}$ is a subset of $F^*(\tau'_{m+1})$ and $F^*(\tau'_{m+1})$ is a subset of $F^*(\tau_{m+1}) \cap \mathbf{f}$, triangulation $F^*(\tau'_{m+1})$ equals $F^*(\tau_{m+1}) \cap \mathbf{f}$.   □

6.3.5   Proof of Isosurface Property 5

The MARCHING HYPERCUBES algorithm retrieves isosurface patches for a grid hypercube $\mathbf{c}$ from the isosurface lookup table for the unit hypercube. The algorithm then embeds those isosurface patches using linear interpolation on the isosurface vertices. Isosurface patches in the lookup table are constructed by applying Algorithm IsoMid4D to the unit hypercube. In place of using the lookup table, we could have constructed isosurface patches by applying IsoMid4D di-

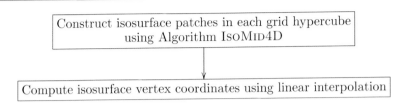

Figure 6.10. Variation of MARCHING HYPERCUBES without a lookup table.

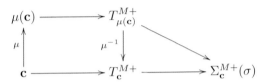

Figure 6.11. The MARCHING HYPERCUBES algorithm constructs the isosurface patch $\Sigma_{\mathbf{c}}^{M+}(\sigma)$ by mapping cube $\mathbf{c}$ to $\mu(\mathbf{c})$, retrieving $T_{\mu(\mathbf{c})}^{M+}$ from the isosurface lookup table and mapping $T_{\mu(\mathbf{c})}^{M+}$ to $\Sigma_{\mathbf{c}}^{M+}(\sigma)$ using linear interpolation. The isosurface patch $\Sigma_{\mathbf{c}}^{M+}(\sigma)$ could be constructed directly from $\mathbf{c}$ by computing $T_{\mathbf{c}}^{M+}$ directly on $\mathbf{c}$ and then mapping $T_{\mathbf{c}}^{M+}$ to $\Sigma_{\mathbf{c}}^{M+}(\sigma)$.

rectly to each grid hypercube $\mathbf{c}$. (See Figure 6.10.) We argue that doing so would give the exact same isosurface.

If $\mu : \mathbb{R}^4 \to \mathbb{R}^4$ is a translation in $\mathbb{R}^4$, then $\mu(\text{conv}(P))$ equals $\text{conv}(\mu(P))$ for any point set $P \subseteq \mathbb{R}^4$. Thus, $\mu(R_{\mathbf{c}}^{M+})$ equals $R_{\mu(\mathbf{c})}^{M+}$ and $\mu(T_{\mathbf{c}}^{M+})$ equals $T_{\mu(\mathbf{c})}^{M+}$. Applying μ^{-1} to $T_{\mu(\mathbf{c})}^{M+}$ gives a triangulation $T_{\mathbf{c}}^{M+}$ of $S_{\mathbf{c}}^{M+}$, which is isomorphic to $T_{\mu(\mathbf{c})}^{M+}$. (See Figure 6.11.)

MARCHING HYPERCUBES embeds $T_{\mu(\mathbf{c})}^{M+}$ in $\mathbf{c}$ using linear interpolation on the isosurface vertices. The location of the isosurface vertices depends upon the isovalue σ. Let $\Sigma_{\mathbf{c}}^{M+}(\sigma)$ (abbreviated $\Sigma_{\mathbf{c}}^{M+}$) represent the resulting isosurface patch for $\mathbf{c}$.

Because $T_{\mu(\mathbf{c})}^{M+}$ and $T_{\mathbf{c}}^{M+}$ are isomorphic, the algorithm could have constructed $\Sigma_{\mathbf{c}}^{M+}$ from $T_{\mathbf{c}}^{M+}$ by reembedding $T_{\mathbf{c}}^{M+}$ in $\mathbf{c}$ using linear interpolation on its vertices. Each vertex v_i of $T_{\mathbf{c}}^{M+}$ is the midpoint m_i of some edge $\mathbf{e}_i$. Vertex v_i is remapped to the interpolated point $w_i \in \mathbf{e}_i$. Map each vertex v_i of $T_{\mathbf{c}}^{M+}$ to $(1 - \alpha)v_i + \alpha w_i$. Linearly extend the mapping to each tetrahedron in $T_{\mathbf{c}}^{M+}$.

The algorithm uses $\mu(\mathbf{c})$ to construct $\Sigma_{\mathbf{c}}^{M+}$ because it is faster to retrieve the precomputed values of $T_{\mu(\mathbf{c})}^{M+}$ from the lookup table than to compute $T_{\mathbf{c}}^{M+}$. However, for the proofs of Properties 5–8 we will view $\Sigma_{\mathbf{c}}^{M+}$ as a homotopy of $T_{\mathbf{c}}^{M+}$.

To prove Property 5, it is necessary to show that there are no cracks between isosurface patches in adjacent hypercubes. We first show that for each cube $\mathbf{c}$, set $|T_{\mathbf{c}}^{M+}|$ separates the positive vertices of $\mathbf{c}$ from the negative ones (Corollary 6.14). ($|T_{\mathbf{c}}^{M+}|$ represents the union of all the tetrahedra of $T_{\mathbf{c}}^{M+}$.) We next show that for two adjacent cubes, $\mathbf{c}_1$ and $\mathbf{c}_2$, region $R_{\mathbf{c}_1}^{M+}$ and $R_{\mathbf{c}_2}^{M+}$ agree on their adjacent boundaries (Corollary 6.17). This implies that there are no "cracks" between $|T_{\mathbf{c}_1}^{M+}|$ and $|T_{\mathbf{c}_2}^{M+}|$ in adjacent cubes and that $\bigcup_{\mathbf{c} \in \Gamma} |T_{\mathbf{c}}^{M+}|$ separates the positive vertices of Γ from the negative ones. We also show that the triangulations $T_{\mathbf{c}_1}^{M+}$ and $T_{\mathbf{c}_2}^{M+}$ agree on their adjacent boundaries, i.e., $\mathbf{t}_1 \cap \mathbf{t}_2$ is a face of $\mathbf{t}_1$ and a face of $\mathbf{t}_2$ for every $\mathbf{t}_1 \in T_{\mathbf{c}_1}^{M+}$ and $\mathbf{t}_2 \in T_{\mathbf{c}_2}^{M+}$ (Lemma 6.18). Thus, $\bigcup_{\mathbf{c} \in \Gamma} T_{\mathbf{c}}^{M+}$ is a triangulation (Lemma 6.22). We define a homotopy map η, which maps this triangulation to the isosurface. Since this homotopy map does not cross any vertices, the resulting isosurface separates the positive vertices of Γ from the negative ones.

As previously noted, we present the lemmas in this section in the more general setting of isosurface vertex sets instead of restricting the lemmas to the midpoints of bipolar edges. Some of the lemmas apply only to isosurface vertex sets that are proper, while others have no such restriction.

We start by showing that if $U_{\mathbf{c}}$ is a proper isosurface vertex set, then $R_{\mathbf{c}}^+(U_{\mathbf{c}})$ is four-dimensional.

Lemma 6.11. *Let $\mathbf{c}$ be a convex polytope in $\mathbb{R}^4$ where each vertex of $\mathbf{c}$ has a positive or a negative label. If $U_{\mathbf{c}}$ is a proper isosurface vertex set of $\mathbf{c}$ and at least one vertex of $\mathbf{c}$ is positive, then $R_{\mathbf{c}}^+(U_{\mathbf{c}})$ is a four-dimensional convex set.*

Proof: Let v be a positive vertex of $\mathbf{c}$. Since v is positive, some interior point of every edge incident on v is in $R_{\mathbf{c}}^+$. Since there are at least four edges incident on v and these edges are not contained in any three-dimensional subspace, $R_{\mathbf{c}}^+$ is four-dimensional. Since $R_{\mathbf{c}}^+$ is the convex hull of the set $U_{\mathbf{c}} \cup V_{\mathbf{c}}^+$, set $R_{\mathbf{c}}^+$ is convex. $\qquad\square$

Corollary 6.12. *Let $\mathbf{c}$ be a convex polytope in $\mathbb{R}^4$ where each vertex of $\mathbf{c}$ has a positive or a negative label. If $U_{\mathbf{c}}$ is a proper isosurface vertex set of $\mathbf{c}$, then $R_{\mathbf{c}}^+(U_{\mathbf{c}})$ is either a four-dimensional convex set or the empty set.*

Proof: By definition, points in $U_{\mathbf{c}}$ do not lie on negative edges of $\mathbf{c}$. Thus, if $V_{\mathbf{c}}^+$ is empty, then $U_{\mathbf{c}}$ must also be empty. By Lemma 6.11, if $V_{\mathbf{c}}^+$ is not empty, then $R_{\mathbf{c}}^+(U_{\mathbf{c}})$ is not empty. $\qquad\square$

Lemma 6.11 and its corollary require $U_{\mathbf{c}}$ to be proper. When $U_{\mathbf{c}}$ is an isosurface vertex set that is not proper, $R_{\mathbf{c}}^+(U_{\mathbf{c}})$ can have any dimension from zero to three.

Let $\mathbf{c}$ be a convex polytope in $\mathbb{R}^4$ where each vertex of $\mathbf{c}$ has a positive or a negative label and let $U_{\mathbf{c}}$ be an isosurface vertex set (not necessarily proper) of $\mathbf{c}$. We define $S_{\mathbf{c}}^+(U_{\mathbf{c}})$ (abbreviated $S_{\mathbf{c}}^+$) to be the set that separates $R_{\mathbf{c}}^+(U_{\mathbf{c}})$ (abbreviated $R_{\mathbf{c}}^+$) from $\mathrm{cl}(\mathbf{c} - R_{\mathbf{c}}^+(U_{\mathbf{c}}))$:

$$S_{\mathbf{c}}^+(U_{\mathbf{c}}) = R_{\mathbf{c}}^+(U_{\mathbf{c}}) \cap \mathrm{cl}(\mathbf{c} - R_{\mathbf{c}}^+(U_{\mathbf{c}})).$$

In abbreviated form, $S_{\mathbf{c}}^+$ is defined as

$$S_{\mathbf{c}}^+ = R_{\mathbf{c}}^+ \cap \mathrm{cl}(\mathbf{c} - R_{\mathbf{c}}^+).$$

(See Section 5.4.3 for a more detailed discussion of separation.)

Algorithm IsoMid4D in Section 6.2.3 returns a set of tetrahedra $T_{\mathbf{c}}^{M+}$ forming an isosurface patch. The corollary to the following lemma shows that this isosurface patch is exactly $S_{\mathbf{c}}^+(M_{\mathbf{c}})$. Proofs of the lemma and its corollaries are exactly the same as the proofs to Lemma 5.10 and its corollaries in Section 5.4.3 and are omitted.

Lemma 6.13. *Let $\mathbf{c}$ be a convex polytope in $\mathbb{R}^4$ where each vertex of $\mathbf{c}$ has a positive or a negative label and let $U_{\mathbf{c}}$ be an isosurface vertex set of $\mathbf{c}$. If $R_{\mathbf{c}}^+(U_{\mathbf{c}})$ has dimension four, then $T_{\mathbf{c}}^+(U_{\mathbf{c}})$ is a triangulation of $S_{\mathbf{c}}^+(U_{\mathbf{c}})$.*

Corollary 6.14. *Let $\mathbf{c}$ be a convex polytope in $\mathbb{R}^4$ where each vertex of $\mathbf{c}$ has a positive or a negative label. If $U_{\mathbf{c}}$ is a proper isosurface vertex set, then $T_{\mathbf{c}}^+(U_{\mathbf{c}})$ is a triangulation of $S_{\mathbf{c}}^+(U_{\mathbf{c}})$.*

Corollary 6.15. *Let $\mathbf{c}$ be a convex polytope in $\mathbb{R}^4$ where each vertex of $\mathbf{c}$ has a positive or a negative label. If $U_{\mathbf{c}}$ is a proper isosurface vertex set, then $|T_{\mathbf{c}}^+(U_{\mathbf{c}})|$ separates the positive vertices of $\mathbf{c}$ from the negative ones.*

We next claim that if $\mathbf{c}_1$ and $\mathbf{c}_2$ are adjacent polytopes, then $R_{\mathbf{c}_1}^+$ and $R_{\mathbf{c}_2}^+$ agree on shared faces of $\mathbf{c}_1$ and $\mathbf{c}_2$. It will follow that there are no "cracks" between adjacent isosurface patches.

We first extend our definitions of $R_{\mathbf{c}}$ from $\mathbf{c}$ to the faces of $\mathbf{c}$. Let $\mathbf{c}$ be a convex polytope in $\mathbb{R}^4$ where each vertex of $\mathbf{c}$ has a positive or a negative label. For each face $\mathbf{f}$ of $\mathbf{c}$, let $V_{\mathbf{f}}^+$ be the set of positive vertices of $\mathbf{f}$. Let $U_{\mathbf{f}}$ be the points of $U_{\mathbf{c}}$ that lie on $\mathbf{f}$. Let $R_{\mathbf{f}}^+(U_{\mathbf{f}})$ (abbreviated as $R_{\mathbf{f}}^+$) equal $\mathrm{conv}(U_{\mathbf{f}} \cup V_{\mathbf{f}}^+)$. In these definitions, face $\mathbf{f}$ can have dimension 0, 1, 2, 3, or 4.

The proofs of the following lemma and its corollary are exactly the same as the proofs of Lemma 5.14 and its corollary in Section 5.4.3 and are omitted.

Lemma 6.16. *Let $\mathbf{c}$ be a convex polytope in $\mathbb{R}^4$ where each vertex of $\mathbf{c}$ has a positive or a negative label and let $U_{\mathbf{c}}$ be an isosurface vertex set of $\mathbf{c}$. For every face $\mathbf{f}$ of polytope $\mathbf{c}$, $R_{\mathbf{c}}^+(U_{\mathbf{c}}) \cap \mathbf{f}$ equals $R_{\mathbf{f}}^+(U_{\mathbf{c}} \cap \mathbf{f})$.*

Corollary 6.17. *Let $\mathbf{c}_1$ and $\mathbf{c}_2$ be adjacent convex polytopes in $\mathbb{R}^4$ where each vertex of $\mathbf{c}_1$ and $\mathbf{c}_2$ has a positive or a negative label. Let $U_{\mathbf{c}_1}$ and $U_{\mathbf{c}_2}$ be isosurface vertex sets of $\mathbf{c}_1$ and $\mathbf{c}_2$, respectively. If $\mathbf{f}$ is a face of both $\mathbf{c}_1$ and $\mathbf{c}_2$, then $R_{\mathbf{c}_1}^+(U_{\mathbf{c}_1}) \cap \mathbf{f}$ equals $R_{\mathbf{c}_2}^+(U_{\mathbf{c}_2}) \cap \mathbf{f}$.*

In the statement of Lemma 6.16 and its corollary, face $\mathbf{f}$ can have dimensions 0, 1, 2, 3 or 4, although the four dimensional case is trivial.

Corollary 6.17 indicates that $R_{\mathbf{c}_1}$ and $R_{\mathbf{c}_2}$ agree on adjacent polytopes $\mathbf{c}_1$ and $\mathbf{c}_2$. We also need to show that triangulations $T_{\mathbf{c}_1}$ and $T_{\mathbf{c}_2}$ agree on their adjacent boundaries.

Lemma 6.18. *Let $\mathbf{c}_1$ and $\mathbf{c}_2$ be adjacent convex polytopes in $\mathbb{R}^4$ where each vertex of $\mathbf{c}_1$ and $\mathbf{c}_2$ has a positive or a negative label. Let $U_{\mathbf{c}_1}$ and $U_{\mathbf{c}_2}$ be isosurface vertex sets of $\mathbf{c}_1$ and $\mathbf{c}_2$, respectively, If $\mathbf{t}_1$ is a tetrahedron in $T_{\mathbf{c}_1}^+(U_{\mathbf{c}_1})$ and $\mathbf{t}_2$ is a tetrahedron in $T_{\mathbf{c}_1}^+(U_{\mathbf{c}_2})$, then $\mathbf{t}_1 \cap \mathbf{t}_2$ is a face of $\mathbf{t}_1$ and a face of $\mathbf{t}_2$.*

Proof: Let $\mathbf{f}$ equal $\mathbf{c}_1 \cap \mathbf{c}_2$, the grid face between $\mathbf{c}_1$ and $\mathbf{c}_2$. Let τ_1 and τ_2 be the lexicographic triangulations of $R_{\mathbf{c}_1}^+$ and $R_{\mathbf{c}_2}^+$, respectively. By Lemma 6.10, $\{\mathbf{t} \cap \mathbf{f} : \mathbf{t} \in \tau_1 \text{ and } \mathbf{t} \cap \mathbf{f} \neq \emptyset\}$ equals the lexicographic triangulation $\tau_{\mathbf{f}}$ of $R_{\mathbf{f}}^+$. Similarly, $\{\mathbf{t} \cap \mathbf{f} : \mathbf{t} \in \tau_2 \text{ and } \mathbf{t} \cap \mathbf{f} \neq \emptyset\}$ equals $\tau_{\mathbf{f}}$.

Let $\mathbf{t}_1$ and $\mathbf{t}_2$ be tetrahedra in $T_{\mathbf{c}_1}^+$ and $T_{\mathbf{c}_2}^+$, respectively. Since $T_{\mathbf{c}_1}^+$ is a subset of τ_1, $\mathbf{t}_1 \cap \mathbf{f}$ is a face $\mathbf{g}_1$ in $\tau_{\mathbf{f}}$. Since $T_{\mathbf{c}_2}^+$ is a subset of τ_2, $\mathbf{t}_1 \cap \mathbf{f}$ is a face $\mathbf{g}_2$ in $\tau_{\mathbf{f}}$. Since $\tau_{\mathbf{f}}$ is a triangulation, the intersection $\mathbf{g}_1 \cap \mathbf{g}_2$ is a face of $\mathbf{g}_1$ and a face of $\mathbf{g}_2$. Since $\mathbf{g}_1 \cap \mathbf{g}_2$ equals $\mathbf{t}_1 \cap \mathbf{t}_2$ and $\mathbf{g}_1$ is a face of $\mathbf{t}_1$ and $\mathbf{g}_2$ is a face of $\mathbf{t}_2$, the intersection $\mathbf{t}_1 \cap \mathbf{t}_2$ is a face of $\mathbf{t}_1$ and a face of $\mathbf{t}_2$. $\qquad\square$

In the statement of Lemma 6.18, $\mathbf{f}$ is a proper face of $\mathbf{c}$ and can have dimensions 0, 1, 2, or 3. Note that Lemmas 6.16 and 6.18 and their corollaries do not require that isosurface vertex sets $U_{\mathbf{c}}$ or $U_{\mathbf{c}_1}$ or $U_{\mathbf{c}_2}$ be proper.

For a convex polytopal mesh Γ, let U be an isosurface vertex set for Γ, and let $U_{\mathbf{c}}$ equals $\mathbf{c} \cap U$ for each polytope $\mathbf{c} \in \Gamma$. Note that $U_{\mathbf{c}}$ is an isosurface vertex set for $\mathbf{c}$ and that if U is proper, then each $U_{\mathbf{c}}$ is proper.

Define M_Γ, $\Sigma_\Gamma^{M+}(\sigma)$, $T_\Gamma^+(U)$, $R_\Gamma^+(U)$ and $S_\Gamma^+(U)$ as

$$M_\Gamma = \bigcup_{\mathbf{c} \in \Gamma} M_{\mathbf{c}},$$

$$\Sigma_\Gamma^{M+}(\sigma) = \bigcup_{\mathbf{c} \in \Gamma} \Sigma_{\mathbf{c}}^{M+}(\sigma),$$

$$T_\Gamma^+(U) = \bigcup_{\mathbf{c} \in \Gamma} T_{\mathbf{c}}^+(U_{\mathbf{c}}),$$

$$R_\Gamma^+(U) = \bigcup_{\mathbf{c} \in \Gamma} R_{\mathbf{c}}^+(U_{\mathbf{c}}),$$

$$S_\Gamma^+(U) = R_\Gamma^+(U) \cap \mathrm{cl}(|\Gamma| - R_\Gamma^+(U)).$$

Set $|\Gamma|$ is the union of all the elements of mesh Γ. Abbreviate $\Sigma_\Gamma^{M+}(\sigma)$, $T_\Gamma^+(U)$, $R_\Gamma^+(U)$ and $S_\Gamma^+(U)$ as Σ_Γ^{M+}, T_Γ^+, R_Γ^+, and S_Γ^+, respectively. Abbreviate $T_\Gamma^+(M_\Gamma)$ and $S_\Gamma^+(M_\Gamma)$ as T_Γ^{M+} and S_Γ^{M+}. Note that S_Γ^+ separates R_Γ^+ from $|\Gamma| - R_\Gamma^+$. (See Lemma B.8 in Appendix B.)

We claim that $S_\Gamma^+(U)$ equals $\bigcup_{\mathbf{c}\in\Gamma} S_\mathbf{c}^+(U\cap\mathbf{c})$ if U is a proper isosurface vertex set. The proofs of the lemma and its two corollaries are exactly the same as the proof of Lemma 5.18 and its corollaries in Section 5.4.3 and are omitted.

Lemma 6.19. *Let Γ be a convex polytopal mesh in $\mathbb{R}^4$ where each vertex of Γ has a positive or negative label. If U is a proper isosurface vertex set of Γ, then set $S_\Gamma^+(U)$ equals $\bigcup_{\mathbf{c}\in\Gamma} S_\mathbf{c}^+(U\cap\mathbf{c})$.*

Corollary 6.20. *Let Γ be a convex polytopal mesh in $\mathbb{R}^4$ where each vertex of Γ has a positive or negative label. If U is a proper isosurface vertex set of Γ, then $|T_\Gamma^+(U)|$ equals $S_\Gamma^+(U)$.*

Corollary 6.21. *Let Γ be a convex polytopal mesh in $\mathbb{R}^4$ where each vertex of Γ has a positive or negative label. If U is a proper isosurface vertex set of Γ, then $|T_\Gamma^+(U)|$ separates the positive vertices of Γ from the negative ones.*

$T_\Gamma^+(U)$ is defined as $\bigcup_{\mathbf{c}\in\Gamma} T_\mathbf{c}^+(U_\mathbf{c})$, the union of sets of tetrahedra. Each $T_\mathbf{c}^+(U_\mathbf{c})$ is a subset of a triangulation of the boundary of $R_\mathbf{c}^+(U_\mathbf{c})$, so each $T_\mathbf{c}^+(U_\mathbf{c})$ is itself a triangulation. The following lemma claims that $T_\Gamma^+(U)$ is a triangulation of $S_\Gamma^+(U)$. The proof is exactly the same as the proof of Lemma 5.21 in Section 5.4.3 and is omitted.

Lemma 6.22. *Let Γ be a convex polytopal mesh in $\mathbb{R}^4$ where each vertex of Γ has a positive or negative label. If U is a proper isosurface vertex set of Γ, then $T_\Gamma^+(U)$ is a triangulation of $S_\Gamma^+(U)$.*

M_Γ is the set of all midpoints of bipolar edges of Γ. Since $T_\Gamma^+(M_\Gamma)$ (abbreviated T_Γ^{M+}) is a triangulation of $S_\Gamma^+(M_\Gamma)$ (abbreviated S_Γ^{M+}), the homotopy η can be extended to all of S_Γ^{M+}. Each vertex v_i of T_Γ^{M+} is the midpoint m_i of some edge $\mathbf{e}_i$. Vertex v_i is remapped to the interpolated point $w_i \in \mathbf{e}_i$. For each vertex v_i of T_Γ^{M+}, define $\eta(v_i,\alpha)$ as $(1-\alpha)v_i + \alpha w_i$. Extend η to all of S_Γ^{M+} by linearly extending η on each tetrahedron in T_Γ^{M+}. Because T_Γ^{M+} is a triangulation of S_Γ^{M+}, homotopy η is well-defined on every point of S_Γ^{M+}. Thus $\eta : S_\Gamma^{M+} \times [0,1]$ is a homotopy from S_Γ^{M+} to $\Sigma_\Gamma^{M+}(\sigma)$ where $\eta(S_\Gamma^{M+},0)$ equals S_Γ^{M+} and $\eta(S_\Gamma^{M+},1)$ equals $\Sigma_\Gamma^{M+}(\sigma)$.

Our last lemma shows that Σ_Γ^{M+} does not contain mesh vertices whose scalar values do not equal the isovalue. The proof is the same as the proof of Lemma 5.22 in Section 5.4.3 and is omitted.

Lemma 6.23. *Let Γ be a convex polytopal mesh in $\mathbb{R}^4$ where each vertex of Γ has a scalar value. If v is a vertex of Γ whose scalar value does not equal $\sigma \in \mathbb{R}$, then $\Sigma_\Gamma^{M+}(\sigma)$ does not contain v.*

Applying Corollary 6.21, the homotopy map η, and Lemma 6.23 gives Property 5.

Property 5.   *The isosurface separates positive grid vertices from negative ones and strictly separates strictly positive grid vertices from negative grid vertices.*

Proof: By Corollary 6.21, set $|T_\Gamma^{M+}|$ strictly separates the positive vertices of Γ from the negative ones. Map η is a homotopy map from S_Γ^{M+} to $\Sigma_\Gamma^{M+}(\sigma)$ that never passes through any vertices of Γ. Vertices of S_Γ^{M+} that lie on the boundary of $|\Gamma|$ remain on the boundary of $|\Gamma|$ throughout the homotopy η. Thus $\eta(S_\Gamma^{M+}) \cap \partial|\Gamma|, \alpha)$ is a subset of $\partial|\Gamma|$ for all $\alpha \in [0,1]$. By Lemma B.18, $\Sigma_\Gamma^{M+}(\sigma)$ separates the positive vertices of Γ from the negative ones.

By Lemma 6.23, $\Sigma_\Gamma^{M+}(\sigma)$ does not contain any mesh vertices whose scalar values equal the isovalue σ. Therefore, $\Sigma_\Gamma^{M+}(\sigma)$ strictly separates vertices with scalar values greater than σ from vertices with scalar value less than σ.   □

6.3.6   Proof of Isosurface Properties 6, 7, and 8

Property 6 states that if no grid scalar value equals the isovalue, then $\Sigma_\Gamma^{M+}(\sigma)$ is a manifold. To prove this property we first show that if $U_{\mathbf{c}_1}$ and $U_{\mathbf{c}_2}$ are proper, then $S_{\mathbf{c}_1}^+(U_{\mathbf{c}_1})$ and $S_{\mathbf{c}_2}^+(U_{\mathbf{c}_2})$ agree on shared faces of $\mathbf{c}_1$ and $\mathbf{c}_2$ (Corollary 6.26). (We already established in Section 6.3.5, Corollary 6.17, that $R_{\mathbf{c}_1}^+(U_{\mathbf{c}_1})$ and $R_{\mathbf{c}_2}^+(U_{\mathbf{c}_2})$ agree on shared faces of $\mathbf{c}_1$ and $\mathbf{c}_2$.) We next show that for every point $p \in S_\Gamma^+(M_\Gamma)$ there is a hyperplane $\mathbf{h}$ containing p such that the orthogonal projection of $S_\Gamma^+(M_\Gamma) \cap \mathbb{N}_p$ onto $\mathbf{h}$ is one-to-one for some sufficiently small neighborhood $\mathbb{N}_p$ of p (Lemma 6.30). From this we show that $S_\Gamma^+(M_\Gamma)$ is an orientable 3-manifold with boundary (Lemma 6.31). Since the mapping from $S_\Gamma^+(M_\Gamma)$ to Σ_Γ^{M+} is one-to-one and onto, Σ_Γ^{M+} is also an orientable 3-manifold with boundary.

We start by extending the definition of $S_{\mathbf{c}}$ from $\mathbf{c}$ to faces of $\mathbf{c}$. Let $\mathbf{f}$ be a face of $\mathbf{c}$. As in Section 5.4.3, set $V_{\mathbf{f}}^+$ is the set of positive vertices of $\mathbf{f}$, set $U_{\mathbf{f}}$ is the points of $U_{\mathbf{c}}$ that lie on $\mathbf{f}$, and $R_{\mathbf{f}}^+(U_{\mathbf{f}})$ (abbreviated $R_{\mathbf{f}}^+$) equals $\mathrm{conv}(U_{\mathbf{f}} \cup V_{\mathbf{f}}^+)$. Define $S_{\mathbf{f}}^+(U_{\mathbf{f}})$ (abbreviated $S_{\mathbf{f}}^+$) as

$$S_{\mathbf{f}}^+(U_{\mathbf{f}}) = R_{\mathbf{f}}^+(U_{\mathbf{f}}) \cap \mathrm{cl}(\mathbf{f} - R_{\mathbf{f}}^+(U_{\mathbf{f}})).$$

As in Section 6.3.5, face $\mathbf{f}$ need not be three-dimensional.

Lemma 6.16 shows that $R_{\mathbf{c}}^+ \cap \mathbf{c}$ equals $R_{\mathbf{f}}^+$. We claim a similar lemma holds for the complements of $R_{\mathbf{c}}^+$ and $R_{\mathbf{f}}^+$. Proofs of Lemma 6.24 and Lemma 6.25 and its corollary are the same as the proofs of Lemma 5.23 and Lemma 5.24 in Section 5.4.4 and are omitted.

Lemma 6.24. *Let $\mathbf{c}$ be a convex polytope in $\mathbb{R}^4$ where each vertex of $\mathbf{c}$ has a positive or a negative label and let $U_{\mathbf{c}}$ be a proper isosurface vertex set of $\mathbf{c}$. For every face $\mathbf{f}$ of polytope $\mathbf{c}$, $\mathrm{cl}(\mathbf{c} - R_{\mathbf{c}}^+(U_{\mathbf{c}})) \cap \mathbf{f}$ equals $\mathrm{cl}(\mathbf{f} - R_{\mathbf{f}}^+(U_{\mathbf{c}} \cap \mathbf{f}))$.*

Lemma 6.25. *Let* **c** *be a convex polytope in* $\mathbb{R}^4$ *where each vertex of* **c** *has a positive or a negative label and let* $U_\mathbf{c}$ *be a proper isosurface vertex set of* **c**. *For every face* **f** *of polytope* **c**, $S_\mathbf{c}^+(U_\mathbf{c}) \cap \mathbf{f}$ *equals* $S_\mathbf{f}^+(U_\mathbf{c} \cap \mathbf{f})$.

Corollary 6.26. *Let* $\mathbf{c}_1$ *and* $\mathbf{c}_2$ *be adjacent convex polytopes in* $\mathbb{R}^4$ *where each vertex of* $\mathbf{c}_1$ *and* $\mathbf{c}_2$ *has a positive or a negative label. Let* $U_{\mathbf{c}_1}$ *and* $U_{\mathbf{c}_2}$ *be isosurface vertex sets for* $\mathbf{c}_1$ *and* $\mathbf{c}_2$, *respectively, where* $U_{\mathbf{c}_1} \cap \mathbf{f}$ *equals* $U_{\mathbf{c}_2} \cap \mathbf{f}$. *If* $U_{\mathbf{c}_1}$ *and* $U_{\mathbf{c}_2}$ *are proper isosurface vertex sets, then* $S_{\mathbf{c}_1}^+(U_{\mathbf{c}_1}) \cap \mathbf{f}$ *equals* $S_{\mathbf{c}_2}^+(U_{\mathbf{c}_2}) \cap \mathbf{f}$.

Lemma 6.25 and its corollary require that the isosurface vertex sets be proper.

To prove Properties 6 and 7, we prove that if isosurface vertex set U is proper, then $S_\Gamma^+(U)$ is a piecewise linear, orientable 3-manifold whose boundary lies on $\partial|\Gamma|$.

In $\mathbb{R}^3$, we proved that S_Γ^+ is a manifold whose boundary lies on $\partial|\Gamma|$. We did this by case analysis of the intersection of S_Γ^+ and edges, facets, and interiors of grid cubes. In $\mathbb{R}^4$, this case analysis becomes rather difficult. Instead, for each point $p \in S_\Gamma^+$, we construct a hyperplane $\mathbf{h}_p$ such that for some neighborhood $\mathbb{N}_p$ of p in S_Γ^+ the orthogonal projection onto $\mathbf{h}_p$ is a homeomorphism from $\mathbb{N}_p$ to $\mathbf{h}_p \cap \mathbb{B}_p \cap |\Gamma|$ for some sufficiently small ball $\mathbb{B}_p$ around p. Since $\mathbf{h}_p \cap \mathbb{B}_p \cap |\Gamma|$ is homeomorphic to $\mathbb{R}^3$ or a half-space in $\mathbb{R}^3$, we have a homeomorphism from $\mathbb{N}_p$ to $\mathbb{R}^3$ or a half-space in $\mathbb{R}^3$.

Lemma 6.11 established that if $U_\mathbf{c}$ is a proper isosurface vertex set then $R_\mathbf{c}^+(U_\mathbf{c})$ is four-dimensional. We need a similar lemma about $R_\mathbf{c}^+ \cap \mathbf{f}$ for each face **f** of **c**.

Lemma 6.27. *Let* **c** *be a convex polytope in* $\mathbb{R}^4$ *where each vertex of* **c** *has a positive or a negative label and let* **f** *be a* j-*dimensional face of* **c**. *If* $U_\mathbf{c}$ *is a proper isosurface vertex set of* **c** *and at least one vertex of* **f** *is positive, then* $R_\mathbf{c}^+ \cap \mathbf{f}$ *has dimension* j.

Proof: Let v be a positive vertex of **f**. Since v is positive, some point of $U_\mathbf{c}$ lies on the interior of every edge incident on v. There are at least j edges incident on v and contained in **f**. Since these edges are edges of **f**, they span a j-dimensional space. Thus the convex hull of $(U_\mathbf{c} \cap \mathbf{f}) \cup v$ has dimension j. Since $R_\mathbf{f}^+(U_\mathbf{c} \cap \mathbf{f})$ contains that convex hull, $R_\mathbf{f}^+(U_\mathbf{c} \cap \mathbf{f})$ is j-dimensional. By Lemma 6.16, set $R_\mathbf{c}^+ \cap \mathbf{f}$ equals $R_\mathbf{f}^+(U_\mathbf{c} \cap \mathbf{f})$, so $R_\mathbf{c}^+ \cap \mathbf{f}$ is j-dimensional. $\square$

By Lemma 6.11, set $R_\mathbf{f}^+$ has the same dimension as **f**. The relative interior of $R_\mathbf{f}^+$, denoted $\mathrm{relint}(R_\mathbf{f}^+)$, is the interior of $R_\mathbf{f}^+$ when $R_\mathbf{f}^+$ is considered as a j-dimensional set lying in $\mathbb{R}^j$. (See Section B.1 in Appendix B for a more rigorous discussion and definition.) We prove that if p and p' are points in $R_\mathbf{f}^+$ and at least one of these points is in $\mathrm{relint}(R_\mathbf{f}^+)$, then the open line segment (p, p') does not intersect $S_\mathbf{f}^+$. (See Figure 6.12(a).)

Lemma 6.28. *Let* **c** *be a convex polytope in* $\mathbb{R}^4$ *where each vertex of* **c** *has a positive or negative label, let* **f** *be a face of* **c**, *let* $U_\mathbf{c}$ *be a proper isosurface vertex*

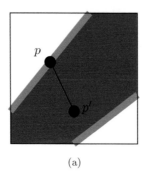
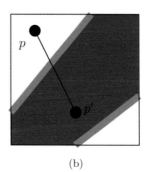

(a)   (b)

Figure 6.12. Red region is $R_{\mathbf{f}}^+$ and the green line segments are $S_{\mathbf{f}}^+$. Point p' is in relint($R_{\mathbf{f}}^+$). (a) Point p is in $R_{\mathbf{f}}^+$. The open line segment (p, p') does not intersect $S_{\mathbf{f}}^+$. (b) Point p is in $\mathbf{f} - R_{\mathbf{f}}^+$. The open line segment (p, p') intersects $S_{\mathbf{f}}^+$ at exactly one point.

set of $\mathbf{c}$, and let $U_{\mathbf{f}}$ equal $U_{\mathbf{c}} \cap \mathbf{f}$. If p is in $R_{\mathbf{f}}^+(U_{\mathbf{f}})$ and p' is in relint($R_{\mathbf{f}}^+(U_{\mathbf{f}})$), then open line segment (p, p') does not intersect $S_{\mathbf{f}}^+(U_{\mathbf{f}})$.

Proof: Since p' is in relint($R_{\mathbf{f}}^+$), some open neighborhood $\mathbb{N}_{p'}$ of p' in $\mathbf{f}$ is contained in relint($R_{\mathbf{f}}^+$). The convex hull of $\mathbb{N}_{p'} \cup \{p\}$ is contained in $R_{\mathbf{f}}^+$. Since open line segment (p, p') is a subset of relint(conv($\mathbb{N}_{p'} \cup \{p\}$)), open line segment (p, p') is a subset of relint($R_{\mathbf{f}}^+$). Thus $(p, p') \cap \text{cl}(\mathbf{f} - R_{\mathbf{f}}^+)$ is the empty set. Since $S_{\mathbf{f}}^+$ is a subset of $\text{cl}(f - R_{\mathbf{f}}^+)$, set $(p, p') \cap S_{\mathbf{f}}^+$ is also the empty set.   □

The following corollary gives a condition for line segment (p, p') to intersect $S_{\mathbf{f}}^+$ at exactly one point.

Corollary 6.29. *Let $\mathbf{c}$ be a convex polytope in $\mathbb{R}^4$ where each vertex of $\mathbf{c}$ has a positive or negative label, let $\mathbf{f}$ be a face of $\mathbf{c}$, let $U_{\mathbf{c}}$ be a proper isosurface vertex set of $\mathbf{c}$ and let $U_{\mathbf{f}}$ equal $U_{\mathbf{c}} \cap \mathbf{f}$. If p is a point in $\mathbf{f} - R_{\mathbf{f}}^+(U_{\mathbf{f}})$ and p' is in relint($R_{\mathbf{f}}^+(U_{\mathbf{f}})$), then open line segment (p, p') intersects $S_{\mathbf{f}}^+(U_{\mathbf{f}})$ at one and only one point.*

Proof: Assume p is in $\mathbf{f} - R_{\mathbf{f}}^+$. Since $p' \in \text{relint}(R_{\mathbf{f}}^+)$ and $p \in f - R_{\mathbf{f}}^+$, open line segment (p, p') intersects $R_{\mathbf{f}}^+ \cap \text{cl}(\mathbf{f} - R_{\mathbf{f}}^+) = S_{\mathbf{f}}^+$. Thus (p, p') intersects $S_{\mathbf{f}}^+$ at least at one point q.

Assume that (p, p') intersected $S_{\mathbf{f}}^+$ at two points q and q'. Without loss of generality, assume that the points on (p, p') lie in the order p, q, q', p'. Since q is in $S_{\mathbf{f}}^+$, point q is in $R_{\mathbf{f}}^+$. By Lemma 6.28, open line segment (q, p') does not intersect $S_{\mathbf{f}}^+$. However, q' is in (q, p') and in $S_{\mathbf{f}}^+$, a contradiction. Thus (p, p') intersects $S_{\mathbf{f}}^+$ at exactly one point.   □

Corollary 6.29 is illustrated in Figure 6.12(b).

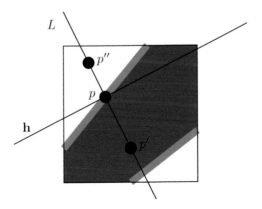

Figure 6.13. Red region is $R_{\mathbf{f}}^+$ and the green line segments are $S_{\mathbf{f}}^+$. Point p is on $S_{\mathbf{f}}^+$, point p' is in $\mathrm{relint}(R_{\mathbf{f}}^+)$, and point p'' is in $\mathbf{f} - \mathrm{relint}(R_{\mathbf{f}}^+)$. Line L passes through points p, p', and p''. The hyperplane $\mathbf{h}$ is orthogonal to L. Only a one-dimensional cross section of $\mathbf{h}$ is shown in the figure.

Let p be a point in $S_{\mathbf{c}}^+$ and let $\mathbf{h}$ be a hyperplane containing p. Under appropriate conditions, the orthogonal projection of $\mathbb{R}^4$ onto $\mathbf{h}$ is a homeomorphism of some neighborhood of p in $S_{\mathbf{c}}^+$ to some neighborhood of p in $\mathbf{h} \cap \mathbf{c}$.

Lemma 6.30. *Let $\mathbf{c}$ be a convex polytope in $\mathbb{R}^4$ where each vertex of $\mathbf{c}$ has a positive or negative label, let $\mathbf{f}$ be a face of $\mathbf{c}$, let $U_{\mathbf{c}}$ be a proper isosurface vertex set of $\mathbf{c}$, and let $U_{\mathbf{f}}$ equal $U_{\mathbf{c}} \cap \mathbf{f}$. Let p be a point in $S_{\mathbf{f}}^+(U_{\mathbf{f}}) \cap \mathrm{relint}(\mathbf{f})$, let p' be a point in $\mathrm{relint}(R_{\mathbf{f}}^+(U_{\mathbf{f}}))$, and let $\mathbf{h}$ be a hyperplane containing p and orthogonal to the line through p and p'. If $\pi : \mathbb{R}^4 \to \mathbf{h}$ is the orthogonal projection of $\mathbb{R}^4$ onto a hyperplane $\mathbf{h}$, then there is some neighborhood $\mathbb{N}_p$ of p in $S_{\mathbf{c}}^+(U_{\mathbf{c}})$ such that π restricted to $\mathbb{N}_p$ is a homeomorphism from $\mathbb{N}_p$ to some neighborhood of p in $\mathbf{h} \cap \mathbf{c}$.*

Proof: By Corollary 6.7, point p does not lie on a vertex of Γ. Let $j \geq 1$ be the dimension of $\mathbf{f}$. By Lemma 6.27, $R_{\mathbf{c}}^+ \cap \mathbf{f}$ is j-dimensional. By Lemma 6.16, $R_{\mathbf{f}}^+$ is also j-dimensional.

Let L be the line through p and p'. Since point p is in $\mathrm{relint}(\mathbf{f})$, there is a point $p'' \in \mathrm{relint}(\mathbf{f}) \cap L$ such that p lies on line segment (p', p''). (See Figure 6.13.) By Lemma 6.28, p'' is not in $R_{\mathbf{f}}^+$ or else p would not be in $S_{\mathbf{f}}^+$. Since $R_{\mathbf{f}}^+$ is a closed set, some open neighborhood of p'' in $\mathbf{f}$ does not intersect $R_{\mathbf{f}}^+$.

Let $\mathbb{B}_p$, $\mathbb{B}_{p'}$, and $\mathbb{B}_{p''}$ be open balls in $\mathbb{R}^4$ around p, p', and p'', respectively, such that

- $\mathbb{B}_p$, $\mathbb{B}_{p'}$, and $\mathbb{B}_{p''}$ have the same radius;
- $\mathbb{B}_{p'} \cap \mathbf{c}$ is contained in $R_{\mathbf{c}}^+$;
- $\mathbb{B}_{p''} \cap \mathbf{c}$ does not intersect $R_{\mathbf{c}}^+$.

Let A equal $\text{conv}(\mathbb{B}_{p'}, \mathbb{B}_{p''})$. Set $S_{\mathbf{c}}^+ \cap A$ is a neighborhood of p in $S_{\mathbf{c}}^+$. Set $\mathbf{h} \cap \mathbb{B}_p \cap \mathbf{c}$ is a neighborhood of p in $\mathbf{h} \cap \mathbf{c}$. Projection π maps $S_{\mathbf{c}}^+ \cap A$ into $\mathbf{h} \cap \mathbb{B}_p \cap \mathbf{c}$. We claim that π restricted to $S_{\mathbf{c}}^+ \cap A$ is a homeomorphism from $S_{\mathbf{c}}^+ \cap A$ to $\mathbf{h} \cap \mathbb{B}_p \cap \mathbf{c}$.

Let q be a point on $S_{\mathbf{c}}^+ \cap A$. Let L_q be the line parallel to L passing through q. Line L_q intersects $\mathbb{B}_{p'}$ at some point q' and intersects $\mathbb{B}_{p''}$ at some point q''. By Corollary 6.29, line segment (q', q'') intersects $S_{\mathbf{c}}^+$ at only one point. Thus π restricted to $S_{\mathbf{c}}^+ \cap A$ is a one-to-one mapping of $S_{\mathbf{c}}^+ \cap A$ into $\mathbf{h} \cap \mathbb{B}_p \cap \mathbf{c}$.

Now, let $\tilde{q}$ be a point in $\mathbf{h} \cap \mathbb{B}_p \cap \mathbf{c}$. Let $L_{\tilde{q}}$ be the line parallel to L passing through $\tilde{q}$, let q' be a point in $L_q \cap \mathbb{B}_{p'} \cap \mathbf{c}$, and let q'' be a point in $L_q \cap \mathbb{B}_{p''} \cap \mathbf{c}$. Point $\tilde{q}$ lies in the relative interior of some face $\mathbf{g}$ of $\mathbf{c}$ (where $\mathbf{g}$ may equal $\mathbf{c}$.) Points q' and q'' must also lie on $\mathbf{g}$. By Corollary 6.29, line segment (q', q'') intersects $S_{\mathbf{g}}^+$ at exactly one point q. Thus $\pi(S_{\mathbf{c}}^+ \cap A)$ covers $\mathbf{h} \cap \mathbb{B}_p \cap \mathbf{c}$. Since π restricted to $S_{\mathbf{c}}^+ \cap A$ is continuous and the inverse mapping taking $\mathbf{h} \cap \mathbb{B}_p \cap \mathbf{c}$ to $S_{\mathbf{c}}^+ \cap A$ is also continuous, π restricted to $S_{\mathbf{c}}^+ \cap A$ is a homeomorphism from $S_{\mathbf{c}}^+ \cap A$ to to $\mathbf{h} \cap \mathbb{B}_p \cap \mathbf{c}$. $\qquad\square$

Lemma 6.31. *Let Γ be a regular grid where each vertex of Γ has a positive or negative label. Set S_{Γ}^+ is a piecewise linear, orientable 3-manifold whose boundary lies on $\partial|\Gamma|$.*

Proof: Let p be a point on S_{Γ}^+. Point p lies in the interior of some face $\mathbf{f}$ of Γ. By Corollary 6.8, point p does not lie on a vertex of Γ. Let $j \geq 1$ be the dimension of $\mathbf{f}$. By Lemma 6.27, $R_{\mathbf{f}}^+ \cap \mathbf{f}$ is j-dimensional.

Let p' be a point in the relative interior of $R_{\mathbf{f}}^+$. By Lemma 6.25, $S_{\Gamma}^+ \cap \mathbf{f}$ equals $S_{\mathbf{f}}^+$. Since point p' is not in $S_{\mathbf{f}}^+$, point p' cannot be in S_{Γ}^+.

Let L be the line through points p and p', and let $\mathbf{h}$ be a hyperplane orthogonal to L. Let $\pi : \mathbb{R}^4 \to \mathbf{h}$ be the orthogonal projection of $\mathbb{R}^4$ onto $\mathbf{h}$.

By Lemma 6.30, for each hypercube $\mathbf{c}$ containing $\mathbf{f}$ there is a neighborhood $\mathbb{N}_{p,\mathbf{c}}$ of p in $S_{\mathbf{c}}^+$ such that π restricted to $\mathbb{N}_{p,\mathbf{c}}$ is a homeomorphism from $\mathbb{N}_{p,\mathbf{c}}$ to some neighborhood of p in $\mathbf{h} \cap \mathbf{c}$. Choose some neighborhood $\mathbb{N}_p$ of p in S_{Γ}^+, such that $\mathbb{N}_p \cap \mathbf{c}$ is a subset of $\mathbb{N}_{p,\mathbf{c}}$ for each $\mathbf{c}$ containing $\mathbf{f}$. By Lemma 6.25, $S_{\mathbf{c}}^+ \cap \mathbf{f}$ equals $S_{\mathbf{c}'}^+ \cap \mathbf{f}$ for every pair of hypercubes $\mathbf{c}, \mathbf{c}'$ containing $\mathbf{f}$. Thus, mapping π restricted to $\mathbb{N}_p$ is a continuous, one-to-one, and onto mapping from $\mathbb{N}_p$ to some neighborhood $\mathbb{N}_p'$ of p in $\mathbf{h} \cap |\Gamma|$. Since the inverse mapping from $\mathbb{N}_p'$ is also continuous, π restricted to $\mathbb{N}_p$ is a homeomorphism from the neighborhood $\mathbb{N}_p$ of p in S_p^+ to some neighborhood $\mathbb{N}_p'$ of p in $\mathbf{h} \cap |\Gamma|$.

Let $\mathbb{B}_p$ be a small four-dimensional ball around p such that $\mathbb{B}_p \cap \mathbf{h} \cap |\Gamma|$ is a subset of $\mathbb{N}_p'$. If p is in the interior of $|\Gamma|$, then set $\mathbb{B}_p \cap \mathbf{h} \cap |\Gamma|$ is homeomorphic to $\mathbb{R}^3$. If p is on the boundary of $|\Gamma|$, then set $\mathbb{B}_p \cap \mathbf{h} \cap |\Gamma|$ is homeomorphic to a half-space in $\mathbb{R}^3$. Thus some neighborhood of p in S_{Γ}^+ is homeomorphic to $\mathbb{R}^3$ or a half-space of $\mathbb{R}^3$, and this neighborhood is homeomorphic to a half-space of $\mathbb{R}^3$ if and only if p is on the boundary of $|\Gamma|$. It follows that S_{Γ}^+ is a manifold with boundary, and the boundary of S_{Γ}^+ lies on the boundary of $|\Gamma|$.

By Lemma 6.22, set Υ is a triangulation of Σ_Γ^{M+}. The tetrahedra in Υ are oriented so that their induced normals point toward the positive region. Thus, their orientation is consistent and Σ_Γ^{M+} is orientable. $\qquad\square$

Since T_Γ^+ is a triangulation of S_Γ^+, the isosurface Σ_Γ^+ is the image of S_Γ^+ under a continuous, piecewise linear mapping. Thus isosurface Σ_Γ^+ is an embedding of S_Γ^+ into $|\Gamma|$. If S_Γ^+ is an orientable 3-manifold with boundary and this embedding is one-to-one, then Σ_Γ^+ is also an orientable 3-manifold with boundary. However, as in $\mathbb{R}^3$, the mapping from S_Γ^+ to Σ_Γ^+ may not be one-to-one. We claim that if $\mathbf{c}$ is a hypercube in $\mathbb{R}^4$, then the embedding is one-to-one.

Lemma 6.32. *If $\mathbf{c}$ is a hypercube in $\mathbb{R}^4$ with scalar values at its vertices such that no vertex value equals the isovalue, then the mapping from $S_{\mathbf{c}}^+$ to $\Sigma_{\mathbf{c}}^+$ is one-to-one.*

The proof for Lemma 6.32 is a computer-based case analysis and is omitted.

Property 6. *The isosurface is a piecewise linear, orientable 3-manifold with boundary.*

Property 7. *The boundary of the isosurface lies on the boundary of the grid.*

Proof of Properties 6 & 7: By Lemma 6.31, the set $S_\Gamma^+(M_\Gamma)$ (abbreviated S_Γ^+) is a piecewise linear 3-manifold with boundary whose boundary lies on $\partial|\Gamma|$. Set $\Sigma_\Gamma^{M+}(\sigma)$ (abbreviated Σ_Γ^{M+}) is an embedding of S_Γ^+ into $|\Gamma|$. We claim that this embedding is one-to-one.

Let p be some point in Σ_Γ^{M+}. Since no grid vertex value equals the isovalue, set Σ_Γ^{M+} does not contain any grid vertices (Lemma 6.23). Moreover, any point of Σ_Γ^{M+} on a grid edge $\mathbf{e}$ is the image $\mathbf{e} \cap S_\Gamma^+$, which can only be the midpoint of $\mathbf{e}$. Thus, if p lies on a grid edge $\mathbf{e}$, then it is the image of only one point in S_Γ^+.

Assume $p \in \Sigma_\Gamma^{M+}$ is on the interior of some hypercube $\mathbf{c}$. By Lemma 6.32, the mapping from $S_{\mathbf{c}}^+$ to $\Sigma_{\mathbf{c}}^+$ is one-to-one. Since only points in $S_{\mathbf{c}}^+$ map to the interior of $\mathbf{c}$, point p is the image of only one point in Σ_Γ^{M+}.

Assume $p \in \Sigma_\Gamma^{M+}$ is in the interior of some proper grid face $\mathbf{f}$. By Lemma 6.32, the mapping from $S_{\mathbf{c}}^+$ to $\Sigma_{\mathbf{c}}^+$ is one-to-one for each hypercube $\mathbf{c}$ containing $\mathbf{f}$. Thus, the mapping from $S_{\mathbf{c}}^+ \cap \mathbf{f}$ to $\Sigma_{\mathbf{c}}^+$ is one-to-one. By Lemma 6.25, $S_{\mathbf{c}}^+ \cap \mathbf{f}$ equals $S_{\mathbf{c}'}^+ \cap \mathbf{f}$ for each pair of hypercubes $\mathbf{c}, \mathbf{c}'$ containing $\mathbf{f}$. Since only points in $\Sigma_\Gamma^{M+} \cap \mathbf{f}$ map to $\mathbf{f}$, only points in hypercubes containing $\mathbf{f}$ map to $\mathbf{f}$. Thus, point p is the image of only one point in Σ_Γ^{M+}.

Since the mapping from Σ_Γ^{M+} to Σ_Γ^{M+} is one-to-one and Σ_Γ^{M+} is an orientable 3-manifold with boundary, set Σ_Γ^{M+} is an orientable 3-manifold with boundary. Since $\partial\Sigma_\Gamma^{M+}$ lies on $\partial|\Gamma|$ and all points on $\partial\Sigma_\Gamma^{M+} \cap \partial|\Gamma|$ are mapped to $\partial|\Gamma|$, the boundary of Σ_Γ^{M+} also lies on $\partial|\Gamma|$. $\qquad\square$

The last property is that Υ does not contain any zero-volume or duplicate tetrahedra and forms a triangulation of the isosurface.

Property 8.   *Set Υ does not contain any zero-volume tetrahedra or duplicate tetrahedra, and the tetrahedra in Υ form a triangulation of the isosurface.*

Proof: Since no grid vertex has scalar value equal to the isovalue, no isosurface vertex lies on a grid vertex. By Property 3, each bipolar grid edge contains only one isosurface vertex. Thus, the linear interpolation on isosurface vertices does not create any zero-volume or duplicate isosurface tetrahedra.

Set Υ is the image of T_Γ^+ under the embedding that takes S_Γ^+ to Σ_Γ^{M+}. Since this embedding is one-to-one and T_Γ^+ defines a triangulation of S_Γ^+ (Lemma 6.22), set Υ is a triangulation of Σ_Γ^{M+}. $\square$

6.4   Marching Simplices

A four-dimensional simplex is called a 4-simplex. A modified version of MARCHING HYPERCUBES constructs an isosurface in a four-dimensional mesh composed of 4-simplices. The basic idea follows the MARCHING HYPERCUBES algorithm. An isosurface lookup table is built on a "generic" simplex, $\mathbf{c}_S$, in $\mathbb{R}^4$. Each simplex in the simplicial mesh is mapped to $\mathbf{c}_S$, the isosurface patch is retrieved from the lookup table, and the vertices of the isosurface patch are positioned using linear interpolation on the mesh edges. The algorithm is called MARCHING SIMPLICES.

As with MARCHING HYPERCUBES, it is extremely important that the triangulation of one isosurface patch match the triangulation of an adjacent one. This can also be a problem for a simplicial mesh. The facet of a 4-simplex is a 3-simplex. If two of the vertices of that 3-simplex are positive and two are negative, then the intersection of the isosurface and that 3-simplex has four vertices and two triangles. If two adjacent 4-simplices, $\mathbf{c}$ and $\mathbf{c}'$, share such a common 3-simplex, then the two triangles generated for $\mathbf{c}$ should be the same two triangles generated for $\mathbf{c}'$. This may or may not happen depending upon the mapping of $\mathbf{c}$ and $\mathbf{c}'$ to the "generic" simplex $\mathbf{c}_S$.

The MARCHING HYPERCUBES algorithm translates each grid cube to the unit hypercube. The lookup table for the unit hypercube contains lexicographic triangulations of the isosurface patches. These lexicographic triangulations translate to lexicographic triangulations of isosurface patches in the mesh element. Since the lexicographic triangulations of two adjacent isosurface patches agree on their common boundaries, the resulting set of tetrahedra form a triangulation of the isosurface. Positioning the isosurface vertices changes the surface geometry but preserves the triangulation.

The problem with using lexicographic triangulations for simplices is that the mapping of a mesh simplex to $\mathbf{c}_S$ is often not a translation but some other affine transformation. Affine transformations do not preserve the lexicographic order of points. Moreover, even an affine transformation of a simplex that preserves the lexicographic order of the simplex vertices may not preserve the lexicographic order of midpoints of simplex edges. We introduce a new type of triangulation, called incremental triangulations, to handle these problems.

6.4.1   Incremental Triangulation

Let $(p_1, \ldots, p_m)$ be a sequence of points in $\mathbb{R}^d$ and let $\mathrm{conv}(p_1, \ldots, p_m)$ be the convex hull of set $\{p_1, p_2, \ldots, p_m\}$. If τ is a triangulation and p is a vertex of some simplex in τ, then $\tau - \{p\}$ is the set of simplices in $F^*(\tau)$ that do not contain vertex p. (set $F^*(\tau)$ is the set τ and all faces of elements of τ.)

Definition 6.33. Triangulation τ is an incremental triangulation of convex set $\mathrm{conv}(p_1, \ldots, p_m)$ induced by the sequence $(p_1, \ldots, p_m)$ if

- triangulation τ is a single vertex, or

- set $\tau - \{p_m\}$ is an incremental triangulation of $\mathrm{conv}(p_1, \ldots, p_{m-1})$ induced by the sequence $(p_1, \ldots, p_{m-1})$.

Note that different sequences may induce different incremental triangulations. The lexicographic triangulation of $\mathrm{conv}(p_1, \ldots, p_m)$ is an incremental triangulation of $\mathrm{conv}(p_1, \ldots, p_m)$ induced by the lexicographic order of $\{p_1, \ldots, p_m\}$.

The dimension of $\mathrm{conv}(p_1, \ldots, p_m)$ may be $k < d$, in which case the largest simplices in τ are k-simplices. Also, the dimension of $\mathrm{conv}(p_1, \ldots, p_{m-1})$ may be less than the dimension of $\mathrm{conv}(p_1, \ldots, p_m)$. Thus, even if τ contains k-simplices, the largest simplices in $\tau - p_m$ could be $(k-1)$-simplices.

The two properties of lexicographic triangulations given in Section 6.3.4 also apply to incremental triangulations. The proofs of the following two lemmas are exactly the same as the proofs of Lemmas 6.9 and 6.10 and are omitted.

Lemma 6.34. *If $(p_1, \ldots, p_m)$ is a sequence of points in $\mathbb{R}^d$, and if τ and τ' are incremental triangulations of $\mathrm{conv}(p_1, p_2, \ldots, p_m)$ induced by the sequence $(p_1, p_2, \ldots, p_m)$, then $F^*(\tau)$ equals $F^*(\tau')$.*

Lemma 6.35. *Let $(p_1, \ldots, p_m)$ be a sequence of points in $\mathbb{R}^d$, let $\mathbf{f}$ be a face of $\mathrm{conv}(P)$, and let $(q_1, q_2, \ldots, q_k)$ be the subsequence of $(p_1, \ldots, p_m)$ consisting of points in $\mathbf{f}$. If τ is an incremental triangulation of $\mathrm{conv}(P)$ induced by $(p_1, p_2, \ldots, p_m)$, and τ' is an incremental triangulation of $\mathrm{conv}(P \cap \mathbf{f})$ induced by $(q_1, q_2, \ldots, q_k)$, then $F^*(\tau) \cap \mathbf{f}$ equals $F^*(\tau')$.*

Incremental triangulations are so named because they are the triangulations constructed by the INCREMENTAL CONVEX HULL algorithm described in Section 6.2.2. If the INCREMENTAL CONVEX HULL algorithm processes the sequence of points $(p_1, \ldots, p_m)$ in the given order, then the algorithm returns the incremental triangulation of $\mathrm{conv}(p_1, \ldots, p_m)$ induced by the sequence $(p_1, \ldots, p_m)$.

6.4.2   Induced Order of Edges

An ordering of polytope vertices induces an ordering on the polytope edges. This ordering will be used to determine the incremental triangulation of isosurface patches.

Let Z_P be a sequence of points. Let E be a set of line segments whose endpoints are the points in Z_P. Let $\mathbf{e} = (p, q)$ and $\mathbf{e}' = (p', q')$ be line segments in E where $p \prec q$ and $p' \prec q'$ in Z_P. Line segment $\mathbf{e} = (p, q)$ precedes $\mathbf{e}' = (p', q')$ if

- $p \prec p'$ in Z_P, or

- $p = p'$ and $q \prec q'$ in Z_P.

Denote "$\mathbf{e}$ precedes $\mathbf{e}'$" by $\mathbf{e} \prec \mathbf{e}'$. The ordering of E induced by Z_P is $(\mathbf{e}_1, \mathbf{e}_2, \ldots, \mathbf{e}_k)$ where $\mathbf{e}_i \prec \mathbf{e}_{i+1}$ for $i = 1, \ldots, k - 1$.

Since polytope vertices are points and polytope edges are line segments, an order on polytope vertices induces an order on the polytope edges.

6.4.3   Isosurface Patch Construction

We present an algorithm, IsoPoly4D, that constructs isosurface patches in any (bounded) four-dimensional convex polytope, not just a simplex. The algorithm is similar to Algorithm IsoMid4D in Section 6.2.3 but it uses incremental triangulations. Input to IsoPoly4D includes an order on the polytope vertices that determines the sequence inducing the incremental triangulation. IsoPoly4D also uses proper isosurface vertex sets in place of edge midpoints. While the application of IsoPoly4D is restricted to edge midpoints in this chapter, we will use IsoPoly4D on other isosurface vertex sets in constructing interval volumes in Chapter 7.

Let $\mathbf{c}$ be a polytope in $\mathbb{R}^4$ where each vertex of $\mathbf{c}$ has a positive or a negative label. The vertices of $\mathbf{c}$ are ordered. Let $U_{\mathbf{c}}$ be a proper isosurface vertex set of $\mathbf{c}$. We show how to construct an isosurface patch in $\mathbf{c}$ using set $U_{\mathbf{c}}$ as its vertices.

Let $V_{\mathbf{c}}^+$ be the vertices of $\mathbf{c}$ with positive labels and let $V_{\mathbf{c}}^-$ be the vertices with negative labels. Let $R_{\mathbf{c}}^+(U_{\mathbf{c}})$ equal $\mathrm{conv}(U_{\mathbf{c}} \cup V_{\mathbf{c}}^+)$, the convex hull of $U_{\mathbf{c}} \cup V_{\mathbf{c}}^+$. Set $R_{\mathbf{c}}^+(U_{\mathbf{c}})$ has a three-dimensional boundary denoted $\partial R_{\mathbf{c}}^+(U_{\mathbf{c}})$.

To build an incremental triangulation of $\partial R_{\mathbf{c}}^+$, we must fix an order for $U_{\mathbf{c}} \cup V_{\mathbf{c}}^+$. The order of $V_{\mathbf{c}}^+$ is simply given by the order on the vertices of $\mathbf{c}$. The order of $U_{\mathbf{c}}$ corresponds to the order on the bipolar edges induced by the order on the vertices of $\mathbf{c}$. More formally, let Z_V^+ be the sequence of positive vertices, $V_{\mathbf{c}}^+$, whose order matches the input order. Let Z_E be the sequence of bipolar edges of $\mathbf{c}$ whose order is induced by the input order on $\mathbf{c}$'s vertices. Each $u_e \in U_{\mathbf{c}}$ lies on some bipolar edge e of $\mathbf{c}$. Let Z_U be the sequence $(u_{e_1}, u_{e_2}, \ldots, u_{e_k})$ where e_i is the ith element of Z_E. Let Z^+ be the sequence $Z_V^+ \circ Z_U$, the concatenation of Z_V^+ with Z_U.

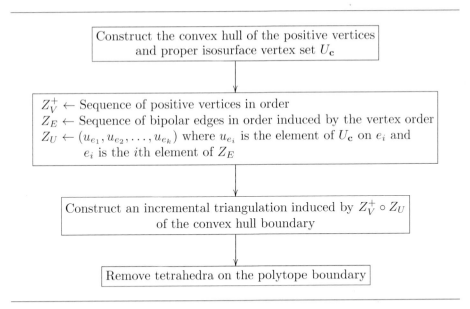

Figure 6.14. Algorithm IsoPoly4D for construction of isosurface patches from proper isosurface set $U_{\mathbf{c}}$. Input includes an order on the vertices of $\mathbf{c}$. Algorithm IsoPoly4D produces an incremental triangulation of the surface patch.

Construct $R_{\mathbf{c}}^+(U_{\mathbf{c}})$ and the incremental triangulation induced by sequence Z^+ of $\partial R_{\mathbf{c}}^+(U_{\mathbf{c}})$. Let $T_{\mathbf{c}}^+(U_{\mathbf{c}})$ be the set of tetrahedra in this triangulation that do not lie on facets of $\mathbf{c}$. Orient the tetrahedra in this triangulation so that the induced normals point toward $R_{\mathbf{c}}^+(U_{\mathbf{c}})$. Although some of the vertices of $\partial R_{\mathbf{c}}^+(U_{\mathbf{c}})$ are from the set $V_{\mathbf{c}}^+$, none of the vertices of $T_{\mathbf{c}}^+$ are from this set. The tetrahedra in $T_{\mathbf{c}}^+(U_{\mathbf{c}})$ form the triangulated isosurface patch whose vertices are $U_{\mathbf{c}}$. This isosurface patch separates the positive vertices from the negative vertices of $\mathbf{c}$.

We call the algorithm IsoPoly4D. Algorithm IsoPoly4D is presented in Figure 6.14.

The combinatorial structure of the isosurface within $\mathbf{c}$ is determined from the configuration of the polytope's vertex labels. The isosurface does not intersect any edge whose endpoints have the same label.

One could use $R_{\mathbf{c}}^-(U_{\mathbf{c}}) = \mathrm{conv}(U_{\mathbf{c}} \cup V_{\mathbf{c}}^-)$ instead of $R_{\mathbf{c}}^+(U_{\mathbf{c}})$. Let Z_V^- be the sequence of negative vertices whose order matches the input order and let Z^- be $Z_V^- \circ Z_U$. Construct an incremental triangulation of $\partial R_{\mathbf{c}}^-(U_{\mathbf{c}})$ induced by Z^-, and let $T_{\mathbf{c}}^-(U_{\mathbf{c}})$ be the set of tetrahedra of that triangulation that does not lie on facets of $\mathbf{c}$. The tetrahedra in $T_{\mathbf{c}}^-(U_{\mathbf{c}})$ form an alternate isosurface patch for $\mathbf{c}$ with vertices $U_{\mathbf{c}}$.

6.4.4   Isosurface Table Construction

Let $\mathbf{c}$ be a convex polytope in $\mathbb{R}^4$. Let n be the number of vertices of $\mathbf{c}$. In the isosurface extraction algorithm, the n vertices of $\mathbf{c}$ receive positive and negative labels representing their relationship to the isovalue. A vertex that has scalar value below the isovalue will receive a negative label, "$-$". A vertex with scalar value above or equal to the isovalue receives a positive label, "$+$".

Since each vertex is either positive or negative, there are 2^n configurations of positive and negative vertex labels. The isosurface lookup table contains 2^n entries, one for each configuration κ. Each entry is a list of quadruples of bipolar edges of $\mathbf{c}$. Each quadruple (e_1, e_2, e_3, e_4) represents an oriented tetrahedron whose vertices lie on e_1 and e_2 and e_3 and e_4. The list of quadruples define the combinatorial structure of the isosurface patch for configuration κ. Let $M_{\mathbf{c}}^{\kappa}$ be the midpoints of the bipolar edges for configuration κ. Lexicographically sort the vertices of $\mathbf{c}$. Construct the isosurface patch by applying Algorithm IsoPoly4D to the isosurface vertex set $M_{\mathbf{c}}^{\kappa}$. The tetrahedra in this isosurface patch are the set $T_{\mathbf{c}}^{+}(M_{\mathbf{c}}^{\kappa})$. The isosurface patch intersects every bipolar edge of $\mathbf{c}$ exactly once and does not intersect any other edges of $\mathbf{c}$.

6.4.5   Algorithm

Input to the MARCHING SIMPLICES algorithm is an isovalue, a four-dimensional simplicial mesh and a set of scalar values at the vertices of the mesh. The algorithm has three steps. (See Figure 6.15.) Read the isosurface lookup table from a preconstructed data file. For each simplex, retrieve from the lookup table a set of isosurface tetrahedra representing the combinatorial structure of the isosurface. The vertices of these tetrahedra form the isosurface vertices. Assign geometric locations to the isosurface vertices based on the scalar values at the simplex endpoints. The basic algorithm is the same as MARCHING HYPERCUBES. The differences are in the construction of the lookup table using IsoPoly4D and in the retrieval of the isosurface tetrahedra from the lookup table.

The isosurface lookup table is built on a "generic" 4-simplex $\mathbf{c}_S$ with vertices and vertex order

$$(0,0,0,0), (1,0,0,0), (0,1,0,0), (0,0,1,0), (0,0,0,1).$$

Any other 4-simplex or any other order on the vertices could also have been used. Algorithm IsoPoly4D is applied to each of the $2^5 = 32$ configurations of positive and negative vertex labels to generate 32 isosurface patches in 32 table entries.

For each mesh simplex $\mathbf{c}$, the isosurface patch triangulation retrieved from the lookup table depends on how the simplex is mapped to $\mathbf{c}_S$. There are 4! ways

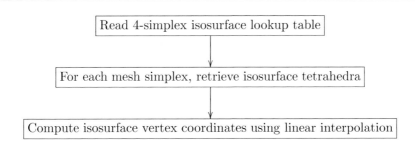

Figure 6.15. MARCHING SIMPLICES.

to map the vertices of **c** to $\mathbf{c}_S$. Each such mapping represents a different affine transformation of **c** to $\mathbf{c}_S$. However, only one such mapping preserves the lexicographic order of the vertices. If v_1, v_2, v_3, v_4, and v_5 are the vertices of **c** in lexicographic order, then map vertex v_1 to $(0,0,0,0)$, vertex v_2 to $(1,0,0,0)$, vertex v_3 to $(0,1,0,0)$, vertex v_4 to $(0,0,1,0)$, and vertex v_5 to $(0,0,0,1)$. This mapping sends the ith vertex of **c** in lexicographic order to the ith vertex of $\mathbf{c}_S$ in lexicographic order.

Consider two adjacent mesh simplices **c** and **c**′ sharing a common face **f**. Sort the vertices of **c** and **c**′ in lexicographical order. Let $Z_{\mathbf{c}}^+$ and $Z_{\mathbf{c'}}^+$ be the sequence of positive vertices and midpoints of bipolar edges of **c** and **c**′, respectively, as described in Section 6.4.3. Restricted to face **f**, the subsequence $Z_{\mathbf{c}}^+ \cap \mathbf{f}$ equals $Z_{\mathbf{c'}}^+ \cap \mathbf{f}$. Because the mapping of **c** to $\mathbf{c}_S$ preserves the lexicographic order of vertices of **c**, the triangulation retrieved from the lookup table for **c** is an incremental triangulation induced by $Z_{\mathbf{c}}^+$. This triangulation restricted to **f** is an incremental triangulation induced by $Z_{\mathbf{c}}^+ \cap \mathbf{f}$. Similarly, the triangulation for **c**′ is an incremental triangulation induced by $Z_{\mathbf{c'}}$. The triangulation induced by $Z_{\mathbf{c'}}$ restricted to **f** is an incremental triangulation induced by $Z_{\mathbf{c'}}^+ \cap \mathbf{f}$. Since $Z_{\mathbf{c}}^+ \cap \mathbf{f}$ equals $Z_{\mathbf{c'}}^+ \cap \mathbf{f}$, these two triangulations match on face **f**.

The isosurface lookup table determines only the edge containing each isosurface vertex. The geometric location of the isosurface vertex along that edge is not defined by the lookup table.

The vertices of the isosurface tetrahedra are the isosurface vertices. To map each isosurface tetrahedra to a geometric tetrahedra, we use linear interpolation to position the isosurface vertices as described in Section 1.7.2. Each isosurface vertex v lies on a grid edge $[p, q]$. If s_p and s_q are the scalar values at p and q and σ is the isovalue, then map v to $(1-\alpha)p + \alpha q$ where $\alpha = (\sigma - s_p)/(s_q - s_p)$.

The MARCHING SIMPLICES algorithm is presented in Algorithm 6.2. Function `LinearInterpolation`, called by this algorithm, is defined in Algorithm 1.1 in Section 1.7.2.

Input   : Γ is a 4D simplicial mesh.
            F is an array of the scalar values of each mesh vertex.
            σ is an isovalue.
Result   : A set Υ of isosurface tetrahedra.

MarchingSimplices(Γ, F, σ, Υ)

1 Read Marching Simplices lookup table into Table;
2 $\mathbf{c}_S \leftarrow$ "generic" simplex from Marching Simplices lookup table;
3 $(v_1^S, v_2^S, v_3^S, v_4^S, v_5^S) \leftarrow$ vertices of $\mathbf{c}_S$ in lexicographic order;
   /* Assign "+" or "−" signs to each vertex   */
4 foreach mesh vertex v do
5 | if F$[v] < \sigma$ then Sign$[v] \leftarrow$ "−";
6 | else Sign$[v] \leftarrow$ "+";   /* F$[v] \geq \sigma$ */
7 end
8 T $\leftarrow \emptyset$;
   /* For each mesh simplex, retrieve isosurface tetrahedra   */
9 foreach mesh simplex $\mathbf{c}$ do
10 | $(v_1, v_2, v_3, v_4, v_5) \leftarrow$ vertices of $\mathbf{c}$ in lexicographic order;
11 | $\kappa \leftarrow$ (Sign$[v_1]$, Sign$[v_2]$, Sign$[v_3]$, Sign$[v_4]$, Sign$[v_5]$);
12 | foreach edge quadruple $(e_1^S, e_2^S, e_3^S, e_4^S) \in$ Table$[\kappa]$ do
13 | | for $i \leftarrow 1$ to 4 do $e_i \leftarrow (v_j, v_k)$ where $e_i^S = (v_j^S, v_k^S)$;
14 | | Insert edge quadruple (e_1, e_2, e_3, e_4) into T;
15 | end
16 end
   /* Compute isosurface vertex coordinates using linear interpolation   */
17 foreach bipolar grid edge e with endpoints v and v' do
18 | $w_e \leftarrow$ LinearInterpolation (v.coord, F$[v]$, v'.coord, F$[v']$, σ);
19 end
   /* Convert T to set of tetrahedra   */
20 $\Upsilon \leftarrow \emptyset$;
21 foreach quadruple of edges $(e_1, e_2, e_3, e_4) \in$ T do
22 | $\Upsilon \leftarrow \Upsilon \cup \{(w_{e_1}, w_{e_2}, w_{e_3}, w_{e_4})\}$;
23 end

Algorithm 6.2. MARCHING SIMPLICES.

6.4.6 Isosurface Properties

The isosurface produced by MARCHING SIMPLICES has the same properties as the one produced by MARCHING HYPERCUBES. Under appropriate conditions, the isosurface is a 3-manifold with boundary.

MARCHING SIMPLICES returns a finite set, Υ, of oriented tetrahedra. The isosurface is the union of these tetrahedra. The vertices of the isosurface are the tetrahedra vertices.

The following properties apply to all isosurfaces produced by the MARCHING SIMPLICES algorithm.

Property 1.   *The isosurface is piecewise linear.*

Property 2.   *The vertices of the isosurface lie on mesh edges.*

Property 3.   *The isosurface intersects every bipolar mesh edge at exactly one point.*

Property 4.   *The isosurface does not intersect any negative or strictly positive mesh edges.*

Property 5.   *The isosurface separates positive mesh vertices from negative ones and strictly separates strictly positive mesh vertices from negative mesh vertices.*

Properties 3 and 4 imply that the isosurface intersects a minimum number of mesh edges. As with MARCHING HYPERCUBES, if both endpoints of a mesh edge have scalar value equal to the isovalue, then the isosurface may intersect the mesh edge zero, one, or two times or may contain the mesh edge.

By Property 3, the isosurface intersects every bipolar mesh edge. However, the bipolar mesh edge may be intersected by zero-volume isosurface tetrahedra.

Consider a simplicial mesh that has the following conditions:

- The isovalue does not equal the scalar value of any mesh vertex.

- The mesh is a partition of a 4-manifold with boundary.

Under these conditions, the isosurface produced by MARCHING SIMPLICES has the following properties:

Property 6.   *The isosurface is a piecewise linear orientable 3-manifold with boundary.*

Property 7.   *The boundary of the isosurface lies on the boundary of the mesh.*

Property 8.   *Set Υ does not contain any zero-volume tetrahedra or duplicate tetrahedra and the tetrahedra in Υ form a triangulation of the isosurface.*

Proofs of the properties for the MARCHING SIMPLICES isosurface are similar to the proofs for the MARCHING HYPERCUBES isosurface and are omitted.

6.5   Marching Polytopes

The three-dimensional MARCHING POLYHEDRA algorithm generalizes to four-dimensional convex meshes. The algorithm is called MARCHING POLYTOPES.

6.5.1   Algorithm

As with MARCHING SIMPLICES, care must be taken to ensure that triangulations of adjacent isosurface patches match. To do so, we require an additional input, an ordering of all the mesh vertices. This ordering will be used to control the mapping to the isosurface lookup tables so that triangulations of adjacent isosurface patches match. This ordering could be a lexicographic ordering on all the mesh vertices. However, as we shall see for a mesh of pyramids, it is sometimes preferable to use a different ordering of mesh vertices.

As in the 3D MARCHING POLYHEDRA, we assume that the mesh is composed of a fixed, predefined set of polytope classes (Section 5.4). Each class a has a precomputed isosurface lookup table built on a "generic" polytope $\mathbf{c}_a$. In 3D, a mesh polytope $\mathbf{c}$ is in class a if there is an affine transformation from $\mathbf{c}$ to $\mathbf{c}_a$. Because triangulations of adjacent isosurface patches must match, this partition into classes needs to be further refined in 4D.

The ordering of mesh vertices input to MARCHING POLYTOPES induces an ordering of the vertices of each mesh element. For each class a, the "generic" polytope $\mathbf{c}_a$ has an ordering on its vertices that is used in constructing the isosurface lookup table for class a. More specifically, this order determines the incremental triangulation of each isosurface patch constructed by Algorithm ISOPOLY4D. A mesh polytope $\mathbf{c}$ is in class a if there is an affine transformation from $\mathbf{c}$ to $\mathbf{c}_a$ that preserves the vertex order. In other words, the affine transformation should map the ith vertex of $\mathbf{c}$ (given by the input ordering of mesh vertices) to the ith vertex of $\mathbf{c}_a$.

If $\mathbf{c}$ and $\mathbf{c}_a$ are 4-simplices, then there is always an affine transformation of $\mathbf{c}$ to $\mathbf{c}_a$ that preserves vertex order. For this reason, the MARCHING SIMPLICES algorithm needs only one isosurface lookup table. However, for other mesh elements, say four-dimensional pyramids whose bases are three-dimensional cubes, there may be no affine transformation preserving vertex order. There may have to be multiple isosurface lookup tables built for such multiple elements, one for each ordering of the vertices.

Input to the MARCHING POLYTOPES algorithm is an isovalue, a four-dimensional convex polytopal mesh, a set of scalar values at the vertices of the mesh, a sequence of all the mesh vertices, and a set of polytope classes.

The MARCHING POLYTOPES algorithm has three steps. (See Figure 6.16.) The first step is reading the isosurface lookup tables for each class from preconstructed data files. Next, for each mesh element, a triangulated isosurface patch

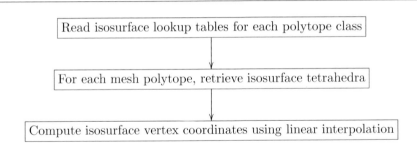

Figure 6.16. MARCHING POLYTOPES.

is retrieved from the appropriate table. Finally, the vertices of the isosurface patch are assigned geometric locations using linear interpolation. We explain the last two steps in more detail.

Let c be a mesh element. The input sequence of mesh vertices represents an ordering of the mesh vertices. Restricted to the vertices of c, this ordering gives an ordering of the vertices of c. Let a be the polytope class of c where some affine transformation preserving vertex order maps c to c_a. To retrieve an isosurface patch for mesh element c from the isosurface lookup table for class a, the ith vertex of c is mapped to the ith vertex of the lookup table polytope. This mapping determines a configuration of positive and negative vertices that identifies the appropriate entry in the lookup table.

The isosurface lookup table determines only the edge containing each isosurface vertex. The geometric location of the isosurface vertex along that edge is not defined by the lookup table.

The vertices of the isosurface tetrahedra are the isosurface vertices. To map each isosurface tetrahedron to a geometric tetrahedron, we use linear interpolation to position the isosurface vertices as described in Section 1.7.2. Each isosurface vertex v lies on a mesh edge $[p, q]$. If s_p and s_q are the scalar values at p and q and σ is the isovalue, then map v to $(1 - \alpha)p + q$ where $\alpha = (\sigma - s_p)/(s_q - s_p)$.

The MARCHING POLYTOPES algorithm is presented in Algorithm 6.3. Function LinearInterpolation, called by this algorithm, is defined in Algorithm 1.1 in Section 1.7.2.

6.5.2 Isosurface Properties

The isosurface produced by MARCHING POLYTOPES has the same properties as the one produced by MARCHING SIMPLICES. Under appropriate conditions the isosurface is a 3-manifold with boundary.

Input   : Γ is a 4D simplicial mesh.
           F is an array of the scalar values of each mesh vertex.
           σ is an isovalue.
           Z_Γ is a sequence of the vertices of Γ.
           TableClasses is a set of isosurface lookup table classes.
Result   : A set Υ of isosurface tetrahedra.

MarchingPolytopes(Γ,F, σ,Z_Γ, TableClasses, Υ)

1 foreach $a \in$ TableClasses do
2 | Read in isosurface lookup table, Table[a], for class a;
3 | $\mathbf{c}_a \leftarrow$ "generic" polytope of Table[a];
4 | $(v_1^a, v_2^a, v_3^a, \ldots) \leftarrow$ ordered sequence of vertices of $\mathbf{c}_a$;
5 end
   /* Assign "+" or "−" signs to each vertex   */
6 foreach mesh vertex v do
7 | if F[v] $< \sigma$ then Sign[v] $\leftarrow$ "−";
8 | else Sign[v] $\leftarrow$ "+";   /* F[v] $\geq \sigma$ */
9 end
10 T $\leftarrow \emptyset$;
   /* For each mesh polytope, retrieve isosurface tetrahedra   */
11 foreach mesh polytope $\mathbf{c}$ do
12 | $(v_1, v_2, v_3, \ldots) \leftarrow$ subsequence of Z_Γ containing vertices of $\mathbf{c}$;
13 | $\kappa \leftarrow$ (Sign[v_1], Sign[v_2], $\ldots$);
14 | $a \leftarrow$ class of polytope $\mathbf{c}$;
   | /* Requirement: There exists an affine transformation from $\mathbf{c}$ to $\mathbf{c}_a$
   | that maps $(v_1, v_2, v_3, \ldots)$ to $(v_1^a, v_2^a, v_3^a, \ldots)$ */
15 | foreach edge quadruple $(e_1^a, e_2^a, e_3^a, e_4^a) \in$ Table[a, κ] do
16 | | for $i \leftarrow 1$ to 4 do $e_i \leftarrow (v_j, v_k)$ where $e_i^a = (v_j^a, v_k^a)$;
17 | | Insert edge quadruple (e_1, e_2, e_3, e_4) into T;
18 | end
19 end
   /* Compute isosurface vertex coordinates using linear interpolation   */
20 foreach bipolar mesh edge e with endpoints v and v' do
21 | $w_e \leftarrow$ LinearInterpolation (v.coord, F[v], v'.coord, F[v'], σ);
22 end
   /* Convert T to set of tetrahedra   */
23 $\Upsilon \leftarrow \emptyset$;
24 foreach quadruple of edges $(e_1, e_2, e_3, e_4) \in$ T do
25 | $\Upsilon \leftarrow \Upsilon \cup \{(w_{e_1}, w_{e_2}, w_{e_3}, w_{e_4})\}$;
26 end

Algorithm 6.3. MARCHING POLYTOPES.

MARCHING POLYTOPES returns a finite set, Υ, of oriented tetrahedra. The isosurface is the union of these tetrahedra. The vertices of the isosurface are the tetrahedra vertices.

The following properties apply to all isosurfaces produced by the MARCHING POLYTOPES algorithm.

Property 1. *The isosurface is piecewise linear.*

Property 2. *The vertices of the isosurface lie on mesh edges.*

Property 3. *The isosurface intersects every bipolar mesh edge at exactly one point.*

Property 4. *The isosurface does not intersect any negative or strictly positive mesh edges.*

Property 5. *The isosurface separates positive mesh vertices from negative ones and strictly separates strictly positive mesh vertices from negative mesh vertices.*

Properties 3 and 4 imply that the isosurface intersects a minimum number of mesh edges. As with MARCHING SIMPLICES, if both endpoints of a mesh edge have scalar value equal to the isovalue, then the isosurface may intersect the mesh edge zero, one, or two times or may contain the mesh edge.

By Property 3, the isosurface intersects every bipolar mesh edge. However, the bipolar mesh edge may be intersected by zero-volume isosurface tetrahedra.

As in 3D, "complex" mesh elements could create isosurface patches that self-intersect, even if the isovalue does not equal the scalar value of any mesh vertex. To ensure a manifold, we restrict the mesh elements to a few simple types, such as cubes, 4-simplices, and pyramids. This list is not meant to be exhaustive and many other convex polytopes could probably be added. Consider a mesh of convex polytopes that has the following conditions:

- The isovalue does not equal the scalar value of any mesh vertex.

- The mesh is a partition of a 3-manifold with boundary.

- The mesh elements are affine transformations of cubes, 4-simplices, and pyramids over a cube base.

Under these conditions, the isosurface produced by MARCHING POLYTOPES has the following properties:

Property 6. *The isosurface is a piecewise linear, orientable 3-manifold with boundary.*

Property 7. *The boundary of the isosurface lies on the boundary of the mesh.*

Property 8.   *Set* Υ *does not contain any zero-volume tetrahedra or duplicate tetrahedra and the tetrahedra in* Υ *form a triangulation of the isosurface.*

Proofs of the properties for the MARCHING POLYTOPES isosurface are similar to the proofs for the MARCHING HYPERCUBES isosurface and are omitted.

6.5.3   Ordering of Mesh Vertices

The MARCHING POLYTOPES algorithm requires some ordering of mesh vertices to determine the mapping of mesh elements to their isosurface lookup tables. This ordering could simply be a lexicographic ordering of mesh vertices. In this section, we give an example of a pyramidal mesh where a different ordering is preferable.

Consider a three-dimensional cube and a four-dimensional hypercube. Adding a vertex v to the center of the cube and connecting vertex v to the six facets of the cube decomposes the cube into six pyramids. Each pyramid has apex at v and a square base. Similarly, adding a vertex v to the center of the hypercube and connecting vertex v to the eight facets of the hypercube decomposes the cube into eight pyramids. Since the facets of a hypercube are three-dimensional cubes, each pyramid has apex at v and a three-dimensional cube base.

Decompose a regular four-dimensional grid of hypercubes into a mesh of pyramids of four-dimensional pyramids by adding a vertex at the center of each hypercube. Any two mesh elements, $\mathbf{c}$ and $\mathbf{c}'$, are affine transformations of each other. However, if $\mathbf{c}$ is a pyramid whose apex is first in lexicographic order and $\mathbf{c}'$ is a pyramid whose apex is last in lexicographic order, then no affine transformation of $\mathbf{c}$ to $\mathbf{c}'$ preserves the lexicographic order of vertices. There are actually three different classes of pyramids based on lexicographic vertex order: one where the apex is first, one where the apex is last, and one where the apex is exactly in the middle of the order. (See Figure 6.17 for a three-dimensional illustration.)

If a lexicographic order of mesh vertices is supplied to MARCHING POLYTOPES, then the algorithm requires three different lookup tables based on the three different locations of the apex in the lexicographic order. Instead, we give a different ordering where only one lookup table is required. Each mesh vertex either is a pyramid apex or is contained in a pyramid base. It cannot be both. Let Z_A be the sequence of mesh vertices at pyramid apices sorted in lexicographic order and let Z_B be the sequence of mesh vertices in pyramid bases sorted in lexicographic order. The sequence composition $Z_A \circ Z_B$ is an ordering on all the mesh vertices. Under this ordering, the apex appears first in the vertex order of each mesh element. Between any two mesh elements, there is an affine transformation that preserves the vertex order. Thus, when $Z_A \circ Z_B$ is used as the input sequence for this mesh, MARCHING POLYTOPES requires only one isosurface lookup table.

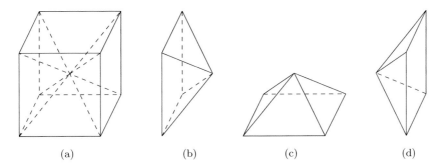

(a)   (b)   (c)   (d)

Figure 6.17. (a) Decomposition of a 3D cube into six pyramids. (b) Pyramid apex is last in lexicographic order. (c) Pyramid apex is in the middle of the lexicographic order of the vertices. (d) Pyramid apex is first in lexicographic order.

Instead of decomposing all hypercubes, one could decomposed only some hypercubes into eight pyramids. The resulting set of pyramids and hypercubes is still a mesh. Let Z_A be the sequence of mesh vertices at pyramid apices sorted in lexicographic order. Let Z_B be the sequence of all other mesh vertices sorted in lexicographic order. Let $Z_A \circ Z_B$ be an ordering on all the mesh vertices. Between any two hypercubes, there is a translation that preserves the vertex order. Between any two pyramids, there is an affine transformation preserving the vertex order. When $Z_A \circ Z_B$ is used as the input sequence for this mesh, Marching Polytopes requires two isosurface lookup tables, one for hypercubes and one for pyramids.

6.6 IsoHull4D

The IsoHull3D algorithm, in Section 5.5, can also be extended to four dimensions. As in IsoHull3D, we first locate the isosurface vertex on mesh edges using linear interpolation and then we construct the isosurface patches directly from the isosurface vertices.

Algorithm IsoHull4D has two steps (Figure 6.18). The first step is computing the location of the isosurface vertices. For each bipolar mesh edge $[p, q]$, apply linear interpolation as in Section 1.7.2 to compute a point on $[p, q]$ representing the intersection of the isosurface with $[p, q]$. If s_p and s_q are the scalar values at p and q and σ is the isovalue, then this point is $(1 - \alpha)p + \alpha q$, where $\alpha = (\sigma - s_p)/(s_q - s_p)$. Let U be the set of all such interpolated points over all bipolar mesh edges.

If the positive endpoint of $[p, q]$ is strictly positive, then $0 < \alpha < 1$ and $\alpha p + (1 - \alpha)q$ is in the interior of $[p, q]$. Thus no point of U lies on a strictly positive vertex of **c**.

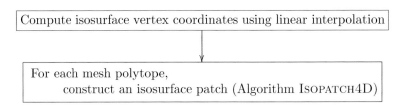

Figure 6.18. Algorithm ISOHULL4D.

Linear interpolation on multiple mesh edges that share the same endpoint could return that endpoint for each such edge. Since U is a set, it contains only a single copy of that endpoint.

The second step is constructing isosurface patches on those vertices. For each mesh polytope $\mathbf{c}$, let $U_{\mathbf{c}}$ be the points of U that lie in $\mathbf{c}$. Each bipolar edge of $\mathbf{c}$ contains one and only one point of $U_{\mathbf{c}}$ and no point of $U_{\mathbf{c}}$ lies on a strictly positive vertex or a negative vertex or a negative edge. Construct a region $R_{\mathbf{c}}^{+}$ that is the convex hull of $U_{\mathbf{c}}$ and the positive vertices of $\mathbf{c}$.

As in 3D, we differentiate three cases. If $R_{\mathbf{c}}^{+}$ is four-dimensional, then we triangulate the boundary of $R_{\mathbf{c}}^{+}$ and remove tetrahedra that lie on facets of $\mathbf{c}$. We orient the remaining tetrahedra so that the induced normals point toward $R_{\mathbf{c}}^{+}$. If $R_{\mathbf{c}}^{+}$ is three-dimensional and $R_{\mathbf{c}}^{+}$ is contained in some facet of $\mathbf{c}$, then we triangulate $R_{\mathbf{c}}^{+}$ and retain all the tetrahedra. We orient the tetrahedra so that the induced normals point toward $\mathbf{c}$. Otherwise, we don't construct any isosurface tetrahedra from $R_{\mathbf{c}}^{+}$.

Algorithm ISOPATCH4D for constructing the 4D isosurface patches is described in Figure 6.19. Apply ISOPATCH4D to each mesh polytope $\mathbf{c}$. The isosurface is represented by all the tetrahedra produced by all the calls to ISOPATCH4D.

6.6.1 Isosurface Properties

The isosurface produced by ISOHULL4D has the following properties:

Property 1. *The isosurface is piecewise linear.*

Property 2. *The vertices of the isosurface lie on mesh edges.*

Property 3. *The isosurface intersects every bipolar mesh edge at exactly one point.*

Property 4. *The isosurface does not intersect any negative or strictly positive mesh edges.*

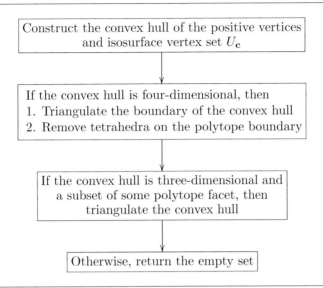

Figure 6.19. Algorithm ISOPATCH4D for isosurface patch construction in $\mathbb{R}^4$ from isosurface vertex set $U_{\mathbf{c}}$.

Property 5. *Set* Υ *does not contain any zero-volume tetrahedra.*

To claim that the isosurface separates strictly positive mesh vertices from strictly negative ones, we need the following condition on the mesh:

• The mesh is a partition of a 4-manifold with boundary.

Under this condition we have the following separation property:

Property 6. *The isosurface strictly separates strictly positive mesh vertices from negative mesh vertices.*

Consider a 4D convex polytopal mesh and isovalue that has the following two conditions:

• The isovalue does not equal the scalar value of any mesh vertex.

• The mesh is a partition of a 4-manifold with boundary.

Under these conditions, the isosurface produced by ISOHULL3D has the following three properties:

Property 7. *The isosurface is a piecewise linear, orientable 3-manifold with boundary.*

Property 8. *The boundary of the isosurface lies on the boundary of the mesh.*

Property 9. *Set Υ does not contain any zero-volume tetrahedra or duplicate tetrahedra and the tetrahedra in Υ form a triangulation of the isosurface.*

6.7   4D Surface Nets

The SURFACE NETS algorithm, described in Section 3.2, extends directly to four dimensions. Let Γ be a four-dimensional regular grid with scalar values assigned to the vertices of Γ. Add an isosurface vertex $w_{\mathbf{c}}$ to each grid hypercube $\mathbf{c}$ with at least one positive and at least one negative vertex. In $\mathbb{R}^4$, eight grid hypercubes can contain a single grid edge. For each bipolar grid edge contained in eight such hypercubes, $\mathbf{c}_1, \mathbf{c}_2, \ldots, \mathbf{c}_8$, add an isosurface hexahedron with vertices $w_{\mathbf{c}_1}, w_{\mathbf{c}_2}, \ldots, w_{\mathbf{c}_8}$ that is the dual of the eight hypercubes. For each pair of hypercubes that share a three-dimensional facet, the isosurface hexahedron has an edge. For each four hypercubes that share a two-dimensional face, the isosurface hexahedron has a quadrilateral.

To position the isosurface vertex within each grid cube, use linear interpolation as described in Section 1.7.2 to approximate the intersection of the isosurface and all the bipolar edges. For each bipolar edge $\mathbf{e} = [p, q]$, let $w_{\mathbf{e}}$ be the point $(1-\alpha)p+\alpha q$ where $\alpha = (\sigma - s_p)/(s_q - s_p)$. Take the centroid, $(w_{\mathbf{e}_1} + \cdots + w_{\mathbf{e}_k})/k$, of all the approximation points as the location of the isosurface vertex.

Dual contouring creates a hexahedral mesh but the vertices of each hexahedral element do not necessarily lie in a single 3-space. In fact, unless isosurface vertices are placed at the center of each grid cube, it is highly unlikely that the eight hexahedron vertices will lie in a single 3-space. To create a piecewise linear surface, one needs to triangulate each hexahedra, subdividing it into tetrahedra.

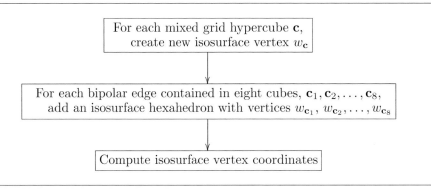

Figure 6.20. SURFACE NETS 4D.

Hexahedra can be triangulated using five or six tetrahedra without adding any new vertices. (See Figure 2.29 in Section 2.4.) The triangulation of the hexahedron induces a triangulation of each of the quadrilateral facets of the hexahedron into two triangles. To produce a triangulation of the isosurface, the triangulation of the quadrilateral facet of a hexahedron must match the triangulation of the quadrilateral facet of the adjacent hexahedron. Figure 2.29 depicts a triangulation into six tetrahedra such that each tetrahedra shares a diagonal edge from one corner to an opposing corner of the cube. Using this triangulation into six tetrahedra guarantees that the facet triangulations will match.

As discussed in Section 3.2.3 for 3D, simply breaking isosurface quadrilaterals produced by SURFACE NETS into two triangles could produce self-intersections in the isosurface. The same problem exists for the triangulation of hexahedra in 4D. Ju et al. [Ju and Udeshi, 2006] solved the problem in 3D by adding an additional isosurface vertex and breaking quadrilaterals into four triangles. It is possible that their approach generalizes to 4D, but no 4D solution has been published.

6.7.1 Isosurface Properties

As in 3D, the boundary of the SURFACE NETS isosurface is not usually on the boundary of the grid. A path connecting positive and negative grid vertices can bypass the isosurface in boundary grid cubes. However, if we restrict the path to interior grid cubes, then the path must intersect the isosurface. Restricted to interior grid cubes, the isosurface separates the positive and negative grid vertices.

If none of the grid vertices have scalar value equal to the isovalue, then the isosurface is minimal in the sense that it intersects interior bipolar grid edges exactly once and does not intersect positive or negative grid edges. The isosurface is not necessarily a manifold, even if none of the grid vertices have scalar value equal to the isovalue.

SURFACE NETS returns a finite set, Υ, of tetrahedra. The isosurface is the union of these triangles. The vertices of the isosurface are the tetrahedra vertices.

The following properties apply to all isosurfaces produced by the SURFACE NETS algorithm.

Property 1. *The isosurface is piecewise linear.*

Property 2. *Set Υ contains at most $6m$ tetrahedra where m is the number of bipolar grid edges.*

Property 3. *The isosurface does not contain any grid vertices whose scalar values do not equal the isovalue.*

Property 4. *The isosurface intersects at exactly one point every interior bipolar grid edge whose positive endpoint is strictly positive.*

Property 5. *The isosurface does not intersect any negative or strictly positive grid edges.*

Property 6. *The isosurface strictly separates the strictly positive vertices of* Γ_{inner} *from the negative vertices of* Γ_{inner}.

Properties 3–5 imply that the isosurface intersects a minimum number of grid edges.

The following properties apply to the SURFACE NETS isosurface whose isovalues do not equal the scalar value of any grid vertex.

Property 7. *Set* Υ *does not contain any zero-volume tetrahedra or duplicate tetrahedra.*

Since tetrahedra in Υ may intersect in their interiors, set Υ is not necessarily a triangulation of the isosurface.

Proofs of these properties are the direct 4D generalizations of the proofs of properties for SURFACE NETS 3D given in Sections 3.2.5, 3.2.6, and 3.2.7 and are omitted.

6.8 Notes and Comments

Weigle and Banks [Weigle and Banks, 1996, Weigle and Banks, 1998] proposed the first algorithm for isosurface construction in four dimensions. Their algorithm decomposes the regular 4D grid into simplices and then constructs isosurface patches in each simplex.

Roberts and Hill [Roberts and Hill, 1999] extended MARCHING CUBES to isosurface construction in 4D using isosurface lookup tables for 4D hypercubes. Their algorithm retrieves isosurface patches from the lookup table for each grid hypercube. The lookup tables were built by case analysis of configurations of "+" and "−" vertex labels.

Lachaud and Montanvert [Lachaud and Montanvert, 2000] and Bhaniramka et al. [Bhaniramka et al., 2000, Bhaniramka et al., 2004a] gave similar algorithms for automatically generating isosurface patches in 4D hypercubes using convex hulls. The algorithms in this chapter are based on their work. The algorithms can be used to generate isosurface patches in any convex polytope, not just 4D hypercubes.

CHAPTER 7

INTERVAL VOLUMES

The region between two level sets is called an interval volume. Formally, the interval volume for a function $\phi : \mathbb{R}^3 \to \mathbb{R}$ is defined as

$$\mathcal{I}_\phi(\sigma_0, \sigma_1) = \{x \in \mathbb{R}^3 : \sigma_0 \le \phi(x) \le \sigma_1\},$$

where $\sigma_0, \sigma_1 \in \mathbb{R}$ and $\sigma_0 < \sigma_1$.

Interval volumes are represented by a set of tetrahedra forming some approximation of $\mathcal{I}_\phi(\sigma_0, \sigma_1)$. Typically, this representation is the triangulation of the region between two piecewise linear isosurface approximations of the level sets $\phi^{-1}(\sigma_0)$ and $\phi^{-1}(\sigma_1)$. Unfortunately, the term *interval volume* is used both for the well-defined set $\mathcal{I}_\phi(\sigma_0, \sigma_1)$ and for the piecewise linear approximation to that set.

To visualize an interval volume $\mathcal{I}_\phi(\sigma_0, \sigma_1)$, the set of tetrahedra representing $\mathcal{I}_\phi(\sigma_0, \sigma_1)$ are volume rendered. Thus, interval volumes represent an intermediate step between direct volume rendering on all the data and rendering the triangulated isosurface. The advantage of interval volumes over direct volume rendering is that a much smaller region of interest is being processed and rendered, leading to much faster rendering times. The advantage of interval volumes over isosurfaces is that a range of values are represented in the image, allowing more complex visualizations of data. Interval volumes also enable visualization of surfaces with unclear boundaries that are not associated with a single isovalue.

Interval volumes were introduced by Guo in [Guo, 1995] and Fujishiro, Maeda, and Takeshima in [Fujishiro et al., 1996]. Nielson and Sung [Nielson and Sung, 1997] gave a MARCHING TETRAHEDRA-type algorithm for constructing interval volume representations using a manually constructed interval volume lookup table.

Bhaniramka et al. in [Bhaniramka et al., 2004b, Bhaniramka et al., 2004a] showed how the problem of constructing an interval volume in $\mathbb{R}^3$ can be transformed into constructing an isosurface in $\mathbb{R}^4$. One could apply this transformation, construct the isosurface in $\mathbb{R}^4$, and then project it back down to $\mathbb{R}^3$.

Instead, we use this transformation to construct an interval volume lookup table for a Marching Cubes-type algorithm for constructing interval volumes. Doing so is more efficient than constructing an isosurface in 4D and then projecting it back to 3D. We first give the algorithm for generating the interval volume using a lookup table and then discuss automatic generation of the lookup table.

7.1   Definitions and Notation

Because an interval volume, $\mathcal{I}_\phi(\sigma_0, \sigma_1)$, is based on two scalar values, σ_0 and σ_1, the definitions of vertex and edge properties for interval volumes are slightly different than in previous chapters.

Definition 7.1. Let v be a mesh vertex with scalar value s_v.

- Vertex v is positive, "+", if $s_v > \sigma_1$.

- Vertex v is negative, "−", if $s_v < \sigma_0$.

- Vertex v has a star label, "∗", if $\sigma_0 \leq s_v \leq \sigma_1$.[1]

Since σ_0 is less than σ_1, every vertex gets one of these three labels. (See Figure 7.1(a).) Note that a vertex with scalar value equal to σ_1 gets a "∗" label, not a positive label.

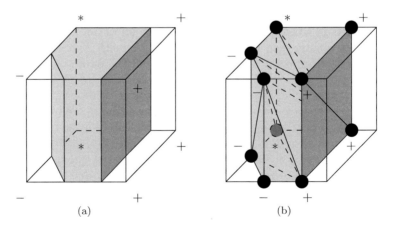

Figure 7.1. (a) Configuration "$- + * + - + * +$" and representative interval volume. (b) Interval volume triangulation vertices and their labels.

---

[1]This differs slightly from the convention in [Nielson and Sung, 1997] where a vertex with scalar value equal to σ_0 receives a positive label and a vertex with scalar value equal to σ_1 receives a negative label.

Let **c** be a convex polyhedron and Γ a convex polyhedral mesh where every vertex of **c** and Γ is labeled negative, positive, or strictly positive. We characterize the edges of **c** and Γ by the labels at their endpoints.

Definition 7.2.

- An edge of **c** or Γ is positive if both its endpoints are positive.

- An edge of **c** or Γ is negative if both its endpoints are negative.

7.2 MCVol

Input to the MCVOL algorithm is two isovalues, σ_0 and σ_1 with $\sigma_0 < \sigma_1$, and a set of scalar values at the vertices of a three-dimensional regular grid. The algorithm has three steps. (See Figure 7.2.) Read in the interval volume lookup table from a preconstructed data file. For each cube, retrieve from the lookup table a set of interval volume tetrahedra representing the combinatorial structure of the interval volume. The vertices of these tetrahedra form the interval volume vertices. Assign geometric locations to the interval volume vertices based on the scalar values at the cube edge endpoints. We first describe the interval volume lookup table.

7.2.1 Lookup Table

Since each vertex is labeled positive, negative, or star, and a cube has eight vertices, there are $3^8 = 6,561$ different configurations of cube vertex labels. The interval volume lookup table contains 6,561 entries, one for each configuration κ.

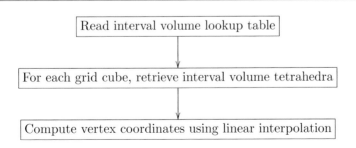

Figure 7.2. MCVOL. Algorithm to construct a piecewise linear approximation to the interval volume.

Each entry contains a list of oriented tetrahedra forming the triangulation of the interval volume for configuration κ. However, unlike vertices of triangles in the isosurface lookup table, vertices of tetrahedra in the interval volume lookup table may be cube vertices. Furthermore, a cube edge $\mathbf{e}$ can contain more than one tetrahedron vertex, although not more than two such vertices.

If a tetrahedron vertex lies in the interior of a cube edge $\mathbf{e}$, then it represents a point that lies on $\phi^{-1}(\sigma_0) \cap e$ or $\phi^{-1}(\sigma_1) \cap e$. If a tetrahedron vertex lies on a cube vertex, then it represents a point that is not on $\phi^{-1}(\sigma_0)$ or $\phi^{-1}(\sigma_1)$. Thus, we can label a tetrahedra vertex, "$-$", "$+$", or "$*$", depending on whether it represents a point on $\phi^{-1}(\sigma_0)$ or $\phi^{-1}(\sigma_1)$ or neither, respectively. (See Figure 7.1(b).) Each tetrahedron vertex is represented by a pair, $(g, \mathcal{L})$, where g is a vertex or edge and $\mathcal{L}$ is a "$-$", "$+$", or "$*$" label.

The entries in the interval volume lookup table are lists of quadruplets, $((g_1, \mathcal{L}_1), (g_2, \mathcal{L}_2), (g_3, \mathcal{L}_3), (g_4, \mathcal{L}_4))$. Each quadruplet represents a tetrahedron whose vertices lie on g_1 and g_2 and g_3 and g_4. The list of quadruplets define the combinatorial structure of the isosurface patch for configuration κ.

An algorithm to construct each of the 6,561 entries in the interval volume lookup table is described in Section 7.3. We first complete the description of algorithm MCVOL.

7.2.2 Algorithm

The first step in algorithm MCVOL is reading in the interval volume lookup table from a preconstructed data file. The next step is retrieving from the lookup table the appropriate list of interval volume tetrahedra for each grid cube.

Each grid vertex is labeled positive, negative, or star. The eight vertices of each grid cube determine a vector of eight vertex labels. Retrieve a list of interval volume tetrahedra from the corresponding entry in the interval volume lookup table.

The interval volume lookup table is constructed on the unit cube with vertices $(0,0,0), (1,0,0), (0,1,0), \ldots, (0,1,1), (1,1,1)$. To construct the interval volume in grid cube (i_x, i_y, i_z), we have to map unit cube edges to edges of cube (i_x, i_y, i_z). Each vertex $v = (v_x, v_y, v_z)$ of the unit cube maps to $v + (i_x, i_y, i_z) = (v_x, v_y, v_z) + (i_x, i_y, i_z) = (v_x + i_x, v_y + i_y, v_z + i_z)$. Each edge $\mathbf{e}$ of the unit square with endpoints (v, v') maps to edge $\mathbf{e} + (i_x, i_y, i_z) = (v + (i_x, i_y, i_z), v' + (i_x, i_y, i_z))$. Finally, each quadruplet $((g_1, \mathcal{L}_1), \ldots, (g_4, \mathcal{L}_4))$ in the interval volume lookup table maps to $((g_1 + (i_x, i_y, i_z), \mathcal{L}_1), \ldots, (g_4 + (i_x, i_y, i_z), \mathcal{L}_4))$.

The interval volume tetrahedra form a triangulation of the interval volume. The triangulation vertices are the vertices of the interval volume tetrahedra. The final step is mapping each triangulation vertex to its geometric location. We use linear interpolation as described in Section 1.7.2 to position the triangulation vertices.

Input : F is a 3D array of scalar values.
   Coord is a 3D array of (x, y, z) coordinates.
   σ_0, σ_1 are scalar values with $\sigma_0 < \sigma_1$.
Result : A set Υ of interval volume tetrahedra.

MCVol(F, Coord, σ_0, σ_1, Υ)

1 Read interval volume lookup table into Table;
   /* Assign "+" or "−" or "∗" labels to each vertex   */
2 foreach grid vertex (i_x, i_y, i_z) do
3 | if $\mathsf{F}[i_x, i_y, i_z] < \sigma_0$ then Label$[i_x, i_y, i_z] \leftarrow$ "−";
4 | else if $F[i_x, i_y, i_z] > \sigma_1$ then Label$[i_x, i_y, i_z] \leftarrow$ "+";
5 | else Label$[i_x, i_y, i_z] \leftarrow$ "∗";   /* $\sigma_0 \leq \mathsf{F}[i_x, i_y, i_z] \leq \sigma_1$ */
6 end
7 $\mathsf{T} \leftarrow \emptyset$;
   /* For each grid cube, retrieve interval volume tetrahedra   */
8 foreach grid cube (i, j, k) do
      /* Cube vertices are $(i_x, i_y, i_z), (i_x{+}1, i_y, i_z), \ldots, (i_x{+}1, i_y{+}1, i_z{+}1)$ */
9 | $\kappa \leftarrow$ (Label$[i_x, i_y, i_z]$, Label$[i_x{+}1, i_y, i_z], \ldots,$ Label$[i_x{+}1, i_y{+}1, i_z{+}1]$);
10 | foreach $((g_1, \mathcal{L}_1), (g_2, \mathcal{L}_2), (g_3, \mathcal{L}_3), (g_4, \mathcal{L}_4)) \in$ Table$[\kappa]$ do
11 | | Insert $((g_1 + (i_x, i_y, i_z), \mathcal{L}_1), \ldots, (g_4 + (i_x, i_y, i_z), \mathcal{L}_4))$ into T;
12 | end
13 end
   /* Compute vertex coordinates using linear interpolation   */
14 foreach grid edge $\mathbf{e} = (i, j)$ labeled $(-, *)$ or $(-, +)$ do
15 | $w_{(\mathbf{e}, -)} \leftarrow$ LinearInterpolation
16 | (Coord$[i_x, i_y, i_z]$, $\mathsf{F}[i_x, i_y, i_z]$, Coord$[j_x, j_y, j_z]$, $\mathsf{F}[j_x, j_y, j_z]$, σ_0);
17 end
18 foreach grid edge $\mathbf{e} = (i, j)$ labeled $(+, *)$ or $(-, +)$ do
19 | $w_{(\mathbf{e}, +)} \leftarrow$ LinearInterpolation
20 | (Coord$[i_x, i_y, i_z]$, $\mathsf{F}[i_x, i_y, i_z]$, Coord$[j_x, j_y, j_z]$, $\mathsf{F}[j_x, j_y, j_z]$, σ_1);
21 end
   /* Store vertex coordinates   */
22 foreach grid vertex v with label "∗" do
23 | $w_{(v, *)} \leftarrow (v_x, v_y, v_x)$;
24 end
   /* Convert T to set of triangles   */
25 $\Upsilon \leftarrow \emptyset$;
26 foreach quadruplet $((g_1, \mathcal{L}_1), (g_2, \mathcal{L}_2), (g_3, \mathcal{L}_3), (g_4, \mathcal{L}_4)) \in \mathsf{T}$ do
27 | $\Upsilon \leftarrow \Upsilon \cup \{(w_{(g_1, \mathcal{L}_1)}, w_{(g_2, \mathcal{L}_2)}, w_{(g_3, \mathcal{L}_3)}, w_{(g_4, \mathcal{L}_4)})\}$;
28 end

Algorithm 7.1. MCVol.

Each pair $(g, \mathcal{L})$ corresponds to a vertex in the triangulation of the interval volume. For each such pair, element g is either a grid vertex or a grid edge. If g is a grid vertex v, then the location of vertex $(g, \mathcal{L})$ is the location of vertex v. Assume g is a grid edge $[p, q]$. Let s_p and s_q be the scalar values at p and q. If $\mathcal{L}$ is "$-$", then map $(g, \mathcal{L})$ to $(1 - \alpha_0)p + \alpha_0 q$ where $\alpha_0 = (\sigma_0 - s_p)/(s_q - s_p)$. If $\mathcal{L}$ is "$+$", then map $(g, \mathcal{L})$ to $(1 - \alpha_1)p + \alpha_1 q$ where $\alpha_1 = (\sigma_1 - s_p)/(s_q - s_p)$. Note that if $\mathcal{L}$ is "$-$" or "$+$", then p and q must have different labels. Thus, scalar s_p does not equal s_q and the denominator $(s_q - s_p)$ is never zero.

Pseudocode for algorithm MCVOL is given in Algorithm 7.1. Function `LinearInterpolation`, called by this algorithm, is defined in Algorithm 1.1 in Section 1.7.2.

7.3   Automatic Table Generation

The interval volume lookup table consists of representative interval volumes, one for each cube configuration. To generate the interval volume for a given configuration, we embed the cube in a four-dimensional hypercube, construct an isosurface in that hypercube, and then project the tetrahedra in that isosurface to the three-dimensional cube to form the interval volume.

We first discuss the relationship between interval volumes in $\mathbb{R}^3$ and isosurfaces in $\mathbb{R}^4$.

7.3.1   Interval Volumes and Isosurfaces

Let

$$\mathcal{I}_\phi(\sigma_0, \sigma_1) = \{x \in \mathbb{R}^3 : \sigma_0 \leq \phi(x) \leq \sigma_1\}$$

be the interval volume for function ϕ for the interval $[\sigma_0, \sigma_1]$ where $\sigma_0 < \sigma_1$. Define the scalar function $\Psi : \mathbb{R}^3 \times [0, 1] \to \mathbb{R}$ as follows:

$$\Psi(p_x, p_y, p_z, \alpha) = \phi(p_x, p_y, p_z) - ((1 - \alpha)\sigma_0 + \alpha\sigma_1),$$

where $0 \leq \alpha \leq 1$. Note that $\{(p_x, p_y, p_z) : \Psi(p_x, p_y, p_z, 0) = 0\}$ equals $\phi^{-1}(\sigma_0)$ and $\{(p_x, p_y, p_z) : \Psi(p_x, p_y, p_z, 1) = 0\}$ equals $\phi^{-1}(\sigma_1)$.

We claim that the isosurface $\Psi^{-1}(0)$ is homeomorphic to the interval volume $\mathcal{I}_\phi(\sigma_0, \sigma_1)$. Let $\pi : \mathbb{R}^4 \to \mathbb{R}^3$ be the orthogonal projection of $\mathbb{R}^4$ onto $\mathbb{R}^3$ where $\pi(p_x, p_y, p_z, p_w) = (p_x, p_y, p_z)$.

Theorem 7.3. *If $\phi : \mathbb{R}^3 \to \mathbb{R}$ is a scalar field in $\mathbb{R}^3$ and $\sigma_0 < \sigma_1$, then π restricted to $\Psi^{-1}(0)$ is a homeomorphism from the isosurface $\Psi^{-1}(0)$ to the interval volume $\mathcal{I}_\phi(\sigma_0, \sigma_1)$.*

Proof: Let π' be the restriction of π to $\Psi^{-1}(0)$. Every point (p_x, p_y, p_z, α) on the isosurface $\Psi^{-1}(0)$ satisfies the equation

$$\phi(p_x, p_y, p_z) - ((1 - \alpha)\sigma_0 + \alpha\sigma_1) = 0$$

or, equivalently,

$$\phi(p_x, p_y, p_z) = (1 - \alpha)\sigma_0 + \alpha\sigma_1.$$

Since α is between 0 and 1, inclusive, $\phi(p_x, p_y, p_z)$ is between σ_0 and σ_1, inclusive, and so (p_x, p_y, p_z) is in $\mathcal{I}_\phi(\sigma_0, \sigma_1)$. Thus, $\pi'(\Psi^{-1}(0))$ is a subset of $\mathcal{I}_\phi(\sigma_0, \sigma_1)$.

For every point $(p_x, p_y, p_z) \in \mathcal{I}_\phi(\sigma_0, \sigma_1)$, scalar $\phi(p_x, p_y, p_z)$ is between σ_0 and σ_1, inclusive. Thus, there is some scalar $\alpha \in [0, 1]$ such that $\phi(p_x, p_y, p_z) = (1 - \alpha)\sigma_0 + \alpha\sigma_1$. Since $\phi(p_x, p_y, p_z) - ((1 - \alpha)\sigma_0 + \alpha\sigma_1)$ equals zero, point (p_x, p_y, p_z, α) is in $\Psi^{-1}(0)$. Since projection $\pi'(p_x, p_y, p_z, \alpha)$ equals (p_x, p_y, p_z), point (p_x, p_y, p_z) is in $\pi'(\Psi^{-1}(0))$. Thus, $\pi'(\Psi^{-1}(0))$ equals $\mathcal{I}_\phi(\sigma_0, \sigma_1)$.

Since σ_0 does not equal σ_1, for every scalar $y \in \mathbb{R}$, there is a unique α such that $y = (1 - \alpha)\sigma_0 + \alpha\sigma_1$. Thus, for each point $(p_x, p_y, p_z) \in \mathcal{I}_\phi(\sigma_0, \sigma_1)$, the scalar $\alpha \in [0, 1]$ such that $\phi(p_x, p_y, p_z) = (1 - \alpha)\sigma_0 + \alpha\sigma_1$ is unique and π' is one-to-one. Since π' is a continuous, one-to-one mapping from $\Psi^{-1}(0)$ onto $\mathcal{I}(\sigma_0, \sigma_1)$, π' is a homeomorphism. $\qquad\square$

7.3.2   Interval Volume Patch Construction

We present an algorithm, IVolPatch3D, that constructs interval volume patches in any three-dimensional convex polyhedron, not just a cube. Input to this algorithm is a convex polyhedron $\mathbf{c}$ in $\mathbb{R}^3$ where each vertex of $\mathbf{c}$ has a positive, negative, or star label, and an ordering of the vertices of $\mathbf{c}$. The ordering of the vertices of $\mathbf{c}$ determines the triangulation of the interval volume.

As in the previous section, $\pi : \mathbb{R}^4 \to \mathbb{R}^3$ is the orthogonal projection of $\mathbb{R}^4$ onto $\mathbb{R}^3$ where $\pi(p_x, p_y, p_z, p_w) = (p_x, p_y, p_z)$. The set $\mathbf{c} \times [0, 1]$ is a convex polytope in $\mathbb{R}^4$. Set $\mathbf{c} \times [0, 1]$ equals $\{(p_x, p_y, p_z, \alpha) : (p_x, p_y, p_z) \in \mathbf{c} \text{ and } \alpha \in [0, 1]\}$. When $\mathbf{c}$ is a cube, set $\mathbf{c} \times [0, 1]$ is a hypercube. Note that $\pi(\mathbf{c} \times [0, 1])$ equals $\mathbf{c}$.

The interval volume in $\mathbf{c}$ is a region that contains all the star vertices and separates the positive from the negative vertices. We show how to construct the interval volume in $\mathbf{c}$ by constructing a three-dimensional isosurface in $\mathbf{c} \times [0, 1]$ and then projecting that isosurface back into $\mathbf{c}$.

The vertices of $\mathbf{c} \times [0, 1]$ lie on the hyperplanes

$$\mathbf{h}_0 = \{(p_x, p_y, p_z, p_w) : p_w = 0\},$$
$$\mathbf{h}_1 = \{(p_x, p_y, p_z, p_w) : p_w = 1\}.$$

Label the vertices of $\mathbf{c} \times [0, 1]$ as follows.

- If $v = (v_x, v_y, v_z)$ has a positive label, then label vertices $(v_x, v_y, v_z, 0)$ and $(v_x, v_y, v_z, 1)$ of $\mathbf{c} \times [0, 1]$ positive.

Label vertices of $\mathbf{c} \times [0,1]$ as follows:
For each vertex $v = (v_x, v_y, v_z)$ of the labeled cube
1. If v is negative, label $(v_x, v_y, v_z, 0)$ and $(v_x, v_y, v_z, 1)$ as negative
2. If v is positive, label $(v_x, v_y, v_z, 0)$ and $(v_x, v_y, v_z, 1)$ as positive
3. If v has a star label,
    label $(v_x, v_y, v_z, 0)$ as positive and $(v_x, v_y, v_z, 1)$ as negative

Construct a proper isosurface vertex set as follows:
For each bipolar edge $\mathbf{e} = (v, v')$ of the unit hypercube
(where v is negative and v' is positive)
1. If $\mathbf{e}$ is in hyperplane $\mathbf{h}_0$, then $u_{\mathbf{e}}^{\mathcal{I}} \leftarrow (2/3)v + (1/3)v'$
2. Else if $\mathbf{e}$ is in hyperplane $\mathbf{h}_1$, then $u_{\mathbf{e}}^{\mathcal{I}} \leftarrow (1/3)v + (2/3)v'$
3. Else $u_{\mathbf{e}}^{\mathcal{I}} \leftarrow (1/2)v + (1/2)v'$

$Z_{\mathbf{c}} \leftarrow$ Sequence of vertices of $\mathbf{c}$ in order
$Z_{\mathbf{c} \times [0,1]} \leftarrow ((v_1, 0), (v_1, 1), (v_2, 0), (v_2, 1), \ldots, (v_i, 0), (v_i, 1), \ldots)$
    where v_i is ith element of $Z_{\mathbf{c}}$

Construct an isosurface patch in $\mathbf{c} \times [0,1] \subset \mathbb{R}^4$
using isosurface vertex set $U^{\mathcal{I}} = \{u_{\mathbf{e}}^{\mathcal{I}}\}$ and order $Z_{\mathbf{c} \times [0,1]}$

Orthogonally project each tetrahedron in the isosurface patch
into $\mathbf{c}$ in $\mathbb{R}^3$

Label interval volume vertices as follows:
1. If $u_{\mathbf{e}}^{\mathcal{I}} \in U^{\mathcal{I}}$ is on $\mathbf{h}_0$, assign label "$-$" to $\pi(u_{\mathbf{e}}^{\mathcal{I}})$
2. Else if $u_{\mathbf{e}}^{\mathcal{I}} \in U^{\mathcal{I}}$ is on $\mathbf{h}_1$, assign label "$+$" to $\pi(u_{\mathbf{e}}^{\mathcal{I}})$
3. Else assign label "$*$" to $\pi(u_{\mathbf{e}}^{\mathcal{I}})$

Figure 7.3. Algorithm IVOLPATCH3D for interval volume construction in a 3D convex polyhedron.

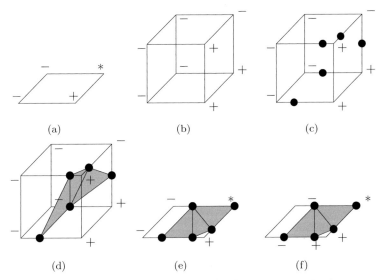

Figure 7.4. Two-dimensional example of interval volume construction in square $\mathbf{c}$ using isosurface in $\mathbf{c} \times [0,1]$. (a) Square $\mathbf{c}$ with vertices labeled "$-$", "$+$", or "$*$". (b) Cube $\mathbf{c} \times [0,1]$ with vertices labeled "$-$" or "$+$" based on the labeling of the vertices of $\mathbf{c}$. (c) Isosurface vertex set $U^{\mathcal{I}}_{\mathbf{c} \times [0,1]}$ for $\mathbf{c} \times [0,1]$. (d) Isosurface patch in $\mathbf{c} \times [0,1]$. (e) Interval volume in $\mathbf{c}$ formed by projecting the isosurface patch in $\mathbf{c} \times [0,1]$ into $\mathbf{c}$. (f) Labels "$-$", "$+$", and "$*$" of interval volume vertices.

- If $v = (v_x, v_y, v_z)$ has a negative label, then label both $(v_x, v_y, v_z, 0)$ and $(v_x, v_y, v_z, 1)$ negative.

- If $v = (v_x, v_y, v_z)$ has a star label, then label $(v_x, v_y, v_z, 0)$ positive and $(v_x, v_y, v_z, 1)$ negative.

See Figure 7.4(b) for a two-dimensional example.

Construct a proper isosurface vertex set $U^{\mathcal{I}}_{\mathbf{c} \times [0,1]}$ for $\mathbf{c} \times [0,1]$ as follows. Edges of $\mathbf{c} \times [0,1]$ are contained in the hyperplane $\mathbf{h}_0$ or are contained in the hyperplane $\mathbf{h}_1$ or have one vertex on $\mathbf{h}_0$ and one vertex on $\mathbf{h}_1$. For each bipolar edge $\mathbf{e} = (v, v')$ of $\mathbf{c} \times [0,1]$, we determine a point $u^{\mathcal{I}}_{\mathbf{e}} \in \mathbf{e}$. Without loss of generality, assume that v is negative and v' is positive.

- If $\mathbf{e}$ is contained in hyperplane $\mathbf{h}_0$, let $u^{\mathcal{I}}_{\mathbf{e}}$ be the point $(2/3)v + (1/3)v'$. (Point $u^{\mathcal{I}}_{\mathbf{e}}$ is the point on $\mathbf{e}$ one-third of the way from v to v'.)

- If $\mathbf{e}$ is contained in hyperplane $\mathbf{h}_1$, let $u^{\mathcal{I}}_{\mathbf{e}}$ be the point $(1/3)v + (2/3)v'$. (Point $u^{\mathcal{I}}_{\mathbf{e}}$ is the point on $\mathbf{e}$ two-thirds of the way from v to v'.)

- If one vertex of $\mathbf{e}$ is on hyperplane $\mathbf{h}_0$ while the other vertex is on hyperplane $\mathbf{h}_1$, let $u^{\mathcal{I}}_{\mathbf{e}}$ be the point $(1/2)v + (1/2)v'$. (Point $u^{\mathcal{I}}_{\mathbf{e}}$ is the midpoint of $\mathbf{e}$.)

See Figure 7.4(c) for a two-dimensional example. Let $U^{\mathcal{I}}_{\mathbf{c} \times [0,1]}$ be the set $\{ u^{\mathcal{I}}_{\mathbf{e}} :$ $\mathbf{e}$ is a bipolar edge of $\mathbf{c} \times [0,1]$ $\}$.

Let $(v_1, v_2, v_3, \ldots)$ be the input ordering on the vertices of $\mathbf{c}$. Construct the isosurface $S^+_{\mathbf{c} \times [0,1]}(U^{\mathcal{I}}_{\mathbf{c} \times [0,1]})$ of $\mathbf{c} \times [0,1]$ and its triangulation using the algorithm IsoPoly4D in Section 6.4.3. Input to IsoPoly4D is the proper isosurface vertex set $U^{\mathcal{I}}_{\mathbf{c} \times [0,1]}$ and polytope $\mathbf{c} \times [0,1]$ with vertices in the order $((v_1, 0), (v_1, 1), (v_2, 0), (v_2, 1), (v_3, 0), (v_3, 1), \ldots)$. Set $U^{\mathcal{I}}_{\mathbf{c} \times [0,1]}$ is the set of vertices of the triangulated isosurface (see Corollary 6.7). The triangulation is three-dimensional and composed of oriented tetrahedra. Orthogonally project each tetrahedron in the triangulation into $\mathbb{R}^3$ (Figure 7.4(e)). The union of the projected tetrahedra form the interval volume in $\mathbf{c}$.

There are three types of vertices in $U^{\mathcal{I}}_{\mathbf{c} \times [0,1]}$: those that lie on the hyperplane $\mathbf{h}_0$, those that lie on the hyperplane $\mathbf{h}_1$, and those that lie between the two hyperplanes. The projections of the first two types lie in the interiors of edges of $\mathbf{c}$. The projections of the last type lie on vertices of $\mathbf{c}$.

Algorithm MCVol determines the geometric location of the interval volume vertices using linear interpolation. For this interpolation step, it is necessary to distinguish between different types of vertices. If $u \in U^{\mathcal{I}}_{\mathbf{c} \times [0,1]}$ lies on hyperplane $\mathbf{h}_0$, then assign a negative label to $\pi(u)$. If $u \in U^{\mathcal{I}}_{\mathbf{c} \times [0,1]}$ lies on hyperplane $\mathbf{h}_1$, then assign a positive label to $\pi(u)$. If $u \in U^{\mathcal{I}}_{\mathbf{c} \times [0,1]}$ is between $\mathbf{h}_0$ and $\mathbf{h}_1$, then assign a star label to $\pi(u)$. (See Figure 7.4(f).) Algorithm IVolPatch3D is presented in Figure 7.3.

7.3.3 Interval Volume Table Construction

As discussed in Section 7.2.1, the interval volume lookup table has $3^8 = 6,561$ entries, one for each configuration of cube vertex labels. For each configuration, construct an interval volume patch using algorithm IVolPatch3D. Input to IVolPatch3D is the cube $(0,0,0), (1,0,0), (0,1,0), \ldots, (1,1,1)$ with vertices sorted in lexicographic order.

Algorithm IVolPatch3D returns a set of oriented tetrahedra representing a triangulation of the interval volume. Note that this is not a lexicographic triangulation of the interval volume but it is an incremental one. (See Section 6.4.1 for the definition and discussion of incremental triangulations.)

An interval volume tetrahedron with vertices (w_1, w_2, w_3, w_4) is represented in the isosurface lookup table by a quadruplet:

$$((g_1, \mathcal{L}_1), (g_2, \mathcal{L}_2), (g_3, \mathcal{L}_3), (g_4, \mathcal{L}_4)).$$

Symbol g_i is the cube edge or cube vertex containing w_i. Symbol $\mathcal{L}_i$ is the label $(-, +, \text{ or } *)$ of w_i.

7.4 MCVol Interval Volume Properties

The MARCHING CUBES isosurface is a 2-manifold with boundary if no grid point has scalar value equal to the isovalue. Similarly, the interval volume produced by MCVOL is a piecewise linear, orientable 3-manifold with boundary if no grid point has scalar value equal to σ_0 or to σ_1. If some grid point has scalar value equal to σ_0 or to σ_1, then the interval volume may not be a 3-manifold.

MCVOL returns a finite set, Υ, of oriented tetrahedra. The interval volume is the union of these tetrahedra. The vertices of the interval volume are the tetrahedra vertices.

The following properties apply to all interval volumes produced by the MCVOL algorithm. The scalar value at a grid vertex v is denoted s_v.

Property 1. *The interval volume is piecewise linear.*

Property 2. *The vertices of the interval volume lie on grid vertices or grid edges.*

Property 3. *The interval volume contains every grid vertex v where $\sigma_0 \le s_v \le \sigma_1$ and intersects each grid edge $(v, \tilde{v})$ where $s_v < \sigma_1$ and $s_{\tilde{v}} > \sigma_0$ (or vice versa) in a line segment.*

Property 4. *The interval volume does not contain any grid vertex v where $s_v < \sigma_0$ or $s_v > \sigma_1$ and does not intersect any grid edges $(v, \tilde{v})$ where both s_v and $s_{\tilde{v}}$ are strictly less than σ_0 or where both s_v and $s_{\tilde{v}}$ are strictly greater than σ_1.*

Property 5. *The interval volume strictly separates vertices with scalar values less than σ_0 from vertices with scalar value greater than σ_1.*

Properties 3 and 4 imply that the interval volume intersects a minimum number of grid edges.

The following property applies to the MCVOL interval volumes when no grid vertex has scalar value equal to σ_0 or σ_1.

Property 6. *The interval volume is a piecewise linear, orientable 3-manifold with boundary.*

Property 7. *Set Υ does not contain any zero-volume tetrahedra or duplicate tetrahedra and the tetrahedra in Υ form a triangulation of the interval volume.*

7.4.1 Proofs of Interval Volume Properties 1–4

We start with a mathematical description of the interval volume constructed by Algorithm IVOLPATCH3D and prove some properties of this interval volume.

These properties apply to interval volume patches constructed in any convex polyhedron, not just in cubes.

Let $\mathbf{c}$ be a convex polyhedron and Γ be a convex polyhedral mesh in $\mathbb{R}^3$ where each vertex of $\mathbf{c}$ and of Γ has the label "$-$", "$+$", or "$*$". Let $\pi : \mathbb{R}^4 \to \mathbb{R}^3$ be the orthogonal projection of $\mathbb{R}^4$ onto $\mathbb{R}^3$ where $\pi(p_x, p_y, p_z, p_w) = (p_x, p_y, p_z)$. Projection π maps $\mathbf{c} \times [0, 1] \subset \mathbb{R}^4$ to $\mathbf{c} \subset \mathbb{R}^3$ and $\Gamma \times [0, 1] \subset \mathbb{R}^4$ to $\Gamma \subset \mathbb{R}^3$. Note that π maps each vertex of $\mathbf{c} \times [0, 1]$ or $\Gamma \times [0, 1]$ to a vertex of $\mathbf{c}$ or Γ, respectively.

Let $\chi(\mathbf{c})$ equal $\mathbf{c} \times [0, 1]$ and $\chi(\Gamma)$ equal $\Gamma \times [0, 1]$, where vertices of $\chi(\mathbf{c})$ and $\chi(\Gamma)$ are labeled as follows:

- A vertex v in $\chi(\mathbf{c})$ or $\chi(\Gamma)$ has a negative label if $\pi(v)$ has a negative label.

- A vertex $v \in \chi(\mathbf{c})$ has a positive label if $\pi(v)$ has a positive label.

- A vertex $v^0 = (v_1, v_2, v_3, 0)$ in $\chi(\mathbf{c})$ or $\chi(\Gamma)$ has a positive label if $\pi(v^0)$ has a star label.

- A vertex $v^1 = (v_1, v_2, v_3, 1)$ in $\chi(\mathbf{c})$ or $\chi(\Gamma)$ has a negative label if $\pi(v^1)$ has a star label.

Define hyperplanes $\mathbf{h}_0, \mathbf{h}_1 \subset \mathbb{R}^4$ as

$$\begin{aligned} \mathbf{h}_0 &= \{(p_x, p_y, p_z, p_w) \in \mathbb{R}^4 : p_w = 0\}, \\ \mathbf{h}_1 &= \{(p_x, p_y, p_z, p_w) \in \mathbb{R}^4 : p_w = 1\}. \end{aligned}$$

For each bipolar edge $\mathbf{e} = (v, v')$ in $\chi(\mathbf{c})$ or $\chi(\Gamma)$, define a point $u_{\mathbf{e}}^{\mathcal{I}}$ as in Section 7.3.2. Assume that v is negative and v' is positive.

- If $\mathbf{e} = (v, v')$ is on hyperplane $\mathbf{h}_0$, then $u_{\mathbf{e}}^{\mathcal{I}}$ equals $(2/3)v + (1/3)v'$.

- If $\mathbf{e} = (v, v')$ is on hyperplane $\mathbf{h}_1$, then $u_{\mathbf{e}}^{\mathcal{I}}$ equals $(1/3)v + (2/3)v'$.

- If $\mathbf{e} = (v, v')$ has one endpoint on $\mathbf{h}_0$ and one endpoint on $\mathbf{h}_1$, then $u_{\mathbf{e}}^{\mathcal{I}}$ equals $(1/2)v + (1/2)v'$.

Let $U_{\chi(\mathbf{c})}^{\mathcal{I}}$ equal $\{u_{\mathbf{e}}^{\mathcal{I}} : \mathbf{e}$ is a bipolar edge of $\chi(\mathbf{c})\}$. Let $U_{\chi(\Gamma)}^{\mathcal{I}}$ equal $\{u_{\mathbf{e}}^{\mathcal{I}} : \mathbf{e}$ is a bipolar edge of $\chi(\Gamma)\}$. Sets $U_{\chi(\mathbf{c})}^{\mathcal{I}}$ and $U_{\chi(\Gamma)}^{\mathcal{I}}$ are proper isosurface vertex sets of $\chi(\mathbf{c})$ and $\chi(\Gamma)$, respectively.

Let $T_{\chi(\mathbf{c})}^{\mathcal{I}}$ be the set of tetrahedra in the incremental triangulation of the isosurface patch constructed by algorithm IsoPoly4D (Chapter 6) using isosurface vertex set $U_{\chi(\mathbf{c})}^{\mathcal{I}}$. Let $T_{\mathbf{c}}^{\mathcal{I}} = \{\pi(t) : t \in T_{\chi(\mathbf{c})}^{\mathcal{I}}\}$ be the orthogonal projection of $T_{\chi(\mathbf{c})}^{\mathcal{I}}$ into $\mathbf{c}$. Set $T_{\mathbf{c}}^{\mathcal{I}}$ is the set of tetrahedra returned by algorithm IVolPatch3D.

Let μ be a translation of convex polyhedron $\mathbf{c}$. Since $\mu(\mathbf{c})$ is a translation of $\mathbf{c}$, polyhedron $\chi(\mu(\mathbf{c}))$ is a translation of $\chi(\mathbf{c})$ and $|T_{\chi(\mu(\mathbf{c}))}^{\mathcal{I}}|$ is a translation of $|T_{\chi(\mathbf{c})}^{\mathcal{I}}|$. (Sets $|T_{\chi(\mu(\mathbf{c}))}^{\mathcal{I}}|$ and $|T_{\chi(\mathbf{c})}^{\mathcal{I}}|$ are the unions of all the tetrahedra in $T_{\chi(\mu(\mathbf{c}))}^{\mathcal{I}}$

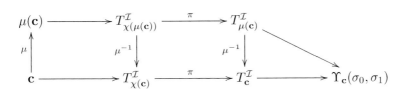

Figure 7.5. The MCVOL algorithm constructs the interval volume patch $\Upsilon_{\mathbf{c}}(\sigma_0, \sigma_1)$ by mapping cube $\mathbf{c}$ to $\mu(\mathbf{c})$, retrieving $T_{\mu(\mathbf{c})}^{\mathcal{I}}$ from the interval volume lookup table and mapping $T_{\mu(\mathbf{c})}^{\mathcal{I}}$ to $\Upsilon_{\mathbf{c}}(\sigma_0, \sigma_1)$ using linear interpolation. The isosurface patch $\Upsilon_{\mathbf{c}}(\sigma_0, \sigma_1)$ could be constructed directly from $\mathbf{c}$ by lifting $\mathbf{c}$ to $\chi(\mathbf{c})$, constructing the isosurface patch $|T_{\chi(\mathbf{c})}^{\mathcal{I}}|$ in $\chi(\mathbf{c})$, projecting $T_{\chi(\mathbf{c})}^{\mathcal{I}}$ into $T_{\mathbf{c}}^{\mathcal{I}}$, and then mapping $T_{\mathbf{c}}^{\mathcal{I}}$ to $\Upsilon_{\mathbf{c}}(\sigma_0, \sigma_1)$.

and $T_{\chi(\mathbf{c})}^{\mathcal{I}}$, respectively.) Since incremental triangulations are invariant under translations, $T_{\chi(\mu(\mathbf{c}))}^{\mathcal{I}}$ and $T_{\mu(\mathbf{c})}^{\mathcal{I}}$ are translations of $T_{\chi(\mathbf{c})}^{\mathcal{I}}$ and $T_{\mathbf{c}}^{\mathcal{I}}$, respectively. (See Figure 7.5.)

Algorithm MCVOL constructs the interval volume for a grid cube $\mathbf{c}$ using the interval volume lookup table for $\mu(\mathbf{c})$, where $\mu(\mathbf{c})$ is the unit cube with vertices $(0, 0, 0), \ldots, (1, 1, 1)$ and μ is a translation of $\mathbf{c}$ to this unit cube. The algorithm retrieves from the lookup table $T_{\mu(\mathbf{c})}^{\mathcal{I}}$ and then embeds $T_{\mu(\mathbf{c})}^{\mathcal{I}}$ in $\mathbf{c}$ using linear interpolation to position the vertices of $T_{\mu(\mathbf{c})}^{\mathcal{I}}$.

Because $T_{\mu(\mathbf{c})}^{\mathcal{I}}$ and $T_{\mathbf{c}}^{\mathcal{I}}$ are isomorphic, the algorithm could have constructed the interval volume from $T_{\mathbf{c}}^{\mathcal{I}}$ by reembedding $T_{\mathbf{c}}^{\mathcal{I}}$ in $\mathbf{c}$ using linear interpolation on its vertices. The algorithm uses $\mu(\mathbf{c})$ to construct the interval volume because it is faster to retrieve the precomputed values of $T_{\mu(\mathbf{c})}^{\mathcal{I}}$ from the lookup table than to compute $T_{\mathbf{c}}^{\mathcal{I}}$. However, for the analysis we will view the interval volume as an embedding of $T_{\mathbf{c}}^{\mathcal{I}}$ into $\mathbf{c}$.

Let $U_{\mathbf{c}}^{\mathcal{I}}$ be the set of vertices of $T_{\mathbf{c}}^{\mathcal{I}}$. Algorithm IVOLPATCH3D assigns a label, "$-$", "$+$", or "$*$" to each vertex in $U_{\mathbf{c}}^{\mathcal{I}}$. For each $u_i \in U_{\mathbf{c}}^{\mathcal{I}}$, define point w_i as follows:

- If u_i has the label "$*$", then $w_i = u_i$.

- If u_i has the label "$-$", then $w_i = (1 - \alpha_0)v + \alpha_0 \tilde{v}$ where $(v, \tilde{v})$ is the edge of $\mathbf{c}$ containing u_i and $\alpha_0 = (\sigma_0 - s_v)/(s_{\tilde{v}} - s_v)$.

- If u_i has the label "$+$", then $w_i = (1 - \alpha_1)v + \alpha_1 \tilde{v}$ where $(v, \tilde{v})$ is the edge of $\mathbf{c}$ containing u_i and $\alpha_1 = (\sigma_1 - s_v)/(s_{\tilde{v}} - s_v)$.

Linearly extend the mapping from u_i to w_i to each tetrahedron in $T_{\mathbf{c}}^{\mathcal{I}}$. Let $\Upsilon_{\mathbf{c}}(\sigma_0, \sigma_1)$ be the resulting set of tetrahedra.

Let Γ be a regular grid with scalar values at its vertices. Define $\Upsilon_\Gamma(\sigma_0, \sigma_1)$ as

$$\Upsilon_\Gamma(\sigma_0, \sigma_1) = \bigcup_{\mathbf{c} \in \Upsilon} \Upsilon_\mathbf{c}(\sigma_0, \sigma_1).$$

Sets $|\Upsilon_\mathbf{c}(\sigma_0, \sigma_1)|$ and $|\Upsilon_\Gamma(\sigma_0, \sigma_1)|$ are the union of all the tetrahedra in $\Upsilon_\mathbf{c}(\sigma_0, \sigma_1)$ and $\Upsilon_\Gamma(\sigma_0, \sigma_1)$, respectively. Algorithm MCVOL returns set $\Upsilon_\Gamma(\sigma_0, \sigma_1)$ representing the interval volume $|\Upsilon_\Gamma(\sigma_0, \sigma_1)|$.

Properties 1 and 2 come directly from the construction of $\Upsilon_\Gamma(\sigma_0, \sigma_1)$.

Property 1.   *The interval volume is piecewise linear.*

Property 2.   *The vertices of the interval volume lie on grid vertices or grid edges.*

Proof of Properties 1 & 2: Let Γ be a regular grid with scalar values at its vertices. Set $|\Upsilon_\Gamma(\sigma_0, \sigma_1)|$ is composed of a finite set of tetrahedra, so it is piecewise linear. By construction, the vertices of these tetrahedra lie on grid edges or grid vertices. $\square$

We need a few lemmas to prove Properties 3 and 4, A vertex of $\mathbf{c}$ is a vertex of $T_\mathbf{c}^\mathcal{I}$ if and only if it has the label "$*$".

Lemma 7.4. *Let $\mathbf{c}$ be a convex polyhedron in $\mathbb{R}^3$ where each vertex of $\mathbf{c}$ has the label "$-$", "$+$", or "$*$". Vertex v of $\mathbf{c}$ is a vertex of some tetrahedron in $T_\mathbf{c}^\mathcal{I}$ if and only if the label of vertex v is "$*$".*

Proof: Vertex v of $\mathbf{c}$ is the projection $\pi(p^0)$ and $\pi(p^1)$ of two vertices p^0 and p^1 of $\mathbf{c} \times [0,1]$. Vertex v is a vertex of some tetrahedron in $T_\mathbf{c}^\mathcal{I}$ if and only if (p^0, p^1) is a bipolar edge in $\mathbf{c} \times [0,1]$. Edge (p^0, p^1) is bipolar if and only if vertex v has the label "$*$". Thus vertex v is a vertex of some tetrahedron in $T_\mathbf{c}^\mathcal{I}$ if and only if vertex v has the label "$*$". $\square$

As defined in Section 7.1, an edge of $\mathbf{c}$ is positive if both endpoints of $\mathbf{e}$ are positive and negative if both endpoints of $\mathbf{e}$ are negative. If an edge is neither positive nor negative, then exactly two vertices and one edge of $T_\mathbf{c}^\mathcal{I}$ lie on $\mathbf{e}$.

Lemma 7.5. *Let $\mathbf{c}$ be a convex polyhedron in $\mathbb{R}^3$ where each vertex of $\mathbf{c}$ has the label "$-$", "$+$", or "$*$" and let $\mathbf{e}$ be an edge of $\mathbf{c}$. If $\mathbf{e}$ is not positive and not negative, then edge $\mathbf{e}$ contains some edge (w, w') of some tetrahedron in $T_\mathbf{c}^\mathcal{I}$ and no tetrahedra vertices of $T_\mathbf{c}^\mathcal{I}$ other than w and w' lie on $\mathbf{e}$.*

Proof: Let $v = (v_1, v_2, v_3)$ and $\tilde{v} = (\tilde{v}_1, \tilde{v}_2, \tilde{v}_3)$ be the endpoints of $\mathbf{e}$. Let $p^0 = (v_1, v_2, v_3, 0)$, $p^1 = (v_1, v_2, v_3, 1)$, $\tilde{p}^0 = (\tilde{v}_1, \tilde{v}_2, \tilde{v}_3, 0)$, and $\tilde{p}^1 = (\tilde{v}_1, \tilde{v}_2, \tilde{v}_3, 1)$ be the corresponding vertices of $\mathbf{c} \times [0,1]$. Square $(p^0, p^1, \tilde{p}^0, \tilde{p}^1)$ is a face $\mathbf{f}$ of $\mathbf{c} \times [0,1]$. Note that $\pi(\mathbf{f})$ equals $\mathbf{e}$. Algorithm IVOLPATCH3D constructs an isosurface patch of $\mathbf{c} \times [0,1]$. The intersection of this isosurface patch and $\mathbf{f}$ is an isocontour in $\mathbf{f}$ separating positive and negative vertices of $\mathbf{f}$. We claim that this intersection is a single line segment in $\mathbf{f}$. We prove this based on case analysis of the labels of v and $\tilde{v}$.

Case I: $(v, \tilde{v})$ have labels $(-, +)$ or $(+, -)$.

Without loss of generality, assume that v has the label "$-$" and $\tilde{v}$ has the label "$+$". By construction, vertices p^0 and p^1 have negative labels and $\tilde{p}^0$ and $\tilde{p}^1$ have positive labels. The isocontour in **f** is a single line segment from edge $(p^0, \tilde{p}^0)$ to edge $(p^1, \tilde{p}^1)$.

Case II: $(v, \tilde{v})$ have labels $(-, *)$ or $(*, -)$.

Without loss of generality, assume that v has the label "$-$" and $\tilde{v}$ has the label "$*$". By construction, vertices p^0, p^1, and $\tilde{p}^1$ have negative labels and $\tilde{p}^0$ has a positive label. The isocontour in **f** is a single line segment from edge $(p^0, \tilde{p}^0)$ to edge $(\tilde{p}^0, \tilde{p}^1)$.

Case III: $(v, \tilde{v})$ have labels $(+, *)$ or $(*, +)$.

This case is similar to Case II, with positive labels interchanged with negative ones. Without loss of generality, assume that v has the label "$+$" and $\tilde{v}$ has the label "$*$". By construction, vertices p^0, p^1, and $\tilde{p}^0$ have positive labels and $\tilde{p}^1$ has a negative label. The isocontour in **f** is a single line segment from edge $(p^1, \tilde{p}^1)$ to edge $(\tilde{p}^0, \tilde{p}^1)$.

Case IV: $(v, \tilde{v})$ have labels $(*, *)$.

By construction, vertices p^0 and $\tilde{p}^0$ have positive labels and p^1 and $\tilde{p}^1$ have negative labels. The isocontour in **f** is a single line segment from edge (p^0, p^1) to edge $(\tilde{p}^0, \tilde{p}^1)$.

Let line segment $(u, \tilde{u})$ be the intersection of **f** and the isosurface patch of $\mathbf{c} \times [0, 1]$. Line segment $(u, \tilde{u})$ is an edge of some tetrahedron in the incremental triangulation of this isosurface patch. Its projection, $(\pi(u), \pi(\tilde{u}))$, is an edge of some tetrahedron in $T_{\mathbf{c}}^{\mathcal{I}}$. Tetrahedra vertices u and $\tilde{u}$ are the only isosurface vertices on **f**. Thus $\pi(u)$ and $\pi(\tilde{u})$ are the only vertices of $T_{\mathbf{c}}^{\mathcal{I}}$ on **e**. $\quad\square$

We are finally ready to prove Properties 3 and 4. Because Property 4 is used in the proof of Property 3, we prove Property 4 first.

Property 4.   *The interval volume does not contain any grid vertex v where $s_v < \sigma_0$ or $s_v > \sigma_1$ and does not intersect any grid edges $(v, \tilde{v})$ where both s_v and $s_{\tilde{v}}$ are strictly less than σ_0 or where both s_v and $s_{\tilde{v}}$ are strictly greater than σ_1.*

Proof: Let Γ be a regular grid with scalar values at its vertices. We prove that $|\Upsilon_\Gamma(\sigma_0, \sigma_1)|$ does not contain any grid vertex where $s_v < \sigma_0$ or $s_v > \sigma_1$. Let v be a grid vertex with scalar value s_v less than σ_0 or greater than σ_1. Such a vertex has the label "$-$" or "$+$". By Lemma 7.4, no interval volume vertex with label "$*$" lies on v. The location of interval volume vertices with labels "$-$" and "$+$" are determined by linear interpolation. Since s_v does not equal σ_0 or σ_1, such vertices do not lie on v. Thus no interval volume vertex lies on v. Since each interval volume tetrahedron is contained in some grid cube, no interval volume tetrahedron contains v. Thus, $|\Upsilon_\Gamma(\sigma_0, \sigma_1)|$ does not contain v.

Finally, consider a grid edge $\mathbf{e} = (v, \tilde{v})$ where both s_v and $s_{\tilde{v}}$ are strictly less than σ_0 or both s_v and $s_{\tilde{v}}$ are strictly greater than σ_1. No interval volume vertices lie on $\mathbf{e}$. Since each interval volume tetrahedron is contained in some grid cube and $\mathbf{e}$ does not intersect the interior of any grid cube, $|\Upsilon_\Gamma(\sigma_0, \sigma_1)|$ does not intersect $\mathbf{e}$. □

Property 3. *The interval volume contains every grid vertex v where $\sigma_0 \le s_v \le \sigma_1$ and intersects each grid edge $(v, \tilde{v})$ where $s_v < \sigma_1$ and $s_{\tilde{v}} > \sigma_0$ (or vice versa) in a line segment.*

Proof: Let Γ be a regular grid with scalar values at its vertices. We first prove that $|\Upsilon_\Gamma(\sigma_0, \sigma_1)|$ contains every grid vertex v where $\sigma_0 \le s_v \le \sigma_1$. Such a grid vertex has the label "$*$". By Lemma 7.4, vertex v is a vertex of some tetrahedron in $T_{\mathbf{c}}^\mathcal{I}$. The linear interpolation step of algorithm MCVOL does not move vertices with label "$*$", so v is in the interval volume constructed by MCVOL.

We now prove that $|\Upsilon_\Gamma(\sigma_0, \sigma_1)|$ intersects each grid edge $s = (v, \tilde{v})$ where $s_v < \sigma_1$ and $s_{\tilde{v}} > \sigma_0$ (or vice versa) in a line segment. Since $s_v < \sigma_1$ and $s_{\tilde{v}} > \sigma_0$, edge $\mathbf{e}$ is neither positive nor negative.

Let $\mathbf{c}$ be a grid cube containing $\mathbf{e}$. By Lemma 7.5, two vertices, u_i and u_j, and one edge, (u_i, u_j), of $T_{\mathbf{c}}^\mathcal{I}$ lie on edge $\mathbf{e}$. The two vertices u_i and u_j map to two interval volume vertices, w_i and w_j, on edge $\mathbf{e}$.

The geometric locations of w_i and w_j depend upon the labels of u_i and u_j, respectively. If u_i has a star label, then w_i is an endpoint of $\mathbf{e}$. If u_i has a negative label, then w_i is the point $(1-\alpha_0)v + \alpha_0\tilde{v}$ where $\alpha_0 = (\sigma_0 - s_v)/(s_{\tilde{v}} - s_v)$. Similarly, if u_i has a positive label, then w_i is the point $(1 - \alpha_1)v + \alpha_1\tilde{v}$ where $\alpha_1 = (\sigma_1 - s_v)/(s_{\tilde{v}} - s_v)$. The same applies to w_j.

We claim that w_i and w_j are different points. The proof is by case analysis on the labels of s_v and $s_{\tilde{v}}$. Since $s_v < \sigma_1$ and $s_{\tilde{v}} > \sigma_0$, vertex v cannot have a positive label and $\tilde{v}$ cannot have a negative label.

Case I: $(v, \tilde{v})$ have labels $(-, +)$.

Since $(v, \tilde{v})$ have labels $(-, +)$, vertices (u_i, u_j) have labels $(-, +)$ or $(+, -)$. Without loss of generality, assume (u_i, u_j) have labels $(-, +)$. Since σ_0 does not equal σ_1, the scalar $\alpha_0 = (\sigma_0 - s_v)/(s_{\tilde{v}} - s_v)$ does not equal $\alpha_1 = (\sigma_1 - s_v)/(s_{\tilde{v}} - s_v)$ and $w_i = (1 - \alpha_0)v + \alpha_0\tilde{v}$ does not equal $w_j = (1 - \alpha_1)v + \alpha_1\tilde{v}$.

Case II: $(v, \tilde{v})$ have labels $(-, *)$.

Since $(v, \tilde{v})$ have labels $(-, *)$, vertices (u_i, u_j) have labels $(-, *)$ or $(*, -)$. Without loss of generality, assume (u_i, u_j) have labels $(-, *)$. By Lemma 7.4, vertex w_j lies on $\tilde{v}$. Since $s_{\tilde{v}} > \sigma_0$, the scalar $\alpha_0 = (\sigma_0 - s_v)/(s_{\tilde{v}} - s_v)$ does not equal one and vertex $w_i = (1 - \alpha_0)v + \alpha_0\tilde{v}$ does not equal $\tilde{v}$.

Case III: $(v, \tilde{v})$ have labels $(*, +)$.

Since $(v, \tilde{v})$ have labels $(*, +)$, vertices (u_i, u_j) have labels $(*, +)$ or $(+, *)$. Without loss of generality, assume (u_i, u_j) have labels $(*, +)$. By Lemma 7.4, vertex w_i lies on v. Since $s_v < \sigma_1$, the scalar $\alpha_1 = (\sigma_1 - s_v)/(s_{\tilde{v}} - s_v)$ does not equal zero and vertex $w_j = (1 - \alpha_1)v + \alpha_1\tilde{v}$ does not equal v.

Case IV: $(v, \tilde{v})$ have labels $(*, *)$.

Since $(v, \tilde{v})$ have labels $(*, *)$, vertices (u_i, u_j) have labels $(*, *)$. By Lemma 7.4, vertex w_i lies on v and vertex w_j lies on $\tilde{v}$ or vice versa. Therefore, w_i does not equal w_j.

Since w_i does not equal w_j, line segment (u_i, u_j) maps to line segment (w_i, w_j). Thus, $|\Upsilon_{\mathbf{c}}(\sigma_0, \sigma_1)| \cap e$ equals line segment (w_i, w_j).

Let $\mathbf{c}'$ be any other grid cube containing $\mathbf{e}$. By the argument given above, $|\Upsilon_{\mathbf{c}'}(\sigma_0, \sigma_1)| \cap e$ is a line segment. The labels of the endpoints of this line segment and hence their geometric locations are determined by the labels of the endpoints of $\mathbf{e}$. Thus, $|\Upsilon_{\mathbf{c}'}(\sigma_0, \sigma_1)| \cap e$ equals $|\Upsilon_{\mathbf{c}}(\sigma_0, \sigma_1)| \cap e$ and their union is a line segment.

We lastly must consider a grid cube $\mathbf{c}'$ that contains only one endpoint of $\mathbf{e}$. By Property 4, if this endpoint has a negative or positive label, then the interval volume patch of $\mathbf{c}'$ does not contain this vertex. By Lemma 7.4, if this endpoint has a star label, then it is a vertex of some tetrahedron in $T_{\mathbf{c}'}^{\mathcal{I}}$. This vertex is in $|\Upsilon_{\mathbf{c}'}(\sigma_0, \sigma_1)|$ but it also is in $|\Upsilon_{\mathbf{c}}(\sigma_0, \sigma_1)|$. Thus, $|\Upsilon_{\mathbf{c}'}(\sigma_0, \sigma_1)| \cap e$ is a subset of $|\Upsilon_{\mathbf{c}}(\sigma_0, \sigma_1)| \cap e$ and again their union is a line segment. $\qquad\square$

7.4.2 Proof of Interval Volume Properties 5-7

Define $T_\Gamma^{\mathcal{I}}$ as the union of all the isosurface patch triangulations $T_{\mathbf{c}}^{\mathcal{I}}$:

$$T_\Gamma^{\mathcal{I}} = \bigcup_{\mathbf{c} \in \Gamma} T_{\mathbf{c}}^{\mathcal{I}}.$$

Set $|T_\Gamma^{\mathcal{I}}|$ is the union of all the tetrahedra in $T_\Gamma^{\mathcal{I}}$.

To prove Property 5, we show that $|T_\Gamma^{\mathcal{I}}|$ separates the positive vertices of Γ from the negative ones. We also show that $T_\Gamma^{\mathcal{I}}$ is a triangulation. We define a homotopy map η that maps this triangulation to the interval volume. Since this triangulation does not cross any vertices, the interval volume separates the positive vertices of Γ from the negative ones. Since the interval volume does not contain any positive or negative vertices (Property 4), the interval volume strictly separates the positive vertices of Γ from the negative ones.

In Sections 6.3.3 and 6.3.5, we gave definitions of positive regions and separating surfaces in 4D polytopes and meshes. We give similar definitions here for the 4D convex polytope $\chi(\mathbf{c})$ and the 4D mesh $\chi(\Gamma)$. We use the proper isosurface vertex sets $U^{\mathcal{I}}_{\chi(\mathbf{c})}$ as defined in the previous section:

$$
\begin{aligned}
V^+_{\chi(\mathbf{c})} &= \text{the positive vertices of } \chi(\mathbf{c}), \\
R^{\mathcal{I}}_{\chi(\mathbf{c})} &= \text{conv}\left(U^{\mathcal{I}}_{\chi(\mathbf{c})} \cup V^+_{\chi(\mathbf{c})}\right), \\
S^{\mathcal{I}}_{\chi(\mathbf{c})} &= R^{\mathcal{I}}_{\chi(\mathbf{c})} \cap \text{cl}\left(\chi(\mathbf{c}) - R^{\mathcal{I}}_{\chi(\mathbf{c})}\right), \\
T^{\mathcal{I}}_{\chi(\Gamma)} &= \bigcup_{\mathbf{c} \in \Gamma} T^{\mathcal{I}}_{\chi(\mathbf{c})}, \\
R^{\mathcal{I}}_{\chi(\Gamma)} &= \bigcup_{\mathbf{c} \in \Gamma} R^{\mathcal{I}}_{\chi(\mathbf{c})}, \\
S^{\mathcal{I}}_{\chi(\Gamma)} &= R^{\mathcal{I}}_{\chi(\Gamma)} \cap \text{cl}\left(|\chi(\Gamma)| - R^{\mathcal{I}}_{\chi(\Gamma)}\right),
\end{aligned}
$$

where $|\chi(\Gamma)|$ is the union of all the polytopes in $\chi(\Gamma)$. Note that the isosurface vertex set $U^{\mathcal{I}}_{\chi(\mathbf{c})}$ is fixed. In Chapter 6, we used a "+" superscript to indicate that we are using positive, not negative, vertices of the polytope. For convenience, we have dropped the "+" superscript in this chapter.

We also need to define positive regions and separating surfaces on the cube facet $\chi(\mathbf{c}) \cap \mathbf{h}_0$ and the subgrid $\Gamma \cap \mathbf{h}_0$:

$$
\begin{aligned}
R^0_{\chi(\mathbf{c})} &= \text{conv}((U^{\mathcal{I}}_{\chi(\mathbf{c})} \cup V^+_{\chi(\mathbf{c})}) \cap \mathbf{h}_0), \\
S^0_{\chi(\mathbf{c})} &= \mathbb{R}^0_{\chi(\mathbf{c})} \cap \text{cl}((\chi(\mathbf{c}) \cap \mathbf{h}_0) - R^0_{\chi(\mathbf{c})}), \\
R^0_{\chi(\Gamma)} &= \bigcup_{\mathbf{c} \in \Gamma} R^0_{\chi(\mathbf{c})}, \\
S^0_{\chi(\Gamma)} &= R^0_{\chi(\Gamma)} \cap \text{cl}(|\chi(\Gamma) \cap \mathbf{h}_0| - R^0_{\chi(\Gamma)}).
\end{aligned}
$$

We now show that $S^0_{\chi(\Gamma)}$ equals $S^{\mathcal{I}}_{\chi(\Gamma)} \cap \mathbf{h}_0$.

Lemma 7.6. *If Γ is a convex polyhedral mesh in $\mathbb{R}^3$ where each vertex of Γ has the label "$-$", "$+$", or "$*$", then $S^0_{\chi(\Gamma)}$ equals $S^{\mathcal{I}}_{\chi(\Gamma)} \cap \mathbf{h}_0$.*

Proof:

$$
\begin{aligned}
S^0_{\chi(\Gamma)} &= \bigcup_{\mathbf{c} \in \Gamma} S^0_{\chi(\mathbf{c})} && \text{(Lemma 6.19)} \\
&= \bigcup_{\mathbf{c} \in \Gamma} (S^{\mathcal{I}}_{\chi(\mathbf{c})} \cap \mathbf{h}_0) && \text{(Lemma 6.25)} \\
&= \mathbf{h}_0 \cap \bigcup_{\mathbf{c} \in \Gamma} S^{\mathcal{I}}_{\chi(\mathbf{c})} \\
&= \mathbf{h}_0 \cap S^{\mathcal{I}}_{\chi(\Gamma)}. && \text{(Lemma 6.19)}
\end{aligned}
$$

Lemmas 6.25 and 6.19 are from Section 6.3.5.   □

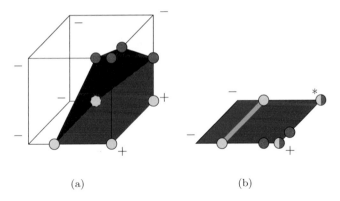

(a)   (b)

Figure 7.6. Two-dimensional example of interval $\pi(R^{\mathcal{I}}_{\chi(\mathbf{c})})$. (a) Positive region $R^{\mathcal{I}}_{\chi(\mathbf{c})}$ (red). (b) Red region is the projection $\pi(R^{\mathcal{I}}_{\chi(\mathbf{c})})$. Blue region is $\pi(\mathbf{f}_0 - R^0_{\chi(\mathbf{c})})$, where $\mathbf{f}_0$ equals $\mathbf{c} \cap \mathbf{h}_0$. Green line segment is $\pi(S^{\mathcal{I}}_{\chi(\mathbf{c})} \cap \mathbf{h}_0)$. Cyan vertices are in $\pi(U_{\mathbf{f}_0} \cup V^+_{\mathbf{f}_0})$. Magenta vertices are in $\pi((U^{\mathcal{I}}_{\chi(\mathbf{c})} \cup V^+_{\chi(\mathbf{c})}) - (U_{\mathbf{f}_0} \cup V^+_{\mathbf{f}_0}))$.

As a corollary to Lemma 7.6, $|T^{\mathcal{I}}_{\Gamma}|$ separates the positive vertices of Γ from the negative ones.

Corollary 7.7. *If Γ is a convex polyhedral mesh in $\mathbb{R}^3$ where each vertex of Γ has the label "$-$", "$+$", or "$*$", then $|T^{\mathcal{I}}_{\Gamma}|$ separates the positive vertices of Γ from the negative ones.*

Proof: Let $\pi : \mathbb{R}^4 \to \mathbb{R}^3$ be the orthogonal projection of $\mathbb{R}^4$ onto $\mathbb{R}^3$. Set $T^{\mathcal{I}}_{\Gamma}$ is the orthogonal projection of $T^{\mathcal{I}}_{\chi(\Gamma)}$ into $\mathbb{R}^3$. By Lemma 6.22 (Section 6.3.5), $T^{\mathcal{I}}_{\chi(\Gamma)}$ is a triangulation of $S^{\mathcal{I}}_{\chi(\Gamma)}$. By Lemma 7.6, $S^0_{\chi(\Gamma)}$ is a subset of $S^{\mathcal{I}}_{\chi(\Gamma)}$. By Lemma B.8 in Appendix B, $S^0_{\chi(\Gamma)}$ separates $R^0_{\chi(\Gamma)}$ from $|\chi(\Gamma) \cap \mathbf{h}_0| - R^0_{\Gamma}$. The positive vertices in $\chi(\Gamma) \cap \mathbf{h}_0$ are contained in $R^0_{\chi(\Gamma)}$ while the negative ones are contained in $|\chi(\Gamma) \cap \mathbf{h}_0| - R^0_{\Gamma}$.

Let v be a vertex in $\chi(\Gamma) \cap \mathbf{h}_0$. If $\pi(v)$ is positive, then v is positive. If $\pi(v)$ is negative, then v is negative. Let π_0 be the restriction of π to the hyperplane $\mathbf{h}_0$. Mapping π_0 is a one-to-one mapping from $\mathbf{h}_0$ to $\mathbb{R}^3$. Since $S^0_{\chi(\Gamma)}$ separates the positive vertices of $\chi(\Gamma) \cap \mathbf{h}_0$ from the negative ones, its projection $\pi(S^0_{\chi(\Gamma)})$ separates the positive vertices of Γ from the negative ones. Since $|T^{\mathcal{I}}_{\Gamma}|$ contains $\pi(S^0_{\chi(\Gamma)})$, set $|T^{\mathcal{I}}_{\Gamma}|$ separates the positive vertices of Γ from the negative ones. $\square$

By Lemma 6.22 (Section 6.3.5), $T^{\mathcal{I}}_{\chi(\Gamma)}$ is a triangulation of $S^{\mathcal{I}}_{\chi(\Gamma)}$. Set $T^{\mathcal{I}}_{\Gamma}$ is the projection of this triangulation into $\mathbb{R}^3$. To show that $T^{\mathcal{I}}_{\Gamma}$ is a triangulation, we show that this projection is one-to-one when restricted to $S^{\mathcal{I}}_{\chi(\Gamma)}$. We first give two properties of the projection $\pi(R^{\mathcal{I}}_{\chi(\mathbf{c})})$ of $R^{\mathcal{I}}_{\chi(\mathbf{c})}$.

Lemma 7.8. *Let* $\mathbf{c}$ *be a convex polyhedron in* $\mathbb{R}^3$ *where each vertex of* $\mathbf{c}$ *has the label "$-$", "$+$", or "$*$". If* $\pi : \mathbb{R}^4 \to \mathbb{R}^3$ *is the orthogonal projection of* $\mathbb{R}^4$ *onto* $\mathbb{R}^3$, *then*

1. *projection* $\pi(R^{\mathcal{I}}_{\chi(\mathbf{c})})$ *equals* $\pi(R^{\mathcal{I}}_{\chi(\mathbf{c})} \cap \mathbf{h}_0)$,

2. *projection* $\pi(R^{\mathcal{I}}_{\chi(\mathbf{c})} - \mathbf{h}_0)$ *does not intersect* $\pi(S^{\mathcal{I}}_{\chi(\mathbf{c})} \cap \mathbf{h}_0)$.

Proof of 1: Let $\mathbf{f}_0$ equal $\mathbf{h}_0 \cap \chi(\mathbf{c})$. Define $U_{\mathbf{f}_0}$ and $V_{\mathbf{f}_0}$ as

$$U_{\mathbf{f}_0} = U^{\mathcal{I}}_{\chi(\mathbf{c})} \cap \mathbf{h}_0,$$
$$V^+_{\mathbf{f}_0} = V^+_{\chi(\mathbf{c})} \cap \mathbf{h}_0.$$

Note that $R^0_{\chi(\mathbf{c})}$ equals $\mathrm{conv}(U_{\mathbf{f}_0} \cup V^+_{\mathbf{f}_0})$. By Lemma 6.16 (Section 6.3.5), $R^{\mathcal{I}}_{\chi(\mathbf{c})} \cap \mathbf{h}_0$ equals $R^0_{\chi(\mathbf{c})}$.

Case analysis (omitted) shows that $\pi(U^{\mathcal{I}}_{\chi(\mathbf{c})} \cup V^+_{\chi(\mathbf{c})})$ is contained in the convex hull of $\pi(U_{\mathbf{f}_0} \cup V^+_{\mathbf{f}_0})$. (See 2D example in Figure 7.6.) Thus,

$$\pi\left(R^{\mathcal{I}}_{\chi(\mathbf{c})}\right) = \pi\left(\mathrm{conv}\left(U^{\mathcal{I}}_{\chi(\mathbf{c})} \cup V^+_{\chi(\mathbf{c})}\right)\right) = \mathrm{conv}\left(\pi\left(U^{\mathcal{I}}_{\chi(\mathbf{c})} \cup V^+_{\chi(\mathbf{c})}\right)\right)$$
$$\subseteq \mathrm{conv}\left(\pi\left(U_{\mathbf{f}_0} \cup V^+_{\mathbf{f}_0}\right)\right) = \pi\left(\mathrm{conv}\left(U_{\mathbf{f}_0} \cup V^+_{\mathbf{f}_0}\right)\right)$$
$$= \pi\left(R^0_{\chi(\mathbf{c})}\right) = \pi\left(R^{\mathcal{I}}_{\chi(\mathbf{c})} \cap \mathbf{h}_0\right).$$

Since $\pi(R^{\mathcal{I}}_{\chi(\mathbf{c})} \cap \mathbf{h}_0)$ is clearly a subset of $\pi(R^{\mathcal{I}}_{\chi(\mathbf{c})})$, the projection $\pi(R^{\mathcal{I}}_{\chi(\mathbf{c})})$ equals $\pi(R^{\mathcal{I}}_{\chi(\mathbf{c})} \cap \mathbf{h}_0)$. $\qquad \square$

Proof of 2: Let $\mathbf{f}_0$ equal $\mathbf{h}_0 \cap \chi(\mathbf{c})$. Case analysis (omitted) shows that $\pi(U^{\mathcal{I}}_{\chi(\mathbf{c})} - U_{\mathbf{f}_0})$ does not intersect $\pi(\mathrm{cl}(\mathbf{f}_0 - R^0_{\chi(\mathbf{c})}))$. (See 2D example in Figure 7.6.) Similarly, $\pi(V^+_{\chi(\mathbf{c})} - V_{\mathbf{f}_0})$ does not intersect $\pi(\mathrm{cl}(\mathbf{f}_0 - R^0_{\chi(\mathbf{c})}))$. Thus $\pi(\mathrm{conv}(U^{\mathcal{I}}_{\chi(\mathbf{c})} \cup V^+_{\chi(\mathbf{c})}) - (U_{\mathbf{f}_0} \cup V^+_{\mathbf{f}_0}))$ does not intersect $\pi(\mathrm{cl}(\mathbf{f}_0 - R^0_{\chi(\mathbf{c})}))$.

Every point $p \in R^{\mathcal{I}}_{\chi(\mathbf{c})}$ can be expressed as a convex combination of the points in $U^{\mathcal{I}}_{\chi(\mathbf{c})} \cup V^+_{\chi(\mathbf{c})}$:

$$p = \sum_{q_i \in U^{\mathcal{I}}_{\chi(\mathbf{c})} \cup V^+_{\chi(\mathbf{c})}} \alpha_i q_i,$$

where $0 \leq \alpha_i \leq 1$. If point p is not on hyperplane $\mathbf{h}_0$, then α_i is greater than zero for some $q_i \notin U_{\mathbf{f}_0} \cup V^+_{\mathbf{f}_0}$. Since $\pi(\mathrm{cl}(\mathbf{f}_0 - R^0_{\chi(\mathbf{c})}))$ does not contain q_i, region $\pi(\mathrm{cl}(\mathbf{f}_0 - R^0_{\chi(\mathbf{c})}))$ does not contain p. Therefore, $\pi(R^{\mathcal{I}}_{\chi(\mathbf{c})} - \mathbf{h}_0)$ does not intersect $\pi(\mathrm{cl}(\mathbf{f}_0 - R^0_{\chi(\mathbf{c})}))$. Since $S^0_{\chi(\mathbf{c})}$ is contained in $\mathrm{cl}(\mathbf{f}_0 - R^0_{\chi(\mathbf{c})})$, region $\pi(R^{\mathcal{I}}_{\chi(\mathbf{c})} - \mathbf{h}_0)$ does not intersect $\pi(S^0_{\chi(\mathbf{c})})$. By Lemma 6.25 (Section 6.3.5), $S^0_{\chi(\mathbf{c})}$ equals $S^{\mathcal{I}}_{\chi(\mathbf{c})} \cap \mathbf{h}_0$, so region $\pi(R^{\mathcal{I}}_{\chi(\mathbf{c})} - \mathbf{h}_0)$ does not intersect $\pi(S^{\mathcal{I}}_{\chi(\mathbf{c})} \cap \mathbf{h}_0)$. $\qquad \square$

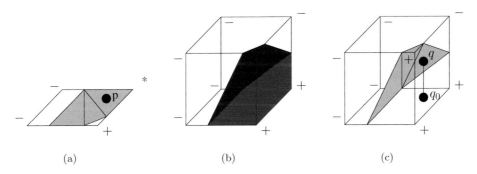

Figure 7.7. Two-dimensional example of $\pi(S^{\mathcal{I}}_{\chi(\mathbf{c})})$. (a) Point $p \in \pi(S^{\mathcal{I}}_{\chi(\mathbf{c})})$. (b) Positive region $R^{\mathcal{I}}_{\chi(\mathbf{c})}$ (red). (c) Point q_0 equals $\pi^{-1}(p) \cap \mathbf{h}_0$. Line segment (q_0, q) equals $\pi^{-1}(p) \cap R^{\mathcal{I}}_{\chi(\mathbf{c})}$.

We show that the projection when restricted to an isosurface patch $S^{\mathcal{I}}_{\chi(\mathbf{c})}$ in a single polyhedron $\chi(\mathbf{c})$ is one-to-one.

Lemma 7.9. *Let* $\mathbf{c}$ *be a convex polyhedron in* $\mathbb{R}^3$ *where each vertex of* $\mathbf{c}$ *has the label "$-$", "$+$", or "$*$". Let* $\pi : \mathbb{R}^4 \to \mathbb{R}^3$ *be the orthogonal projection of* $\mathbb{R}^4$ *onto* $\mathbb{R}^3$. *The projection* π *when restricted to* $S^{\mathcal{I}}_{\chi(\mathbf{c})}$ *is one-to-one.*

Proof: Let p be a point in $\pi(S^{\mathcal{I}}_{\chi(\mathbf{c})})$. The preimage, $\pi^{-1}(p)$, of p is a line in $\mathbb{R}^4$. By Lemma 7.8, $\pi(R^{\mathcal{I}}_{\chi(\mathbf{c})})$ equals $\pi(R^{\mathcal{I}}_{\chi(\mathbf{c})} \cap \mathbf{h}_0)$. Thus, line $\pi^{-1}(p) \cap R^{\mathcal{I}}_{\chi(\mathbf{c})}$ intersects $\mathbf{h}_0$ at some point $q_0 \in \mathbf{h}_0$. (See 2D example in Figure 7.7.)
We consider two cases, depending upon whether q_0 is in $S^0_{\chi(\mathbf{c})}$.

Case I: $S^0_{\chi(\mathbf{c})}$ contains q_0.

By Lemma 6.25 (Section 6.3.5), point q_0 is in $S^{\mathcal{I}}_{\chi(\mathbf{c})} \cap \mathbf{h}_0$ and p is in $\pi(S^{\mathcal{I}}_{\chi(\mathbf{c})} \cap \mathbf{h}_0)$. By Lemma 7.8, $\pi(R^{\mathcal{I}}_{\chi(\mathbf{c})} - \mathbf{h}_0)$ does not intersect $\pi(S^{\mathcal{I}}_{\chi(\mathbf{c})} \cap \mathbf{h}_0)$. Thus, q_0 is the only point on $\pi^{-1}(p) \cap R^{\mathcal{I}}_{\chi(\mathbf{c})}$ and thus the only point in $\pi^{-1}(p) \cap S^{\mathcal{I}}_{\chi(\mathbf{c})}$.

Case II: $S^0_{\chi(\mathbf{c})}$ does not contain q_0.

By Lemma 6.25, point q_0 is not in $S^{\mathcal{I}}_{\chi(\mathbf{c})}$. By definition of $S^{\mathcal{I}}_{\chi(\mathbf{c})}$, point q_0 is not in $\mathrm{cl}(\chi(\mathbf{c}) - R^{\mathcal{I}}_{\chi(\mathbf{c})})$.

Since $R^{\mathcal{I}}_{\chi(\mathbf{c})}$ is convex, line $\pi^{-1}(p)$ intersects $R^{\mathcal{I}}_{\chi(\mathbf{c})}$ in a point or a line segment. Since q_0 is in $\pi^{-1}(p) \cap R^{\mathcal{I}}_{\chi(\mathbf{c})}$ but not in $\pi^{-1}(p) \cap S^{\mathcal{I}}_{\chi(\mathbf{c})}$, the intersection $\pi^{-1} \cap R^{\mathcal{I}}_{\chi(\mathbf{c})}$ is not a single point. Thus line $\pi^{-1}(p)$ intersects $R^{\mathcal{I}}_{\chi(\mathbf{c})}$ in a line segment. One endpoint of this line segment is the point q_0. Let q be the other endpoint.

Since $R^{\mathcal{I}}_{\chi(\mathbf{c})}$ is convex and q_0 is not in $\mathrm{cl}(\chi(\mathbf{c}) - R^{\mathcal{I}}_{\chi(\mathbf{c})})$, no point of $\pi^{-1}(p) \cap R^{\mathcal{I}}_{\chi(\mathbf{c})}$ is in $\mathrm{cl}(\chi(\mathbf{c}) - R^{\mathcal{I}}_{\chi(\mathbf{c})})$ other than q. Thus q is the only point on $S^{\mathcal{I}}_{\chi(\mathbf{c})}$ whose projection $\pi(q)$ equals p. □

As a corollary, the mapping π, restricted to $S^{\mathcal{I}}_{\chi(\Gamma)}$, is one-to-one.

Corollary 7.10. *Let Γ be a convex polyhedral mesh in $\mathbb{R}^3$ where each vertex of Γ has the label "$-$", "$+$", or "$*$". Let $\pi : \mathbb{R}^4 \to \mathbb{R}^3$ be the orthogonal projection of $\mathbb{R}^4$ onto $\mathbb{R}^3$. The projection π, when restricted to $S^{\mathcal{I}}_{\chi(\Gamma)}$, is one-to-one.*

Proof: By Lemma 6.19 (Section 6.3.5), $S^{\mathcal{I}}_{\chi(\Gamma)}$ equals $\bigcup_{\mathbf{c} \in \Gamma} S^{\mathcal{I}}_{\chi(\mathbf{c})}$. By Lemma 7.9, the mapping π is one-to-one for every isosurface patch $S^{\mathcal{I}}_{\chi(\mathbf{c})}$ in a polytope $\chi(\mathbf{c})$. By Corollary 6.26 (Section 6.3.5), $S_{\chi(\mathbf{c})} \cap \mathbf{f}$ equals $S^{\mathcal{I}}_{\chi(\mathbf{c}')} \cap \mathbf{f}$ for every pair of polytopes $\chi(\mathbf{c})$ and $\chi(\mathbf{c}')$ sharing a face $\mathbf{f}$. Thus π is one-to-one when restricted to $S^{\mathcal{I}}_{\chi(\Gamma)}$. □

As another corollary to Lemma 7.9, set $T^{\mathcal{I}}_\Gamma$ is a triangulation.

Corollary 7.11. *If Γ is a convex polyhedral mesh in $\mathbb{R}^3$ where each vertex of Γ has the label "$-$", "$+$", or "$*$", then $T^{\mathcal{I}}_\Gamma$ is a triangulation.*

Proof: $T^{\mathcal{I}}_\Gamma$ is the image of $T^{\mathcal{I}}_{\chi(\Gamma)}$ under the projection π. By Lemma 6.22 (Section 6.3.5), $T^{\mathcal{I}}_{\chi(\Gamma)}$ is a triangulation of $S^{\mathcal{I}}_{\chi(\Gamma)}$. By Corollary 7.10, π is one-to-one when restricted to $S^{\mathcal{I}}_{\chi(\Gamma)}$. Thus $T^{\mathcal{I}}_\Gamma$ is a triangulation of $\pi(S^{\mathcal{I}}_{\chi(\Gamma)})$. □

Let Γ be a regular grid with scalar values at each of the grid vertices. Each vertex v_i of $T^{\mathcal{I}}_\Gamma$ is mapped to some vertex w_i of $\Upsilon_\Gamma(\sigma_0, \sigma_1)$. For each vertex v_i of $T^{\mathcal{I}}_\Gamma$, define $\eta(v_i, \alpha)$ as $(1 - \alpha)v_i + \alpha w_i$. Extend η to all of $|T^{\mathcal{I}}_\Gamma|$ by linearly extending η on each tetrahedron in $T^{\mathcal{I}}_\Gamma$. Thus $\eta : T^{\mathcal{I}}_\Gamma \times [0,1]$ is a homotopy from $T^{\mathcal{I}}_\Gamma$ to $|\Upsilon_\Gamma(\sigma_0, \sigma_1)|$ where $\eta(T^{\mathcal{I}}_\Gamma, 0)$ equals $|T^{\mathcal{I}}_\Gamma|$ and $\eta(T^{\mathcal{I}}_\Gamma, 1)$ equals $|\Upsilon_\Gamma(\sigma_0, \sigma_1)|$.

We are finally ready to show that $|\Upsilon_\Gamma(\sigma_0, \sigma_1)|$ strictly separates grid vertices with scalar values less than σ_0 from grid vertices with scalar values greater than σ_1.

Property 5. *The interval volume strictly separates vertices with scalar values less than σ_0 from vertices with scalar value greater than σ_1.*

Proof: Let v_1 and v_2 be vertices of Γ where v_1 has scalar value less than σ_0 and v_2 has scalar value greater than σ_1. By Corollary 7.7, $|T^{\mathcal{I}}_\Gamma|$ separates v_1 from v_2.

Map η is a homotopy from $|T^{\mathcal{I}}_\Gamma|$ to $|\Upsilon_\Gamma(\sigma_0, \sigma_1)|$ that never passes through v_1 or v_2. Vertices of $|T^{\mathcal{I}}_\Gamma|$ that lie on the boundary of $|\Gamma|$ remain on the boundary of $|\Gamma|$ throughout the homotopy η. Thus $\eta(T^{\mathcal{I}}_\Gamma \cap \partial|\Gamma|, \alpha)$ is a subset of $\partial|\Gamma|$ for all $\alpha \in [0,1]$. By Lemma B.18, $|\Upsilon_\Gamma(\sigma_0, \sigma_1)|$ separates v_1 from v_2.

By Property 4, neither v_1 nor v_2 are in $|\Upsilon_\Gamma(\sigma_0, \sigma_1)|$. Thus $|\Upsilon_\Gamma(\sigma_0, \sigma_1)|$ strictly separates v_1 from v_2. □

Set $|\Upsilon_\Gamma(\sigma_0,\sigma_1)|$ is an embedding of $|T_\Gamma^{\mathcal{I}}|$ into $|\Gamma|$. Even restricted to a single cube, this embedding is not necessarily one-to-one. However, this embedding is one-to-one when no vertex has scalar value equal to σ_0 or σ_1.

Define $\eta^1 : \mathbb{R}^3 \to \mathbb{R}^3$ as

$$\eta^1(p) = \eta(p, 1),$$

where η is the homotopy map from $|T_\Gamma^{\mathcal{I}}|$ to $|\Upsilon_\Gamma(\sigma_0,\sigma_1)|$. Mapping η^1 is the embedding of $|T_\Gamma^{\mathcal{I}}|$ into $|\Gamma|$ that maps $|T_\Gamma^{\mathcal{I}}|$ to $|\Upsilon_\Gamma(\sigma_0,\sigma_1)|$.

Lemma 7.12. *Let σ_0 and σ_1 be scalar values with $\sigma_0 < \sigma_1$. Let $\mathbf{c}$ be a cube in $\mathbb{R}^3$ where each vertex v of $\mathbf{c}$ has a scalar value, s_v, and each vertex v has a negative label if $s_v < \sigma_0$, a positive label if $s_v > \sigma_1$, or a star label if $\sigma_0 \leq s_v \leq \sigma_1$. If no vertex of $\mathbf{c}$ has scalar value equal to σ_0 or σ_1, then $\eta^1 : |T_{\mathbf{c}}^{\mathcal{I}}| \to |\Upsilon_{\mathbf{c}}(\sigma_0,\sigma_1)|$ is one-to-one and orientation preserving.*

The proof for Lemma 7.12 is a computer-based analysis and is omitted.

We claim that if no vertex of regular grid Γ has scalar value σ_0 or σ_1, then η^1 is one-to-one. The proof is a corollary to the following lemma.

Lemma 7.13. *Let σ_0 and σ_1 be scalar values with $\sigma_0 < \sigma_1$. Let $\mathbf{c}$ be a convex polyhedron in $\mathbb{R}^3$ where each vertex v of $\mathbf{c}$ has a scalar value, s_v, and each vertex v has a negative label if $s_v < \sigma_0$, a positive label if $s_v > \sigma_1$, or a star label if $\sigma_0 \leq s_v \leq \sigma_1$. Let $\mathbf{f}$ be a vertex, edge, or face of $\mathbf{c}$. If no vertex of $\mathbf{f}$ has scalar value σ_0 or σ_1, then $\eta^1(|T_\Gamma^{\mathcal{I}}|) \cap \mathbf{f}$ equals $\eta^1(|T_\Gamma^{\mathcal{I}}| \cap \mathbf{f})$.*

Proof: Since $T_{\mathbf{c}}^{\mathcal{I}}$ is the triangulation of $|T_{\mathbf{c}}^{\mathcal{I}}|$,

$$\eta^1(|T_{\mathbf{c}}^{\mathcal{I}}|) = \bigcup_{\mathbf{t} \in T_{\mathbf{c}}^{\mathcal{I}}} \eta^1(\mathbf{t}).$$

Let $\mathbf{t}$ be a tetrahedron in $T_{\mathbf{c}}^{\mathcal{I}}$. Since no vertex of $\mathbf{f}$ has scalar value σ_0 or σ_1, a vertex v of $\mathbf{t}$ lies on $\mathbf{f}$ if and only if $\eta^1(\mathbf{t})$ lies on $\mathbf{f}$. Since η^1 is linear on $\mathbf{t}$ and $\mathbf{t}$ is the convex hull of its vertices, $\eta^1(\mathbf{t}) \cap \mathbf{f}$ equals $\eta^1(\mathbf{t} \cap \mathbf{f})$. Thus,

$$\eta^1(|T_{\mathbf{c}}^{\mathcal{I}}|) \cap \mathbf{f} \;=\; \bigcup_{\mathbf{t} \in T_{\mathbf{c}}^{\mathcal{I}}} (\eta^1(\mathbf{t}) \cap \mathbf{f}) = \bigcup_{\mathbf{t} \in T_{\mathbf{c}}^{\mathcal{I}}} \eta^1(\mathbf{t} \cap \mathbf{f}) = \eta^1 \left(\bigcup_{\mathbf{t} \in T_{\mathbf{c}}^{\mathcal{I}}} (\mathbf{t} \cap \mathbf{f}) \right)$$

$$= \; \eta^1(|T_{\mathbf{c}}^{\mathcal{I}}| \cap \mathbf{f}). \qquad\qquad \square$$

Corollary 7.14. *Let σ_0 and σ_1 be scalar values with $\sigma_0 < \sigma_1$. Let Γ be a regular grid in $\mathbb{R}^3$ where each vertex v of $\mathbf{c}$ has a scalar value, s_v, and each vertex v has a negative label if $s_v < \sigma_0$, a positive label if $s_v > \sigma_1$, or a star label if $\sigma_0 \leq s_v \leq \sigma_1$. If no vertex of $\mathbf{f}$ has scalar value σ_0 or σ_1, then $\eta^1 : |T_\Gamma^{\mathcal{I}}| \to |\Upsilon_\Gamma(\sigma_0,\sigma_1)|$ is one-to-one.*

Proof: Assume that there are distinct points $p, p' \in |T_\Gamma^\mathcal{I}|$ such that $\eta^1(p)$ equals $\eta^1(p')$. By Lemma 7.12, function η^1 is one-to-one when restricted to $|T_\mathbf{c}^\mathcal{I}|$ for a single cube $\mathbf{c}$. Thus points p and p' must belong to different grid cubes $\mathbf{c}$ and $\mathbf{c}'$. Since $\mathbf{c}$ contains $\eta^1(|T_\mathbf{c}^\mathcal{I}|)$ and $\mathbf{c}'$ contains $\eta^1(|T_{\mathbf{c}'}^\mathcal{I}|)$, point $\eta^1(p) = \eta^1(p')$ must lie in $f = \mathbf{c} \cap \mathbf{c}'$. Since Γ is a regular grid, $\mathbf{f}$ is a face of $\mathbf{c}$.

By assumption, no vertex of $\mathbf{f}$ has value σ_0 or σ_1. By Lemma 7.13, $\eta^1(|T_\mathbf{c}^\mathcal{I}|) \cap \mathbf{f}$ equals $\eta^1(|T_\mathbf{c}^\mathcal{I}| \cap \mathbf{f})$. Thus $\eta^1(p) \in \eta^1(|T_\mathbf{c}^\mathcal{I}|) \cap \mathbf{f}$ must be equal $\eta^1(q)$ for some $q \in |T_\mathbf{c}^\mathcal{I}| \cap \mathbf{f}$. Since η^1 is one-to-one on $|T_\mathbf{c}^\mathcal{I}|$, point p must equal q and so $p \in |T_\mathbf{c}^\mathcal{I}| \cap \mathbf{f}$. Since $\mathbf{f}$ is a face of $\mathbf{c}'$, point p is also in $\mathbf{c}'$. Since η^1 is one-to-one on $|T_{\mathbf{c}'}^\mathcal{I}|$ and $p, p' \in |T_{\mathbf{c}'}^\mathcal{I}|$, point p must equal p', a contradiction. We conclude that $\eta^1 : |T_\Gamma^\mathcal{I}| \to |\Upsilon_\Gamma(\sigma_0, \sigma_1)|$ is one-to-one. $\square$

Finally, we show that if no grid vertex has scalar value equal to σ_0 or σ_1, then $|\Upsilon_\Gamma(\sigma_0, \sigma_1)|$ is a three-dimensional manifold with boundary.

Property 6. *The interval volume is a piecewise linear, orientable 3-manifold with boundary.*

Proof: By Lemma 6.31 (Section 6.3.5), the set $S_{\chi(\Gamma)}^\mathcal{I}$ is a piecewise linear, orientable 3-manifold with boundary. By Lemma 7.9, the projection of $S_{\chi(\Gamma)}^\mathcal{I}$ into $\mathbb{R}^3$ is one-to-one and so $|T_\Gamma^\mathcal{I}|$ is an orientable 3-manifold with boundary. Set $|\Upsilon_\Gamma(\sigma_0, \sigma_1)|$ is an embedding of $|T_\Gamma^\mathcal{I}|$ into $|\Gamma|$. By Corollary 7.14, this embedding is one-to-one. Thus, $|\Upsilon_\Gamma(\sigma_0, \sigma_1)|$ is an orientable 3-manifold with boundary. $\square$

Property 7. *Set Υ does not contain any zero-volume tetrahedra or duplicate tetrahedra and the tetrahedra in Υ form a triangulation of the interval volume.*

Proof: By Corollary 7.11, $T_\Gamma^\mathcal{I}$ is a triangulation. Function η^1 maps tetrahedra in $T_\Gamma^\mathcal{I}$ to tetrahedra in $\Upsilon_\Gamma(\sigma_0, \sigma_1)$. Since η^1 is continuous and one-to-one, $\Upsilon_\Gamma(\sigma_0, \sigma_1)$ is a triangulation of $|\Upsilon_\Gamma(\sigma_0, \sigma_1)|$ and has no duplicate tetrahedra. Since no tetrahedron in $T_\Gamma^\mathcal{I}$ has zero volume, no tetrahedron in $\Upsilon_\Gamma(\sigma_0, \sigma_1)$ has zero volume. $\square$

7.5 Tetrahedral Meshes

Nielson and Sung [Nielson and Sung, 1997] gave a version of MCVOL for tetrahedral meshes. Their algorithm stores triangulated interval volume patches in an interval volume lookup table and retrieves the patches to build interval volumes within each tetrahedra. We call the algorithm MTETVOL. Just as with cubes, it is important that the triangulation of an interval volume patch matches the triangulation of neighboring patches. Nielson and Sung constructed the lookup table by hand, but we use algorithm IVOLPATCH3D to generate the interval volume patches in the lookup table.

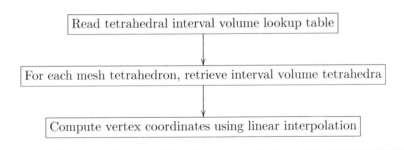

Figure 7.8. MTETVOL. Algorithm to construct a piecewise linear approximation to the interval volume in a tetrahedral mesh.

7.5.1 Algorithm

Input to the MTETVOL algorithm is two scalar values, σ_0 and σ_1, a three-dimensional tetrahedral mesh and a set of scalar values at the vertices of the mesh. The algorithm has three steps. (See Figure 7.8.) Read an interval volume lookup table from a preconstructed data file. For each mesh tetrahedron, retrieve from the lookup table a set of isosurface tetrahedra representing the combinatorial structure of the isosurface. The vertices of the isosurface tetrahedra form the isosurface vertices. Assign geometric locations to the isosurface vertices based on the scalar values at the simplex endpoints.

The isosurface lookup table is built on the "generic" tetrahedron $\mathbf{c}_T$ whose vertices are

$$(0,0,0), (1,0,0), (0,1,0), (0,0,1).$$

Any other tetrahedron could also have been used. Algorithm IVOLPATCH3D is applied to each of the $3^4 = 81$ configurations of positive, negative, and star vertex labels to generate 81 interval volume patches in 81 table entries.

For each mesh tetrahedron $\mathbf{c}$, the the isosurface patch triangulation retrieved from the lookup table depends on how the simplex is mapped to $\mathbf{c}_T$. There are 4! ways to map the vertices of $\mathbf{c}$ to $\mathbf{c}_T$. Each such mapping represents a different affine transformation of $\mathbf{c}$ to $\mathbf{c}_T$. However, only one such mapping preserves the lexicographic order of the vertices. If v_1, v_2, v_3, and v_4 are the vertices of $\mathbf{c}$ in lexicographic order, then map vertex v_1 to $(0,0,0)$, vertex v_2 to $(1,0,0)$, vertex v_3 to $(0,1,0)$, and vertex v_4 to $(0,0,1)$. This mapping sends the ith vertex of $\mathbf{c}$ in lexicographic order to the ith vertex of $\mathbf{c}_T$ in lexicographic order.

The final step is mapping each triangulation vertex to its geometric location. We use linear interpolation as described in Section 1.7.2 to position the triangulation vertices.

Each pair $(g, \mathcal{L})$ corresponds to a vertex in the triangulation of the interval volume. For each such pair, element g is either a grid vertex or a grid edge. If g is a grid vertex v, then the location of vertex $(g, \mathcal{L})$ is the location of vertex v. Assume g is a grid edge $[p, q]$. Let s_p and s_q be the scalar values at p and q. If $\mathcal{L}$ is "$-$", then map $(g, \mathcal{L})$ to $(1 - \alpha_0)p + \alpha_0 q$ where $\alpha_0 = (\sigma_0 - s_p)/(s_q - s_p)$. If $\mathcal{L}$ is "$+$", then map $(g, \mathcal{L})$ to $(1 - \alpha_1)p + \alpha_1 q$ where $\alpha_1 = (\sigma_1 - s_p)/(s_q - s_p)$. Note that if $\mathcal{L}$ is "$-$" or "$+$", then p and q must have different labels. Thus, scalar s_p does not equal s_q and the denominator $(s_q - s_p)$ is never zero.

7.5.2 Interval Volume Properties

The interval volume produced by MTETVOL has the same properties as the one produced by MCVOL. Under appropriate conditions the interval volume is a 3-manifold with boundary

MTETVOL returns a finite set, Υ, of orientable tetrahedra. The interval volume is the union of these tetrahedra. The vertices of the interval volume are the tetrahedra vertices.

The following properties apply to all interval volumes produced by the MTET-VOL algorithm. The scalar value at a mesh vertex v is denoted s_v.

Property 1. *The interval volume is piecewise linear.*

Property 2. *The vertices of the interval volume lie on mesh vertices or mesh edges.*

Property 3. *The interval volume contains every mesh vertex v where $\sigma_0 \leq s_v \leq \sigma_1$ and intersects each mesh edge $(v, \tilde{v})$ where $s_v < \sigma_1$ and $s_{\tilde{v}} > \sigma_0$ (or vice versa) in a line segment.*

Property 4. *The interval volume does not contain any mesh vertex v where $s_v < \sigma_0$ or $s_v > \sigma_1$ and does not intersect any mesh edges $(v, \tilde{v})$ where both s_v and $s_{\tilde{v}}$ are strictly less than σ_0 or both s_v and $s_{\tilde{v}}$ are strictly greater than σ_1.*

Property 5. *The interval volume strictly separates vertices with scalar values less than σ_0 from vertices with scalar value greater than σ_1.*

Properties 3 and 4 imply that the interval volume intersects a minimum number of mesh edges.

Consider a simplicial mesh which has the following conditions:

- The isovalue does not equal the scalar value of any mesh vertex.

- The mesh is a partition of a 4-manifold with boundary.

Under these conditions, the interval volume produced by MTETVOL has the following properties:

Property 6. *The interval volume is a piecewise linear, orientable 3-manifold with boundary.*

Property 7. *Set* Υ *does not contain any zero-volume tetrahedra or duplicate tetrahedra and the tetrahedra in* Υ *form a triangulation of the interval volume.*

Proofs of the properties for the MTETVOL interval volume are similar to the proofs for the MCVOL interval volume and are omitted.

7.6   Convex Polyhedral Meshes

Algorithm MCVOL also generalizes to meshes of convex polyhedral elements.

The three-dimensional MARCHING POLYHEDRA algorithm generalizes to four-dimensional convex meshes. The algorithm is called MPOLYVOL.

7.6.1   Algorithm

As with MTETVOL, care must be taken to ensure that triangulations of adjacent interval volume patches match. To do so, we require an additional input, an ordering of all the mesh vertices. This ordering will be used to control the mapping to the isosurface lookup tables so that triangulations of adjacent isosurface patches match. This ordering could be a lexicographic ordering on all the mesh vertices but other orders may be preferable.

As in the MARCHING POLYHEDRA algorithm for isosurface construction, we assume that the mesh is composed of a fixed, predefined set of polytope classes (Section 5.4.) Each class a has a precomputed isosurface lookup table built on a "generic" polytope $\mathbf{c}_a$. For isosurface construction, a mesh polytope $\mathbf{c}$ is in class a if there is an affine transformation from $\mathbf{c}$ to $\mathbf{c}_a$. Because triangulations of adjacent interval volume patches must match, this partition into classes needs to be further refined for interval volume construction.

The ordering of mesh vertices input to MPOLYVOL induces an ordering of the vertices of each mesh element. For each class a, the "generic" polytope $\mathbf{c}_a$ has an ordering on its vertices that is used in constructing the isosurface lookup table for class a. More specifically, this order determines the incremental triangulation of each interval volume patch constructed by algorithm IVOLPATCH3D. A mesh polytope $\mathbf{c}$ is in class a if there is an affine transformation from $\mathbf{c}$ to $\mathbf{c}_a$ that preserves the vertex order. In other words, the affine transformation should map the ith vertex of $\mathbf{c}$ (given by the input ordering of mesh vertices) to the ith vertex of $\mathbf{c}_a$.

Input to MPOLYVOL algorithm is two scalar values, σ_0 and σ_1, a three-dimensional convex polyhedral mesh, a set of scalar values at the vertices of the mesh, a sequence of all the mesh vertices, and a set of polyhedron classes.

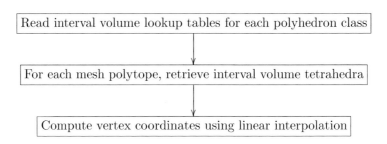

Figure 7.9. MPOLYVOL. Algorithm to construct a piecewise linear approximation to the interval volume in a convex polyhedral mesh.

The MPOLYVOL algorithm has three steps. The first step is reading the interval volume lookup tables for each class from preconstructed data files. Next, for each mesh element, a triangulated interval volume patch is retrieved from the appropriate table. Finally, the vertices of the interval volume patch are assigned geometric locations using linear interpolation. We explain the last two steps in more detail.

Let $\mathbf{c}$ be a mesh element. The input sequence of mesh vertices represents an ordering of the mesh vertices. Restricted to the vertices of $\mathbf{c}$, this ordering gives an ordering of the vertices of $\mathbf{c}$. Let a be the polytope class of $\mathbf{c}$ where some affine transformation preserving vertex order maps $\mathbf{c}$ to $\mathbf{c}_a$. To retrieve an isosurface patch for mesh element $\mathbf{c}$ from the isosurface lookup table for class a, the ith vertex of $\mathbf{c}$ is mapped to the ith vertex of the lookup table polytope. This mapping determines a configuration of positive, negative, and star vertices that identifies the appropriate entry in the lookup table.

The final step is mapping each triangulation vertex to its geometric location. We use linear interpolation as described in Section 1.7.2 to position the triangulation vertices.

Each pair $(g, \mathcal{L})$ corresponds to a vertex in the triangulation of the interval volume. For each such pair, element g is either a mesh vertex or a mesh edge. If g is a mesh vertex v, then the location of vertex $(g, \mathcal{L})$ is the location of vertex v. Assume g is a mesh edge $[p, q]$. Let s_p and s_q be the scalar values at p and q. If $\mathcal{L}$ is "$-$", then map $(g, \mathcal{L})$ to $\alpha_0 p + (1 - \alpha_0)q$ where $\alpha_0 = (s_p - \sigma_0)/(s_q - s_p)$. If $\mathcal{L}$ is "$+$", then map $(g, \mathcal{L})$ to $\alpha_1 p + (1 - \alpha_1)q$ where $\alpha_1 = (s_p - \sigma_1)/(s_q - s_p)$. Note that if $\mathcal{L}$ is "$-$" or "$+$", then p and q must have different labels. Thus, scalar s_p does not equal s_q and the denominator $(s_q - s_p)$ is never zero.

MPOLYVOL requires some ordering of mesh vertices to determine the mapping of mesh elements to their isosurface lookup tables. This ordering could simply be a lexicographic ordering of mesh vertices. However, as noted in Section 6.5.3, alternate orderings may be preferable. For instance, the lexicographic ordering of vertices in a pyramidal decomposition of the regular grid requires three pyramid isosurface lookup tables, one for each possible location of the pyramid apex in the order. The ordering given in Section 6.5.3 requires only a single pyramid isosurface vertex lookup table.

7.6.2   Interval Volume Properties

The interval volume produced by MPOLYVOL has the same properties as the one produced by MCVOL. Under appropriate conditions the interval volume is a 3-manifold with boundary

MPOLYVOL returns a finite set, Υ, of oriented tetrahedra. The interval volume is the union of these tetrahedra. The vertices of the interval volume are the tetrahedra vertices.

The following properties apply to all interval volumes produced by the MPOLY-VOL algorithm. The scalar value at a mesh vertex v is denoted s_v.

Property 1.   *The interval volume is piecewise linear.*

Property 2.   *The vertices of the interval volume lie on mesh vertices or mesh edges.*

Property 3.   *The interval volume contains every mesh vertex v where $\sigma_0 \leq s_v \leq \sigma_1$ and intersects each mesh edge $(v, \tilde{v})$ where $s_v < \sigma_1$ and $s_{\tilde{v}} > \sigma_0$ (or vice versa) in a line segment.*

Property 4.   *The interval volume does not contain any mesh vertex v where $s_v < \sigma_0$ or $s_v > \sigma_1$ and does not intersect any mesh edges $(v, \tilde{v})$ where both s_v and $s_{\tilde{v}}$ are strictly less than σ_0 or both s_v and $s_{\tilde{v}}$ are strictly greater than σ_1.*

Property 5.   *The interval volume strictly separates vertices with scalar values less than σ_0 from vertices with scalar value greater than σ_1.*

Properties 3 and 4 imply that the interval volume intersects a minimum number of mesh edges.

As with MARCHING POLYTOPES, "complex" mesh elements could create interval volume patches that self-intersect, even if the isovalue does not equal the scalar value of any mesh vertex. To ensure a manifold, we restrict the mesh elements to a few simple types, such as cubes, tetrahedra, and pyramids. This list is not meant to be exhaustive and many other convex polytopes could probably be added. Consider a mesh of convex polytopes that has the following conditions:

- The isovalue does not equal the scalar value of any mesh vertex.

- The mesh is a partition of a 3-manifold with boundary.

- The mesh elements are affine transformations of cubes, tetrahedra, and pyramids over a square base.

Under these conditions, the interval volume produced by MPOLYVOL has the following properties:

Property 6. *The interval volume is a piecewise linear, orientable 3-manifold with boundary.*

Property 7. *Set Υ does not contain any zero-volume tetrahedra or duplicate tetrahedra and the tetrahedra in Υ form a triangulation of the interval volume.*

Proofs of the properties for the MPOLYVOL interval volume are similar to the proofs for the MCVOL interval volume and are omitted.

7.7   Notes and Comments

Guo [Guo, 1995] and Fujishiro et al. [Fujishiro et al., 1996] independently introduced interval volumes. Guo constructed the interval volumes using Delaunay triangulations. Fujishiro et al. constructed interval volumes as the difference of two solid regions bounded by isosurfaces. Nielson and Sung [Nielson and Sung, 1997] gave the algorithm presented in Section 7.5 for constructing an interval volume from a tetrahedral mesh. They built the lookup table by hand instead of automatically generating it using algorithm IVOLPATCH3D.

The interval volume algorithms in Sections 7.2, 7.3, and 7.6 are described in [Bhaniramka et al., 2004b] and are based on techniques from [Bhaniramka et al., 2000, Bhaniramka et al., 2004a].

Labelle and Schewchuk use interval volumes in [Labelle and Shewchuk, 2007] to construct tetrahedral meshes with good tetrahedral mesh elements. They give guaranteed lower and upper bounds on the dihedral angles of the tetrahedra forming the mesh.

CHAPTER 8

DATA STRUCTURES FOR FASTER ISOSURFACE CONSTRUCTION

Much of the running time of the MARCHING CUBES algorithm and its variants is spent processing grid cubes that are not intersected by the isosurface. While it is theoretically possible for an isosurface to intersect every grid cube in the input, in practice a single isosurface intersects only a small fraction of the input grid cubes. A grid cube that is not intersected by a given isosurface is called an empty cube. Of course, whether a cube is empty depends upon the given isosurface.

To improve the running time of isosurface extraction algorithms, data structures have been designed for quickly identifying the grid cubes intersected by isosurfaces. These data structures can be divided into three categories: partitions of the domain, partitions of the range, and seed sets. The domain is the volume covered by the input grid. The range is the interval from the smallest to the largest scalar value at any grid vertex.

The span of a set $\mathbb{C}$ of grid cubes or mesh elements is the interval from the minimum scalar value $s_{\min}(\mathbb{C})$ to the maximum scalar value $s_{\max}(\mathbb{C})$ of any cube vertex or element vertex. For instance, the cube in Figure 8.1 has span [2,7]. If an isovalue lies in the span of a cube, then the corresponding isosurface intersects the cube. If the isovalue lies in the span of a set of cubes, then the isosurface intersects at least one cube in that set.

Both domain and range data structures compute the spans of sets of cubes and compare those spans to the isovalue to determine if the isosurface intersects the set. The difference is that domain-based structures group together cubes that are geometrically close to one another while the range data structures group together cubes that have similar spans.

Uniform partitioning (Section 8.1) and octrees (Section 8.2) are both domain-based methods. In uniform partitioning, the grid is divided into a set of uniform

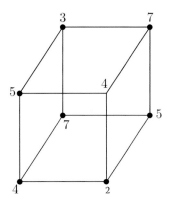

Figure 8.1. Cube with span [2,7].

$k \times k \times k$ regions while octrees represent multiple layers of subdivisions of the grid. The root of the octree represents the entire grid while the leaves represent individual grid cubes.

The span space priority tree (Section 8.3) is a range-based method. The grid cubes are organized in a priority tree whose keys are based on the spans of the grid cubes.

The third approach to faster isosurface retrieval is based on "growing" isosurfaces from a small set of cubes. Starting from an initial "seed set" of cubes, one determines which cubes in the seed set are intersected by the isosurface. One then grows the isosurface from each such seed cube, checking whether neighboring cubes are intersected by the isosurface. If a neighboring cube is intersected by the isosurface, then it is processed and its neighbors are checked.

If no cube in the seed set intersects the isosurface, then seed growing will not find any of the cubes intersecting the isosurface. Even if the seed set intersects the isosurface, seed growing may find only a subset of the cubes intersecting an isosurface with more than one connected component. The challenge is to find a small seed set such that for any isovalue seed growing finds all the cubes intersecting the isosurface.

Each data structure is associated with two algorithms, one to build the data structure and one to search the data structure and retrieve cubes intersected by the isosurface. The data structures reduce isosurface extraction time by avoiding the processing of empty grid cubes that do not intersect the isosurface. However, building a data structure requires processing each cube in the grid. If only a single isosurface is to be extracted from a grid, then the time saved in the extraction phase is outweighed by the time to create the data structure. Thus, using data structures is only productive if more than one isosurface is to be extracted from a scalar grid.

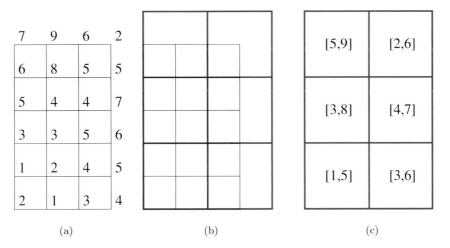

7	9	6	2
6	8	5	5
5	4	4	7
3	3	5	6
1	2	4	5
2	1	3	4

(a)   (b)   (c)

Figure 8.2. Example of uniform partitioning into 2×2 regions. (a) 2D scalar grid; (b) Uniform partition into 2×2 regions. Note that some regions extend beyond the grid boundary. (c) Span associated with each grid region.

8.1   Uniform Grid Partitions

Uniform partitioning is the partitioning of a grid into regions of cube dimensions $k \times k \times k$, i.e., every region edge is composed of k grid edges. (See Section 1.7.1 for the definition and an example of cube dimensions.) The span of each region is computed and stored. To identify the cubes intersected by the isosurface, the span of each region is compared with the isovalue. If the isovalue is contained in the span, then isosurface patches are extracted from the k^3 cubes in the region. Of course, some cubes may be empty, in which case their isosurface patches will be empty. If the isovalue is not contained in the region's span, then the cubes in the region are ignored, saving the time to compare k^3 cubes with the isovalue. The method is simple, yet results in significant reductions of isosurface extraction time.

For each $k \times k \times k$ region, we store the span of the region, the minimum and maximum values of all grid vertices in the region. We also store a reference to the grid cube with the lowest coordinates in the region. All the other grid cubes in the region can be computed from this reference grid cube.

If the grid has cube dimensions $m_x \times m_y \times m_z$ and some m_i is not a multiple of k, then the grid cannot be exactly partitioned into $k \times k \times k$ regions. Instead, we let some regions extend beyond the grid boundary. (See Figure 8.2 for a two-dimensional example.) These partial regions contain fewer than k^3 cubes. Because it is easier and faster to process full regions than partial ones, the partial regions should be kept in a list separate from the full ones.

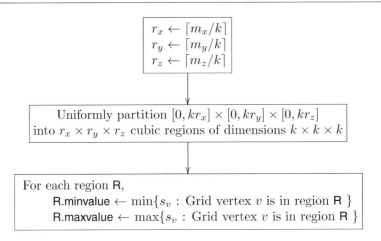

Figure 8.3. BUILDUNIFORM. Input is a positive integer k and a scalar grid with cube dimensions $m_x \times m_y \times m_z$.

```
Input   : Isovalue is the isovalue.
          Regions is a set of regions.
          Each region has a minvalue and a maxvalue.

SearchUniformPartition(Isovalue, Regions)
1 foreach R ∈ Regions do
2  │  if (R.minvalue ≤ Isovalue ≤ R.maxvalue) then
3  │  │  foreach cube c ∈ R do
4  │  │  │  if (c.minvalue ≤ Isovalue ≤ c.maxvalue) then
5  │  │  │  │  Report(c);
6  │  │  end
7  │  end
8  end
9 end
```

Algorithm 8.1. SEARCHUNIFORMPARTITION.

8.2 Octrees

While uniform grid partitions are simple, they have two drawbacks. First, users must decide on a region size k for the $k \times k \times k$ regions. Different region sizes may be better for different scalar data sets. Even for a fixed data set, the optimal

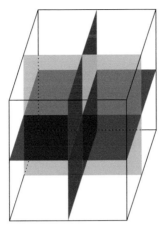

Figure 8.4. Partition of a cube into eight octants.

region sizes may depend upon the isovalue. Second, even for a fixed data set and scalar value, some portions of the data set may have large empty regions while others have smaller ones. Using a single region size for the entire set may also result in suboptimal performance.

8.2.1 Full Octrees

An octree is a tree representing a decomposition of some three-dimensional volume. Each node in the tree represents some subregion of the volume. A region is partitioned into eight octants by dividing the region with three axis-parallel planes through a "center" point. (See Figure 8.4.) What constitutes the "center" point for a region depends upon the application. Each internal node has eight children representing the eight octants, hence the name "octree." The root represents the entire volume.

A similar data structure in two dimensions is a quadtree where each region is partitioned into four quadrants by two axis-parallel lines through a center point. (See Figure 8.6.) The data structure can also be generalized to d dimensions, where each region is partitioned into 2^d regions by d axis-parallel planes. Unfortunately, the data structure is misleadingly still called an octree in higher dimensions, even though each node has 2^d, not eight, children.

In isosurface construction, octrees are used to partition regular grids into subregions. The leaves of the octree represent individual grid cubes. The parents of leaves represent $2 \times 2 \times 2$ regions, each containing eight grid cubes. The parents of those nodes represent $4 \times 4 \times 4$ regions, each containing eight $2 \times 2 \times 2$ regions and 64 grid cubes. Nodes at height h in the tree represent regions with dimension $2^h \times 2^h \times 2^h$, each containing 2^{3h} cubes. The root represents the entire grid.

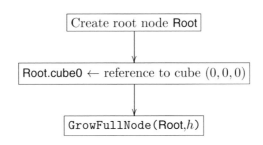

Figure 8.5. BUILDISOOCTREE. Input is a scalar grid with cube dimensions $2^h \times 2^h \times 2^h$.

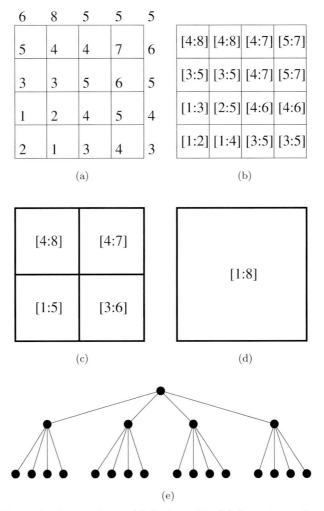

Figure 8.6. Example of a quadtree. (a) Scalar grid. (b) Span intervals at grid cubes. (c) Span intervals at 2×2 regions. (d) Span intervals at 4×4 regions. (e) Full quadtree.

A full octree is an octree whose internal nodes have exactly eight children. If a grid has cube dimensions $2^h \times 2^h \times 2^h$, then it can be represented by a full octree, all of whose leaves are at depth h in the tree. A $2^h \times 2^h \times 2^h$ grid divides perfectly into eight $2^{h-1} \times 2^{h-1} \times 2^{h-1}$ regions. Each $2^i \times 2^i \times 2^i$ regions subdivides perfectly into eight $2^{i-1} \times 2^{i-1} \times 2^{i-1}$ and the $2 \times 2 \times 2$ regions subdivide perfectly into eight cubes. However, if a grid has cube dimensions $m \times m \times m$ and m is not a power of two, or if a grid has cube dimensions $m_x \times m_y \times m_z$ and the m_i have different values, then there is no such perfect subdivision. We shall discuss octrees for such grids in the next section. For now, we restrict ourselves to the idealized case of grids with cube dimensions $2^h \times 2^h \times 2^h$.

At each node in the octree, we store the span of the corresponding region, by storing the minimum and maximum values of all grid vertices in the region. We also store a reference to the grid cube with the lowest coordinates in the region.

Each internal octree node has eight children. Instead of storing eight references, we allocate the nodes representing the eight children in a single array of eight nodes and store a reference to the array. The jth child of the node is the jth node in this array.

The octree is built recursively, starting at the root. At the root we store a reference to cube (0,0,0), the cube with lowest coordinates in the grid. To add children to the root we call the procedure GROWFULLNODE (Algorithm 8.2). GROWFULLNODE takes two parameters, a node from the octree and the height of the node. The height of the octree root is h where $2^h \times 2^h \times 2^h$ are the dimensions of the input grid.

GROWFULLNODE processes its input node as follows. If the height of the node is zero, then the node is a leaf of the tree corresponding to a single cube. The node contains a reference to that cube. GROWFULLNODE computes the minimum and maximum scalar values over all vertices of that cube and stores those values to represent the span of the node.

If the height of the node is not zero, then the node is an internal one. The node represents a region of dimensions $2^{\mathsf{Height}} \times 2^{\mathsf{Height}} \times 2^{\mathsf{Height}}$ starting at the cube referenced by the node. GROWFULLNODE creates an array of eight child nodes and stores a reference to that array in the parent node.

Let L equal $2^{\mathsf{Height}-1}$. Each child node represents an $\mathsf{L} \times \mathsf{L} \times \mathsf{L}$ region and contains a reference to the cube with lowest coordinates in that region. The cube for the first child node is the same as the cube for its parent. The region and cube for the second child node are translates along the x-axis of the region and cube for the first child node. Let (x_0, y_0, z_0) be the coordinates of the cube for the first child node. The cube for the second child node has coordinates $(x_0, y_0, z_0) + (\mathsf{L}, 0, 0) = (x_0 + \mathsf{L}, y_0, z_0)$. The cubes for the third through eighth child nodes are translates of the cube for the first child node in directions $(0, \mathsf{L}, 0)$, $(\mathsf{L}, \mathsf{L}, 0)$, $(0, 0, \mathsf{L})$, $(\mathsf{L}, 0, \mathsf{L})$, $(0, \mathsf{L}, \mathsf{L})$, and $(\mathsf{L}, \mathsf{L}, \mathsf{L})$, respectively. For each child node, GROWFULLNODE stores a reference to the cube with the appropriate coordinates.

Input : **Node** is an octree node.

Height is the height of **Node** in the octree.

GrowFullNode(**Node**, **Height**)

1   if **Height** $= 0$ then
2   **Node**.minvalue $\leftarrow$ minimum scalar value of any vertex of **Node**.cube0;
3   **Node**.maxvalue $\leftarrow$ maximum scalar value of any vertex of **Node**.cube0;
4   return;
5   end
6   $L \leftarrow 2^{\text{Height}-1}$;
7   $(x_0, y_0, z_0) \leftarrow$ coordinates of **Node**.cube0;
8   Create array, **NodeArray**, of eight new octree nodes;
9   **Node**.child $\leftarrow$ reference to array **NodeArray**;
10   $j \leftarrow 0$;
11   for $i_x = 0$ to 1 do
12   for $i_y = 0$ to 1 do
13   for $i_z = 0$ to 1 do
14         $(x, y, z) \leftarrow (x_0 + i_x \times L, y_0 + i_y \times L, z_0 + i_z \times L)$;
15   **NodeArray**[j].cube0 $\leftarrow$ reference to cube (x, y, z);
16         $j \leftarrow j + 1$;
17   end
18   end
19   end
20   for $i = 0, \ldots, 7$ do
21   GrowFullNode(**NodeArray**[i], **Height**-1);
22   end
23   **Node**.minvalue $\leftarrow$ minimum of $\{$**NodeArray**[i].minvalue $\}_{i=0,\ldots,7}$;
24   **Node**.maxvalue $\leftarrow$ maximum of $\{$**NodeArray**[i].maxvalue $\}_{i=0,\ldots,7}$;

Algorithm 8.2. GROWFULLNODE.

To create subtrees at each child node, GROWFULLNODE recursively calls itself on each child node. The recursive call sets the minimum and maximum scalar values for each child node. These values represent the minimum and maximum scalar values over all vertices in the corresponding region. GROWFULLNODE sets the minimum and maximum scalar values for the parent node by computing the minimum over all the minimum scalar values and the maximum over all maximum scalar values of the child nodes.

The algorithm to search the octree takes an isovalue σ as input. Starting at the octree root, the search algorithm processes an internal octree node by determining if σ lies in the interval between the nodes minimum and maximum values. If σ lies in the interval, then the search algorithm recursively processes

---

Input : Isovalue is the isovalue.
            Node is an octree node.
            Height is the height of Node in the octree.

SearchBON (Isovalue, Node, Height)

  1   if (Node.minvalue $\leq$ Isovalue $\leq$ Node.maxvalue) then
  2   if Height $= 0$ then
  3   Report (Node.cube0);
  4   else
  5         $k \leftarrow$ Node.numChildren;
  6   for $j = 0, \ldots, k - 1$ do
  7   SearchBON (Isovalue, Node.child (j), Height-1);
  8   end
  9   end
10   end

Algorithm 8.3. SEARCHBON.

the node's children. If σ does not lie in the interval, then the search algorithm does not intersect the region represented by the node and the entire subtree rooted at that node is ignored.

The search algorithm recursively processes internal nodes until it reaches the octree leaves. It again determines if σ lies in the interval between the nodes minimum and maximum values. If σ lies in the interval, then the cube referenced at that node is added to a list of intersected cubes.

8.2.2 Branch on Need Octrees

When a grid does not have cube dimensions $2^h \times 2^h \times 2^h$, then its subdivisions cannot be represented by a full octree. One solution might be to pad the grid with extra cubes until it has dimensions $2^h \times 2^h \times 2^h$. Unfortunately, such a padding of an $m \times m \times m$ grid could increase the number of cubes by a factor of eight. If the grid has cube dimensions $m_x \times m_y \times m_z$ where not all the m_i are equal, the increase could be much larger.

The solution presented here is the branch-on-need (BON) octree proposed by Wilhelms and Van Gelder [Wilhelms and Gelder, 1992]. Let $m_x \times m_y \times m_z$ be the cube dimensions of the input grid and let $(0, 0, 0)$ be its vertex with lowest coordinates. Let 2^h be the smallest power of two that is greater than $\max(m_x, m_y, m_z)$. The root of the octree represents a $2^h \times 2^h \times 2^h$ region extending from $(0, 0, 0)$ to $(2^h, 2^h, 2^h)$. This region covers the grid. If m_i does not equal 2^h, then this region will extend beyond the edge of the grid.

The root region is partitioned into eight congruent octants by dividing it with three axis-parallel planes through the point $(2^{h-1}, 2^{h-1}, 2^{h-1})$. Some of these octants will intersect the grid but some may not. Add a child to the root for each octant that contains one or more grid cubes. Octants that do not contain any grid cubes are not represented in the tree, hence the name "Branch on Need."

Each internal node in the tree represents a cubical region that intersects the input grid. The three axis-parallel planes through the center of the region partition the region into eight congruent octants. As with the root, some of the octants will intersect the input grid but some may not. The node contains a child for each octant that contains one or more grid cubes. Each internal nodes has one to eight children.

The leaves of the octree are individual grid cubes. Because every node of the tree at depth d represents a $2^{h-d} \times 2^{h-d} \times 2^{h-d}$ region, every leaf has depth h. Thus all leaves are at the same level in the tree.

The number of leaves in the branch-on-need octree is $m_x \times m_y \times m_z$, the number of cubes in the input grid. The nodes at height one in the tree represent cubical regions of dimension $2 \times 2 \times 2$. They form a grid of dimensions $\lceil m_x/2 \rceil \times \lceil m_y/2 \rceil \times \lceil m_z/2 \rceil$ where the cubes in the grid are $2 \times 2 \times 2$ regions. If some m_i are odd, then some of these $2 \times 2 \times 2$ regions will extend beyond the end of the input grid.

Nodes at height j of the tree represent cubical regions of dimension $2^j \times 2^j \times 2^j$. These regions form a grid of dimensions $\lceil m_x/2^j \rceil \times \lceil m_y/2^j \rceil \times \lceil m_z/2^j \rceil$ where the cubes in the grid are $2^j \times 2^j \times 2^j$ regions. If some m_i is not divisible by j, then some of these $2^j \times 2^j \times 2^j$ regions will extend beyond the end of the input grid.

As in the full octree, each node in the branch-on-need octree contains the minimum and maximum values of all grid vertices in its corresponding region and a reference to the grid cube with the lowest coordinates in the region. Each internal node contains a reference to a single array of nodes that stores its children. The kth child of the node is the kth node in this array. However, the number of children is not necessarily eight as it is in the full octree. Each internal node in the branch-on-need octree explicitly stores its number of children.

To build the branch-on-need octree on an $m_x \times m_y \times m_z$ input grid, we first compute the smallest h such that $2^h \geq \max(m_x, m_y, m_z)$. The root of the octree represents a $2^h \times 2^h \times 2^h$ region.

The octree is built recursively, starting at the root. At the root we store a reference to cube $(0,0,0)$, the cube with the lowest coordinates in the grid. To add children to the root we call the procedure GROWBON (Algorithm 8.4). GROWBON takes two parameters, a node from the octree and the height of the node. The height of the octree root is h.

GROWBON processes its input node as follows. If the height of the node is zero, then the node is a leaf of the tree corresponding to a single cube. The node contains a reference to that cube. GROWBON computes the minimum and

---

Input : **Node** is an octree node.
　　　　　 Height is the height of **Node** in the octree.

GrowBON (**Node**, **Height**)

1 if **Height** $= 0$ then
2 $\quad$ **Node**.minvalue $\leftarrow$ minimum scalar value of any vertex of **Node**.cube0;
3 $\quad$ **Node**.maxvalue $\leftarrow$ maximum scalar value of any vertex of **Node**.cube0;
4 $\quad$ return;
5 end
6 $\mathsf{L} \leftarrow 2^{\mathsf{Height}-1}$;
7 $(x_0, y_0, z_0) \leftarrow$ coordinates of **Node**.cube0;
8 if $(x_0 + \mathsf{L} < m_x)$ then $k_x \leftarrow 2$; else $k_x \leftarrow 1$;
9 if $(y_0 + \mathsf{L} < m_y)$ then $k_y \leftarrow 2$; else $k_y \leftarrow 1$;
10 if $(z_0 + \mathsf{L} < m_z)$ then $k_z \leftarrow 2$; else $k_z \leftarrow 1$;
11 $k \leftarrow k_x \times k_y \times k_z$;
12 **Node**.numChildren $\leftarrow k$;
13 Create array, **NodeArray**, of k new octree nodes;
14 **Node**.child $\leftarrow$ reference to array **NodeArray**;
15 $j \leftarrow 0$;
16 for $i_x = 0$ to $k_x - 1$ do
17 $\quad$ for $i_y = 0$ to $k_y - 1$ do
18 $\quad\quad$ for $i_z = 0$ to $k_z - 1$ do
19 $\quad\quad\quad$ $(x, y, z) \leftarrow (x_0 + i_x \times \mathsf{L}, y_0 + i_y \times \mathsf{L}, z_0 + i_z \times \mathsf{L})$;
20 $\quad\quad\quad$ **NodeArray**[j].cube0 $\leftarrow$ reference to cube (x, y, z);
21 $\quad\quad\quad$ $j \leftarrow j + 1$;
22 $\quad\quad$ end
23 $\quad$ end
24 end
25 for $j = 0$ to $k - 1$ do
26 $\quad$ GrowBON (**NodeArray**[j], **Height**-1);
27 end
28 **Node**.minvalue $\leftarrow$ minimum of $\{$**NodeArray**[j].minvalue $\}_{j=0,\dots,k-1}$;
29 **Node**.maxvalue $\leftarrow$ maximum of $\{$**NodeArray**[j].maxvalue $\}_{j=0,\dots,k-1}$;

Algorithm 8.4. GROWBON.

maximum scalar values over all vertices of that cube and stores those values to represent the span of the node.

If the height of the node is not zero, then the node is an internal one. We need to determine the number of children of the node. Let (x_0, y_0, z_0) be the coordinates of the cube referenced by the node. The node represents a region of dimensions $2^{\mathsf{Height}} \times 2^{\mathsf{Height}} \times 2^{\mathsf{Height}}$ starting at cube (x_0, y_0, z_0). Let L equal

$2^{\mathsf{Height}-1}$. Let k_x equal one if $x_0 + \mathsf{L}$ is less than m_x and let k_x equal two, otherwise. Similarly, let k_y equal one if $y_0 + \mathsf{L}$ is less than m_y and equal two, otherwise, and let k_z equal one if $z_0 + \mathsf{L}$ is less than m_z and equal two, otherwise. The node has $k = k_x \times k_y \times k_z$ children. Create an array of m child nodes and store a reference to this array and its length at the parent node.

To set the cubes for each of the child nodes we index the child nodes by (i_x, i_y, i_z) where $0 \le i_x < k_x$ and $0 \le i_y < k_y$ and $0 \le i_z < k_z$. Note that i_x, i_y, and i_z have values zero or one. There are m such indices corresponding to the m children. The cube for child (i_x, i_y, i_z) is $(x_0 + i_x \times \mathsf{L}, y_0 + i_y \times \mathsf{L}, z_0 + i_z \times \mathsf{L})$. We store a reference to that cube in the child node. The cube represents an octant of the parent node which intersects the input grid.

The rest of the algorithm follows the algorithm GROWFULLOCTREE for the full octree. To create subtrees at each child node, GROWBON recursively calls itself on each child node. The recursive call sets the minimum and maximum scalar values for each child node. These values represent the minimum and maximum scalar values over all vertices in the corresponding region. GROWBON sets the minimum and maximum scalar values for the parent node by computing the minimum over all the minimum scalar values and the maximum over all maximum scalar values of the child nodes.

The algorithm to search the branch-on-need octree takes an isovalue σ as input. Starting at the octree root, the search algorithm processes an internal octree node by determining if σ lies in the interval between the node's minimum and maximum values. If σ lies in the interval, then the search algorithm recursively processes the node's children. If σ does not lie in the interval, then the search algorithm does not intersect the region represented by the node and the entire subtree rooted at that node is ignored.

The search algorithm recursively processes internal nodes until it reaches the octree leaves. It again determines if σ lies in the interval between the nodes minimum and maximum values. If σ lies in the interval, then the cube referenced at that node is added to a list of intersected cubes.

8.2.3   Space and Time Analysis

Full octree. We first give an analysis of the space and time complexity for a full octree of cube dimensions $2^h \times 2^h \times 2^h$. The total number of cubes in the octree is 2^{3h}. The number of nodes in the octree is

$$2^{3h} + 2^{3h}/8 + 2^{3h}/8^2 + \ldots + 8^2 + 8 + 1 < (8/7)2^{3h}.$$

Thus the space used by the octree is $\Theta(N)$ where $N = 2^{3h}$ is the number of cubes in the octree. The height of the full octree is $h = \log_8(N)$.

The time to build the octree (algorithm BUILDISOOCTREE, Figure 8.5) is proportional to the size of the octree. Thus, the time to build the octree is also $\Theta(N)$.

```
Input   : Isovalue is the isovalue.
          Node is an octree node.
          Height is the height of Node in the octree.

SearchFullNode(Isovalue, Node, Height)
1  if (Node.minvalue ≤ Isovalue ≤ Node.maxvalue) then
2  |    if Height = 0 then
3  |    |    Report(Node.cube0);
4  |    else
5  |    |    for i = 0,…,7 do
6  |    |    |    SearchFullNode(Isovalue, Node.child (i), Height-1);
7  |    |    end
8  |    end
9  end
```

Algorithm 8.5. SEARCHFULLNODE.

We search the octree by recursively calling Algorithm 8.5 SEARCHFULLNODE. For each cube whose span contains the input isovalue, σ, we make at most $\log_8(N)$ recursive calls to SEARCHFULLNODE. Thus searching the octree takes $O(1 + M \log(N))$ time where M is the number of cubes whose span contains σ.

We summarize the space and running times in the following proposition.

Proposition 8.1. *Let* $N = 2^{3h}$ *be the number of cubes in the input grid with cube dimensions* $2^h \times 2^h \times 2^h$.

- *Algorithm* BUILDISOOCTREE *creates an octree of size* $\Theta(N)$ *and height* $\lfloor \log_8(N) \rfloor$.

- *Algorithm* BUILDISOOCTREE *runs in* $\Theta(N)$ *time.*

- *Algorithm* SEARCHFULLNODE *runs in* $O(1 + M \log(N))$ *time where* M *is the number of cubes whose span contains the input isovalue.*

If the span of every cube contains σ, then searching the octree takes only $O(M)$ time since we can make at most one recursive call to SEARCHFULLNODE for each node in the octree. Thus $O(1 + M \log(N))$ time is not a tight upper bound.

Branch-on-need (BON) octree. Consider a grid Γ with cube dimensions $m_x \times m_y \times m_z$. Without loss of generality, assume that $m_x \geq m_y \geq m_z$. Let $m'_x = 2^{h_x}$ be the smallest power of two that is greater than m_x. Similarly, let $m'_y = 2^{h_y}$ and $m'_z = 2^{h_z}$ be the smallest powers of two that are greater than m_y and m_z, respectively. Note that $m'_x < 2m_x$ and $m'_y < 2m_y$ and $m'_z < 2m_z$. Let Γ' be a grid with cube dimensions $m'_x \times m'_y \times m'_z$.

The octree for grid Γ' contains the octree for grid Γ. We bound the size and height of the octree for grid Γ by bounding the size and height of the octree for grid Γ'.

Nodes in the first $h_x - h_y - 1$ levels of the octree for grid Γ' each have only two nonempty children. The total number of nodes in these levels is

$$1 + 2 + 2^2 + 2^3 + \ldots + 2^{h_x - h_y - 1} < 2^{h_x - h_y} \leq m'_x.$$

The nodes in the next $h_y - h_z - 1$ levels of the octree each have four nonempty children. There are exactly $2^{h_x - h_y}$ nodes at level $h_x - h_y$ and the number of nodes increases by a factor of four at each level. The total number of nodes in these levels is

$$\begin{aligned}
2^{h_x - h_y}(1 + 4 + 4^2 + 4^3 + \ldots + 4^{h_y - h_z - 1}) &< 2^{h_x - h_y} 4^{h_y - h_z} \\
&= 2^{h_x - h_y} 2^{2h_y - 2h_z} \\
&\leq 2^{h_x} 2^{h_y} \leq m'_x m'_y.
\end{aligned}$$

The interior nodes in the next $h_z + 1$ levels of the octree each have eight nonempty children. There are exactly $2^{h_x - h_y} 4^{h_y - h_z}$ nodes at level $h_x - h_z$. The total number of nodes in these levels is

$$\begin{aligned}
2^{h_x - h_y} 4^{h_y - h_z}(1 + 8 + 8^2 + 8^3 + \ldots + 8^{h_z}) &< 2^{h_x - h_y} 4^{h_y - h_z} 8^{h_z + 1} \\
&= 2^{h_x - h_y} 2^{2h_y - 2h_z} 2^{3h_z + 1} \\
&= 2 \times 2^{h_x} 2^{h_y} 2^{h_z} = 2m'_x m'_y m'_z.
\end{aligned}$$

Thus total number of nodes in the BON octree for grid Γ' is $\Theta(m'_x m'_y m'_z)$. Since $m'_x/2 < m_x \leq m'_x$ and $m'_y/2 < m_y \leq m'_y$ and $m'_z/2 < m_z \leq m'_z$, the total number of nodes in the BON octree for grid Γ is $\Theta(m_x m_y m_z)$ or $\Theta(N)$ where $N = m_x m_y m_z$ is the number of cubes in the grid.

The height of the BON octree for both grid Γ and grid Γ' is $h_x = \log_2(m'_x)$. Since $m'_x/2 < m_x \leq m'_x$, the height of the BON octree for grid Γ is $\Theta(\log_2(m_x))$. Since $m_x \leq N \leq m_x^3$, the height of the BON octree for grid Γ is $\Theta(\log_2(N))$.

The time to build the octree (Algorithm BUILDBONOCTREE, Figure 8.7) is proportional to the size of the octree. Thus, the time to build the octree is also $O(N)$.

We search the octree by recursively calling Algorithm 8.5 SEARCHFULLNODE. For each cube whose span contains the input isovalue, σ, we make at most $\log_2(N)$ recursive calls to SEARCHFULLNODE. Thus searching the octree takes $O(1 + M\log(N))$ time where M is the number of cubes whose span contains σ.

We summarize the space and running times in the following proposition.

Proposition 8.2. *Let N be the number of cubes in the input grid.*

- *Algorithm* BUILDBONOCTREE *creates an octree of size $\Theta(N)$ and height at most $\Theta(\log(N))$.*

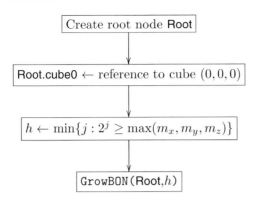

Figure 8.7. BuildBONOctree. Input is a scalar grid with cube dimensions $m_x \times m_y \times m_z$.

- *Algorithm* BuildBONOctree *runs in* $\Theta(N)$ *time.*

- *Algorithm* SearchBON *runs in* $O(1 + M \log(N))$ *time where* M *is the number of cubes whose span contains the input isovalue.*

8.3   Span Space Priority Trees

Octrees cluster grid cubes together based on their geometric location in space. Span space data structures cluster grid cubes together based on the proximity of cube spans. In this section, we present a data structure similar to one given with Livnat et al. in [Livnat et al., 1996] but replace the kd-tree in their paper by McCreight's priority tree [McCreight, 1985]. The modification using priority trees was suggested by Bajaj et al. in [Bajaj et al., 1996].

8.3.1   Span Space

As previously defined, the span of a grid cube $\mathbf{c}$ is the interval from the minimum scalar value $s_{\min}(\mathbf{c})$ to the maximum scalar value $s_{\max}(\mathbf{c})$ of the cube. The span space of a set of cubes $\mathbf{c}_1, \mathbf{c}_2, \mathbf{c}_3, \ldots$ is the set of points $P = \{(x_i, y_i) : [x_i, y_i]$ is the span of cube $\mathbf{c}_i\}$. Note that P is actually a multiset, i.e., the same point can occur multiple times in P.

Given an isovalue σ, we want to find the cubes $\mathbf{c}_i$ such that $x_i \leq \sigma \leq y_i$. Plot the points (x_i, y_i) in the plane. The horizontal and vertical lines through the

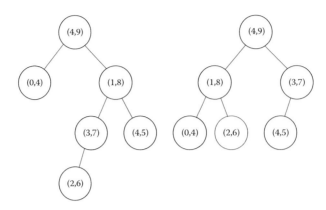

Figure 8.8. Two priority trees on the same set of points. The priority tree on the right is an almost complete binary tree.

point (σ, σ) divide the plane into four quadrants. The points of P in the upper-left quadrant correspond exactly to the cubes $\mathbf{c}_i$ where $x_i \leq \sigma \leq y_i$. Finding these cubes becomes the problem of finding all points in this upper-left quadrant.

In [McCreight, 1985], McCreight presented an efficient data structure for reporting all points $(x, y) \in P$ that lie in a query region $\{(x, y) : x_{\min} \leq x \leq x_{\max}$ and $y \geq y_{\min}\}$. If we let $x_{\min}$ equal $-\infty$ and $x_{\max} = y_{\min} = \sigma$, then this region is the upper-left quadrant defined by the horizontal and vertical lines through (σ, σ). The reported set is exactly the points (x_i, y_i) such that $x_i \leq \sigma \leq y_i$. Thus, we can use McCreight's priority trees to solve the problem of reporting cubes based on their span.

Instead of McCreight's priority trees, octrees or kd-trees could be used to report the points (x_i, y_i) such that $x_i \leq \sigma \leq y_i$. (See [Samet, 1990b] for a description of an octree for reporting points. See [Livnat et al., 1996] for definition and discussion of kd-trees.) We use priority trees because their worst-case query time is $O(\log n + k)$ where n is the number of grid elements and k is the number of reported points. Octrees and kd-trees have worst-case query time of $O(\sqrt{n} + k)$. (The octree for reporting points in the span space should not be confused with the octree in Section 8.2 that recursively partitions the grid into octants. Because grid cubes are evenly distributed across the grid, the octree in Section 8.2 has an $O(k \log(n))$ upper bound on its query time. The span space points have no structure and can be distributed in any manner.)

8.3.2   McCreight Priority Trees

A priority tree is a binary tree with a point (x, y) assigned to each node such that each node satisfies the following two conditions:

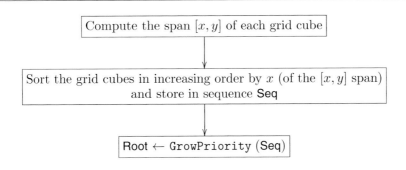

Figure 8.9. BUILDPRIORITY. Input is a scalar grid.

1. Points in the left subtree of the node have an x-coordinate less than or equal to points in the right subtree of the node.

2. Coordinate y is greater than or equal to the y-coordinates of any descendants of the node.

A priority tree can be viewed as a combination of a binary search tree built on the x-coordinates and a heap built on the y-coordinates. There can be many different priority trees for a given set of points.

Our description of priority trees is specialized for the problem of isosurface extraction. Each node in a priority tree contains a reference to some cube in the search grid. It contains an x- and a y-coordinate representing the span $[x, y]$ of the referenced cube. It also contains a value minX, representing the minimum x value of any node in the subtree rooted at the node. Finally, it contains a reference to its left and right child.

To build the span space priority tree, construct a sequence of all the grid cubes and their $[x, y]$ spans. Each element of the sequence contains a reference to a cube and the x and y values of its span. Sort the sequence in increasing order by the x-coordinate. Call the subroutine GROWPRIORITY, which creates a priority tree and returns the root of that tree. (See Figure 8.9.)

GROWPRIORITY creates the priority tree as follows. Let h equal $\lfloor \log_2(n) \rfloor$. As explained in the next section, h will be the height of the priority tree. Create a new tree node and store the minimum x value of any element of the sequence at that node. Find the maximum y value of any element of the sequence and a sequence element with that maximum y value. (Break ties by choosing the last element of the sequence with the maximum y value.) Copy the x, y, and cube reference of the sequence element into new node. Delete this element from the sequence.

Let n_L equal $\min\{2^h - 1, n' - (2^{h-1} - 1)\}$ where n' is the number of elements remaining in the sequence. Let LeftSeq be the first n_L elements of the sequence

---

Input   : Seq is a sequence of points sorted by increasing x value.

GrowPriority(Seq)

1   if Seq $= \emptyset$ then return NULL;
2   numElem $\leftarrow$ number of elements of Seq;
3   Height $\leftarrow \lfloor \log_2(\textsf{numElem}) \rfloor$;
4   Create new tree node Node;
5   Node.minX $\leftarrow$ Seq$[0].x$;
6   maxY $\leftarrow \max_i \{\textsf{Seq}[i].y\}$;
7   $j \leftarrow \max\{i : \textsf{Seq}[i].y = \textsf{maxY}\}$;
8   Node.cube $\leftarrow$ Seq$[j]$.cube;
9   Node.$x \leftarrow$ Seq$[j].x$;
10 Node.$y \leftarrow$ Seq$[j].y$;
11 Delete element j from Seq;
12 numElem $\leftarrow$ numElem $- 1$;
13 numLeft $\leftarrow \min\{2^{\textsf{Height}} - 1, \textsf{numElem} - (2^{\textsf{Height}-1} - 1)\}$;
14 numRight $\leftarrow$ numElem $-$ numLeft;
15 LeftSeq $\leftarrow$ first numLeft elements of Seq;
16 RightSeq $\leftarrow$ last numRight elements of Seq;
17 Node.left $\leftarrow$ GrowPriority(LeftSeq);
18 Node.right $\leftarrow$ GrowPriority(RightSeq);
19 return reference to Node;

Algorithm 8.6. GROWPRIORITY.

and RightSeq be the last $n' - n_L$ elements of the sequence. Value n_L is chosen so that GROWPRIORITY creates an almost complete binary tree as will be defined in Section 8.3.4. Either the nodes in LeftSeq will form a complete binary subtree or the nodes in RightSeq will form a complete binary subtree.

Recursively call GROWPRIORITY on sequence LeftSeq. GROWPRIORITY returns the root of a priority tree on the nodes in LeftSeq. Set the left child of the current node to this root. Similarly, recursively call GROWPRIORITY on RightSeq and set the right child of the current node to this root.

The left and right subtrees at each node satisfy the first condition on priority tree nodes while the node's y value satisfies the second condition. Thus GROW-PRIORITY creates a priority tree. In the next section, we show that this priority tree has height $h = \lfloor \log_2(n) \rfloor$.

The algorithm to search the priority tree takes an isovalue as input. Starting at the priority tree root, the algorithm processes node v by comparing the isovalue to v.minX. The v.minX is the minimum x value over all nodes in the subtree rooted at v. If the isovalue is less than or equal to v.minX, then the isovalue is less than or equal to x for every node in the subtree rooted at v. The algorithm

```
Input   : Isovalue is the isovalue.
          Node is a priority tree node.

SearchPriority(Isovalue, Node)
1  if (Isovalue ≤ Node.minX) then return;
2  if (Isovalue > Node.y) then return;
3  if (Node.x ≤ Isovalue) then
4  |   Report(Node.cube);
5  end
6  if Node.left ≠ NULL then
7  |   SearchPriority(Isovalue,Node.left);
8  end
9  if Node.right ≠ NULL then
10 |   SearchPriority(Isovalue,Node.right);
11 end
```

Algorithm 8.7. SEARCHPRIORITY.

skips the entire subtree. If the isovalue is greater than $v.y$, then the isovalue is greater than the y value of every node in the subtree rooted at v. Again, the algorithm skips the entire subtree.

The remaining case is that the isovalue is in the range $[v.\mathsf{minX}, v.y]$. If the isovalue is in the range $[v.x, v.y]$, then the algorithm reports the cube at node v. Whether or not the isovalue is in the range $[v.x, v.y]$, the algorithm recursively processes the left and right subtrees of node v.

8.3.3 Space and Time Analysis

Let N be the number of cubes in the input grid. Each recursive call to algorithm GROWPRIORITY creates one tree node for each element of the input sequence. Therefore, the size of the priority tree is $\Theta(N)$.

BUILDPRIORITY takes $\Theta(N)$ time to compute the span of each cube and $\Theta(N \log(N))$ time to sort the cubes. To compute the running time of GROW-PRIORITY, we represent the recursive calls to GROWPRIORITY in a recursion tree. Each node in the tree represents a recursive call and the two children of the node represent the two recursive calls in Statements 17 and 18. The first call to GROWPRIORITY is the root of the tree. The leaves of the recursion tree are calls to GROWPRIORITY with input Seq equal to the empty set. The internal nodes of the recursion tree represents a call to GROWPRIORITY, which creates a new node in the priority tree. Therefore, the internal nodes correspond to nodes in the priority tree. As shown in the next section, the priority tree has height $\lfloor \log_2(N) \rfloor$. Thus, the recursion tree has height $\lfloor \log_2(N) \rfloor + 1$.

Let N_v be the size of Seq in the recursive call associated with node v of the recursion tree. The recursive call takes $\Theta(N_v)$ time to find maxY and $\max\{i : \text{Seq}[i].x = \text{maxY}\}$ in steps 6 and 7. Splitting Seq into the subsequence LeftSeq and RightSeq takes $\Theta(N_v)$ time. Thus, the running time for this recursive call of GrowPriority is $\Theta(N_v)$.

The total running time is $\Theta(\sum_v N_v)$. Let V_k be the set of nodes at depth k in the tree. Rewrite $\sum_v N_v$ as $\sum_{k=0}^{h} \sum_{v \in V_k} N_v$ where h is the height of the tree. Because LeftSeq and RightSeq share no elements, each cube is in at most one sequence at a recursive call of depth k. Thus, $\sum_{v \in V_k} N_v$ is at most N and

$$\sum_v N_v = \sum_{k=0}^{h} \sum_{v \in V_k} N_v \leq \sum_{k=0}^{h} N = Nh.$$

Since the height h of the tree is $\lfloor \log_2(N) \rfloor + 1$, the running time of GROWPRIORITY is $\Theta(N \log(N))$. Thus, the running time of BUILDPRIORITY is $\Theta(N \log(N))$.

Finally, we analyze the running time of SEARCHPRIORITY. Let M be the number of elements reported by SEARCHPRIORITY. Each recursive call to SEARCHPRIORITY takes $\Theta(1)$ time. Let v be the current node in the priority tree. If the input isovalue is at most v.minX or if the isovalue is greater than $v.y$, then the recursive call returns immediately. We charge this time to the parent routine, i.e., the routine that called the current one. If the isovalue is in the range $[v.x, v.y]$, then we associate this time with the element being reported. There are M such recursive calls and they take a total of $\Theta(M)$ time. We are left with recursive calls where the isovalue is in the range $[v.\text{minX}, v.x]$.

Because nodes in the priority tree are ordered from left to right by x value, if the isovalue is less than v.right.minX, then the isovalue is less than v'.minX for every node v' in the subtree rooted at v.right. If the isovalue is greater than v.right.minX, then the isovalue is greater than $v'.x$ for every node v' in the subtree rooted at v.left. Thus, if the isovalue is in the ranges $[v'.\text{minX}, v'.x]$ and $[v''.\text{minX}, v''.x]$ for two nodes v' and v'', then one of these nodes must be a descendant of the other. The nodes v' where the isovalue is in the range $[v'.\text{minX}, v'.x]$ must lie on a path from the root to some leaf. Since the priority tree has height $\lfloor \log_2(N) \rfloor$, there are at most $\lfloor \log_2(N) \rfloor$ such nodes. Processing such nodes takes a total of $\Theta(\log(N))$ time. Thus, the running time of SEARCHPRIORITY is $\Theta(\log(N) + M)$.

We summarize the space and running times in the following proposition.

Proposition 8.3. *Let N be the number of cubes in the input grid.*

- *Algorithm* BUILDPRIORITY *creates a priority of size $\Theta(N)$ and height $\lfloor \log_2(N) \rfloor$.*

- *Algorithm* BUILDPRIORITY *runs in $\Theta(N \log(N))$ time.*

- *Algorithm* SEARCHPRIORITY *runs in $\Theta(\log(N) + M)$ time where M is the number of cubes whose span contains the input isovalue.*

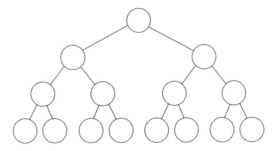

Figure 8.10. Complete binary tree of height three.

8.3.4 Almost Complete Binary Trees

The running time of the search algorithm depends upon the height of the priority tree. As previously noted, there are many possible priority trees on a given set of nodes. We claim that Algorithm BUILDPRIORITY creates a very specialized tree called an almost complete binary tree. In Section 8.3.5, we will discuss how almost complete binary trees have very simple, compact implementations.

A complete binary tree is a binary tree where all internal nodes have degree two and all leaves have the same depth. (See Figure 8.10.) Alternately, a complete binary tree of height $h > 0$ has the following recursive property:

> The left and right subtrees of the root of a complete binary tree of height h are complete binary trees of height $h - 1$.

A single node is a complete binary tree of height zero.

A complete binary tree of height h has exactly $2^{h+1} - 1$ nodes and all its leaves are at depth h. Thus there is no complete binary tree on n nodes if $n + 1$ is not a power of two. An almost complete binary tree is a relaxation of the requirements of a complete binary tree to accommodate any number of nodes.

A single node is an almost complete binary tree of height zero. An almost complete binary tree of height one is either a complete binary tree of height one or a root node and a leaf node with the leaf as the left child of the root. An almost complete binary tree of height $h > 1$ has one of the following two recursive properties:

(a) The left subtree of the root is a complete binary tree of height $h - 1$ and the right subtree is an almost complete binary tree of height $h - 1$.

(b) The left subtree of the root is an almost complete binary tree of height $h - 1$ and the right subtree is a complete binary tree of height $h - 2$.

See Figure 8.11 for an example of an almost complete binary tree.

We claim the following properties of almost complete binary trees.

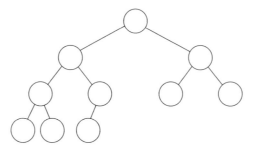

Figure 8.11. Almost complete binary tree of height three. The left subtree of the root is an almost complete binary tree of height two while the right subtree of the root is a complete binary tree of height one.

Proposition 8.4.

1. *If n is the number of nodes and h is the height of an almost complete binary tree, then $2^h \leq n < 2^{h+1}$, or, equivalently, $h = \lfloor \log_2(n) \rfloor$.*

2. *An almost complete binary tree on $2^{h+1} - 1$ nodes is a complete binary tree of height h.*

Proof of 1: The proof is by induction on n. Assume that every almost complete binary tree on $k < n$ nodes has height $\lfloor \log_2(k) \rfloor$.

Consider an almost binary tree on n nodes with height h. Let n_L be the number of elements in the left subtree and n_R be the number of elements in the right subtree. Note that n equals $n_L + n_R + 1$. We consider two cases based on whether the tree has recursive property (a) or (b).

If the tree has recursive property (a), then the left subtree is a complete binary tree of height $h - 1$ and n_L equals $2^h - 1$. The right subtree is an almost complete binary tree of height $h - 1$. By induction, $2^{h-1} \leq n_R < 2^h$. Thus, $2^h \leq 2^h - 1 + 2^{h-1} + 1 \leq n_L + n_R + 1 < 2^{h+1}$ and $2^h \leq n < 2^{h+1}$.

If the tree has recursive property (b), then the right subtree is a complete binary tree of height $h - 2$ and n_R equals $2^{h-1} - 1$. The left subtree is an almost complete binary tree of height $h - 1$. By induction, $2^{h-1} \leq n_L < 2^h$. Thus, $2^h \leq n_L + n_R + 1 < 2^h + 2^{h-1} < 2^{h+1}$ and $2^h \leq n < 2^{h+1}$. □

Proof of 2: The proof is by induction on h. When h equals zero, the tree is a single leaf node at depth zero. When h equals one, the tree is the complete binary tree on three nodes. Assume that for all $h' < h$, an almost complete binary tree on $2^{h'+1} - 1$ nodes is a complete binary tree of height h'.

Consider an almost complete binary tree on $2^h - 1$ nodes. The left and right subtrees of the root of an almost complete binary tree of height h are almost complete binary trees of height $h-1$ or $h-2$. By Statement 1 in this proposition, the left and right subtrees each have at most $2^h - 1$ nodes. Since the original tree has $2^{h+1} - 1$ nodes, the left and right subtrees must each have exactly $2^h - 1$ nodes. By the inductive assumption, the left and right subtrees are complete binary trees of height $h - 1$. Thus, the original tree is a complete binary tree of height h.   □

We claim that algorithm BUILDPRIORITY constructs an almost complete binary tree.

Proposition 8.5. *Let N be the number of cubes in the input grid. Algorithm* BUILDPRIORITY *constructs an almost complete binary tree of height* $\lfloor \log_2(N) \rfloor$.

Proof: Algorithm BUILDPRIORITY calls GROWPRIORITY to build the priority tree. We prove that GROWPRIORITY constructs an almost complete binary tree. The proof is by induction on the number of elements in the input sequence. When there are no elements or a single element, GROWPRIORITY creates the empty tree or a tree with one element, both of which are almost complete binary trees. Assume that on sequences of $k < N$ elements, GROWPRIORITY creates an almost complete binary tree. Let h equal $\lfloor \log_2(N) \rfloor$. By Steps 12 and 13 in Algorithm 8.6, the number of nodes in the left subtree is $\min(2^h - 1, N - 2^{h-1})$. We consider two cases based on the values of h and N.

Case I: $2^h - 1 < N - 2^{h-1}$:

Since $(2^h - 1)$ is less than $(N - 2^{h-1})$, GROWPRIORITY creates a left subtree with $(2^h - 1)$ nodes and a right subtree with $(N - 2^h)$ nodes. By induction, both subtrees are almost complete binary trees.

By Proposition 8.4, the left subtree is a complete binary tree of height $(h - 1)$ and the right subtree has height $\lfloor \log_2(N - 2^h) \rfloor$. Since the initial tree had height h, the right subtree has height less than h. By assumption, $N - 2^{h-1} > 2^h - 1$ or $N - 2^{h-1} \geq 2^h$. Thus,

$$\lfloor \log_2(N - 2^h) \rfloor = \lfloor \log_2(N - 2^{h-1} - 2^{h-1}) \rfloor$$
$$\geq \lfloor \log_2(2^h - 2^{h-1}) \rfloor = \lfloor \log_2(2^{h-1}) \rfloor = h - 1.$$

Thus, $\lfloor \log_2(N - 2^h) \rfloor$ equals $(h - 1)$. The left subtree is a complete binary tree of height $(h - 1)$ and the right subtree is an almost complete binary tree of height $(h - 1)$, so the tree is an almost complete binary tree of height $h = \lfloor \log_2(N) \rfloor$.

Case II: $2^h - 1 \geq N - 2^{h-1}$:

Since $(2^h - 1)$ is at least $(N - 2^{h-1})$, GROWPRIORITY creates a left subtree with $(N - 2^{h-1})$ nodes and a right subtree with $(2^{h-1} - 1)$ nodes. By induction, both subtrees are almost complete binary trees.

By Proposition 8.4, the left subtree is a nearly complete binary tree of height $\lfloor \log_2(N - 2^{h-1}) \rfloor$ and the right subtree is a complete binary tree of

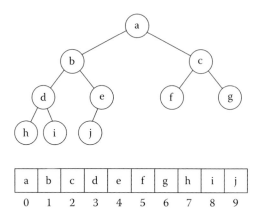

Figure 8.12. Array representation of an almost complete binary tree. Left child of A[i] is A[$2i + 1$]. Right child of A[i] is A[$2i + 2$].

height $(h - 2)$. Since the initial tree had height h, the left subtree has height less than h. Since $N \geq 2^h$,

$$\lfloor \log_2(N - 2^{h-1}) \rfloor \geq \lfloor \log_2(2^h - 2^{h-1}) \rfloor \geq \lfloor \log_2(2^{h-1}) \rfloor = h - 1.$$

Thus, $\lfloor \log_2(N - 2^{h-1}) \rfloor$ equals $(h-1)$. The left subtree is an almost complete binary tree of height $(h - 1)$ and the right subtree is a complete binary tree of height $(h - 2)$, so the tree is an almost complete binary tree of height $h = \lfloor \log_2(N) \rfloor$. □

8.3.5 Priority Tree Implementation

As stated in Proposition 8.5, BUILDPRIORITY creates an almost complete binary tree. An almost complete binary tree on N nodes has a very efficient representation as an array, A, with N elements. Each element of the array from A[0] to A[$N - 1$] represents a node of the tree. (See Figure 8.12.)

The root of the tree is A[0]. The left child of the root is A[1] and its right child is A[2]. More generally, the left child of A[i] is A[$2i + 1$] and the right child

```
Left(i)     /* Return index of left child of ith node */
return 2*i+1;


Right(i)    /* Return index of right child of ith node */
return 2*i+2;
```

Algorithm 8.8. LEFT and RIGHT.

Input   : A is an array storing the tree nodes.
        A[i] is the current node.
        Seq is a sequence of points sorted by increasing x value.

GrowPriority2(A, i, Seq)

1 if Seq $= \emptyset$ then return;
2 numElem $\leftarrow$ number of elements of Seq;
3 Height $\leftarrow \lfloor \log_2(\text{numElem}) \rfloor$;
4 A[i].minX $\leftarrow$ Seq[0].x;
5 maxY $\leftarrow \max_i\{\text{Seq}[i].y\}$;
6 $j \leftarrow \max\{i : \text{Seq}[i].x = \text{maxY}\}$;
7 A[i].cube $\leftarrow$ Seq[j].cube;
8 A[i].x $\leftarrow$ Seq[j].x;
9 A[i].y $\leftarrow$ Seq[j].y;
10 Delete element j from Seq;
11 numElem $\leftarrow$ numElem $- 1$;
12 numLeft $\leftarrow \min\{2^{\text{Height}} - 1, \text{numElem} - (2^{\text{Height}-1} - 1)\}$;
13 numRight $\leftarrow$ numElem $-$ numLeft;
14 LeftSeq $\leftarrow$ first numLeft elements of Seq;
15 RightSeq $\leftarrow$ last numRight elements of Seq;
16 GrowPriority2(A, Left(i), LeftSeq);
17 GrowPriority2(A, Right(i), RightSeq);
18 return;

Algorithm 8.9. GROWPRIORITY2.

is A[$2i + 2$]. If $2i + 1$ is greater than or equal to N, then A[i] has no left child. Similarly, if $2i + 2$ is greater than or equal to N, then A[i] has no right child.

The modified versions of GROWPRIORITY and SEARCHPRIORITY are in Algorithms 8.9 and 8.10.

8.3.6 Homogeneous Cubes

A cube is homogeneous if the scalar values of all its vertices are identical. Let $s_{\mathbf{c}}$ be the scalar values of all the vertices of homogeneous cube $\mathbf{c}$. If $s_{\mathbf{c}}$ is not the isovalue, then the isosurface does not intersect $\mathbf{c}$ and cube $\mathbf{c}$ can be ignored. Even if $s_{\mathbf{c}}$ equals the isovalue, cube $\mathbf{c}$ does not generate any isosurface triangles so $\mathbf{c}$ can again be ignored. Thus homogeneous cubes need not be inserted in the span space priority tree.

Many data sets have a significant number of homogeneous cubes. Duffy et. al. [Duffy et al., 2012] analyzed benchmark data sets from www.volvis.org and www.stereofx.org. They reported that on average 30% of the cubes in the 8-bit data sets and 9% of the cubes in the 12-bit data sets were homogeneous.

Input   : Isovalue is the isovalue.
          A is an array storing the tree nodes.
          N is the length of array A.
          A[i] is the current node.

SearchPriority2(Isovalue, A, N, i)

1 if (Isovalue $\leq$ A[i].minX) then return;
2 if (Isovalue $>$ A[i].y) then return;
3 if (A[i].x $<$ Isovalue) then
4 | Report(A[i].cube);
5 end
6 if Left(i) $< N$ then
7 | SearchPriority2(Isovalue,A,N,Left(i));
8 end
9 if Right(i) $< N$ then
10 | SearchPriority2(Isovalue,A,N,Right(i));
11 end

Algorithm 8.10. SEARCHPRIORITY2.

8.4 Seed Sets

Our last approach to speeding up isosurface retrieval is a method for identifying the cubes intersected by the isosurface from a small initial set of grid cubes, called a seed set. We identify those grid cubes in the seed set whose spans contain the isovalue, i.e., those cubes that intersect the isosurface. We "grow" this set of identified cubes by adding adjacent grid cubes that intersect the isosurface, repeating the process until no adjacent cubes can be added. The technique is called isosurface propagation and produces a set of cubes that intersect the isosurface.

In order for isosurface propagation to find the cubes intersecting the isosurface, some cube in the seed set must intersect the isosurface. If the isosurface has multiple connected components, then each component needs to be intersected by the seed set.[1] An isosurface component that is not intersected by the seed set may be missed by the isosurface propagation algorithm.

In Section 8.4.1, we discuss isosurface propagation in greater detail. In Section 8.4, we give a formal definition of seed sets and discuss how to compute small ones. Finally, in Section 8.4.3, we discuss how isosurface propagation can

---

[1] This is not quite true. If some grid cube intersects two connected components, then the seed set need only intersect one of the two components.

be combined with the span space priority trees of Section 8.3. Methods for computing small seed sets based on contour trees are described in Section 12.8.

8.4.1   Isosurface Propagation

Grid cubes can be adjacent in different ways. They can share facets, edges, or vertices. For isosurface propagation, we use facet adjacency. Two grid cubes are facet adjacent if they share a common facet.

In order to avoid processing a cube more than once, we need a marker bit for each cube. This bit is used to mark the cubes that have been processed. We preprocess the cubes by setting each marker bit to false. (See Algorithm 8.11, IsoPropagateInit.) These bits can either be stored with the individual cubes or in a separate array.

The isosurface propagation algorithm proceeds as follows. For each cube c_0 in the seed set, we call SeedGrow to find adjacent cubes intersecting the isosurface. Algorithm SeedGrow first checks if the isovalue is in the span of c_0 and if c_0 is not marked. If both conditions are true, then SeedGrow reports and marks c_0 and adds c_0 to a first-in-first-out queue Q. While Q is not empty, SeedGrow removes and processes the first element of the queue. Let c be the first element of Q. For each grid cube c' that shares a facet f with c, SeedGrow checks if the isovalue is in the span of f and if c' is marked. If both conditions are true, then SeedGrow marks and reports c' and adds c' to the end of the queue. The span of facet f is the interval from the minimum to the maximum scalar value of any vertex on f.

After SeedGrow is applied to all elements of the seed set, we unmark all the reported (i.e., marked) cubes. If the reported cubes are stored in some list, then the time to unmark the cubes is proportional to the length of the list, not the size of the grid. By unmarking the reported cubes, we can apply algorithm IsoPropagate to a different isovalue without the costly step of reinitializing all the marker bits.

The running time of the initialization algorithm IsoPropagateInit and the reporting algorithm IsoPropagate are as follows:

Proposition 8.6.

- *Algorithm* IsoPropagateInit *runs in* $\Theta(N)$ *time where* N *is the total number of grid cubes.*

- *Algorithm* IsoPropagate *runs in* $\Theta(M_1 + M_2)$ *time where* M_1 *is the number of cubes reported by the algorithm and* M_2 *is the size of the seed set.*

The output of algorithm IsoPropagate depends upon the seed set provided to the algorithm. If that seed set misses the isosurface or some connected component of the isosurface, then IsoPropagate will fail to produce all the cubes

---

Input : G is a regular grid.

`IsoPropagateInit(G)`

1 Unmark all cubes in G;

Algorithm 8.11. IsoPropagateInit.

Input : Isovalue is the isovalue.
SeedSet is a set of seed cubes.

Precondition: All grid cubes are unmarked.

`IsoPropagate(Isovalue, SeedSet)`

1 foreach cube c_0 in SeedSet do
2 | SeedGrow (Isovalue, c_0);
3 end
4 Unmark all reported cubes;

Algorithm 8.12. IsoPropagate.

that intersect the isosurface. In the next section, we discuss seed set properties that will guarantee that IsoPropagate produces all the cubes intersecting the isosurface.

Note that our algorithm IsoPropagate processes all the cubes reachable from one seed cube before it processes the next seed cube. This makes it more likely that adjacent cubes are reported together in the output. Algorithm Iso-Propagate processes cubes in a first-in-first-out order. This gives a breadth-first search, as opposed to depth-first search, of the cubes intersecting the isosurface. A breadth-first search also increases the number of adjacent cubes reported together in the output. Reporting adjacent cubes together may help isosurface extraction algorithms identify isosurface vertices shared across cubes. It also helps build triangle strips, strips of contiguous triangles, which can be rendered faster than isolated triangles.

8.4.2 Constructing Small Seed Sets

Algorithm IsoPropagate starts from a set of cubes called a seed set. This seed set is precomputed from the scalar grid and is independent of any specific isovalue. An obvious requirement is that the seed set intersects every possible isosurface. However, if an isosurface has more than one connected component, then even a seed set that intersects one component of the isosurface may fail to find the cubes that intersect the other components.

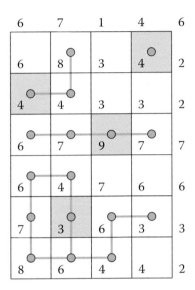

Figure 8.13. Two-dimensional example of graph $\mathcal{G}_\sigma$ (isovalue 5). Vertices (green) of $\mathcal{G}_\sigma$ are squares whose span contains isovalue 5. Edges (green) of $\mathcal{G}_\sigma$ are $(\mathbf{c}, \mathbf{c}')$ where $\mathbf{c}$ and $\mathbf{c}'$ share an edge $\mathbf{e}$ and the span of $\mathbf{e}$ contains isovalue 5. Graph $\mathcal{G}_\sigma$ has four connected components. (One component is a single, isolated vertex.) Yellow cubes form a seed set intersecting all the connected components of $\mathcal{G}_\sigma$.

Connected isosurface components are outputs of isosurface construction algorithms, not intrinsic properties of the scalar grid. Even the MARCHING CUBES algorithm can produce topologically different isosurfaces with different numbers of connected components, depending on the lookup table used. (See Section 2.2.5, Figure 2.13.) Bajaj et al. [Bajaj et al., 1996] gave seed set requirements based on the spans of facet adjacent cubes. We give a slightly modified but equivalent presentation of those requirements.

For each isovalue σ, define a graph $\mathcal{G}_\sigma$ as follows. The vertices of $\mathcal{G}_\sigma$ are the set of grid cubes whose spans contain the isovalue σ. If cubes $\mathbf{c}_1$ and $\mathbf{c}_2$ share a facet $\mathbf{f}$ and the span of $\mathbf{f}$ contains σ, then $(\mathbf{c}_1, \mathbf{c}_2)$ is an edge of graph $\mathcal{G}_\sigma$. A connected component of $\mathcal{G}_\sigma$ is a maximal connected subset of $\mathcal{G}_\sigma$. Figure 8.13 contains a 2D example of $\mathcal{G}_\sigma$.

Note that two cubes $\mathbf{c}$ and $\mathbf{c}'$ could share a facet $\mathbf{f}$, and both have spans containing σ, and yet be in different connected components of $\mathcal{G}_\sigma$. Correspondingly, if the span of facet $\mathbf{f}$ does not contain σ, then algorithm IsoPROPAGATE may report $\mathbf{c}$ but not $\mathbf{c}'$.

If algorithm IsoPROPAGATE on input isovalue σ reports grid cube $\mathbf{c}$, then it will also report all the grid cubes in the connected component of $\mathcal{G}_\sigma$ containing $\mathbf{c}$. Thus, if the seed set intersects every connected component of $\mathcal{G}_\sigma$, then IsoPROPAGATE will report all cubes whose spans contain σ.

Input : Isovalue is the isovalue.
           c_0 is a seed cube.

SeedGrow(Isovalue, c_0)

1 if (Isovalue is not in the span of c_0) or (c_0 is marked) then return;
2 Mark c_0;
3 Report(c_0);
   /* Q is a first-in-first-out queue   */
4 Q.Enqueue(c_0);
5 while Q is not empty do
6      $c \leftarrow Q$.Dequeue();
7   foreach cube c' facet-adjacent to c do
8          $f \leftarrow$ common facet of c and c';
9   if (Isovalue is in the span of f) and (c' is not marked) then
10   Mark c';
11   Report(c');
12   Q.Enqueue(c');
13   end
14   end
15 end

Algorithm 8.13. SEEDGROW.

A complete seed set is a set of cubes that intersect every component of $\mathcal{G}_\sigma$ for every isovalue σ.

Proposition 8.7. *For every isovalue σ, if the seed set input to algorithm* ISO-PROPAGATE *is complete, then* ISOPROPAGATE *will report all the cubes whose span contains σ.*

Proof: Algorithm SEEDGROW (Algorithm 8.13) processes every cube in the connected component of $\mathcal{G}_\sigma$ intersected by c_0. Since a complete seed set intersects every connected component of $\mathcal{G}_\Sigma$, every cube in $\mathcal{G}_\sigma$ is reported by algorithm ISOPROPAGATE. □

Now that we have a definition of a complete seed set, the next problem is to find a "small" complete seed set. Van Kreveld et al. [van Kreveld et al., 1997] describe a polynomial time algorithm for finding a minimum size complete seed set, but the polynomial is large. Fortunately, we do not really need to find a minimum size complete seed set, just a small one.

A number of heuristics have been proposed for finding small, complete seed sets [Bajaj et al., 1996, Itoh and Koyamada, 1994, Itoh and Koyamada, 1995, Itoh et al., 1996, Itoh et al., 2001, van Kreveld et al., 1997]. We present a simple heuristic based on ideas in [Itoh and Koyamada, 1995] by Itoh and Koyamada.

The minimum and maximum scalar values of any vertex of cube $\mathbf{c}$ are denoted $s_{\min}(\mathbf{c})$ and $s_{\max}(\mathbf{c})$, respectively. A grid vertex v is a local minimum if its scalar value equals $s_{\min}(\mathbf{c})$ for every cube $\mathbf{c}$ containing v. A grid vertex v is a local maximum if its scalar value equals $s_{\max}(\mathbf{c})$ for every cube $\mathbf{c}$ containing v. A grid vertex is an extremum if it is a local minimum or local maximum.

As previously defined, set $\mathbb{Y} \subseteq \mathbb{X}$ separates point $p \in \mathbb{X}$ from point $q \in \mathbb{X}$ if every path in $\mathbb{X}$ connecting p to q intersects $\mathbb{Y}$. We claim that any connected isosurface component separates some local minimum grid vertex from some local maximum grid vertex. Because isosurface components are algorithm dependent, we formally state this as a property of connected components of $\mathcal{G}_\sigma$.

Connected components of $\mathcal{G}_\sigma$ are sets of cubes. A sequence $(\mathbf{c}_1, \ldots, \mathbf{c}_k)$ of cubes from point p to point q is a vertex-connected path of cubes from p to q if $\mathbf{c}_1$ contains p and $\mathbf{c}_k$ contains q and every $\mathbf{c}_i$ shares at least one vertex with $\mathbf{c}_{i+1}$. A set $\mathbb{C}$ of cubes cube separates point p from point q in a grid if every vertex-connected path of grid cubes from p to q contains some cube in $\mathbb{C}$.

Proposition 8.8. *For every isovalue σ, every connected component of $\mathcal{G}_\sigma$ cube separates some local minimum grid vertex from some local maximum grid vertex.*

The proof of Proposition 8.8 is given in Section 8.4.4.

Corollary 8.9. *A set $\mathbb{C}$ of cubes is a complete seed set if for every local minimum vertex v and every local maximum vertex v' there is a vertex-connected path of cubes in $\mathbb{C}$ from v to v'.*

Proof: Let λ be a connected component of $\mathcal{G}_\sigma$. By Proposition 8.8, there is some local minimum vertex v and some local maximum vertex v' such that λ cube separates v from v'. Since $\mathbb{C}$ contains a vertex-connected path from v to v', some cube in $\mathbb{C}$ must also lie in λ. Thus, $\mathbb{C}$ contains some cube in every connected component of $\mathcal{G}_\sigma$ for every σ. Therefore, $\mathbb{C}$ is a complete seed set. $\qquad\square$

Note that the corollary gives sufficient but not necessary conditions for a seed set to be complete.

Finding the minimum size set of cubes satisfying the conditions in Corollary 8.9 is a difficult problem. It is similar to the NP-complete Geometric Steiner Tree Problem [Garey and Johnson, 1979] and is probably also NP-complete. Itoh and Koyamada in [Itoh and Koyamada, 1995] and [Itoh et al., 2001] give heuristics for solving this problem.

Instead of presenting a heuristic to find a small set of cubes connecting the extremum vertices, we give a simple algorithm that performs well when extrema are uniformly distributed across the grid. Let N be the number of grid cubes and K be the number of extrema vertices in the grid. If the extrema are uniformly distributed across the grid, then the smallest vertex-connected path of cubes

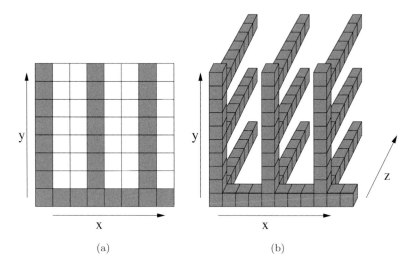

Figure 8.14. (a) 2D example of spine. (b) 3D seed set spine.

connecting an extremum to its closest neighbor will have on average $(N/K)^{1/3}$ cubes. Thus a set of cubes connecting all K extrema grid vertices will have $K \times (N/K)^{1/3}$ or $K^{2/3}N^{1/3}$ cubes. We show how to build a complete seed set with approximately $2K^{2/3}N^{1/3}$ cubes.

Let $m_x \times m_y \times m_z$ be the cube dimensions of the grid. Relabel the axis of the grid so that $m_x \leq m_y \leq m_z$. Note that $m_x \times m_y \times m_z$ equals N.

Label the cubes (i_x, i_y, i_z) where i_x, i_y, i_z are integers starting at 0 and let k equal $\lceil (N/K)^{1/3} \rceil$. The seed set consists of two parts, a set of cubes forming a spine and a set of cubes connecting the extrema grid vertices to the spine. The spine consists of all cubes $(i_x, 0, 0)$, all cubes $(i_x, i_y, 0)$ where $(i_x \bmod k)$ equals zero, and all cubes (i_x, i_y, i_z) where $(i_x \bmod k)$ and $(i_y \bmod k)$ equal zero (Figure 8.14).

Connect each extremum grid vertex v to the spine as follows. Let (i_x, i_y, i_z) be the cube with the smallest i_x, i_y, and i_z coordinates containing v. Let j_x equal $i_x - (i_x \bmod k)$ and j_y equal $i_y - (i_y \bmod k)$. The cube (j_x, j_y, i_z) is in the spine. Connect cube (j_x, j_y, i_z) to cube (i_x, i_y, i_z) by a diagonal set of cubes followed by a stack of cubes in either the x or the y direction (Figure 8.15). More specifically, let d_x equal $(i_x - j_x)$ and d_y equal $(i_y - j_y)$. If d_x is greater than or equal to d_y, then connect (j_x, j_y, i_z) to (i_x, i_y, i_z) by the cubes $(j_x + h, j_y + h, i_z)$ for $h = 1, \ldots, d_y$ and $(j_x + h, i_y, i_z)$ for $h = d_y, \ldots, d_x$. If d_y is greater than d_x, then connect (j_x, j_y, i_z) to (i_x, i_y, i_z) by the cubes $(j_x + h, j_y + h, i_z)$ for $h = 1, \ldots, d_x$ and $(i_x, j_y + d_y, i_z)$ for $h = d_x, \ldots, d_y$. The number of grid cubes used to connect (i_x, i_y, i_z) to the spine is $\max(d_x, d_y) \leq m$. The algorithm for constructing the seed set is presented in Algorithm 8.14.

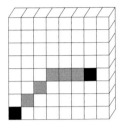

Figure 8.15. Cubes connecting (j_x, j_y, i_z) to (i_x, i_y, i_z) where $j_x = i_x - (i_x \bmod 8)$ and $j_y = i_y - (i_y \bmod 8)$.

Input : G is a regular scalar grid.

k determines spacing between rows and columns in the spine.

Output : A complete seed set for G.

ExtremumSeedSet(G, k)

1   SeedSet$_x$ $\leftarrow$ cubes $(i_x, 0, 0)$;

2   SeedSet$_y$ $\leftarrow$ cubes $(i_x, i_y, 0)$ where $i_x \bmod k = 0$;

3   SeedSet$_z$ $\leftarrow$ cubes (i_x, i_y, i_z) where $i_x \bmod k = 0$ and $i_y \bmod k = 0$;

4   SeedSet $\leftarrow$ SeedSet$_x$ $\bigcup$ SeedSet$_y$ $\bigcup$ SeedSet$_z$;

5   V_E $\leftarrow$ extremum grid vertices ;

6   foreach $v \in V_E$ do

7   (i_x, i_y, i_z) $\leftarrow$ cube with lowest indices containing v;

8     $d_x \leftarrow i_x \bmod k$;

9     $d_y \leftarrow i_y \bmod k$;

10     $j_x \leftarrow i_x - d_x$;

11     $j_y \leftarrow i_y - d_y$;

      /* Add cubes connecting (j_x, j_y, i_z) to (i_x, i_y, i_z) to *SeedSet*   */

      /* Add cubes along diagonal   */

12   Add cubes $\{(j_x + h, j_y + h, i_z) : 1 \le h \le \min(d_x, dy)\}$ to SeedSet;

13   if $dx \ge dy$ then

        /* Add cubes in x direction   */

14   Add cubes $\{(j_x + h, i_y, i_z) : dy < h \le d_x\}$ to SeedSet;

15   else

        /* Add cubes in y direction   */

16   Add cubes $\{(i_x, j_y + h, i_z) : d_x < h \le d_y\}$ to SeedSet;

17   end

18   end

19   return (SeedSet);

Algorithm 8.14. ExtremumSeedSet.

The spine has $m_x + (m_x/k) \times m_y + (m_x/k) \times (m_y/k) \times m_z$ cubes. Since $m_x \leq m_y \leq m_z$, length m_x is at most $N^{1/3}$ and $m_x \times m_y$ is at most $N^{2/3}$. Thus,

$$m_x + (m_x/k) \times m_y + (m_x/k) \times (m_y/k) \times m_z \leq N^{1/3} + N^{2/3}/k + N/k^2.$$

At most k cubes connect each extremum to the spine for a total of kK cubes. Thus the total number of cubes in the seed set is $N^{1/3} + N^{2/3}/k + N/k^2 + kK$. Since k equals $\lceil (N/K)^{1/3} \rceil$, this is approximately $N^{1/3} + N^{1/3}K^{1/3} + 2N^{1/3}K^{2/3}$ cubes. Dropping the low-order terms gives approximately $2N^{1/3}K^{2/3}$ cubes in the seed set.

Algorithm EXTREMUMSEEDSET may add the same cube more than once to the seed set. Having multiple references to the same cube in the seed set is not a problem for algorithmIsoPROPAGATE. After the first reference to a cube is processed, the cube will be marked, and subroutine SEEDGROW immediately returns if it receives an already marked cube. However, multiple references to the same cube can be eliminated by sorting the cube references in **SeedSet** and scanning the sorted list for repeated references.

If a cube has two vertices with the same scalar value, then both could be extrema vertices. Large clusters of adjacent vertices with the same scalar value can create large clusters of adjacent cubes in the seed set. Clusters of adjacent vertices could be identified in a preprocessing step and at most one vertex marked as an extremum in the cluster.

A simpler solution is to symbolically perturb the scalar values of vertices so that no two vertices have precisely the same scalar value. Let (v_x, v_y, v_z) be the coordinates of grid vertex v and let s_v be its scalar value. For each grid vertex v, define a new scalar value,

$$s'_v = s_v + \epsilon v_x + \epsilon^2 v_y + \epsilon^3 v_z,$$

for some extremely small $\epsilon > 0$. The same ϵ is used to define all s'_v. If the value of ϵ is very small, then $s'_u < s'_v$ whenever $s_u < s_v$ and no two values s'_v are equal.

Instead of defining a precise ϵ, we symbolically perturb the scalar values as follows. For each pair of grid vertices u and v, define $u \prec v$ if

- $s_u < s_v$, or

- $s_u = s_v$ and $u_x < v_x$, or

- $s_u = s_v$ and $u_x = v_x$ and $u_y < v_y$, or

- $s_u = s_v$ and $u_x = v_x$ and $u_y = v_y$ and $u_z < v_z$.

Note that $u \prec v$ whenever $s_u < s_v$. If u and v have different coordinates, then either $u \prec v$ or $v \prec u$. A grid vertex u is a local minimum for this symbolically perturbed grid if $u \prec v$ for every vertex $v \neq u$ in every cube containing u. A grid vertex u is a local maximum if $v \prec u$ for every vertex $v \neq u$ in every cube containing u. A grid vertex is an extremum if it is either a local minimum or a local maximum.

Proposition 8.8 and Corollary 8.9 still hold for the local minima and maxima of symbolically perturbed grids. We use the extrema of the symbolically perturbed grid in Step 5 of algorithm EXTREMUMSEEDSET to compute the complete seed set.

8.4.3 Seed Sets and Span Space Priority Trees

Seed sets require that the entire seed set be searched, even if the isosurface only intersects a few grid cubes. For large seed sets, this searching can be costly. Bajaj et al. [Bajaj et al., 1996] suggest storing the seed set in a span space priority tree, such as the one described in Section 8.3. The time to report the cubes intersecting the isosurface $\Theta(\log(K) + M)$ where K is the number of cubes in the seed set and M is the number of reported cubes. The priority tree uses $\Theta(K)$ space. Algorithm ISOPROPAGATE still requires $\Theta(N)$ space for cube markers.

8.4.4 Seed Set Proofs

In this section, we prove Proposition 8.8, which states that every connected component of $\mathcal{G}_\sigma$ cube separates some local minimum grid vertex from some local maximum grid vertex. The vertices of $\mathcal{G}_\sigma$ are the set of grid cubes whose spans contain the isovalue σ. The edges of $\mathcal{G}_\sigma$ are pairs of cubes that share a facet $\mathbf{f}$ whose span contains σ.

As previously defined, $s_{\min}(\mathbf{c})$ and $s_{\max}(\mathbf{c})$ are the minimum and maximum scalar values, respectively, of any vertex of grid cube $\mathbf{c}$. Minimum $s_{\min}(\mathbb{C})$ and maximum $s_{\max}(\mathbb{C})$ are the minimum and maximum of $s_{\min}(\mathbf{c})$ and $s_{\max}(\mathbf{c})$, respectively, over all cubes $\mathbf{c} \in \mathbb{C}$. Define $s_{\min}(\mathbf{f})$ and $s_{\max}(\mathbf{f})$ are the minimum and maximum scalar values, respectively, of any vertex of grid facet $\mathbf{f}$.

Let $\mathcal{G}_\sigma^-$ be a graph whose vertices are the set of grid cubes $\{\mathbf{c} : s_{\min}(\mathbf{c}) \leq \sigma\}$. If cubes $\mathbf{c}_1$ and $\mathbf{c}_2$ share a facet $\mathbf{f}$ and $s_{\min}(\mathbf{f})$ is less than or equal to σ, then $(\mathbf{c}_1, \mathbf{c}_2)$ is an edge of graph $\mathcal{G}_\sigma^-$. Figure 8.16(a) contains a 2D example of $\mathcal{G}_\sigma^-$.

Similarly, let $\mathcal{G}_\sigma^+$ be a graph whose vertices are the set of grid cubes $\{\mathbf{c} : s_{\max}(\mathbf{c}) \geq \sigma\}$. If cubes $\mathbf{c}_1$ and $\mathbf{c}_2$ share a facet $\mathbf{f}$ and $s_{\max}(\mathbf{f})$ is greater than or equal to σ, then $(\mathbf{c}_1, \mathbf{c}_2)$ is an edge of graph $\mathcal{G}_\sigma^+$. Figure 8.16(b) contains a 2D example of $\mathcal{G}_\sigma^+$.

To understand the connected components of $\mathcal{G}_\sigma$, we need to look at facets of cubes in $\mathcal{G}_\sigma$ whose span does not contain σ. Define sets of facets F_σ, F_σ^+, and F_σ^- as follows:

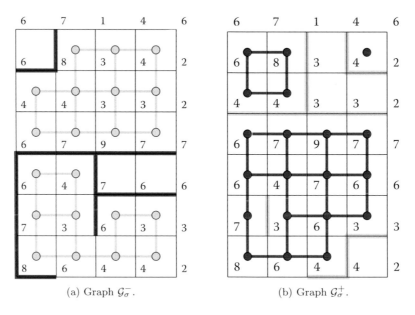

(a) Graph $\mathcal{G}_\sigma^-$.   (b) Graph $\mathcal{G}_\sigma^+$.

Figure 8.16. Two-dimensional example. (a) Graph $\mathcal{G}_\sigma^-$ (isovalue 5). Vertices (blue) of $\mathcal{G}_\sigma^-$ are squares $\mathbf{c}$ where $s_{\min}(\mathbf{c}) \leq 5$. Edges (blue) of $\mathcal{G}_\sigma^-$ are $(\mathbf{c}, \mathbf{c}')$ where $\mathbf{c}$ and $\mathbf{c}'$ share an edge $\mathbf{e}$ and $s_{\min}(\mathbf{e}) \leq 5$. Red grid edges are F_σ^-. (b) Graph $\mathcal{G}_\sigma^+$ (isovalue 5). Vertices (red) of $\mathcal{G}_\sigma^+$ are squares $\mathbf{c}$ where $s_{\max}(\mathbf{c}) \geq 5$. Edges (red) of $\mathcal{G}_\sigma^+$ are $(\mathbf{c}, \mathbf{c}')$ where $\mathbf{c}$ and $\mathbf{c}'$ share an edge $\mathbf{e}$ and $s_{\max}(\mathbf{e}) \geq 5$. Blue grid edges are F_σ^+.

$$F_\sigma = \{\mathbf{f} : \mathbf{f} \text{ is a facet of some cube of } \mathcal{G}_\sigma \text{ and } s_{\max}(\mathbf{f}) < \sigma \text{ or } \sigma < s_{\min}(\mathbf{f})\};$$
$$F_\sigma^- = \{\mathbf{f} : \mathbf{f} \text{ is a facet of some cube of } \mathcal{G}_\sigma^- \text{ and } s_{\min}(\mathbf{f}) > \sigma\};$$
$$F_\sigma^+ = \{\mathbf{f} : \mathbf{f} \text{ is a facet of some cube of } \mathcal{G}_\sigma^+ \text{ and } s_{\max}(\mathbf{f}) < \sigma\}.$$

Figure 8.17 contains a 2D example of F_σ, F_σ^-, and F_σ^+.

Each facet in F_σ is either in F_σ^- or in F_σ^+.

Lemma 8.10. $F_\sigma = F_\sigma^- \cup F_\sigma^+$.

Proof: Let $\mathbf{f}$ be a facet in F_σ. By definition, $\mathbf{f}$ is a facet of some cube $\mathbf{c}$ whose span contains σ. Thus cube $\mathbf{c}$ is a vertex in $\mathcal{G}_\sigma^-$ and a vertex in $\mathcal{G}_\sigma^+$. Since $\mathbf{f}$ is in F_σ, either $s_{\max}(\mathbf{f}) < \sigma$ or $s_{\min}(\mathbf{f}) > \sigma$. If $s_{\max}(\mathbf{f}) < \sigma$, then $\mathbf{f}$ is in F_σ^+. If $s_{\min}(\mathbf{f}) > \sigma$, then $\mathbf{f}$ is in F_σ^-.

Now assume that $\mathbf{f}$ is a facet in F_σ^+. By definition, $\mathbf{f}$ is a facet of some cube $\mathbf{c}$ where $s_{\max}(\mathbf{c})$ is greater than or equal to σ. Since $\mathbf{f}$ is in F_σ^+, the vertices of $\mathbf{f}$ all have scalar value less than σ. Thus the span of $\mathbf{c}$ contains σ and so $\mathbf{c}$ is a vertex of $\mathcal{G}_\sigma$. Since $\mathbf{f}$ is the facet of $\mathbf{c}$ and $s_{\max}(\mathbf{f})$ is less than σ, facet $\mathbf{f}$ is in F_σ. By similar reasoning, if $\mathbf{f}$ is a facet in F_σ^-, then $\mathbf{f}$ is in F_σ.

Since $\mathbf{f}$ is in F_σ if and only if $\mathbf{f}$ is in $F_\sigma^- \cup F_\sigma^+$, set F_σ equals $F_\sigma^- \cup F_\sigma^+$. $\qquad\square$

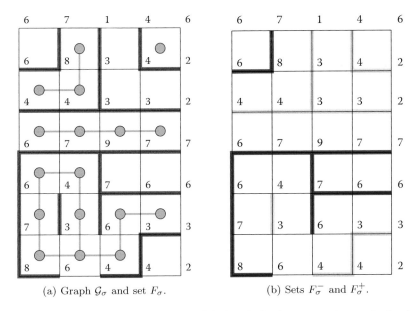

(a) Graph $\mathcal{G}_\sigma$ and set F_σ.   (b) Sets F_σ^- and F_σ^+.

Figure 8.17. Two-dimensional example. (a) Graph $\mathcal{G}_\sigma$ (green vertices and edges) and set F_σ (magenta grid edges). (b) Sets F_σ^- (red grid edges) and F_σ^+ (blue grid edges). Note that F_σ equals $F_\sigma^- \cup F_\sigma^+$ (Lemma 8.10) and that $|F_\sigma^-|$ does not intersect $|F_\sigma^+|$ (Lemma 8.11).

Let $|F_\sigma|$, $|F_\sigma^-|$, and $|F_\sigma^+|$ be the union of the facets in F_σ, F_σ^-, and F_σ^+, respectively. The following lemma, that $|F_\sigma^-|$ does not intersect $|F_\sigma^+|$, is central to the proof of Proposition 8.8.

Lemma 8.11. $|F_\sigma^-| \cap |F_\sigma^+| = \emptyset$.

Proof: Sets $|F_\sigma^-|$ and $|F_\sigma^+|$ are unions of grid facets that intersect only at their edges and vertices. Thus, $|F_\sigma^-|$ intersects $|F_\sigma^+|$ if and only if some grid vertex is in $|F_\sigma^-| \cap |F_\sigma^+|$.

Consider any grid vertex v. If v had scalar value less than or equal to σ, then every facet incident on v would have minimum scalar value less than or equal to σ. Thus, no facet incident on v would be in F_σ^- and v would not be in $|F_\sigma^-|$. On the other hand, if v had scalar value greater than or equal to σ, then no facet incident on v would be in F_σ^+ and v would not be in $|F_\sigma^+|$. Thus no vertex v is in $|F_\sigma^-| \cap |F_\sigma^+|$, proving the lemma.   □

We now consider a connected component of $\mathcal{G}_\sigma$. Let λ be a connected component of $\mathcal{G}_\sigma$ (Figure 8.18(a)). Since both $\mathcal{G}_\sigma^-$ and $\mathcal{G}_\sigma^+$ contain $\mathcal{G}_\sigma$, both $\mathcal{G}_\sigma^-$ and $\mathcal{G}_\sigma^+$ contain λ. Let λ^- be the connected component of $\mathcal{G}_\sigma^-$ containing λ (Figure 8.18(b)). Let λ^+ be the connected component of $\mathcal{G}_\sigma^+$ containing λ

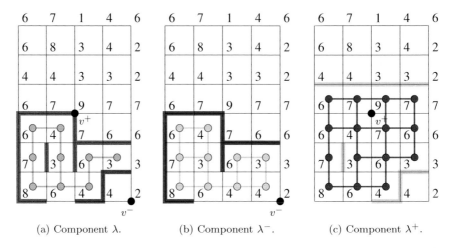

(a) Component λ.   (b) Component λ^-.   (c) Component λ^+.

Figure 8.18. Two-dimensional example. (a) Connected component λ of $\mathcal{G}_\sigma$. Set $\mathbb{C}_\lambda$ is the set of squares containing (green) vertices of $\mathcal{G}_\sigma$. Any vertex-connected path from grid vertex v^- to grid vertex v^+ must contain some cube in $\mathbb{C}_\lambda$. (b) Connected component λ^- of $\mathcal{G}_\sigma^-$ containing λ. Set $\mathbb{C}_{\lambda^-}$ is the set of squares containing (blue) vertices of $\mathcal{G}_\sigma^-$. Grid vertex v^- has minimum scalar value over all vertices of $\mathbb{C}_{\lambda^-}$. (c) Connected component λ^+ of $\mathcal{G}_\sigma^+$ containing λ. Set $\mathbb{C}_{\lambda^+}$ is the set of squares containing (red) vertices of $\mathcal{G}_\sigma^+$. Grid vertex v^+ has maximum scalar value over all vertices of $\mathbb{C}_{\lambda^+}$.

(Figure 8.18(c)). Let $\mathbb{C}_\lambda$, $\mathbb{C}_{\lambda^-}$ and $\mathbb{C}_{\lambda^+}$ be the cubes corresponding to the vertices of λ, λ^-, and λ^+, respectively.

By Lemma B.8 in Appendix B, set $|\mathbb{C}_\lambda| \cap \mathrm{cl}(|\Gamma| - |\mathbb{C}_\lambda|)$ separates $|\mathbb{C}_\lambda|$ from the points of $|\Gamma|$ that are not in $|\mathbb{C}_\lambda|$. Similarly, $|\mathbb{C}_{\lambda^-}| \cap \mathrm{cl}(|\Gamma| - |\mathbb{C}_{\lambda^-}|)$ and $|\mathbb{C}_{\lambda^+}| \cap \mathrm{cl}(|\Gamma| - |\mathbb{C}_{\lambda^+}|)$ separate $|\mathbb{C}_{\lambda^-}|$ from $(|\Gamma| - |\mathbb{C}_{\lambda^-}|)$ and $|\mathbb{C}_{\lambda^+}|$ from $(|\Gamma| - |\mathbb{C}_{\lambda^+}|)$, respectively. The following lemma relates these separation sets to F_σ, F_σ^-, and F_σ^+.

Lemma 8.12.

1. $|\mathbb{C}_\lambda| \cap \mathrm{cl}(|\Gamma| - |\mathbb{C}_\lambda|) \subseteq |F_\sigma|$.

2. $|\mathbb{C}_{\lambda^-}| \cap \mathrm{cl}(|\Gamma| - |\mathbb{C}_{\lambda^-}|) \subseteq |F_\sigma^-|$.

3. $|\mathbb{C}_{\lambda^+}| \cap \mathrm{cl}(|\Gamma| - |\mathbb{C}_{\lambda^+}|) \subseteq |F_\sigma^+|$.

Proof: Since grid cubes intersect only on their facets, edges, and vertices, set $|\mathbb{C}_\lambda| \cap \mathrm{cl}(|\Gamma| - |\mathbb{C}_\lambda|)$ is the union of a set of grid facets. Let $\mathbf{f}$ be a facet of $|\mathbb{C}_\lambda| \cap \mathrm{cl}(|\Gamma| - |\mathbb{C}_\lambda|)$. Since $\mathbf{f}$ is in $|\mathbb{C}_\lambda|$, facet $\mathbf{f}$ is incident on some cube $\mathbf{c} \in \mathbb{C}_\lambda$. Since $\mathbf{f}$ is in $\mathrm{cl}(|\Gamma| - |\mathbb{C}_\lambda|)$, facet $\mathbf{f}$ is incident on some grid cube $\mathbf{c}' \notin \mathbb{C}_\lambda$. Since $\mathbf{c}$ is in $\mathbb{C}_\lambda$ and $\mathbf{c}'$ is not, the span of $\mathbf{f}$ does not contain σ. Thus $\mathbf{f}$ is in F_σ.

Similar arguments prove Statements 2 and 3. $\square$

The proof of the next lemma (Lemma 8.13) relies upon the proposition that if $\mathbb{Y}_1$ is connected and $\mathbb{Y}_2$ is connected and the boundaries of $\mathbb{Y}_1$ and $\mathbb{Y}_2$ do not intersect, then $\mathbb{Y}_1 \cap \mathbb{Y}_2$ is connected. The proof of Lemma 8.13 requires a slightly more general form of the proposition that replaces the boundaries by separating sets. The formal statement of the proposition and its proof are Lemma B.12 in Appendix B.

Lemma 8.13. $\mathbb{C}_\lambda = \mathbb{C}_{\lambda+} \cap \mathbb{C}_{\lambda-}$.

Proof: By definition, $\mathbb{C}_\lambda$ is a subset of $\mathbb{C}_{\lambda-}$ and a subset of $\mathbb{C}_{\lambda+}$, so $\mathbb{C}_\lambda$ is a subset of $\mathbb{C}_{\lambda-} \cap \mathbb{C}_{\lambda+}$. We show that $\mathbb{C}_{\lambda-} \cap \mathbb{C}_{\lambda+}$ is a subset of $\mathbb{C}_\lambda$.

The facets of F_σ^- do not intersect $|\mathbb{C}_{\lambda-}|$ or they lie on the boundary of $|\mathbb{C}_{\lambda-}|$ or they separate two cubes in $\mathbb{C}_{\lambda-}$ that do not share an edge in λ^-. Since λ^- is connected, the set $\mathbb{Y}^- = |\mathbb{C}_{\lambda-}| - |F_\sigma^-|$ is connected. Similarly, $\mathbb{Y}^+ = |\mathbb{C}_{\lambda+}| - |F_\sigma^+|$ is connected.

By Lemma 8.12, set $|F_\sigma^-|$ contains $\mathrm{cl}(\mathbb{Y}^-) \cap \mathrm{cl}(|\Gamma| - \mathbb{Y}^-)$ and set $|F_\sigma^+|$ contains $\mathrm{cl}(\mathbb{Y}^+) \cap \mathrm{cl}(|\Gamma| - \mathbb{Y}^+)$. By Lemma 8.11, $|F_\sigma^-|$ does not intersect $|F_\sigma^+|$. By Lemma B.12 in Appendix B, $\mathbb{Y}^- \cap \mathbb{Y}^+$ is connected.

Let p be a point in the interior of $|\mathbb{C}_\lambda|$. Let q be a point in $\mathbb{Y}^- \cap \mathbb{Y}^+$. Since $\mathbb{Y}^- \cap \mathbb{Y}^+$ is connected, there is a path ζ in $\mathbb{Y}^- \cap \mathbb{Y}^+$ from p to q. This path does not intersect $|F_\sigma^-|$ or $|F_\sigma^+|$. By Lemma 8.12, $|\mathbb{C}_\lambda| \cap \mathrm{cl}(|\Gamma| - |\mathbb{C}_\lambda|)$ is contained in $|F_\sigma|$. Since $|F_\sigma|$ is a subset of $|F_\sigma^-| \cup |F_\sigma^+|$ (Lemma 8.10), $|\mathbb{C}_\lambda| \cap \mathrm{cl}(|\Gamma| - |\mathbb{C}_\lambda|)$ is contained in $|F_\sigma^-| \cup |F_\sigma^+|$. Thus ζ does not intersect $|\mathbb{C}_\lambda| \cap \mathrm{cl}(|\Gamma| - |\mathbb{C}_\lambda|)$. By Lemma B.8 in Appendix B, set $|\mathbb{C}_\lambda| \cap \mathrm{cl}(|\Gamma| - |\mathbb{C}_\lambda|)$ separates $|\mathbb{C}_\lambda|$ from $|\Gamma| - |\mathbb{C}_\lambda|$. Since ζ does not intersect $|\mathbb{C}_\lambda| \cap \mathrm{cl}(|\Gamma| - |\mathbb{C}_\lambda|)$, point q is also in $|\mathbb{C}_\lambda|$. Thus $|\mathbb{Y}^- \cap \mathbb{Y}^+|$ is a subset of $|\mathbb{C}_\lambda|$. Thus $\mathbb{C}_{\lambda-} \cap \mathbb{C}_{\lambda+}$ is a subset of $\mathbb{C}_\lambda$.

Since $\mathbb{C}_\lambda$ is a subset of $\mathbb{C}_{\lambda-} \cap \mathbb{C}_{\lambda+}$ and $\mathbb{C}_{\lambda-} \cap \mathbb{C}_{\lambda+}$ is a subset of $\mathbb{C}_\lambda$, set $\mathbb{C}_\lambda$ equals $\mathbb{C}_{\lambda-} \cap \mathbb{C}_{\lambda+}$. □

Let v^- be a vertex of $\mathbb{C}_{\lambda-}$ with minimum value and let v^+ be a vertex of $\mathbb{C}_{\lambda+}$ with maximum value. (See Figures 8.18(b) and 8.18(c).) We show that v^- is a local minimum and v^+ is a local maximum. In the proof of Proposition 8.8, we will show that any vertex-connected path of cubes from v^- and v^+ intersects $\mathbb{C}_\lambda$. (See Figure 8.18(a) for a 2D example.)

Lemma 8.14.

1. *If v^- is a vertex of a cube $\mathbb{C}_{\lambda-}$ with scalar value $s_{\min}(\mathbb{C}_{\lambda-})$, then v^- is a local minimum.*

2. *If v^+ is a vertex of a cube in $\mathbb{C}_{\lambda+}$ with scalar value $s_{\max}(\mathbb{C}_{\lambda+})$, then v^+ is a local maximum.*

Proof: We prove Statement 1.

Let v^- be a vertex of some cube in $\mathbb{C}_{\lambda-}$ with scalar value $s_{\min}(\mathbb{C}_{\lambda-})$. Vertex v^- has scalar value less than or equal to σ. Thus, every cube and facet containing

v^- has minimum scalar value less than or equal to σ and all cubes containing v^- are in $\mathbb{C}_{\lambda-}$. Since vertex v^- has scalar value less than or equal to the minimum value of each such grid cube, vertex v^- is a local minimum vertex.

A similar argument proves Statement 2.   □

Finally, we come to the proof of Proposition 8.8.

Proposition 8.8. *For every isovalue σ, every connected component of $\mathcal{G}_\sigma$ cube separates some local minimum grid vertex from some local maximum grid vertex.*

Proof: Let λ be a connected component of $\mathcal{G}_\sigma$. Let λ^- be the connected component of $\mathcal{G}_\sigma^-$ containing λ, and let λ^+ be the connected component of $\mathcal{G}_\sigma^+$ containing λ, as defined above. Let $\mathbb{C}_\lambda$, $\mathbb{C}_{\lambda-}$, and $\mathbb{C}_{\lambda+}$ be the cubes corresponding to the vertices of λ, λ^-, and λ^+, respectively.

Let v^- be a vertex of a cube $\mathbb{C}_{\lambda-}$ whose scalar value is $s_{\max}(\mathbb{C}_{\lambda-})$. Similarly, let v^+ be a vertex of a cube in $\mathbb{C}_{\lambda+}$ whose scalar value is $s_{\min}(\mathbb{C}_{\lambda+})$. By Lemma 8.14, vertex v^- is a local minimum vertex and vertex v^+ is a local maximum vertex.

By Lemmas 8.10 and 8.12, set $|F_\sigma^-| \cup |F_\sigma^+|$ separates $|\mathbb{C}_\lambda|$ from $|\Gamma| - |\mathbb{C}_\lambda|$. Set $\mathrm{cl}(|\Gamma| - |\mathbb{C}_\lambda|)$ is composed of connected components. By Lemma 8.11, $|F_\sigma^-|$ does not intersect $|F_\sigma^+|$. Thus each connected component of $\mathrm{cl}(|\Gamma| - |\mathbb{C}_\lambda|)$ is either separated from $|\mathbb{C}_\lambda|$ by $|F_\sigma^-|$ or by $|F_\sigma^+|$.

Let $\zeta_\mathbf{c} = (\mathbf{c}_1, \ldots, \mathbf{c}_k)$ be a vertex-connected path of cubes from v^- to v^+. If v^- is not in $\mathrm{cl}(|\Gamma| - |\mathbb{C}_\lambda|)$, then $\mathbf{c}_1$ is in $\mathbb{C}_\lambda$ and $\zeta_\mathbf{c}$ intersects $\mathbb{C}_\lambda$. If v^+ is not in $\mathrm{cl}(|\Gamma| - |\mathbb{C}_\lambda|)$, then $\mathbf{c}_k$ is in $\mathbb{C}_\lambda$ and again $\zeta_\mathbf{c}$ intersects $\mathbb{C}_\lambda$.

Assume neither v^- nor v^+ are in $\mathrm{cl}(|\Gamma| - |\mathbb{C}_\lambda|)$. Since v^- is in $\mathbb{C}_{\lambda-}$, vertex v^- is in a component of $\mathrm{cl}(|\Gamma| - |\mathbb{C}_\lambda|)$ separated from $|\mathbb{C}_\lambda|$ by F_σ^+. (Remember that F_σ^+ consists of facets $\mathbf{f}$ such that $s_{\max}(\mathbf{f}) < \sigma$. These facets are on the boundary of $|\mathbb{C}_{\lambda+}|$.) Since v^+ is in $\mathbb{C}_{\lambda+}$, vertex v^+ is in a component of $\mathrm{cl}(|\Gamma| - |\mathbb{C}_\lambda|)$ separated from $|\mathbb{C}_\lambda|$ by F_σ^-. Thus v^- and v^+ are in separate components of $\mathrm{cl}(|\Gamma| - |\mathbb{C}_\lambda|)$ and $\zeta_\mathbf{c}$ intersects $\mathbb{C}_\lambda$. Thus any vertex-connected path of cubes from v^- to v^+ intersects $\mathbb{C}_\lambda$ and $\mathbb{C}_\lambda$ cube separates v^- from v^+.   □

8.5   Notes and Comments

Samet's books [Samet, 1990a, Samet, 1990b, Samet, 2005] on spatial data structures contain a full treatment of quadtrees and octrees. Wilhelms and Van Gelder [Wilhelms and Gelder, 1992] proposed the branch-on-need (BON) octree. Shi and Jaja [Shi and JaJa, 2006] give a modification of the BON octree that supports fast, view-dependent isosurface extraction and fast extraction of the 3D isosurface formed by slicing a 4D isosurface with a hyperplane. Their algorithm runs in $O(M + \log(N))$ time where N is the total number of grid cubes and M is the number of cubes whose span contains the isovalue.

Span (range) space algorithms are given in [Giles and Haimes, 1990, Gallagher, 1991, Shen and Johnson, 1995]. The span space algorithm in Section 8.3.1 is based on the one by Livnat et al. [Livnat et al., 1996] but with the kd-tree replaced by McCreight's priority tree [McCreight, 1985] as suggested in [Bajaj et al., 1996]. Shen et al. [Shen et al., 1996] proposed a variation of [Livnat et al., 1996] where the kd-tree is replaced by a uniform partition. Cignoni et al. [Cignoni et al., 1997] use interval trees over the span space to select cubes intersected by the isosurface.

Bloomenthal [Bloomenthal, 1988] and Howie and Blake [Howie and Blake, 1994] present seed growing algorithms to construct isosurfaces but do not describe how to find or precompute seed sets. The algorithm for finding small seed sets in Section 8.4.2 is based on [Itoh and Koyamada, 1995] by Itoh and Koyamada. Van Kreveld et al. [van Kreveld et al., 1997] and Carr and Snoeyink [Carr and Snoeyink, 2003] construct small seed sets using contour trees. (See Section 12.8.) Other approaches to seed set construction are in [Bajaj et al., 1996, Itoh and Koyamada, 1994, Itoh et al., 1996, Itoh et al., 2001].

In [Sutton et al., 2000], Sutton et al. compare data structures for isosurface extraction, including the branch-on-need (BON) octree, three span space data structures, and isosurface propagation from a seed set. They found that all approaches were two to three times faster than Marching Cubes. Cache misses caused one of the span space data structures, the interval tree [Cignoni et al., 1997], to have a noticeably worse performance than the other data structures.

View-dependent isosurface extraction is discussed in [Livnat and Hansen, 1998, Shi and JaJa, 2006]. Data structures for isosurface extraction from time-varying data fields are discussed in [Shen, 1998, Shen et al., 1999, Vrolijk et al., 2004, Gregorski et al., 2004, Wang and Chiang, 2009].

The data structures described in this chapter are not suitable for large data sets that will not fit into core memory. Out of core techniques for faster isosurface extraction are described in [Chiang and Silva, 1997, Chiang et al., 1998, Chiang and Silva, 1999, Sutton and Hansen, 2000, Gregorski et al., 2002, Chiang et al., 2001, Bordoloi and Shen, 2003, Waters et al., 2006].

CHAPTER 9

MULTIRESOLUTION
TETRAHEDAL MESHES

The MARCHING CUBES algorithm and its variants create isosurfaces composed of tremendous numbers of vertices and triangles. Such "large" isosurfaces require considerable memory and are time-consuming to process and render.

The size of a mesh of simplices is the number of vertices and simplices in the mesh. There are two different approaches to reducing the size of an isosurface mesh. The first approach is to post-process the isosurface mesh, collapsing mesh edges and triangles to create a smaller isosurface. The smaller isosurface should approximate the original one, both geometrically and topologically. Reducing the number of mesh elements by eliminating or collapsing elements is called mesh decimation.

Mesh decimation is not unique to isosurface meshes and can be applied to any surface mesh, regardless of how it was originally constructed. For this reason, this book does not cover mesh decimation algorithms. Interested readers are encouraged to see the book *Tutorials on Multiresolution in Geometric Modelling* [Iske et al., 2002] or the survey papers [Cignoni et al., 1998], [Garland, 1999], or [Luebke, 2001].

The second approach to reducing isosurface size is to use different resolution grids or meshes in different portions of the scalar field. In regions of high interest, the isosurface is extracted from high-resolution subgrids while low-resolution subgrids are used in areas of less interest. Special mesh elements are used to connect the high-resolution and low-resolution subgrids. Extracting isosurfaces using multiple levels of resolution is called multiresolution isosurface extraction.

Figure 9.1 contains two images of multiresolution isosurfaces. Figure 9.1(a) is an image of a torus whose upper-left and lower-right regions are at high resolution. Figure 9.1(b) contains a close-up of a tooth with one root at high resolution. The torus isosurface has 3,608 triangles and the entire tooth isosurface (not shown) has 5,768 triangles. If the entire torus and tooth isosurfaces

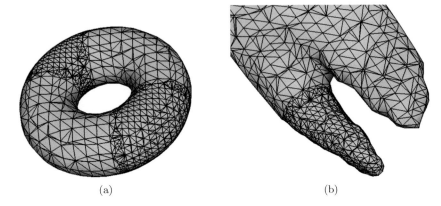

(a)   (b)

Figure 9.1. Multiresolution isosurfaces. (a) Multiresolution torus. Upper-left and lower-right regions on torus are at higher resolution. (b) Multiresolution tooth (close-up). Data set created by GE Aircraft Engines.

were constructed at high resolution, they would have 5,344 and 13,888 triangles, respectively.

In this chapter, we present an algorithm for multiresolution isosurface extraction using multiresolution tetrahedral meshes. Our description follows the presentation by Zhou, Chen, and Kaufmann in [Zhou et al., 1997] and by Gerstner and Pajarola in [Gerstner and Pajarola, 2000]. In the following chapter, we describe other algorithms for multiresolution isosurface extraction.

9.1 Bisection of Tetrahedra

9.1.1 2D Recursive Bisection

We start with a two-dimensional illustration of a multiresolution triangle mesh. The diagonal (p_0, p_3) of a square (p_0, p_1, p_2, p_3) splits the square into two congruent triangles. (See Figure 9.2.) Each triangle contains the diagonal and two square edges. The center q_0 of the diagonal is the center of the square. Bisect the two triangles at q_0, creating four congruent triangles. Each such triangle contains one square edge. Bisect each triangle at the center of its square edge, creating eight congruent triangles. Each of these eight triangles is a replica of one of the two original triangles scaled by a factor of one-half.

The eight triangles form a triangle subdivision of the square. Since they are similar to the original triangles, the two bisections can be applied to these triangles, creating a subdivision of the square into thirty-two triangles. (See

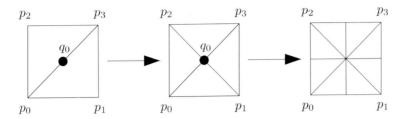

Figure 9.2. Triangle bisection.

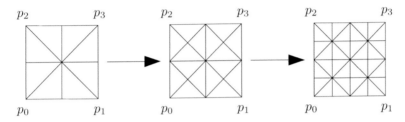

Figure 9.3. Recursive triangle bisection.

Figure 9.3.) Each of these triangles is similar to the original triangles. Repeated application of the bisection steps can lead to triangles of arbitrarily small size.

The triangles form a binary tree whose root is the original square. (See Figure 9.4.) Each tree node other than the root represents a triangle. The two children of a node representing triangle **t** are the two triangles created by splitting **t** into two congruent subtriangles. All the triangles represented at a given level in the tree are congruent.

Note that each triangle edge is horizontal, vertical, or at 45 degrees. All triangles are isosceles right triangles. After two bisections, the triangles form a subdivision of the four subsquares of the original square. After four bisections, the triangles form a subdivision of sixteen subsquares. Bisecting the triangles $2k$ times gives a subdivision of the 4^k squares in a $(2^k + 1) \times (2^k + 1)$ grid.

By choosing to bisect only some of the triangles, we can construct a multiresolution triangulation. For instance, in Figure 9.5 the original two triangles are bisected into four triangles but only three of those four triangles are split again. The next bisection is applied to only four triangles, the following bisection is applied to three triangles, and the final bisection to only two. The resulting triangulation has triangles of various sizes, although all are either similar to triangle (p_0, p_1, p_3) or to triangle (p_0, p_1, q_0).

In bisecting some but not all of the triangles, we still require that the resulting set of triangles form a triangulation, i.e., that the intersection of every two triangles is the empty set or a vertex of each triangle or an edge of each triangle. (See Appendix B.5.) To ensure this property, we apply the following rule:

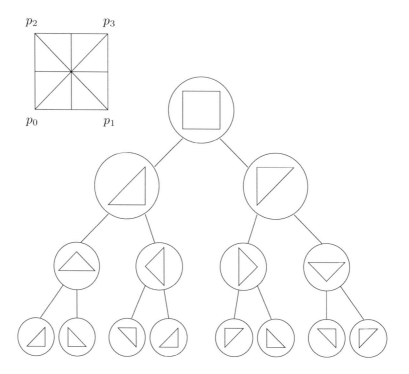

Figure 9.4. Binary tree representation.

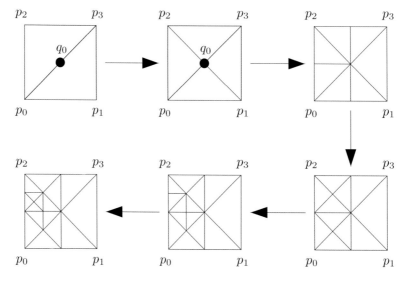

Figure 9.5. Multiresolution triangulation.

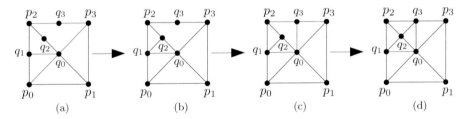

Figure 9.6. (a) Initial triangulation. (b) Split of triangle (q_0, q_1, p_2) at q_2 forces split of triangle (q_0, p_2, p_3). (c) Split of triangle (q_0, q_1, p_2) at q_2 forces split of triangle (q_0, q_3, p_2). (d) Multiresolution triangulation. The intersection of every two triangles is the empty set or a vertex of each triangle or an edge of each triangle.

Bisection Rule (2D, N×N). *If triangle* **t** *is split at point q and q lies in the interior of an edge of triangle* **t**′, *then triangle* **t**′ *is split.*

For instance, in Figure 9.6, splitting triangle (q_0, q_1, p_2) at q_2 forces the split of triangle (q_0, p_2, p_3). Note that triangle (q_0, p_2, p_3) is split at q_3, not at q_2. Triangle (q_0, p_2, p_3) splits into two triangles, (q_0, q_3, p_2) and (q_0, q_3, p_3). Point q_2 lies in the interior of edge (q_0, p_2) of triangle (q_0, q_3, p_2). Therefore, triangle (q_0, q_3, p_2) is split into two triangles. This time the split is at q_2. The final set of triangles form a triangulation of the square.

Proposition 9.1. *Let τ^c be a triangulation of the square into two triangles sharing a square diagonal. If a sequence of splits of τ^c follows the Bisection Rule, then the resulting set τ of triangles is a triangulation of the square.*

Proof: Assume that the nonempty intersection of two triangles **t**, **t**′ ∈ τ is not a vertex or edge of **t**. Since the triangles in τ partition the square, the interiors of **t** and **t**′ do not intersect. Thus the intersection must be a point or line segment on an edge **e** of **t**. If the intersection is a point, then this point must be a vertex q of **t**′. If the intersection is a line segment on edge **e** of **t**, then, by assumption, this line segment does not equal **e**. Some endpoint of this line segment lies in the interior of **e**. This endpoint is again a vertex q of **t**′.

Vertex q was created by splitting some triangle **t**″ at q. Point q is in the interior of an edge of **t** but **t** is not split, violating the Bisection Rule. We conclude that the intersection of every two triangles of τ is the empty set or a vertex or edge of **t**. Thus τ is a triangulation of the square.   □

9.1.2 3D Recursive Bisection

A three-dimensional cube can be subdivided into six tetrahedra as in Figure 9.7. Each of the tetrahedra is the convex hull of cube diagonal (p_0, p_7) and one of the six cube edges not incident on p_0 or p_7.

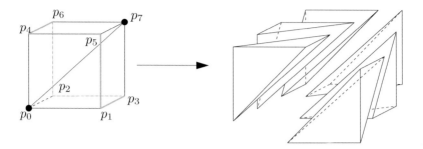

Figure 9.7. Subdivision of a cube into six tetrahedra. Each tetrahedron is the convex hull of the red diagonal (p_0, p_7) and one of the green cube edges. All six tetrahedra are congruent.

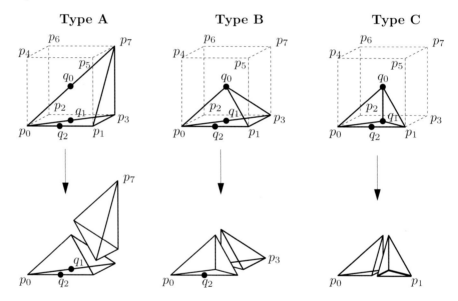

Figure 9.8. Recursive tetrahedral bisection.

All six of the tetrahedra are congruent. Each tetrahedron has three edges that are cube edges, two edges that are facet diagonals, and one edge (p_0, p_7) that is a diagonal of the cube. The center q_0 of diagonal (p_0, p_7) is the center of the cube. Consider tetrahedron (p_0, p_1, p_3, p_7). Bisecting this tetrahedron at point q_0 splits the tetrahedron into two congruent tetrahedra. (See Figure 9.8.) Each of these tetrahedra contains one edge that is the diagonal of a cube facet. For instance, tetrahedron (p_0, p_1, p_3, q_0) has an edge (p_0, p_3), which is the diagonal of the bottom facet of the cube. Bisecting tetrahedron (p_0, p_1, p_3, q_0) at the center q_1 of diagonal (p_0, p_3) splits it into two congruent tetrahedra. Each of these tetrahedra contains one edge that is an edge of the original cube. Bisecting

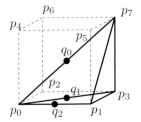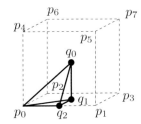

Figure 9.9. Tetrahedron (p_0, p_1, p_3, p_7) is similar to tetrahedron (p_0, q_2, q_1, q_0). Tetrahedron (p_0, q_2, q_1, q_0) is the result of three bisections of tetrahedron (p_0, p_1, p_3, p_7).

tetrahedron (p_0, p_1, q_1, q_0) at the midpoint q_2 of edge (p_0, p_1) split it into two congruent tetrahedra. These two tetrahedra are similar to the original tetrahedron, (p_0, p_1, p_3, p_7), i.e., they are congruent to the tetrahedron formed by shrinking (p_0, p_1, p_3, p_7) by a factor of one-half. (See Figure 9.9.)

The original six congruent tetrahedra sharing the diagonal (p_0, p_7) divide into twelve congruent tetrahedra, each containing one facet diagonal. These twelve tetrahedra subdivide into twenty-four congruent tetrahedra, each containing exactly one cube edge. These twenty-four tetrahedra subdivide into forty-eight tetrahedra. Since these forty-eight tetrahedra are similar to the original six tetrahedra, the three bisection steps can be repeated on these forty-eight tetrahedra, creating tetrahedra whose edges are one-fourth the original size. Repeated application of the bisection steps can lead to tetrahedra of arbitrarily small size.

The tetrahedra form a tree whose root is the original cube. The root has six children, one for each of the six tetrahedra in the triangulation of the cube. Each tree node other than the root represents a tetrahedron. The two children of a node representing tetrahedron **t** are the two tetrahedra created by splitting **t** into two congruent tetrahedra. All the tetrahedra represented at a given level in the tree are congruent.

Let Γ be a $(2^k + 1) \times (2^k + 1) \times (2^k + 1)$ grid containing 2^{3k} cubes. Triangulate the cubic region covered by Γ with six tetrahedra sharing the diagonal from vertex $(0, 0, 0)$ to vertex $(2^k, 2^k, 2^k)$. Recursively bisect these tetrahedra $3k$ times. The set of all tetrahedra produced by $3k$ or fewer bisections form a tree. We call this a full-resolution tree of Γ.

We note a few things about this recursive bisection scheme. First, all tetrahedra can be divided into three types of similar tetrahedra. The three types are shown in Figure 9.8. Second, after three bisection steps, each of the resulting forty-eight tetrahedra has an edge from one of the cube vertices p_i to q_0, the center of the cube. Conversely, for each cube vertex p_i, exactly six of the forty-eight tetrahedra contain the edge (p_i, q_0) and these six tetrahedra form a triangulation of the cube with opposite vertices p_i and q_0. Thus the three bisection steps convert a triangulation of a cube into a triangulation of the eight subcubes of that cube. Repeating the bisection steps another three times gives a triangulation of

8^2 subcubes of the original cube. Repeating the bisection steps $3k$ times gives a triangulation of the 2^{3k} cubes in a $(2^k + 1) \times (2^k + 1) \times (2^k + 1)$ grid.

By choosing to bisect only some of the tetrahedra, we can construct a multiresolution triangulation. As in 2D, we require that the resulting set of tetrahedra form a triangulation, i.e., that the intersection of every two tetrahedra is either the empty set or a vertex, edge, or face of each tetrahedron. (See Appendix B.5.) We apply the following rule to enforce this property:

Bisection Rule (3D, N×N×N). *If tetrahedron* **t** *is split at point q, and if q lies in the interior of an edge of tetrahedron* **t**′, *then tetrahedron* **t**′ *is split.*

A sequence of splits that follows the Bisection Rule creates a triangulation of the cube.

Proposition 9.2. *Let* $\tau^{\mathbf{c}}$ *be a triangulation of the cube into six tetrahedra sharing a cube diagonal. If a sequence of splits of* $\tau^{\mathbf{c}}$ *follows the Bisection Rule, then the resulting set* τ *of tetrahedra is a triangulation of the cube.*

Proof is left for Section 9.1.3.

For each tetrahedron **t** in the full-resolution tree of Γ, let $q_{\text{split}}(\mathbf{t})$ be the point used to split **t** into two congruent tetrahedra. Point $q_{\text{split}}(\mathbf{t})$ is the midpoint of some edge of **t**.

Let $\mathcal{P}_j(\Gamma)$ be the partition of grid Γ into cubes with edge length 2^j. Let $\mathcal{P}_*(\Gamma)$ be the set of all the cubes in $\mathcal{P}_j(\Gamma)$ for $j = 2$ to k. The vertices, edges, and facets in $\mathcal{P}_*(\Gamma)$ are the set of vertices, edges, and facets, respectively, of all the cubes in $\mathcal{P}_j(\Gamma)$ for $j = 2$ to k. Let $q_{\text{center}}(\mathbf{c})$, $q_{\text{center}}(\mathbf{f})$ and $q_{\text{center}}(\mathbf{e})$ represent the centers of cube **c**, facet **f**, and edge **e**, respectively.

The following propositions relates the tetrahedra in the full-resolution tree to the centers of edges, facets, and cubes in $\mathcal{P}_j(\Gamma)$.

Proposition 9.3. *Let* Γ *be a* $(2^k + 1) \times (2^k + 1) \times (2^k + 1)$ *grid containing* 2^{3k} *cubes. If* $\mathbf{t}_A$, $\mathbf{t}_B$, *and* $\mathbf{t}_C$ *are type A, type B, and type C tetrahedra, respectively, in the full-resolution tree of* Γ, *then there exists cube* **c**, *a facet* **f**, *and an edge* **e** *in* $\mathcal{P}_*(\Gamma)$ *such that*

- $q_{\text{split}}(\mathbf{t}_A)$ *equals* $q_{\text{center}}(\mathbf{c})$,

- $q_{\text{split}}(\mathbf{t}_B)$ *equals* $q_{\text{center}}(\mathbf{f})$,

- $q_{\text{split}}(\mathbf{t}_C)$ *equals* $q_{\text{center}}(\mathbf{e})$.

Moreover,

  1. if bisecting $\mathbf{t}_A$ *creates* $\mathbf{t}_B$, *then* **f** *is a facet of* **c**;

  2. if bisecting $\mathbf{t}_B$ *creates* $\mathbf{t}_C$, *then* **e** *is an edge of* **f**;

  3. if bisecting $\mathbf{t}_C$ *creates* $\mathbf{t}_A$, *then* $q_{\text{center}}(\mathbf{e})$ *is a vertex of* **c**.

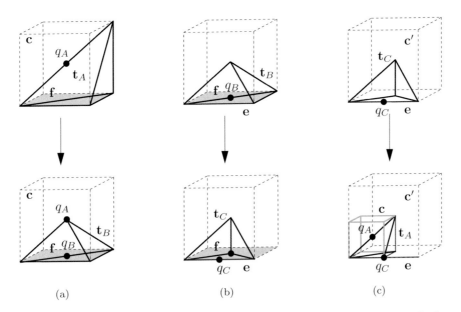

Figure 9.10. Bisecting tetrahedra. (a) Bisecting tetrahedron $\mathbf{t}_A$ creates tetrahedron $\mathbf{t}_B$. Facet $\mathbf{f}$ is a facet of cube $\mathbf{c}$. Point q_A is $q_{\text{split}}(\mathbf{c})$. Point q_B is $q_{\text{split}}(\mathbf{f})$. (b) Bisecting tetrahedron $\mathbf{t}_B$ creates tetrahedron $\mathbf{t}_C$. Edge $\mathbf{e}$ is an edge of facet $\mathbf{f}$. Point q_B is $q_{\text{split}}(\mathbf{f})$. Point q_C is $q_{\text{split}}(\mathbf{e})$. (c) Bisecting tetrahedron $\mathbf{t}_C$ creates tetrahedron $\mathbf{t}_A$. Edge $\mathbf{e}$ is an edge of cube $\mathbf{c}'$. Cube $\mathbf{c}$ is a subcube of cube $\mathbf{c}'$. Point q_C is $q_{\text{split}}(\mathbf{e})$ and a vertex of cube $\mathbf{c}$. Point q_A is $q_{\text{split}}(\mathbf{c})$.

Note that statements 1, 2, and 3 in the proposition are independent. In the first statement, $\mathbf{f}$ is a facet of $\mathbf{c}$. In the second, $\mathbf{e}$ is an edge of $\mathbf{f}$. In the third, $\mathbf{c}$ is a cube whose edge length is half the length of $\mathbf{e}$. Obviously, all three of these properties cannot be simultaneously true.

Proof of Proposition 9.3: Let 2^j be the length of the edges of $\mathbf{t}_A$ that are parallel to the coordinate axes. Tetrahedron $\mathbf{t}_A$ shares its vertices with some cube $\mathbf{c}$ in $\mathcal{P}_j(\Gamma)$. Tetrahedron $\mathbf{t}_A$ is split at the center of $\mathbf{c}$. (See Figure 9.10(a).)

Tetrahedron $\mathbf{t}_B$ is created by bisecting some type A tetrahedron. Without loss of generality, assume this tetrahedron is $\mathbf{t}_A$. Tetrahedron $\mathbf{t}_B$ is split at some point at the center of a facet $\mathbf{f}$ of $\mathbf{c}$.

Tetrahedron $\mathbf{t}_C$ is created by bisecting some type B tetrahedron. Without loss of generality, assume this tetrahedron is $\mathbf{t}_B$. Tetrahedron $\mathbf{t}_C$ is split at the center of some edge $\mathbf{e}$ of $\mathbf{f}$. (See Figure 9.10(b).)

Now assume tetrahedron $\mathbf{t}_A$ is created by bisecting tetrahedron $\mathbf{t}_C$. Cube $\mathbf{c}$ is one of the eight subcubes of a cube $\mathbf{c}' \in \mathcal{P}_{j+1}(\Gamma)$. Tetrahedron $\mathbf{t}_C$ is split at the center of some edge $\mathbf{e}$ of $\mathbf{c}'$ and $q_{\text{center}}(\mathbf{e})$ is a vertex of $\mathbf{c}$. (See Figure 9.10(c).)□

The following proposition is the converse of the previous one.

Proposition 9.4. *Let* Γ *be a* $(2^k + 1) \times (2^k + 1) \times (2^k + 1)$ *grid containing* 2^{3k} *cubes.*

1. *If* **c** *is a cube in* $\mathcal{P}_*$ *and* **f** *is a facet of* **c** *and* **e** *is an edge of* **f**, *then the full-resolution tree contains a type A tetrahedron* $\mathbf{t}_A$, *a type B tetrahedron* $\mathbf{t}_B$, *and a type C tetrahedron* $\mathbf{t}_C$ *such that*

   - $q_{\mathrm{split}}(\mathbf{t}_A)$ *equals* $q_{\mathrm{center}}(\mathbf{c})$,
   - $q_{\mathrm{split}}(\mathbf{t}_B)$ *equals* $q_{\mathrm{center}}(\mathbf{f})$,
   - $q_{\mathrm{split}}(\mathbf{t}_C)$ *equals* $q_{\mathrm{center}}(\mathbf{e})$,
   - *splitting* $\mathbf{t}_A$ *creates* $\mathbf{t}_B$,
   - *splitting* $\mathbf{t}_B$ *creates* $\mathbf{t}_C$.

2. *If* **c** *is a cube in* $\mathcal{P}_*$ *and* v *is a vertex of* **c** *and is not a corner of grid* Γ, *then the full-resolution tree contains a type A tetrahedron* $\mathbf{t}_A$ *and a tetrahedron* **t** *such that*

   - $q_{\mathrm{split}}(\mathbf{t}_A)$ *equals* $q_{\mathrm{center}}(\mathbf{c})$,
   - $q_{\mathrm{split}}(\mathbf{t})$ *equals* v,
   - $\mathbf{t}_A$ *is contained in* **t**.

Tetrahedron **t** in Statement 2 is not necessarily a type C tetrahedron.

Proof: If **c** is a cube in $\mathcal{P}_*(\Gamma)$, then the full-resolution tree contains six type A tetrahedra that share a diagonal of **c** and that triangulate **c**. Each of these tetrahedra are split at the center of **c**.

The six type A tetrahedra triangulating **c** split into twelve type B tetrahedra. Since **f** is a facet of **c**, two of these share the diagonal of **f**. These two are split at the center of **f**.

Since **e** is an edge of **f**, one of the twelve type B tetrahedra contains both **e** and the diagonal of **f**. The bisection of this type B tetrahedra creates a type C tetrahedra $\mathbf{t}_C$ containing **e**. Tetrahedron $\mathbf{t}_C$ is split at the center of **e**.

Assume v is a vertex of **c** and is not a corner of grid Γ. Since v is not a corner of Γ, vertex v was created by the bisection of some tetrahedron **t** in the full-resolution tree. Tetrahedron **t** was split at v.

Tetrahedron **t** intersects the interior of **c**. Thus at least one of the six type A tetrahedra triangulating **c** is contained in **t**.   □

9.1.3   3D Bisection Proof

We present the proof of Proposition 9.2. For convenience, we restate the proposition.

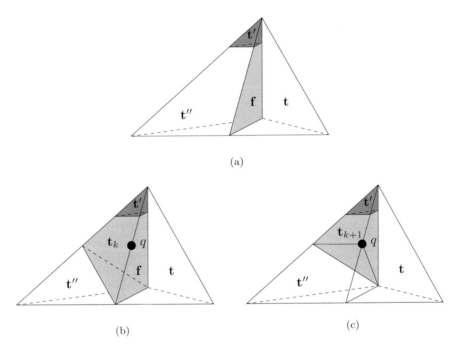

Figure 9.11. Intersecting tetrahedra. (a) Tetraheda $\mathbf{t}''$ and $\mathbf{t}$ are congruent and share a face $\mathbf{f}$. Tetrahedron $\mathbf{t}'$ is a subset of $\mathbf{t}''$. (b) Tetrahedron $\mathbf{t}_k$ where $\mathbf{t}' \subset \mathbf{t}_k \subset \mathbf{t}''$ and $\mathbf{t}_k \cap \mathbf{t} = \mathbf{f}$. (c) Tetrahedron $\mathbf{t}_{k+1}$ produced by splitting $\mathbf{t}_k$ at point q. Note $\mathbf{t}' \subset \mathbf{t}_{k+1} \subset \mathbf{t}_k \subset \mathbf{t}''$. The intersection $\mathbf{t}_{k+1} \cap \mathbf{t}$ is not a face of $\mathbf{t}$.

Proposition 9.2.   *Let $\tau^{\mathbf{c}}$ be a triangulation of the cube into six tetrahedra sharing a cube diagonal. If a sequence of splits of $\tau^{\mathbf{c}}$ follows the Bisection Rule, then the resulting set τ of tetrahedra is a triangulation of the cube.*

Proof: For a tetrahedron $\mathbf{t}$ in the full-resolution tree, let $\tau_{\mathrm{cong}}(\mathbf{t})$ be the set of all tetrahedra congruent to $\mathbf{t}$ in the full-resolution tree. Set $\tau_{\mathrm{cong}}(\mathbf{t})$ forms a triangulation of the cube. Every tetrahedron $\mathbf{t}'$ that is not congruent to $\mathbf{t}$ is either contained in some tetrahedron congruent to $\mathbf{t}$ or contains some tetrahedron congruent to $\mathbf{t}$.

Let $\mathbf{t}$ and $\mathbf{t}'$ be two intersecting tetrahedra in τ. Since the tetrahedra in τ partition the cube, $\mathbf{t}$ is not a subset of $\mathbf{t}'$ and $\mathbf{t}'$ is not a subset of $\mathbf{t}$. Without loss of generality, assume that $\mathbf{t}'$ is contained in a tetrahedron $\mathbf{t}'' \in \tau_{\mathrm{cong}}(\mathbf{t})$ congruent to $\mathbf{t}$. (See Figure 9.11(a).)

Since $\tau_{\mathrm{cong}}(\mathbf{t})$ is a triangulation, $\mathbf{t} \cap \mathbf{t}''$ is a face $\mathbf{f}$ of $\mathbf{t}''$. Since $\mathbf{t} \cap \mathbf{t}'$ is a subset of $\mathbf{t} \cap \mathbf{t}''$, intersection $\mathbf{t} \cap \mathbf{t}'$ is a face of $\mathbf{t}'$.

We need to show that $\mathbf{t} \cap \mathbf{t}'$ is a face of $\mathbf{t}$. Assume that $\mathbf{t} \cap \mathbf{t}'$ is not a face of $\mathbf{t}$. Since $\mathbf{t} \cap \mathbf{t}''$ is a face of $\mathbf{t}$, tetrahedron $\mathbf{t}'$ is not equal to $\mathbf{t}''$. Thus tetrahedron $\mathbf{t}''$ must be split.

Let $\mathbf{t}_0, \mathbf{t}_1, \ldots, \mathbf{t}_m$ be a sequence of splits where

1. tetrahedron $\mathbf{t}_0$ equals $\mathbf{t}''$,

2. tetrahedron $\mathbf{t}_m$ equals $\mathbf{t}'$,

3. tetrahedron $\mathbf{t}_{i+1}$ is one of the two tetrahedra created by splitting $\mathbf{t}_i$.

Since $\mathbf{t}_0$ equals $\mathbf{t}''$ and $\mathbf{t}_m$ equals $\mathbf{t}'$, it follows that $\mathbf{t}' \subseteq \mathbf{t}_i \subseteq \mathbf{t}''$ for all i.

Let $\mathbf{t}_k$ be the smallest tetrahedron in the sequence such that $\mathbf{t} \cap \mathbf{t}_k$ equals $\mathbf{t} \cap \mathbf{t}''$ (Figure 9.11(b)). Since $\mathbf{t} \cap \mathbf{t}'$ does not equal $\mathbf{t} \cap \mathbf{t}''$, tetrahedron $\mathbf{t}_k$ does not equal $\mathbf{t}_m$. Thus $\mathbf{t}_k$ is split at some point q. Since $\mathbf{t} \cap \mathbf{t}_k$ is face of $\mathbf{t}$ and $\mathbf{t} \cap \mathbf{t}_{k+1}$ is not, point q lies on the interior of an edge of $\mathbf{t}$. (See Figure 9.11(c).)

Point q is in the interior of an edge of $\mathbf{t}$ but $\mathbf{t}$ is not split, violating the Bisection Rule. We conclude that $\mathbf{t} \cap \mathbf{t}'$ is a face of $\mathbf{t}$. Since $\mathbf{t} \cap \mathbf{t}'$ is a face of $\mathbf{t}$ and of $\mathbf{t}'$ for every two intersecting tetrahedra $\mathbf{t}, \mathbf{t}' \in \tau$, set τ is a triangulation of the cube. $\square$

9.2   Multiresolution Isosurfaces

9.2.1   Multiresolution Grid

We need a quick, efficient way to determine if a tetrahedron is split in a multiresolution triangulation. Let Γ be a $(2^k + 1) \times (2^k + 1) \times (2^k + 1)$ grid containing 2^{3k} cubes.

Let $\mathbf{t}$ be any tetrahedron in an internal node of the full-resolution tree. Because the grid vertices are the coordinates from $(0, 0, 0)$ to $(2^k, 2^k, 2^k)$, point $q_{\text{split}}(\mathbf{t})$ always lies on a grid vertex. For instance, if $\mathbf{t}$ is one of initial six tetrahedra, then $q_{\text{split}}(\mathbf{t})$ is the midpoint of the diagonal $((0, 0, 0), (2^k, 2^k, 2^k))$. This midpoint is the grid vertex $(2^{k-1}, 2^{k-1}, 2^{k-1})$. After the first split, two of the tetrahedra, $\mathbf{t}'$ and $\mathbf{t}''$, contain the edge $((0, 0, 0), (2^k, 2^k, 0))$. These two tetrahedra are split at $(2^{k-1}, 2^{k-1}, 0)$, the midpoint of edge $((0, 0, 0), (2^k, 2^k, 0))$, and thus $q_{\text{split}}(\mathbf{t}') = q_{\text{split}}(\mathbf{t}'') = (2^{k-1}, 2^{k-1}, 0)$.

If tetrahedron $\mathbf{t}$ is a leaf of the full-resolution tree, then it lies entirely in one grid cube and its vertices lie on the corners of that cube. Point $q_{\text{split}}(\mathbf{t})$ is the center of the cube, which is not a grid vertex.

We represent the multiresolution triangulation by a multiresolution grid Γ^M of $(2^k + 1) \times (2^k + 1) \times (2^k + 1)$ Boolean values. The grid vertices represent the points $q_{\text{split}}(\mathbf{t})$ for all tetrahedra $\mathbf{t}$ at internal nodes of the full-resolution tree. The points $q_{\text{split}}(\mathbf{t})$ for tetrahedra $\mathbf{t}$ at the leaves of the tree are not represented but tetrahedra at the leaves are never split into smaller tetrahedra.

The multiresolution grid Γ^M defines a partition of the grid into tetrahedra. Starting from the initial six tetrahedra, bisect tetrahedron $\mathbf{t}$ if $\Gamma^M(q_{\mathrm{split}}(\mathbf{t}))$ is true. Repeat this process on the resulting tetrahedra until either $\mathbf{t}$ is a leaf of the full-resolution tree or $\Gamma^M(q_{\mathrm{split}}(\mathbf{t}))$ is false. The resulting tetrahedra partition the grid into tetrahedra. However, the sequence of splits does not necessarily satisfy the Bisection Rule from Section 9.1.2, and the resulting set of tetrahedra is not necessarily a triangulation. To ensure a triangulation, we must impose a condition on the multiresolution grid.

In order for the multiresolution grid to represent a multiresolution triangulation, we require that the multiresolution grid satisfy the following condition:

Multiresolution Condition (N×N×N). *Let tetrahedron $\mathbf{t}$ be an internal node of the full-resolution tree. If $\mathbf{t}$ is a child of tetrahedron $\mathbf{t}'$ in the full-resolution tree and $\Gamma^M(q_{\mathrm{split}}(\mathbf{t}))$ is true, then $\Gamma^M(q_{\mathrm{split}}(\mathbf{t}'))$ is true.*

Note that the Multiresolution Condition implies that if $\mathbf{t}$ is a descendant of $\mathbf{t}'$ in the full-resolution tree and $\Gamma^M(q_{\mathrm{split}}(\mathbf{t}))$ is true, then $\Gamma^M(q_{\mathrm{split}}(\mathbf{t}'))$ is true. We claim that the multiresolution grid satisfying the Multiresolution Condition defines a triangulation of the grid.

Proposition 9.5. *If a multiresolution grid of a cubic region satisfies the Multiresolution Condition, then the set of tetrahedra defined by the multiresolution grid is a triangulation of the cubic region.*

Proof: Consider two tetrahedra, $\mathbf{t}$ and $\mathbf{t}'$, such that $q_{\mathrm{split}}(\mathbf{t})$ lies on the interior of an edge of $\mathbf{t}'$. Assume $\mathbf{t}$ is split. We wish to show that $\mathbf{t}'$ is split.

Let τ be the triangulation of the grid via bisection into tetrahedra that are all congruent to $\mathbf{t}$. Some tetrahedron $\mathbf{t}''$ in τ contains $\mathbf{t} \cap \mathbf{t}'$ and is contained in $\mathbf{t}'$. Since $q_{\mathrm{split}}(\mathbf{t})$ lies in the interior of an edge of $\mathbf{t}$ and $\mathbf{t} \cap \mathbf{t}''$ is an edge or facet of $\mathbf{t}$, point $q_{\mathrm{split}}(\mathbf{t})$ lies in the interior of an edge of $\mathbf{t}''$. Since $\mathbf{t}''$ is congruent to $\mathbf{t}$, point $q_{\mathrm{split}}(\mathbf{t}'')$ equals $q_{\mathrm{split}}(\mathbf{t})$ and $\Gamma^M(q_{\mathrm{split}}(\mathbf{t}''))$ equals $\Gamma^M(q_{\mathrm{split}}(\mathbf{t}))$. Since $\mathbf{t}$ is split, $\Gamma^M(q_{\mathrm{split}}(\mathbf{t}))$ is true and so $\Gamma^M(q_{\mathrm{split}}(\mathbf{t}'')) = \Gamma^M(q_{\mathrm{split}}(\mathbf{t}))$ is true. Tetrahedron $\mathbf{t}'$ contains $\mathbf{t}''$. By the Multiresolution Condition, $\Gamma^M(q_{\mathrm{split}}(\mathbf{t}'))$ is also true. Thus tetrahedron $\mathbf{t}'$ is split, satisfying the Bisection Rule. By Proposition 9.2, the resulting set of tetrahedra is a triangulation of the grid. □

Given a grid of Boolean values, we wish to turn it into a multiresolution grid satisfying the Multiresolution Condition. We do so by changing some of the false values to true. Of course, we could satisfy the Multiresolution Condition by setting all values to true, but then we would have a full-resolution triangulation, not a multiresolution one. The challenge is to set values to true only where necessary.

A multiresolution grid Γ^M violates the Multiresolution Condition if tetrahedron $\mathbf{t}$ is a child of tetrahedron $\mathbf{t}'$ in the full-resolution tree and $\Gamma^M(q_{\mathrm{split}}(\mathbf{t}))$ is true while $\Gamma^M(q_{\mathrm{split}}(\mathbf{t}'))$ is false. To fix this violation, we need to set $\Gamma^M(q_{\mathrm{split}}(\mathbf{t}'))$

to true. We could create and process a full-resolution tree but it is much more efficient to work directly on the multiresolution grid Γ^M.

As in Section 9.1.2, let $\mathcal{P}_j(\Gamma^M)$ be the partition of grid Γ^M into cubes of edge length 2^j. Let $\mathbf{c}$ be a cube in $\mathcal{P}_j(\Gamma^M)$, $\mathbf{f}$ a facet of $\mathbf{c}$, and $\mathbf{e}$ an edge of $\mathbf{f}$. By Proposition 9.4, there are tetrahedra $\mathbf{t}_A$, $\mathbf{t}_B$, and $\mathbf{t}_C$ such that $q_{\mathrm{split}}(\mathbf{t}_A)$, $q_{\mathrm{split}}(\mathbf{t}_B)$, and $q_{\mathrm{split}}(\mathbf{t}_C)$ are the centers of $\mathbf{c}$, $\mathbf{f}$, and $\mathbf{e}$, respectively, and splitting $\mathbf{t}_A$ creates $\mathbf{t}_B$ and splitting $\mathbf{t}_B$ creates $\mathbf{t}_C$. Thus, if $\Gamma^M(q_{\mathrm{center}}(\mathbf{f}))$ is true and $\Gamma^M(q_{\mathrm{center}}(\mathbf{c}))$ is false, then the multiresolution grid violates the Multiresolution Condition. Similarly, if $\Gamma^M(q_{\mathrm{center}}(\mathbf{e}))$ is true and $\Gamma^M(q_{\mathrm{center}}(\mathbf{f}))$ is false, then there is also a violation of the multiresolution condition. To fix such violations, set $\Gamma^M(q_{\mathrm{center}}(\mathbf{c}))$ to true whenever $\Gamma^M(q_{\mathrm{center}}(\mathbf{f}))$ is true for any facet $\mathbf{f}$ of $\mathbf{c}$, and set $\Gamma^M(q_{\mathrm{center}}(\mathbf{f}))$ to true whenever $\Gamma^M(q_{\mathrm{center}}(\mathbf{e}))$ is true for any edge $\mathbf{e}$ of $\mathbf{f}$.

Let v be a vertex of $\mathbf{c}$ that is not a corner of the grid Γ^M. By Proposition 9.4, there is a type A tetrahedron $\mathbf{t}_A$ and a tetrahedron $\mathbf{t}$ such that $q_{\mathrm{split}}(\mathbf{t}_A)$ is the center of $\mathbf{c}$ and $q_{\mathrm{split}}(\mathbf{t})$ equals v and $\mathbf{t}_A$ is contained in $\mathbf{t}$. If $\Gamma^M(q_{\mathrm{center}}(\mathbf{c}))$ is true and $\Gamma^M(v)$ is false, then the multiresolution grid violates the Multiresolution Condition. Set $\Gamma^M(v)$ to true whenever $q_{\mathrm{center}}(\mathbf{c})$ is true for any edge vertex v of $\mathbf{c}$.

Changing $\Gamma^M(q_{\mathrm{split}}(\mathbf{t}))$ from false to true for tetrahedron $\mathbf{t}$ can create new violations of the Multiresolution Condition. If tetrahedron $\mathbf{t}'$ is the parent of $\mathbf{t}$ in the full-resolution grid and $\Gamma^M(q_{\mathrm{split}}(\mathbf{t}'))$ is false, then setting $\Gamma^M(q_{\mathrm{split}}(\mathbf{t}))$ to true creates a violation at $\mathbf{t}$ and $\mathbf{t}'$. To eliminate all violations of the Multiresolution Condition, we need to first process cubic regions of edge length 2, then cubic regions of edge length 4, etc. Within each cubic region $\mathbf{c}$, we first check edge midpoints to see if facet centers must be set to true. We then check facet centers to see if $q_{\mathrm{center}}(\mathbf{c})$ must be set to true. Finally, we check $q_{\mathrm{center}}(\mathbf{c})$ to see if the corners of $\mathbf{c}$ must be set to true. Pseudocode for procedure SATISFYMUL-TIRESN×N×N is presented in Algorithm 9.1.

We claim that the multiresolution grid produced by SATISFYMULTIRES-N×N×N satisfies the Multiresolution Condition.

Proposition 9.6. *Let Γ^M be an $n \times n \times n$ grid of Boolean values where $n = 2^k + 1$. After the application of procedure* SATISFYMULTIRESN×N×N *to Γ^M, grid Γ^M satisfies the Multiresolution Condition.*

Proof: Let $\mathbf{t}$ be a child of tetrahedron $\mathbf{t}'$ in the full-resolution tree. We need to show that if $\Gamma^M(q_{\mathrm{split}}(\mathbf{t}))$ is true, then $\Gamma^M(q_{\mathrm{split}}(\mathbf{t}'))$ is true.

Case I: Tetrahedron $\mathbf{t}$ is a type A tetrahedron.

Let $\mathbf{c_t}$ be the cube in $\mathcal{P}_*(\mathbf{g})$ such that $q_{\mathrm{center}}(\mathbf{c_t})$ equals $q_{\mathrm{split}}(\mathbf{t})$. Since $\mathbf{t}$ is type A, tetrahedron $\mathbf{t}'$ must be type C. By Proposition 9.3, point $q_{\mathrm{split}}(\mathbf{t}')$ is a vertex of $\mathbf{c_t}$.

Input : Γ^M is an $n \times n \times n$ array of Boolean values where $n = 2^k + 1$.
        k determines the dimensions of Γ^M.

SatisfyMultiresNxNxN(Γ^M, k)

1 for $j = 1$ to k do
     /* $\mathcal{P}_j(\Gamma^M)$ is the partition of Γ^M into cubes of edge length 2^j */
2   foreach edge $\mathbf{e}$ in $\mathcal{P}_j(\Gamma^M)$ do
        /* $q_{\mathrm{center}}(\mathbf{e})$ is the midpoint of edge $\mathbf{e}$ */
3   if ($\Gamma^M(q_{\mathrm{center}}(\mathbf{e})) =$ true) then
4   foreach facet $f \in \mathcal{P}_j(\Gamma^M)$ containing $\mathbf{e}$ do
            /* $q_{\mathrm{center}}(\mathbf{f})$ is the center of facet $\mathbf{f}$ */
5             $\Gamma^M(q_{\mathrm{center}}(\mathbf{f})) \leftarrow$ true;
6   end
7   end
8   end
9   foreach facet $\mathbf{f}$ in $\mathcal{P}_j(\Gamma^M)$ do
10   if ($\Gamma^M(q_{\mathrm{center}}(\mathbf{f})) =$ true) then
11   foreach cube $\mathbf{c} \in \mathcal{P}_j(\Gamma^M)$ containing $\mathbf{f}$ do
            /* $q_{\mathrm{center}}(\mathbf{c})$ is the center of cube $\mathbf{c}$ */
12            $\Gamma^M(q_{\mathrm{center}}(\mathbf{c})) \leftarrow$ true;
13   end
14   end
15   end
16   foreach cube $\mathbf{c}$ in $\mathcal{P}_j(\Gamma^M)$ do
17   if ($G(q_{\mathrm{center}}(\mathbf{c})) =$ true) then
18   foreach corner v of $\mathbf{c}$ do
19            $\Gamma^M(v) \leftarrow$ true;
20   end
21   end
22   end
23 end

Algorithm 9.1. SATISFYMULTIRESN×N×N for $n \times n \times n$ grids where $n = 2^k + 1$.

Cube $\mathbf{c_t}$ is in $\mathcal{P}_j$ for some j. Consider the jth iteration of the main loop of SATISFYMULTIRESN×N×N. In Steps 16–22, if $\Gamma^M(q_{\mathrm{center}}(\mathbf{c_t}))$ is true, then $\Gamma^M(v)$ is set to true for every vertex v of $\mathbf{c}$. Note that Steps 16–22 do not modify $\Gamma^M(q_{\mathrm{center}}(\mathbf{c_t}))$. The value of $\Gamma^M(q_{\mathrm{center}}(\mathbf{c_t}))$ can never be modified after Step 22, since j increases at the next iteration of the main loop. Thus, if $\Gamma^M(q_{\mathrm{split}}(\mathbf{t}))$ is true, then $\Gamma^M(q_{\mathrm{split}}(\mathbf{t}'))$ is true.

Case II: Tetrahedron $\mathbf{t}$ is a type B tetrahedron.

Since $\mathbf{t}$ is type B, tetrahedron $\mathbf{t}'$ must be type A. Let $\mathbf{f_t}$ be the facet and $\mathbf{c_{t'}}$ be the cube in $\mathcal{P}_*(G)$ such that $q_{\text{center}}(\mathbf{f_t})$ equals $q_{\text{split}}(\mathbf{t})$ and $q_{\text{center}}(\mathbf{c_{t'}})$ equals $q_{\text{split}}(\mathbf{t}')$. By Proposition 9.3, $\mathbf{f_t}$ is a facet of $\mathbf{c_{t'}}$.

Cube $\mathbf{c_{t'}}$ is in $\mathcal{P}_j$ for some j. Consider the jth iteration of the main loop of SATISFYMULTIRESN×N×N. In Steps 9–15, if $\Gamma^M(q_{\text{center}}(\mathbf{f_t}))$ is true, then the value of $\Gamma^M(q_{\text{center}}(\mathbf{c_{t'}}))$ is set to true. Note that Steps 9–22 do not modify $\Gamma^M(q_{\text{center}}(\mathbf{f_t}))$. The value of $\Gamma^M(q_{\text{center}}(\mathbf{f_t}))$ can never be modified after Step 22, since j increases at the next iteration of the main loop. Thus, if $\Gamma^M(q_{\text{split}}(\mathbf{t}))$ is true, then $\Gamma^M(q_{\text{split}}(\mathbf{t}'))$ is true.

Case III: Tetrahedron $\mathbf{t}$ is a type C tetrahedron.

Since $\mathbf{t}$ is type C, tetrahedron $\mathbf{t}'$ must be type B. Let $\mathbf{e_t}$ be the edge and $\mathbf{f_{t'}}$ be the facet in $\mathcal{P}_*(G)$ such that $q_{\text{center}}(\mathbf{e_t})$ equals $q_{\text{split}}(\mathbf{t})$ and $q_{\text{center}}(\mathbf{f_{t'}})$ equals $q_{\text{split}}(\mathbf{t}')$. By Proposition 9.3, $\mathbf{e_t}$ is an edge of $\mathbf{f_{t'}}$.

Facet $\mathbf{f_{t'}}$ is in $\mathcal{P}_j$ for some j. Consider the jth iteration of the main loop of SATISFYMULTIRESN×N×N. In Steps 2–8, if $\Gamma^M(q_{\text{center}}(\mathbf{e_t}))$ is true, then the value of $\Gamma^M(q_{\text{center}}(\mathbf{f_{t'}}))$ is set to true. Note that Steps 2–22 do not modify $\Gamma^M(q_{\text{center}}(\mathbf{e_t}))$. The value of $\Gamma^M(q_{\text{center}}(\mathbf{e_t}))$ can never be modified after Step 22, since j increases at the next iteration of the main loop. Thus, if $\Gamma^M(q_{\text{split}}(\mathbf{t}))$ is true, then $\Gamma^M(q_{\text{split}}(\mathbf{t}'))$ is true. □

9.2.2   Isosurface Extraction

Algorithm MULTIRESOLUTION MARCHING TETRAHEDRA N×N×N extracts a multiresolution isosurface from a multiresolution triangulation of an $n \times n \times n$ grid where $n = 2^k + 1$. Input to the algorithm is a scalar grid, Γ^S; an array, **Coord**, of grid coordinates; a set, **RegionList**, of high-resolution regions; and a scalar value σ. (See Figure 9.12 and Algorithm 9.2.)

The algorithm has four steps. The first step is to convert the high-resolution regions to a multiresolution grid, Γ^M (Algorithm 9.3, CREATEMULTIRESGRID-N×N×N). Every vertex within the high-resolution regions is set to true. All other vertices are set to false. Procedure SATISFYMULTIRESN×N×N (Algorithm 9.1) is applied to the multiresolution grid so that it satisfies the Multiresolution Condition.

The second step in the algorithm is to read the isosurface lookup table from a preconstructed data file. Entries in the table correspond to configurations of ("+","−") labels of a tetrahedron.

The multiresolution grid represents a multiresolution triangulation. The third algorithm step is to extract isosurface patches from the tetrahedra in this triangulation. Instead of explicitly constructing the triangulation, we recursively descend the full-resolution tree, identifying and processing the tetrahedra in the

---

Input : Γ^S is an $n \times n \times n$ array of scalar values where $n = 2^k + 1$ for
some integer k.
Coord is a 3D array of (x, y, z) coordinates.
RegionList is a list of regions in Γ^S.
σ is an isovalue.

MultiresMarchingTetNxNxN(Γ^S, Coord, RegionList, σ)

1   CreateMultiresGridNxNxN(Γ^S,RegionList,Γ^M);
2   Read Marching Cubes lookup table into Table;
3   $T \leftarrow \emptyset$;
4   $\tau^c \leftarrow$ triangulation of grid Γ^M into six tetrahedra sharing edge
$((0, 0, 0), (2^k, 2^k, 2^k))$;
5   foreach tetrahedron $\mathbf{t} \in \tau^c$ do
6   Let (p_0, p_1, p_2, p_3) be the vertices of $\mathbf{t}$ where:
      $p_0 = (0, 0, 0)$ and
      $p_1 \in \{(2^k, 0, 0), (0, 2^k, 0), (0, 0, 2^k)\}$ and
      $p_2 \in \{(2^k, 2^k, 0), (2^k, 0, 2^k), (0, 2^k, 2^k)\}$ and
      $p_3 = (2^k, 2^k, 2^k)$.
7   MultiresRecursiveA(Γ^S,Γ^M,σ,p_0, p_1, p_2, p_3, 2^k,T, Table);
8   end
    /* Compute isosurface vertex coordinates using linear interpolation   */
9   $E \leftarrow \bigcup_{(\mathbf{e}_1,\mathbf{e}_2,\mathbf{e}_3) \in T}\{\mathbf{e}_1, \mathbf{e}_2, \mathbf{e}_3\}$;
10   foreach edge $e \in E$ with endpoints (i_1, j_1, k_1) and (i_2, j_2, k_2) do
11     $u_e \leftarrow$ LinearInterpolation
12   (Coord$[i_1, j_1, k_1]$, $\Gamma^S[i_1, j_1, k_1]$, Coord$[i_2, j_2, k_2]$, $\Gamma^S[i_2, j_2, k_2]$, σ);
13   end
    /* Convert T to set of triangles   */
14   $\Upsilon \leftarrow \emptyset$;
15   foreach triple of edges $(\mathbf{e}_1, \mathbf{e}_2, \mathbf{e}_3) \in T$ do
16     $\Upsilon \leftarrow \Upsilon \cup \{(u_{\mathbf{e}_1}, u_{\mathbf{e}_2}, u_{\mathbf{e}_3})\}$;
17   end

Algorithm 9.2. MULTIRESOLUTION MARCHING TETRAHEDRA N×N×N for
$n \times n \times n$ grids where $n = 2^k + 1$.

tree. Upon reaching a tetrahedron $\mathbf{t}$, if $\mathbf{t}$ is a leaf of the tree or if $\Gamma^M(q_{\text{split}}(\mathbf{t}))$
is false, then an isosurface patch is extracted from $\mathbf{t}$. Otherwise, $\mathbf{t}$ is split into
two tetrahedra, $\mathbf{t}'$ and $\mathbf{t}''$, and we repeat the procedure on $\mathbf{t}'$ and $\mathbf{t}''$.

The isosurface vertices lie on bipolar edges of a multiresolution triangulation
of the grid. The final step in the algorithm is to assign geometric locations to the
isosurface vertices based on the scalar values at the endpoints of these bipolar
edges.

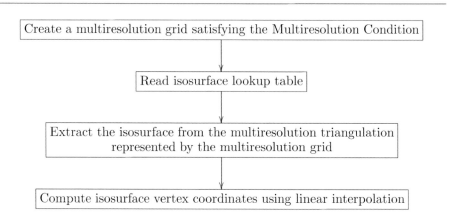

Figure 9.12. MULTIRESOLUTION MARCHING TETRAHEDRA N×N×N for $n \times n \times n$ grids where $n = 2^k + 1$.

Input   : Γ^S is an $n \times n \times n$ array of scalar values where $n = 2^k + 1$ for
           some integer k.
           RegionList is a list of regions in Γ^S.
Output   : Γ^M is an $n \times n \times n$ array of Boolean values satisfying the
           Multiresolution Condition.

`CreateMultiresGridNxNxN(`Γ^S`, RegionList, `Γ^M`)`

1 Initialize $\Gamma^M(v)$ to false for all vertices v;
2 foreach region R in RegionList do
3 | foreach vertex v in R do
4 | | $\Gamma^M(v) \leftarrow$ true;
5 | end
6 end
7 `SatisfyMultiresNxNxN` (Γ^M, k);

Algorithm 9.3. Procedure CREATEMULTIRESGRIDN×N×N for $n \times n \times n$ grids
where $n = 2^k + 1$.

We describe the third algorithm step, extracting the isosurface, in more detail. Grid Γ^M has a triangulation comprised of six tetrahedra sharing edge $((0,0,0),(2^k,2^k,2^k))$. We call procedure MULTIRESRECURSIVEA (Algorithm 9.4) on each of the six tetrahedra.

Procedure MULTIRESRECURSIVEA takes as input the scalar grid Γ^S, the multiresolution grid Γ^M, the isovalue σ, the four vertices, p_0, p_1, p_2, p_3, of a type

Input : Γ^S is an $n_1 \times n_2 \times n_3$ array of scalar values.
        Γ^M is an $n_1 \times n_2 \times n_3$ array of Boolean values satisfying
              the Multiresolution Condition.
        σ is an isovalue.
        p_0, p_1, p_2, p_3 are vertices of a type A tetrahedron.
        L is the length of line segment (p_0, p_1).
        T is a set of edge triples representing isosurface triangles.
        Table is the isosurface lookup table.

MultiresRecursiveA(Γ^S, Γ^M, σ, p_0, p_1, p_2, p_3, L, T, Table)

1 if $(L = 1)$ then
2 | ExtractIsoPatch(Γ^S, σ, p_0, p_1, p_2, p_3, T, Table);
3 else
4 | $q_0 \leftarrow$ midpoint of line segment (p_0, p_3);
5 | if $(\Gamma^M(q_0) = $ false) then
6 | | ExtractIsoPatch(Γ^S, σ, p_0, p_1, p_2, p_3, T, Table);
7 | else
8 | | MultiresRecursiveB(Γ^S, Γ^M, p_0, p_1, p_2, q_0, L, T, Table);
9 | | MultiresRecursiveB(Γ^S, Γ^M, p_3, p_2, p_1, q_0, L, T, Table);
10 | end
11 end

****

Input : Γ^S, Γ^M, σ, L and T are as defined above.
        p_0, p_1, p_2, q_0 are vertices of a type B tetrahedron.

MultiresRecursiveB(Γ^S, Γ^M, σ, p_0, p_1, p_2, q_0, L, T, Table)

1 $q_1 \leftarrow$ midpoint of line segment (p_0, p_2);
2 if $(\Gamma^M(q_1) = $ false) then
3 | ExtractIsoPatch(Γ^S, σ, p_0, p_1, p_2, q_0, T, Table);
4 else
5 | MultiresRecursiveC(Γ^S, Γ^M, p_0, p_1, q_1, q_0, L, T, Table);
6 | MultiresRecursiveC(Γ^S, Γ^M, p_2, p_1, q_1, q_0, L, T, Table);
7 end

****

Input : Γ^S, Γ^M, σ, L and T are as defined above.
        p_0, p_1, q_1, q_0 are vertices of a type C tetrahedron.

MultiresRecursiveC(Γ^S, Γ^M, σ, p_0, p_1, q_1, q_0, L, T, Table)

1 $q_2 \leftarrow$ midpoint of line segment (p_0, p_1);
2 if $(\Gamma^M(q_2) = $ false) then
3 | ExtractIsoPatch(Γ^S, σ, p_0, p_1, q_1, q_0, T, Table);
4 else
5 | MultiresRecursiveA(Γ^S, Γ^M, p_0, q_2, q_1, q_0, $L/2$, T, Table);
6 | MultiresRecursiveA(Γ^S, Γ^M, p_2, q_2, q_1, q_0, $L/2$, T, Table);
7 end

Algorithm 9.4. MULTIRES RECURSIVE.

Input : Γ^S is an $n_1 \times n_2 \times n_3$ array of scalar values.

σ is an isovalue.

p_0, p_1, p_2, p_3 are vertices of a tetrahedron.

T is a set of edge triples representing isosurface triangles.

Table is the isosurface lookup table.

ExtractIsoPatch(Γ^S, σ, p_0, p_1, p_2, p_3, T, Table)

/* *Assign "+" or "−" signs to each vertex* */

1   for $i = 0$ to 3 do

2   if $\Gamma^S(p_i) < \sigma$ then Sign$[p_i] \leftarrow$ "−";

3   else Sign$[p_i] \leftarrow$ "+";   /* $\Gamma^S(p_i) \geq \sigma$ */

4   end

5   $\kappa \leftarrow$ (Sign$[p_0]$, Sign$[p_1]$, Sign$[p_2]$, Sign$[p_3]$);

6   foreach edge triple $(\mathbf{e}_1, \mathbf{e}_2, \mathbf{e}_3) \in$ Table$[\kappa]$ do

7   for $j = 1$ to 3 do

8       $\mathbf{e}'_i \leftarrow$ edge of tetrahedron (p_0, p_1, p_2, p_3) corresponding to $\mathbf{e}_i$;

9   end

10   Insert edge triple $(\mathbf{e}'_1, \mathbf{e}'_2, \mathbf{e}'_3)$ into T;

11   end

Algorithm 9.5. EXTRACTISOPATCH.

A tetrahedron, and the length L of line segment (p_0, p_1). (See Figure 9.8 for an illustration of the three types of tetrahedra.) The vertices of the tetrahedron should be a subset of the vertices of an $L \times L \times L$ cubic region with p_0 and p_3 on opposite corners of the cubic region. Line segments (p_0, p_1), (p_1, p_2), and (p_2, p_3) should be edges of the cubic region.

If length L equals one, then tetrahedron $\mathbf{t} = (p_0, p_1, p_2, p_3)$ is as small as possible and can be split no further. MULTIRESRECURSIVEA calls procedure EXTRACTISOPATCH (Algorithm 9.5) to extract the isosurface patch from tetrahedron $\mathbf{t}$. Otherwise, MULTIRESRECURSIVEA computes the midpoint, q_0, of line segment (p_0, p_3). Point q_0 is a grid vertex. If $\Gamma^M(q_0)$ is false, then tetrahedron $\mathbf{t}$ is not split and MULTIRESRECURSIVEA calls EXTRACTISOPATCH on $\mathbf{t}$. If $\Gamma^M(q_0)$ is true, then tetrahedron $\mathbf{t}$ is split into two tetrahedra, (p_0, p_1, p_2, q_0) and (p_3, p_2, p_1, q_0), of type B. Procedure MULTIRESRECURSIVEA calls Procedure MULTIRESRECURSIVEB on these two tetrahedra.

Procedure MULTIRESRECURSIVEB takes as input the scalar grid Γ^S, the multiresolution grid Γ^M, the isovalue σ, the four vertices, p_0, p_1, p_2, q_0, of a type B tetrahedron, and the length L of line segment (p_0, p_1). Vertices (p_0, p_1, p_2) should be a subset of the vertices of an $L \times L \times L$ cubic region with q_0 at the cube center. Line segments (p_0, p_1) and (p_1, p_2) should be edges of that cubic region.

Procedure MULTIRESRECURSIVEB computes the midpoint, q_1, of line segment (p_0, p_1). Point q_1 is a grid vertex. If $\Gamma^M(q_1)$ is false, then tetrahedron $\mathbf{t} = (p_0, p_1, p_2, q_0)$ is not split and MULTIRESRECURSIVEB calls procedure EXTRACTISOPATCH on $\mathbf{t}$. If $\Gamma^M(q_1)$ is false, then tetrahedron $\mathbf{t}$ is split into two tetrahedra, (p_0, p_1, q_1, q_0) and (p_2, p_1, q_1, q_0), of type C. Procedure MULTIRESRECURSIVEB calls procedure MULTIRESRECURSIVEC on these two tetrahedra.

Procedure MULTIRESRECURSIVEC takes as input the scalar grid Γ^S, the multiresolution grid Γ^M, the isovalue σ, the four vertices, p_0, p_1, q_1, q_0, of a type B tetrahedron, and the length L of line segment (p_0, p_1). Line segment (p_0, p_1) should be an edge of an $L \times L \times L$ cubic region with q_0 at the cube center and q_1 at the center of a cube facet.

Procedure MULTIRESRECURSIVEC computes the midpoint, q_2, of line segment (p_0, p_1). Point q_2 is a grid vertex. If $\Gamma^M(q_2)$ is false, then tetrahedron $\mathbf{t} = (p_0, p_1, q_1, q_0)$ is not split and MULTIRESRECURSIVEC calls Procedure EXTRACTISOPATCH on $\mathbf{t}$. If $\Gamma^M(q_2)$ is true, then tetrahedron $\mathbf{t}$ is split into two tetrahedra, (p_0, q_2, q_1, q_0) and (p_2, q_2, q_1, q_0), of type A. Procedure MULTIRESRECURSIVEC calls Procedure MULTIRESRECURSIVEA on these two tetrahedra. Note that line segments (p_0, q_2) and (p_2, q_2) have length $L/2$.

9.2.3 $n_1 \times n_2 \times n_3$ Grids

So far we have assumed that the input grid Γ^S has vertex dimensions $(2^k + 1) \times (2^k + 1) \times (2^k + 1)$ for some integer k. This condition on Γ^S is highly restrictive. In this section we show how to modify the algorithms from Sections 9.2.1 and 9.2.2 to apply to grids with any dimensions.

Let Γ be a scalar grid with vertex dimensions $n_1 \times n_2 \times n_3$. Let $|\Gamma|$ be the union of all the grid cubes in Γ. Partition $|\Gamma|$ into cubic regions as follows. Let k be the smallest integer such that $2^k + 1 \geq \max(n_1, n_2, n_3)$. Let $\mathbf{c}^*$ be a cubic region with edge length 2^k and with opposing vertices at $(0, 0, 0)$ and $(2^k, 2^k, 2^k)$. Region $\mathbf{c}^*$ contains $|\Gamma|$. If $|\Gamma|$ does not equal $\mathbf{c}^*$, divide $\mathbf{c}^*$ into eight cubic subregions. Recursively subdivide each of the subregions until each subregion is either contained in $|\Gamma|$ or is disjoint from the interior of $|\Gamma|$. After k partitions, each subregion has edge length one. Thus the recursive subdivision stops after at most k steps. Let $\mathcal{P}_{oct}$ be the set of cubic subregions that are contained in $|\Gamma|$. Set $\mathcal{P}_{oct}$ forms a partition of $|\Gamma|$.

Region $\mathbf{c}^*$ and its subdivisions form an octree containing $|\Gamma|$. (See Section 8.2.1 for a definition of octrees.) Set $\mathcal{P}_{oct}$ is the leaves of this octree that represent regions contained in $|\Gamma|$. We call $\mathcal{P}_{oct}$ the octree partition of $|\Gamma|$.

For each cubic subregion $\mathbf{c} \in \mathcal{P}_{oct}$, triangulate $\mathbf{c}$ using six congruent tetrahedra sharing a common edge. This common edge connects opposing vertices of $\mathbf{c}$. There are four possible ways to create such a triangulation of $\mathbf{c}$, corresponding to the four pairs of opposing vertices of $\mathbf{c}$. If $\mathbf{c}$ does not equal $\mathbf{c}^*$, then $\mathbf{c}$ was created by subdividing a cubic region $\mathbf{c}'$ into eight octants. The center of $\mathbf{c}'$ is a vertex q_0 of $\mathbf{c}$. Let $d_{\mathrm{split}}(\mathbf{c})$ be the diagonal of $\mathbf{c}$ connecting q_0 to the opposite vertex in $\mathbf{c}$. If $\mathbf{c}$ equals $\mathbf{c}^*$, then let $d_{\mathrm{split}}(\mathbf{c})$ be the diagonal from $(0,0,0)$ to $(2^k, 2^k, 2^k)$. (Cube $\mathbf{c} \in \mathcal{P}_{oct}$ can only equal $\mathbf{c}^*$ if $\mathbf{c}^*$ equals $|\Gamma|$. In that case, $\mathcal{P}_{oct}$ contains only $\mathbf{c}^*$.) Choose the triangulation of $\mathbf{c}$ whose tetrahedra share edge $d_{\mathrm{split}}(\mathbf{c})$. Label this triangulation $\tau^{\mathbf{c}}$.

The union of the triangulations $\tau^{\mathbf{c}}$ form a subdivision of $|\Gamma|$ into tetrahedra, although not necessarily a triangulation of $|\Gamma|$. We show how to turn this subdivision into a multiresolution triangulation of $|\Gamma|$ by applying recursive bisection from Section 9.1.2 to the tetrahedra.

Each tetrahedron $\mathbf{t} \in \bigcup_{r \in \mathcal{P}_{oct}} \tau^{\mathbf{c}}$ is of type A. (See Figure 9.8.) Tetrahedron $\mathbf{t}$ splits into two tetrahedra of type B and these two split into four tetrahedra of type C. Splitting the type C tetrahedra creates type A tetrahedra. Splitting continues until the tetrahedra vertices are a subset of the vertices of a unit grid cube.

The subdivisions of $\tau^{\mathbf{c}}$ form a full-resolution tree as described in Section 9.2.1. The root of this tree contains the cubic region $\mathbf{c}$. The children of the root are the six tetrahedra in $\tau^{\mathbf{c}}$. Every other internal node of the tree has exactly two children formed by splitting its tetrahedron. Leaves of the tree represent tetrahedra whose vertices are a subset of the vertices of the unit cube.

The collection of all full-resolution trees over all $\mathbf{c} \in \mathcal{P}_{oct}$ form a full-resolution forest. Each node in this forest represents a cubic region $\mathbf{c} \in \mathcal{P}_{oct}$ or a tetrahedron. Different trees in the forest have different heights. The height of a tree depends upon the size of the cubic region $\mathbf{c}$ at its root.

As with the $(2^k+1) \times (2^k+1) \times (2^k+1)$ grid, we represent the multiresolution triangulation of the $n_1 \times n_2 \times n_3$ grid by an $n_1 \times n_2 \times n_3$ grid Γ^M of Boolean values. The grid vertices represent the points $q_{\mathrm{split}}(\mathbf{t})$ for all tetrahedra $\mathbf{t}$ at internal nodes of the full-resolution forest. The points $q_{\mathrm{split}}(\mathbf{t})$ for tetrahedra $\mathbf{t}$ at leaves of the forest are not represented but they are never split into smaller tetrahedra.

The multiresolution grid Γ^M represents a multiresolution subdivision of $|\Gamma|$ as follows. Starting from the triangulations $\tau^{\mathbf{c}}$ for all $\mathbf{c} \in \mathcal{P}_{oct}$, bisect tetrahedron $\mathbf{t} \in \tau^{\mathbf{c}}$ if $q_{\mathrm{split}}(\mathbf{t})$ is true. Repeat this process on the resulting tetrahedra until either $\mathbf{t}$ is either a leaf of the full-resolution forest or $\Gamma^M(q_{\mathrm{split}}(\mathbf{t}))$ is false. The resulting tetrahedra form a multiresolution subdivision of $|\Gamma|$.

To ensure that this multiresolution subdivision is a triangulation, we need additional conditions on the multiresolution grid. The Multiresolution Condition in Section 9.2.1 generalizes to the following condition on full-resolution forests:

Multiresolution Conditions (Forest). *Let tetrahedron $\mathbf{t}$ be an internal node, and let cubic region $\mathbf{c}$ be a root node in the full-resolution forest.*

1. *If $\mathbf{t}$ is a child of tetrahedron $\mathbf{t}'$ in the full-resolution tree and $q_{\mathrm{split}}(\mathbf{t})$ is true, then $q_{\mathrm{split}}(\mathbf{t}')$ is true.*

2. *Every corner v of region $\mathbf{c}$ is true.*

Input : Γ^M is an $n_1 \times n_2 \times n_3$ array of Boolean values.

SatisfyMultires(Γ^M)

1    $\mathcal{P}_{oct} \leftarrow$ octree partition of Γ^M;

2   foreach cubic region $\mathbf{c} \in \mathcal{P}_{oct}$ do

3   foreach corner v of $\mathbf{c}$ do

4         $\Gamma^M(v) \leftarrow$ true;

5   end

6   end

7   Let k be the smallest integer such that $2^k + 1 \geq \max(n_1, n_2, n_3)$;

8   for $j = 1$ to k do

      /* $\mathcal{P}_j(\Gamma^M)$ is the partition of subgrid Γ_j^M of Γ^M into cubes of length 2^j where Γ_j^M is the largest subgrid of Γ^M containing $(0,0,0)$ that can be partitioned into cubes of length 2^j. */

9   foreach edge $\mathbf{e}$ in $\mathcal{P}_j(\Gamma^M)$ do

        /* $q_{\text{center}}(\mathbf{e})$ is the midpoint of edge $\mathbf{e}$ */

10   if $(\Gamma^M(q_{\text{center}}(\mathbf{e})) =$ true$)$ then

11   foreach facet $f \in \mathcal{P}_j(\Gamma^M)$ containing $\mathbf{e}$ do

            /* $q_{\text{center}}(\mathbf{f})$ is the center of facet $\mathbf{f}$ */

12             $\Gamma^M(q_{\text{center}}(\mathbf{f})) \leftarrow$ true;

13   end

14   end

15   end

16   foreach facet $\mathbf{f}$ in $\mathcal{P}_j(\Gamma^M)$ do

17   if $(\Gamma^M(q_{\text{center}}(\mathbf{f})) =$ true$)$ then

18   foreach cube $\mathbf{c} \in \mathcal{P}_j(\Gamma^M)$ containing $\mathbf{f}$ do

            /* $q_{\text{center}}(\mathbf{c})$ is the center of cube $\mathbf{c}$ */

19             $\Gamma^M(q_{\text{center}}(\mathbf{c})) \leftarrow$ true;

20   end

21   end

22   end

23   foreach cube $\mathbf{c}$ in $\mathcal{P}_j(\Gamma^M)$ do

24   if $(\Gamma^M(q_{\text{center}}(\mathbf{c})) =$ true$)$ then

25   foreach corner v of $\mathbf{c}$ do

26             $\Gamma^M(v) \leftarrow$ true;

27   end

28   end

29   end

30   end

Algorithm 9.6. SATISFYMULTIRES for $n_1 \times n_2 \times n_3$ grids.

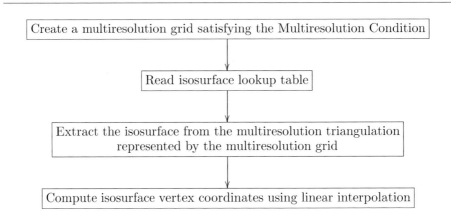

Figure 9.13. MULTIRESOLUTION MARCHING TETRAHEDRA (RECURSIVE) for $n_1 \times n_2 \times n_3$ grids.

We claim that a multiresolution grid that satisfies the Multiresolution Conditions defines a multiresolution triangulation.

Proposition 9.7. *If an $n_1 \times n_2 \times n_3$ multiresolution grid Γ^M satisfies the Multiresolution Conditions, then the set of tetrahedra defined by Γ^M is a triangulation of the region covered by Γ^M.*

The proof is left for the next section.

The algorithm to convert an array of Boolean values into a multiresolution grid satisfying the Multiresolution Conditions is almost exactly the same as procedure SATISFYMULTIRES N×N×N in Section 9.2.1. One difference is that the corners of each cubic region in the octree partition of $|\Gamma|$ must be set to true. The only other difference is that it may not be possible to partition an $n_1 \times n_2 \times n_3$ grid Γ^M into cubic regions of edge length 2^j for some values of j. Instead, let Γ_j^M be the largest subgrid of Γ^M containing $(0, 0, 0)$ that can be partitioned into cubic regions of edge length 2^j. For instance, if Γ^M has cube dimensions $5 \times 7 \times 8$, then Γ_1^M has cube dimensions $4 \times 6 \times 8$ and Γ_2^M has cube dimensions $4 \times 4 \times 8$. (Cube dimensions are the number of edges along each axis. See Section 1.7.1.) Let $\mathcal{P}_j(\Gamma^M)$ be the partition of subgrid Γ_j^M into cubic regions of edge length 2^j.

Within each cubic region $\mathbf{c}$ of $\mathcal{P}_j(\Gamma^M)$, first check edge centers to see if facet centers must be set to true. Next, check facet centers to see if $q_{\mathbf{c}}$ must be set to true. Finally, check $q_{\mathbf{c}}$ to see if the vertices of $\mathbf{c}$ must be set to true. Pseudocode for procedure SATISFYMULTIRES is presented in Algorithm 9.6.

The algorithm for extracting an isosurface from the multiresolution triangulation is called MULTIRESOLUTION MARCHING TETRAHEDRA (Algorithm 9.7).

Input : Γ^S is an $n_1 \times n_2 \times n_3$ array of scalar values.
Coord is a 3D array of (x, y, z) coordinates.
RegionList is a list of regions in Γ^S.
σ is an isovalue.

MultiresMarchingTetRecursive(Γ^S, Coord, RegionList, σ)

1 CreateMultiresGrid(Γ^S,RegionList,Γ^M);
2 Read tetrahedron isosurface lookup table into Table;
3 $\mathsf{T} \leftarrow \emptyset$;
4 $\mathcal{P}_{oct} \leftarrow$ octree partition of Γ^S;
5 foreach cubic region $\mathbf{c} \in \mathcal{P}_{oct}$ do
6 $\quad$ $\tau^{\mathbf{c}} \leftarrow$ triangulation of cubic region $\mathbf{c}$ into six tetrahedra sharing edge $d_{\mathrm{split}}(\mathbf{c})$;
7 $\quad$ $L \leftarrow$ length of the edges of $\mathbf{c}$;
8 $\quad$ foreach tetrahedron $\mathbf{t} \in \tau^{\mathbf{c}}$ do
9 $\quad\quad$ Let (p_0, p_1, p_2, p_3) be the vertices of $\mathbf{t}$ where:
$\quad\quad\quad$ p_0 and p_3 are endpoints of $d_{\mathrm{split}}(\mathbf{c})$ and
$\quad\quad\quad$ $|(p_0, p_1)| = |(p_1, p_2)| = |(p_2, p_3)|$.
10 $\quad\quad$ MultiresRecursiveA(Γ^S,Γ^M,σ,p_0, p_1, p_2, p_3, L,T,Table);
11 $\quad$ end
12 end
$\quad$ /* Compute isosurface vertex coordinates using linear interpolation $\qquad$ */
13 $\mathsf{E} \leftarrow \bigcup_{(\mathbf{e}_1, \mathbf{e}_2, \mathbf{e}_3) \in \mathsf{T}} \{\mathbf{e}_1, \mathbf{e}_2, \mathbf{e}_3\}$;
14 foreach edge $e \in \mathsf{E}$ with endpoints (i_1, j_1, k_1) and (i_2, j_2, k_2) do
15 $\quad$ $u_e \leftarrow$ LinearInterpolation
16 $\quad\quad$ (Coord$[i_1, j_1, k_1]$, $\Gamma^S[i_1, j_1, k_1]$, Coord$[i_2, j_2, k_2]$, $\Gamma^S[i_2, j_2, k_2]$, σ);
17 end
$\quad$ /* Convert T to set of triangles $\qquad$ */
18 $\Upsilon \leftarrow \emptyset$;
19 foreach triple of edges $(\mathbf{e}_1, \mathbf{e}_2, \mathbf{e}_3) \in \mathsf{T}$ do
20 $\quad$ $\Upsilon \leftarrow \Upsilon \cup \{(u_{\mathbf{e}_1}, u_{\mathbf{e}_2}, u_{\mathbf{e}_3})\}$;
21 end

Algorithm 9.7. MULTIRESOLUTION MARCHING TETRAHEDRA (RECURSIVE) for $n_1 \times n_2 \times n_3$ grids.

It is almost the same as the algorithm MULTIRESOLUTION MARCHING TETRAHEDRA N×N×N (Algorithm 9.2) for $n \times n \times n$ grids where $n = 2^k + 1$. The only difference is the construction and use of an octree partition on the grid.

Input to MULTIRESOLUTION MARCHING TETRAHEDRA is a scalar grid, Γ^S; an array, Coord, of grid coordinates; a set, RegionList, of high-resolution regions; and a scalar value σ. Start by converting the high-resolution regions to a mul-

---

Input   : Γ^S is an $n_1 \times n_2 \times n_3$ grid.
          RegionList is a list of regions in Γ^S.
Output   : Γ^M is an $n_1 \times n_2 \times n_3$ array of Boolean values satisfying the
          Multiresolution Condition.

CreateMultiresGrid(Γ^S, RegionList, Γ^M)

1 Initialize $\Gamma^M(v)$ to false for all vertices v;
2 foreach region R in RegionList do
3   foreach vertex v in R do
4       $\Gamma^M(v) \leftarrow$ true;
5   end
6 end
7 SatisfyMultires (Γ^M);

Algorithm 9.8. CREATEMULTIRESGRID for $n_1 \times n_2 \times n_3$ grids.

tiresolution grid Γ^M (Algorithm 9.8). Set every vertex within the high-resolution regions to true. Set all other vertices to false. Apply procedure SATISFYMUL-TIRES (Algorithm 9.6) to the multiresolution grid so that it satisfies the Multiresolution Conditions.

Read the isosurface lookup table from a preconstructed data file. To extract the isosurface, create an octree partition $\mathcal{P}_{oct}$ of Γ^S. For each cubic region $\mathbf{c}$ in $\mathcal{P}_{oct}$, construct the triangulation $\tau^{\mathbf{c}}$ consisting of six tetrahedra sharing the edge $d_{\mathrm{split}}(\mathbf{c})$. Call procedure MULTIRESRECURSIVEA (Algorithm 9.4) on each tetrahedron $\mathbf{t} \in \tau^{\mathbf{c}}$. If $\Gamma^M(q_{\mathrm{split}}(\mathbf{t}))$ is false, then MULTIRESRECURSIVEA splits $\mathbf{t}$ and calls MULTIRESRECURSIVEB, which may call MULTIRESRECURSIVEC. These procedures descend the full-resolution tree rooted at $\mathbf{c}$, extracting iso-surface patches from the multiresolution triangulation defined by the multireso-lution grid. Note MULTIRESOLUTION MARCHING TETRAHEDRA and MULTIRES-OLUTION MARCHING TETRAHEDRA N×N×N use exactly the same procedures, MULTIRESRECURSIVEA, MULTIRESRECURSIVEB, and MULTIRESRECURSIVEC, to extract isosurface patches from the multiresolution triangulation.

In the final step, assign geometric locations to the isosurface vertices based on the scalar values at the endpoints of these bipolar edges.

9.2.4   Isosurface Properties

By Proposition 9.5, the set of tetrahedra represented by the multiresolution grid form a tetrahedral mesh. Algorithm MULTIRESOLUTION MARCHING TETRAHE-DRA is equivalent to applying MARCHING TETRAHEDRA (Section 2.4) to that tetrahedral mesh. Thus, the properties of the MULTIRESOLUTION MARCHING

TETRAHEDRA isosurface are exactly the properties listed in Section 2.4.2 for the MARCHING TETRAHEDRA isosurface.

MULTIRESOLUTION MARCHING TETRAHEDRA constructs an isosurface from a tetrahedral mesh covering a rectangular region. Thus, the condition in Section 2.4.2 about the tetrahedral mesh covering a 3-manifold with boundary is implicit and can be omitted.

As with MARCHING TETRAHEDRA, algorithm MULTIRESOLUTION MARCHING TETRAHEDRA has no ambiguous configurations. However, the topology of the constructed isosurface often depends upon the resolution used in its construction.

9.2.5 Triangulation Proof

We present a proof that multiresolution grids that satisfy the Multiresolution Conditions define triangulations (Proposition 9.7.) We start by modifying the Bisection Rule to apply to octree partitions for $n_1 \times n_2 \times n_3$ grids.

Bisection Rules (3D Octree Partition).

1. *If tetrahedron $\mathbf{t}$ is split at point q and q lies in the interior of an edge of tetrahedron $\mathbf{t}'$, then tetrahedron $\mathbf{t}'$ is split.*

2. *If q is the corner of some region in the octree partition and q lies in the interior of an edge of tetrahedron $\mathbf{t}'$, then tetrahedron $\mathbf{t}'$ is split.*

To prove that a sequence of splits that follows the Bisection Rule creates a triangulation, we need the following two lemmas.

Lemma 9.8. *Let $\mathcal{P}_{oct}$ be an octree partition of an $n_1 \times n_2 \times n_3$ grid Γ and let $\mathbf{t}$ be a tetrahedron in the full-resolution forest of $\mathcal{P}_{oct}$. If $\tau_{\mathrm{cong}}(\mathbf{t})$ is the set of all tetrahedron congruent to $\mathbf{t}$ in the full-resolution forest, then $\tau_{\mathrm{cong}}(\mathbf{t})$ is a triangulation.*

Note that some small regions in $\mathcal{P}_{oct}$ may not contain any tetrahedra congruent to $\mathbf{t}$, so τ may not cover the entire grid.

Proof of Lemma 9.8: Let k be the smallest integer such that

$$2^k + 1 \geq \max(n_1, n_2, n_3), \tag{9.1}$$

and let n equal $2^k + 1$. Let Γ' be an $n \times n \times n$ grid containing Γ with origin $(0, 0, 0)$. Let $\tau'_{\mathrm{cong}}(\mathbf{t})$ be the set of all tetrahedron in the full-resolution tree of Γ' that are congruent to $\mathbf{t}$. By Lemma 9.9, every tetrahedron in the full-resolution forest of Γ is in the full-resolution tree of Γ'. Thus, every tetrahedron in $\tau_{\mathrm{cong}}(\mathbf{t})$ is in $\tau'_{\mathrm{cong}}(\mathbf{t})$. Since $\tau'_{\mathrm{cong}}(\mathbf{t})$ is a triangulation and $\tau_{\mathrm{cong}}(\mathbf{t})$ is a subset of $\tau'_{\mathrm{cong}}(\mathbf{t})$, set $\tau'_{\mathrm{cong}}(\mathbf{t})$ is also a triangulation. □

Lemma 9.9. *Let Γ be an $n_1 \times n_2 \times n_3$ grid, let k be the smallest integer such that $2^k + 1 \geq \max(n_1, n_2, n_3)$, and let n equal $2^k + 1$. Let Γ' be an $n \times n \times n$ grid containing Γ with origin $(0,0,0)$. If τ is the set of all tetrahedra in the full-resolution forest of Γ and τ' is the set of all tetrahedra in the full-resolution tree of Γ', then τ is a subset of τ'.*

Proof: If Γ equals Γ', then τ is a full-resolution tree on Γ and so τ equals τ'. Assume Γ is a subgrid of Γ'.

Let $\mathbf{t}$ be a tetrahedron in τ. Let $\mathbf{c}$ be the cubic region containing $\mathbf{t}$ in the octree partition $\mathcal{P}_{oct}$ of Γ. Tetrahedron $\mathbf{t}$ is contained in some tetrahedron $\mathbf{t}' \in \tau^{\mathbf{c}}$.

Since Γ does not equal Γ', region $\mathbf{c}$ is not at the root of this octree. Thus region $\mathbf{c}$ has some parent, $\mathbf{c}'$, in the octree on Γ'. Region $\mathbf{c}'$ is split into eight octants sharing the point $q_{\text{center}}(\mathbf{c}')$ at the center of $\mathbf{c}'$.

Region $\mathbf{c}'$ is triangulated by six tetrahedra from τ' sharing a common edge. After three bisection steps, these six tetrahedra become forty-eight tetrahedra. All forty-eight of these tetrahedra have a vertex at $q_{\text{center}}(\mathbf{c}')$. Six of these forty-eight tetrahedra form a triangulation of $\mathbf{c}$. Diagonal $d_{\text{split}}(\mathbf{c})$ is the diagonal of $\mathbf{c}$ containing $q_{\text{center}}(\mathbf{c}')$. Since the six tetrahedra in $\mathbf{c}$ share a vertex at $q_{\text{center}}(\mathbf{c}')$, they share a common edge $d_{\text{split}}(\mathbf{c})$. By definition of triangulation $\tau^{\mathbf{c}}$, these six tetrahedra also form the triangulation $\tau^{\mathbf{c}}$ of $\mathbf{c}$. Thus, one of these tetrahedra is $\mathbf{t}'$ and tetrahedron $\mathbf{t}'$ is in τ'. All tetrahedra derived $\mathbf{t}'$ are also in τ'. In particular, tetrahedron $\mathbf{t}$ is in τ'. Thus, τ is a subset of τ'. $\square$

We prove that a sequence of splits that follows the Bisection Rule creates a triangulation.

Proposition 9.10. *Let $\mathcal{P}_{oct}$ be an octree partition of an $n_1 \times n_2 \times n_3$ grid. For each cubic region $\mathbf{c} \in \mathcal{P}_{oct}$, let $\tau^{\mathbf{c}}$ be the triangulation of $\mathbf{c}$ into six tetrahedra sharing the cube diagonal $d_{\text{split}}(\mathbf{c})$. If a sequence of splits of $\tau = \bigcup_{r \in \mathcal{P}_{oct}} \tau^{\mathbf{c}}$ follows the Bisection Rule, then the resulting set τ' of tetrahedra is a triangulation of the grid.*

Proof: Let $\mathbf{t}$ and $\mathbf{t}'$ be tetrahedra in τ such that $\mathbf{t}$ intersects $\mathbf{t}'$. If $\mathbf{t}$ and $\mathbf{t}'$ lie in the same cubic region $\mathbf{c} \in \mathcal{P}_{oct}$, then, by Proposition 9.2 in Section 9.1.2, $\mathbf{t} \cap \mathbf{t}'$ is a vertex, edge, or facet of $\mathbf{t}$ and of $\mathbf{t}'$.

By Lemma 9.8, the set of all tetrahedra congruent to $\mathbf{t}$ form a triangulation. If tetrahedron $\mathbf{t}'$ is congruent to $\mathbf{t}$, then $\mathbf{t}'$ is part of this triangulation and $\mathbf{t} \cap \mathbf{t}'$ is a vertex, edge, or face of $\mathbf{t}$ and of $\mathbf{t}'$.

Assume that $\mathbf{t}$ and $\mathbf{t}'$ lie in different regions of $\mathcal{P}_{oct}$ and that tetrahedron $\mathbf{t}'$ is not congruent to $\mathbf{t}$. Tetrahedron $\mathbf{t}'$ is either contained in some tetrahedron congruent to $\mathbf{t}$ or contains some tetrahedron congruent to $\mathbf{t}$. Without loss of generality, assume that tetrahedron $\mathbf{t}'$ is contained in some tetrahedron congruent to $\mathbf{t}$.

Let $\tau_{\text{cong}}(\mathbf{t})$ be the set of all tetrahedra congruent to $\mathbf{t}$ in the full-resolution forest. We first show that $\mathbf{t}'$ is contained in some tetrahedron in $\tau_{\text{cong}}(\mathbf{t})$. Assume that it is not. Let $\mathbf{c}$ and $\mathbf{c}'$ be the cubic regions in $\mathcal{P}_{oct}$ containing $\mathbf{t}$ and $\mathbf{t}'$, respectively. Consider an octree on the region $\mathbf{c}$. Let $\mathbf{c_t}$ be the smallest cubic region in this octree that contains $\mathbf{t}$. Since $\mathbf{t}'$ is not congruent to $\mathbf{t}$ and not contained in any tetrahedron in $\tau_{\text{cong}}(\mathbf{t})$, region $\mathbf{c}'$ is smaller than region $\mathbf{c_t}$.

Let $\tau^{\mathbf{c_t}}$ be the six tetrahedra in the full-resolution forest of $\mathcal{P}_{oct}$ that form a triangulation of $\mathbf{c_t}$. Tetrahedron $\mathbf{t}$ is not necessarily one of these six tetrahedra. It may be one of the twelve or twenty-four tetrahedra created by splitting an element of $\tau^{\mathbf{c_t}}$ one or two times. However, $\mathbf{t}$ either equals or is contained in one of these six tetrahedron $\tilde{\mathbf{t}}$.

Let k be the smallest integer such that $2^k + 1 \geq \max(n_1, n_2, n_3)$. Let $\mathbf{c}^*$ be a cubic region with edge length 2^k and with opposing vertices at $(0,0,0)$ and $(2^k, 2^k, 2^k)$. Region $\mathbf{c}'$ is contained in some region $\mathbf{c}''$ congruent to $\mathbf{c_t}$ in the octree on $\mathbf{c}^*$. Since tetrahedron $\mathbf{t}$ intersects $\mathbf{t}'$, tetrahedron $\tilde{\mathbf{t}}$ contains $\mathbf{t}$, and region $\mathbf{c}''$ contains $\mathbf{t}'$, tetrahedron $\tilde{\mathbf{t}}$ intersects $\mathbf{c}''$.

Tetrahedron $\tilde{\mathbf{t}}$ intersects $\mathbf{c}''$ at a vertex, edge, or facet of $\tilde{\mathbf{t}}$. If tetrahedron $\tilde{\mathbf{t}}$ intersects $\mathbf{c}''$ at a vertex, then $\mathbf{t} \cap \mathbf{t}'$ is a vertex of $\mathbf{t}$ and of $\mathbf{t}'$. Otherwise, $\tilde{\mathbf{t}}$ and $\mathbf{c}''$ share some edge $\mathbf{e}$. Edge $\mathbf{e}$ is also an edge of $\mathbf{c}$ and therefore an edge of $\mathbf{t}$ as well.

Since $\mathbf{c}'$ is smaller than $\mathbf{c}''$, region $\mathbf{c}''$ must be split into subregions in the octree partition $\mathcal{P}_{oct}$. At least one of these subregions has a vertex on the interior of $\mathbf{e}$. Thus some vertex of a subregion lies on the interior of an edge of $\mathbf{t}$, violating the Bisection Rule for 3D Octree Partitions. We conclude that $\mathbf{t}'$ is contained in some tetrahedron $\mathbf{t}'' \in \tau_{\text{cong}}(\mathbf{t})$.

The rest of the proof proceeds as in Proposition 9.2. Let $\mathbf{t}''$ be the tetrahedron in $\tau_{\text{cong}}(\mathbf{t})$ that contains $\mathbf{t}'$. By Lemma 9.8, $\tau_{\text{cong}}(\mathbf{t})$ is a triangulation. Thus, $\mathbf{t} \cap \mathbf{t}''$ is a face of $\mathbf{t}''$. Since $\mathbf{t} \cap \mathbf{t}'$ is a subset of $\mathbf{t} \cap \mathbf{t}''$, intersection $\mathbf{t} \cap \mathbf{t}'$ is a face of $\mathbf{t}'$.

We need to show that $\mathbf{t} \cap \mathbf{t}'$ is a face of $\mathbf{t}$. Assume that it is not. Since $\mathbf{t} \cap \mathbf{t}''$ is a face of $\mathbf{t}$, tetrahedron $\mathbf{t}'$ is not equal to $\mathbf{t}''$. Thus tetrahedron $\mathbf{t}''$ must be split.

Let $\mathbf{t}_0, \mathbf{t}_1, \ldots, \mathbf{t}_m$ be a sequence of splits where

1. tetrahedron $\mathbf{t}_0$ equals $\mathbf{t}''$,

2. tetrahedron $\mathbf{t}_m$ equals $\mathbf{t}'$,

3. tetrahedron $\mathbf{t}_{i+1}$ is one of the two tetrahedra created by splitting $\mathbf{t}_i$.

Since $\mathbf{t}_0$ equals $\mathbf{t}''$ and $\mathbf{t}_m$ equals $\mathbf{t}'$, it follows that $\mathbf{t}' \subseteq \mathbf{t}_i \subseteq \mathbf{t}''$ for all i.

Let $\mathbf{t}_k$ be the smallest tetrahedron in the sequence such that $\mathbf{t} \cap \mathbf{t}_k$ equals $\mathbf{t} \cap \mathbf{t}''$. Since $\mathbf{t} \cap \mathbf{t}'$ does not equal $\mathbf{t} \cap \mathbf{t}''$, tetrahedron $\mathbf{t}_k$ does not equal $\mathbf{t}_m$. Thus $\mathbf{t}_k$ is split at some point q. Since $\mathbf{t} \cap \mathbf{t}_k$ is face of $\mathbf{t}$ and $\mathbf{t} \cap \mathbf{t}_{k+1}$ is not, point q lies on the interior of an edge of $\mathbf{t}$.

Point q is in the interior of an edge of $\mathbf{t}$ but $\mathbf{t}$ is not split, violating the Bisection Rule. We conclude that $\mathbf{t} \cap \mathbf{t}'$ is a face of $\mathbf{t}$. Since $\mathbf{t} \cap \mathbf{t}'$ is a face of $\mathbf{t}$ and of $\mathbf{t}'$ for every two intersecting tetrahedra $\mathbf{t}, \mathbf{t}' \in \tau'$, set τ' is a triangulation of the grid. $\qquad \square$

We give the proof of Proposition 9.7. For convenience, the proposition is restated here.

Proposition 9.7. *If an $n_1 \times n_2 \times n_3$ multiresolution grid Γ^M satisfies the Multiresolution Conditions, then the set of tetrahedra defined by Γ^M is a triangulation of the region covered by Γ^M.*

Proof: We wish to show that the splits defined by the multiresolution grid satisfy the Bisection Condition for 3D Octree Partitions.

Case I: Tetrahedron $\mathbf{t}$ is split at point $q_{\mathrm{split}}(\mathbf{t})$ and $q_{\mathrm{split}}(\mathbf{t})$ lies in the interior of an edge of tetrahedron $\mathbf{t}'$.

Let $\tau_{\mathrm{cong}}(\mathbf{t})$ be the tetrahedra in the full-resolution forest that are congruent to $\mathbf{t}$. Since $q_{\mathrm{split}}(\mathbf{t})$ lies in the interior of an edge of $\mathbf{t}$, tetrahedron $\mathbf{t}'$ is triangulated by a subset of $\tau_{\mathrm{cong}}(\mathbf{t})$. Some tetrahedron $\mathbf{t}''$ in $\tau_{\mathrm{cong}}(\mathbf{t})$ contains $\mathbf{t} \cap \mathbf{t}'$ and is contained in $\mathbf{t}'$. Since $q_{\mathrm{split}}(\mathbf{t})$ lies in the interior of an edge of $\mathbf{t}$ and $\mathbf{t} \cap \mathbf{t}''$ is an edge or facet of $\mathbf{t}$, point $q_{\mathrm{split}}(\mathbf{t})$ lies in the interior of an edge of $\mathbf{t}'$. Since $\mathbf{t}''$ is congruent to $\mathbf{t}$, point $q_{\mathrm{split}}(\mathbf{t}'')$ equals $q_{\mathrm{split}}(\mathbf{t})$. Since $\mathbf{t}$ is split, $q_{\mathrm{split}}(\mathbf{t})$ is true and so $q_{\mathrm{split}}(\mathbf{t}'') = q_{\mathrm{split}}(\mathbf{t})$ is true. Tetrahedron $\mathbf{t}'$ contains $\mathbf{t}''$. By the Multiresolution Conditions, $q_{\mathrm{split}}(\mathbf{t}')$ is also true. Thus tetrahedron $\mathbf{t}'$ is split, satisfying the first condition of the Bisection Rule.

Case II: Point q is the corner of some region $\mathbf{c}$ in the octree partition of Γ^M and lies in the interior of an edge of tetrahedron $\mathbf{t}'$.

Let $\mathbf{c}'$ be the region in the octree partition containing $\mathbf{t}'$. Since q lies in the interior of an edge of $\mathbf{t}'$, region $\mathbf{c}'$ is larger than region $\mathbf{c}$. Consider the octree partition of $\mathbf{c}'$. Let $\mathbf{c}''$ be the smallest region in this octree partition containing $\mathbf{t}'$.

Let k be the smallest integer such that $2^k + 1 \geq \max(n_1, n_2, n_3)$. Let $\mathbf{c}^*$ be a cubic region with edge length 2^k and with opposing vertices at $(0, 0, 0)$ and $(2^k, 2^k, 2^k)$. Region $\mathbf{c}$ is contained in some region $\tilde{\mathbf{c}}$ congruent to $\mathbf{c}''$ in the octree on $\mathbf{c}^*$. Since point q lies in the interior of an edge $\mathbf{e}$ of $\mathbf{t}'$ and point q lies in $\tilde{\mathbf{c}}$, edge $\mathbf{e}$ lies on the boundary of $\tilde{\mathbf{c}}$.

Since $\mathbf{c}$ is smaller than $\tilde{\mathbf{c}}$, region $\tilde{\mathbf{c}}$ is split into eight octants in constructing the octree partition of Γ^M. Thus the midpoint q' of $\mathbf{e}$ is a corner of some region in the octree partition of $\mathcal{P}_{\mathrm{oct}}$. Point q' may not be $q_{\mathrm{split}}(\mathbf{t}')$ but it is equal to $q_{\mathrm{split}}(\mathbf{t}'')$ for some descendant $\mathbf{t}''$ of $\mathbf{t}'$. By the Multiresolution Conditions, $\Gamma^M(q')$ is true. By the Multiresolution Conditions, since $\Gamma^M(q_{\mathrm{split}}(\mathbf{t}''))$ is true, $\Gamma^M(q_{\mathrm{split}}(\mathbf{t}'))$ is true. Therefore, $\mathbf{t}'$ is split, satisfying the second condition of the Bisection Rule.

Input   : Γ^S is an $n_1 \times n_2 \times n_3$ array of scalar values.
          RegionList is a list of regions in Γ^S.
Output   : Γ^M is an $n_1 \times n_2 \times n_3$ array of Boolean values satisfying the
          Multiresolution Conditions.

CreateMultiresGrid(Γ^S, RegionList, Γ^M)

1 Initialize $\Gamma^M(v)$ to false for all vertices v;
2 foreach region R in RegionList do
3     $k \leftarrow$ R.Subsample;
4     $V \leftarrow$ vertices $(i_x, i_y, i_z) \in \Gamma^S$ where $(i_x \equiv i_y \equiv i_z \equiv 0) \bmod k$;
5   foreach vertex $v \in V$ that lies in R do
6         $\Gamma^M(v) \leftarrow$ true;
7   end
8 end
9 SatisfyMultires (Γ^M);

Algorithm 9.9. CREATEMULTIRESGRID2.

Since the sequence of splits defined by the multiresolution grid satisfy the Bisection Rule, the resulting set of tetrahedra form a triangulation of the grid.□

9.2.6 More on Creating Multiresolution Grids

Procedure CREATEMULTIRESGRID (Algorithm 9.8) from Section 9.2.3 uses full-resolution tetrahedra in the regions listed in RegionList. Instead of using the maximum resolution for every such region, it is possible to specify different resolutions for each such region. For each region R ∈ RegionList, associate a subsampling rate, R.Subsample. This subsample rate should be a nonnegative power of two. Let V be the grid vertices (i_x, i_y, i_z) where i_x, i_y, and i_z are congruent to zero modulo R.Subsample. The vertices of V form a subgrid of the original grid where each element of the subgrid contains k^3 cubes from the original grid. Every vertex of V that lies in R is set to true. Procedure SATISFYMULTIRES (Algorithm 9.6) is applied to the multiresolution grid so that it satisfies the Multiresolution Conditions. Pseudocode for procedure CREATEMULTIRESGRID2 is presented in Algorithm 9.9.

Regions in RegionList may overlap. By setting one of the regions to be the entire region covered by the grid, we can specify a minimum resolution for the entire grid.

9.2.7 Iterative Isosurface Extraction

Algorithm 9.7 makes approximately two recursive calls for each element in the multiresolution triangulation. For low-resolution isosurfaces, this is not a prob-

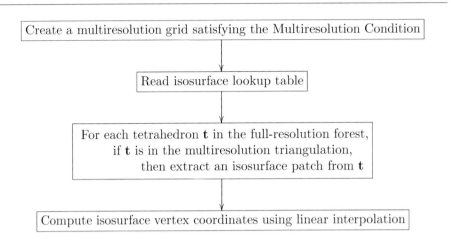

Figure 9.14. MULTIRESOLUTION MARCHING TETRAHEDRA (ITERATIVE) for $n_1 \times n_2 \times n_3$ grids.

lem. However, for full- and near full-resolution isosurfaces, the numerous recursive calls take significant time. We present an iterative version of MULTIRESOLUTION MARCHING TETRAHEDRA, which runs faster than the recursive version when extracting high-resolution isosurfaces.

The first two steps of the iterative algorithm are the same as the first two steps of the recursive algorithm. (See Figure 9.14.) The iterative algorithm creates a multiresolution grid satisfying the Multiresolution Condition and then reads the isosurface lookup table. The difference between the recursive and iterative algorithm is in the third step. Instead of recursively splitting tetrahedra, the iterative algorithm enumerates and processes all tetrahedra in the full-resolution forest. For each tetrahedron **t** in the full-resolution forest, if **t** is in the multiresolution triangulation, the algorithm extract an isosurface patch from **t**. The fourth step of the iterative algorithm, computing isosurface vertex coordinates, is again the same as the fourth step of the recursive algorithm. We explain the third step in more detail. We start by showing how to nonrecursively enumerate tetrahedra in the full-resolution forest.

Let $|\Gamma|$ be the region covered by grid Γ^M. Let $\mathbf{c}^*$ be a cubic region with edge length 2^k containing $|\Gamma|$ and with opposing vertices $(0, 0, 0)$ and $(2^k, 2^k, 2^k)$. Consider the octree of height k on $\mathbf{c}^*$. Every node in this octree represents a cubic region. Each tetrahedron **t** in the full-resolution forest lies inside and shares at least one edge with a cubic region in the octree.

A cubic region **c** divides into six type A tetrahedra sharing a diagonal of **c**. There are four distinct diagonals connecting opposite vertices in a cubic region. These four diagonals correspond to four different subdivisions of the cubic region

---

Input : Γ^S is an $n_1 \times n_2 \times n_3$ array of scalar values.
    Coord is a 3D array of (x, y, z) coordinates.
    RegionList is a list of regions in Γ^S.
    σ is an isovalue.

MultiresMarchingTetIterative(Γ^S, Coord, RegionList, σ)

1 CreateMultiresGrid(Γ^S,RegionList,Γ^M);
2 Read tetrahedron isosurface lookup table into Table;
3 T $\leftarrow \emptyset$;
4 foreach unit cube **c** in Γ^M do
5   foreach tetrahedron **t** $\in \tau_A^{\mathbf{c}}$ do
6   if ($\Gamma^M(q_{\text{prevsplit}}(\mathbf{t})) = $ true) then
7        $(p_0, p_1, p_2, p_3) \leftarrow$ vertices of **t**;
8   ExtractIsoPatch(Γ^S, σ, p_0, p_1, p_2, p_3,T,Table);
9   end
10   end
11 end
12 Let k be the smallest integer such that $2^k + 1 \geq \max(n_1, n_2, n_3)$;
13 for $j = 1$ to k do
    /* $\mathcal{P}_j(\Gamma^M)$ is the partition of subgrid Γ_j^M of Γ^M into cubes of length 2^j
    where Γ_j^M is the largest subgrid of Γ^M containing $(0,0,0)$ that can be
    partitioned into cubes of length 2^j.   */
14   foreach cubic region **c** in $\mathcal{P}_j(\Gamma^M)$ do
15   foreach tetrahedron **t** $\in \tau_A^{\mathbf{c}} \cup \tau_B^{\mathbf{c}} \cup \tau_C^{\mathbf{c}}$ do
16   if ($\Gamma^M(q_{\text{prevsplit}}(\mathbf{t})) = $ true and $\Gamma^M(q_{\text{split}}(\mathbf{t})) = $ false) then
17          $(p_0, p_1, p_2, p_3) \leftarrow$ vertices of **t**;
18   ExtractIsoPatch(Γ^S, σ, p_0, p_1, p_2, p_3,T,Table);
19   end
20   end
21   end
22 end
    /* Compute isosurface vertex coordinates using linear interpolation   */
23 E $\leftarrow \bigcup_{(\mathbf{e}_1, \mathbf{e}_2, \mathbf{e}_3) \in \mathsf{T}} \{\mathbf{e}_1, \mathbf{e}_2, \mathbf{e}_3\}$;
24 foreach edge $e \in$ E with endpoints (i_1, j_1, k_1) and (i_2, j_2, k_2) do
25    $u_e \leftarrow$ LinearInterpolation
26   (Coord$[i_1, j_1, k_1]$, $\Gamma^S[i_1, j_1, k_1]$, Coord$[i_2, j_2, k_2]$, $\Gamma^S[i_2, j_2, k_2]$, σ);
27 end
    /* Convert T to set of triangles   */
28 $\Upsilon \leftarrow \emptyset$;
29 foreach triple of edges $(\mathbf{e}_1, \mathbf{e}_2, \mathbf{e}_3) \in$ T do
30    $\Upsilon \leftarrow \Upsilon \cup \{u_{\mathbf{e}_1}, u_{\mathbf{e}_2}, u_{\mathbf{e}_3}\}$;
31 end

Algorithm 9.10. Multiresolution Marching Tetrahedra (Iterative). for $n_1 \times n_2 \times n_3$ grids

into six type A tetrahedra sharing the diagonal. The tetrahedra in only one of these subdivisions is in the full-resolution forest.

Every region $\mathbf{c}$ other than $\mathbf{c}^*$ is created by splitting a region $\mathbf{c}'$ into eight octants. As in Section 9.2.3, let $d_{\mathrm{split}}(\mathbf{c})$ be the diagonal of $\mathbf{c}$ containing $q_{\mathrm{center}}(\mathbf{c}')$. If $\mathbf{c}$ equals $\mathbf{c}^*$, let $d_{\mathrm{split}}(\mathbf{c})$ be the diagonal containing $(0,0,0)$. Let $\tau_A^{\mathbf{c}}$ be the set of six type A tetrahedra sharing edge $d_{\mathrm{split}}(\mathbf{c})$. The full-resolution forest contains $\tau_A^{\mathbf{c}}$.

Each tetrahedron in $\tau_A^{\mathbf{c}}$ splits into two type B tetrahedra. Let $\tau_B^{\mathbf{c}}$ be the set of twelve type B tetrahedra formed by splitting each tetrahedron in $\tau_A^{\mathbf{c}}$. Let $\tau_C^{\mathbf{c}}$ be the set of twenty-four type C tetrahedra formed by splitting each tetrahedron in $\tau_B^{\mathbf{c}}$. The tetrahedra in the full-resolution forest are enumerated by processing each octree region $\mathbf{c}$ contained in $|\Gamma|$, and generating $\tau_A^{\mathbf{c}}$ and $\tau_B^{\mathbf{c}}$ and $\tau_C^{\mathbf{c}}$.

We enumerate the octree regions contained in $|\Gamma|$ without explicitly constructing the octree. Let Γ_j^M be the largest subgrid of Γ^M containing $(0,0,0)$ that can be partitioned into cubic regions of edge length 2^j. Let $\mathcal{P}_j(\Gamma^M)$ be the partition of Γ_j^M into cubic regions of edge length 2^j. These cubic regions are the octree regions that are contained in $|\Gamma|$.

We determine if a tetrahedron from the full-resolution forest is in the multiresolution triangulation as follows. Let Γ^M be a multiresolution grid satisfying the Multiresolution Condition. Let $\mathbf{t}$ be a tetrahedron in the full-resolution forest on grid Γ^M. If $\mathbf{t}$ is the child of a tetrahedron $\mathbf{t}'$, then $\mathbf{t}$ is in the multiresolution mesh if $\mathbf{t}'$ is split and $\mathbf{t}$ is not. Equivalently, $\mathbf{t}$ is in the multiresolution mesh if $\Gamma^M(q_{\mathrm{split}}(\mathbf{t}'))$ is true and $\Gamma^M(q_{\mathrm{split}}(\mathbf{t}))$ is false. If $\mathbf{t}$ is at a root node, then $\mathbf{t}$ is in the multiresolution mesh if $\mathbf{t}$ is not split, i.e., $\Gamma^M(q_{\mathrm{split}}(\mathbf{t}))$ is false.

We can determine $q_{\mathrm{split}}(\mathbf{t}')$ without explicitly constructing $\mathbf{t}'$. Let $\mathbf{c}$ be the cubic region sharing at least one edge with $\mathbf{t}$. Let 2^j be the edge length of $\mathbf{c}$. If $\mathbf{t}$ has type B, let $q_{\mathrm{prevsplit}}(\mathbf{t})$ be the vertex of $\mathbf{t}$ at the center of $\mathbf{c}$. If $\mathbf{t}$ has type C, let $q_{\mathrm{prevsplit}}(\mathbf{t})$ be the vertex of $\mathbf{t}$ at the center of some facet of $\mathbf{c}$. If $\mathbf{t}$ has type A and $\mathbf{c}$ is the child of region $\mathbf{c}'$ in an octree, let $q_{\mathrm{prevsplit}}(\mathbf{t})$ be the vertex of $\mathbf{t}$ at the midpoint of an edge of $\mathbf{c}'$. Equivalently, if $\mathbf{t}$ has type A and $\mathbf{c}$ is contained in a $2^{j+1} \times 2^{j+1} \times 2^{j+1}$ region $\mathbf{c}'$ in the partition $P_{j+1}(G)$, then $q_{\mathrm{prevsplit}}(\mathbf{t})$ is the vertex of $\mathbf{t}$ at the midpoint of an edge of $\mathbf{c}'$. If $\mathbf{t}$ has type A and $\mathbf{c}$ is at the root of an octree, let $q_{\mathrm{prevsplit}}(\mathbf{t})$ be the point the endpoint of $d_{\mathrm{split}}(\mathbf{c})$ furthest from the origin. In all cases, if $\mathbf{t}$ is the child of tetrahedron $\mathbf{t}'$, then $q_{\mathrm{prevsplit}}(\mathbf{t})$ equals $q_{\mathrm{split}}(\mathbf{t}')$.

The third step in the iterative algorithm is as follows. Let Γ^M be the $n_1 \times n_2 \times n_3$ multiresolution grid constructed in the first step. Let k be the smallest integer such that $2^k + 1 \geq \max(n_1, n_2, n_3)$. For each unit cube $\mathbf{c}$ in Γ^M and each type A tetrahedron $\mathbf{t}$ sharing edge $d_{\mathrm{split}}(\mathbf{c})$, if $\Gamma^M(q_{\mathrm{prevsplit}}(\mathbf{t}))$ is true, then extract an isosurface patch from $\mathbf{t}$. For j equals one to k, let Γ_j^M be the largest scalar grid of Γ^M containing $(0,0,0)$ that can be partitioned into cubic regions of edge length 2^j. Let $\mathcal{P}_j(\Gamma^M)$ be the partition of Γ_j^M into cubic regions of edge length 2^j. For each cubic region $\mathbf{c}$ in $\mathcal{P}_j(\Gamma^M)$, construct $\tau_A^{\mathbf{c}} \cup \tau_B^{\mathbf{c}} \cup \tau_C^{\mathbf{c}}$. For each tetrahedron $\mathbf{t}$ in $\tau_A^{\mathbf{c}} \cup \tau_B^{\mathbf{c}} \cup \tau_C^{\mathbf{c}}$, if $\Gamma^M(q_{\text{prevsplit}}(\mathbf{t}))$ is true and $\Gamma^M(q_{\text{split}}(\mathbf{t}))$ is true, then extract the isosurface patch from $\mathbf{t}$. Pseudocode is given in Algorithm 9.10.

MULTIRESOLUTION MARCHING TETRAHEDRA (ITERATIVE) produces the exact same sets of triangles and vertices as MULTIRESOLUTION MARCHING TETRAHEDRA (RECURSIVE). Thus, all the properties in Section 2.4.2 apply to an isosurface produced by MULTIRESOLUTION MARCHING TETRAHEDRA (ITERATIVE).

9.3 Notes and Comments

Initial work on multiresolution isosurfaces partitioned the domain into cubes or rectangular regions of varying sizes. Because adjacent cubes did not necessarily have the same size, the partition did not form a convex polyhedral mesh. Isosurface patches in different cubes did not necessarily align on their boundaries, creating "cracks" in the isosurface. These cracks were eliminated either by adding polygons to fill the cracks [Shu et al., 1995] or by moving isosurface vertices [Müller and Stark, 1993, Shekhar et al., 1996, He et al., 1996] to align the patch boundaries. An alternative approach [Bloomenthal, 1988, Poston et al., 1998, Westermann et al., 1999, Bai et al., 2006] first constructs the isosurface patch boundaries, ensuring that adjacent boundaries align, and then adds isosurface triangles within the boundaries.

Cignoni et al. [Cignoni et al., 1994] extract isosurfaces from multiresolution tetrahedral meshes where the mesh is a Delaunay triangulation at every level of resolution. Chiang and Lu [Chiang and Lu, 2003] give an algorithm for simplifying tetrahedral meshes while ensuring that isosurface topology is preserved.

The recursive tetrahedral bisection in Section 9.1.2 and the resulting multiresolution grid was described by Maubach in [Maubach, 1995]. Zhou et al. [Zhou et al., 1997] were the first to apply this subdivision to multiresolution isosurface extraction. Gerstner and Pajarola in [Gerstner and Pajarola, 2000] present a simple recursive algorithm for traversing the tetrahedral mesh that is given in Section 9.2.2. Gerstner and Rumpf [Gerstner and Rumpf, 2000] give criteria for choosing the amount of refinement at a location in the isosurface. They also discuss load balancing when extracting a multiresolution isosurface using multiple processors.

Pascucci and Bajaj [Pascucci and Bajaj, 2000] use multiresolution meshes for fast interactive rendering of isosurfaces. Weiss and De Floriani [Weiss and De Floriani, 2008] and [Weiss and De Floriani, 2010] use recursive tetrahedral bisection to produce compact representations of isosurfaces and interval volumes. Gerstner [Gerstner, 2002] shows how to quickly sort the multiresolution mesh produced from recursive tetrahedral bisection in a view dependent ordering. This ordering can be used to efficiently render transparent isosurfaces using alpha blending.

Balmelli et al. [Balmelli et al., 2002] warp the regular grid to concentrate grid cubes in regions of interest. The isosurface extracted from this warped grid has higher resolution in regions with high concentrations of grid cubes.

Bischoff and Kobbelt [Bischoff and Kobbelt, 2002] simplify isosurface topology by using morphological operators such as erosion and dilation. Szymczak and Vanderhyde [Szymczak and Vanderhyde, 2003] simplify topology by starting with a crude genus-0 approximation to the isosurface and adding tunnels. Wood et al. [Wood et al., 2004] start with a full-resolution isosurface and fill in tunnels and handles by modifying scalar values in the underlying grid.

Surveys on mesh decimation for polygonal meshes can be found in [Cignoni et al., 1998, Garland, 1999, Luebke, 2001, Iske et al., 2002]. An algorithm that simultaneously constructs and decimates an isosurface is presented in [Attali et al., 2005].

CHAPTER 10

MULTIRESOLUTION POLYHEDRAL MESHES

In the previous chapter, we presented an algorithm for constructing an isosurface using a multiresolution tetrahedral mesh. This algorithm requires the entire mesh to be broken into tetrahedra, even if most of the mesh is uniform and only a small portion is at high resolution.

Tetrahedral meshes have two drawbacks compared with regular grids of cubes. First, the isosurface created using tetrahedral meshes tends to be less smooth and to contain more visual artifacts compared with the isosurface from a regular grid. Second, a full-resolution tetrahedral mesh creates an isosurface with about three times the number of triangles created by MARCHING CUBES. Unless the multiresolution isosurface has large areas at low resolution, the multiresolution isosurface created from a tetrahedral mesh can have more triangles than the full-resolution isosurface created by MARCHING CUBES. When this happens, one is better off using MARCHING CUBES instead of a multiresolution tetrahedral mesh.

In this chapter, we present an algorithm to construct a mesh of cubes, pyramids, and tetrahedra and to extract the isosurface from that mesh. The uniform regions in the mesh are covered by cubes with pyramids and tetrahedra only used to connect regions at different resolution.

The SURFACE NETS algorithm in Chapter 3 generalizes easily to construct multiresolution isosurfaces from a multiresolution partition of the grid into cubic regions. The multiresolution algorithm is presented in Section 10.2. Section 10.3 gives algorithms for multiresolution isosurfaces in 4D.

10.1   Multiresolution Convex Polyhedral Mesh

In this section, we show how to create multiresolution meshes that include cubic grid elements. Pyramid and tetrahedral elements are used to connect cubes of different resolutions.

10.1.1   Subdivisions of Cubes, Pyramids, and Tetrahedra

Just as the multiresolution tetrahedral mesh is based on bisections of tetrahedra, the multiresolution mesh of cubes, pyramids, and tetrahedra is based on subdivisions of these polyhedra.

A cube is divided in one of two ways. A cube is sometimes divided into eight congruent subcubes sharing a vertex at the cube center. A cube can also be divided into six congruent square pyramids whose apices are the cube center and whose bases are the six square cube facets. (A square pyramid is a pyramid whose base is a square.)

A square pyramid is also divided in two ways. A square pyramid can be divided into four square subpyramids by splitting its square base into four squares and connecting its apex to each of the four squares. It can also be divided into four tetrahedra by splitting its square base into four triangles and connecting its apex to each of the four triangles.

If a cube is divided into six square pyramids and these square pyramids are divided into tetrahedra, then the resulting tetrahedra are similar to the type C tetrahedra in Figure 9.8 in the multiresolution tetrahedra. These tetrahedra share one edge with the original cube. They are bisected at the midpoint of this edge into two smaller type A tetrahedra.

Note that the angle between a square pyramid or tetrahedral edge and a coordinate axis is zero, forty-five, or ninety degrees.

Creation of the multiresolution mesh starts by creating an octree partition as described in Section 9.2.3. Each region in this octree partition is a cube. Each cubic region in the octree partition either is not subdivided, is subdivided into eight subcubes, or is subdivided into six square pyramids. Similarly any cubic region created through subdivision is either not subdivided or subdivided into eight subcubes or subdivided into six square pyramids.

Each square pyramid either is not subdivided, is subdivided into four square subpyramids, or is subdivided into four type C tetrahedra. Subpyramids of a square pyramid are never subdivided. Each type C tetrahedron either is not subdivided or is subdivided into two type A tetrahedra. Type A tetrahedra are never subdivided.

A convex polyhedral mesh is a set of convex polytopes such that the intersection of any two polytopes is a face of each. (See Appendix B.5.) The recursive application of the subdivisions listed above creates a set of cubes, square pyramids, and tetrahedra that partition the grid. To ensure that these polyhedra form a convex polyhedral mesh, one could formulate a list of rules similar to the Bisection Rules in Section 9.2.5. However, such a list is fairly long and complicated. Instead, we define a multiresolution grid to determine when and how the polyhedra are subdivided and show that the resulting polyhedral subdivision is a convex polyhedral mesh.

10.1.2 Multiresolution Grids for a Convex Polyhedral Mesh

As with tetrahedral meshes, we rely upon a multiresolution grid, Γ^M, to specify the multiresolution convex polyhedral mesh. However, the multiresolution grid for the polyhedral mesh indicates not only whether to split a mesh element but also the split type.

For each vertex q of grid Γ^M, grid element $\Gamma^M(q)$ has a Boolean value, $\Gamma^M(q).b$, and a split type, $\Gamma^M(q).t$. We say that $\Gamma^M(q)$ is true or false or has type X or Y to mean that $\Gamma^M(q).b$ is true or false or that $\Gamma^M(q).t$ equals X or Y.

For each polyhedron ψ, we define a point $q_{\mathrm{split}}(\psi)$ that determines if and how ψ is split into smaller polyhedra. For a cube $\mathbf{c}$, point $q_{\mathrm{split}}(\mathbf{c})$ is the center of $\mathbf{c}$. For a square pyramid $\mathbf{p}$, point $q_{\mathrm{split}}(\mathbf{p})$ is the center of the square base of $\mathbf{p}$. For a tetrahedron $\mathbf{t}_C$ of type C, point $q_{\mathrm{split}}(\mathbf{t}_C)$ is the midpoint of the edge that $\mathbf{t}_C$ shares with the cubic region containing $\mathbf{t}_C$. A tetrahedron $\mathbf{t}_A$ of type A is never split and $q_{\mathrm{split}}(\mathbf{t}_A)$ is undefined.

Point $q_{\mathrm{split}}(\psi)$ is always the center of some cubic region or the center of some face of a cubic region. Because these cubic regions have edge length 2^j, point $q_{\mathrm{split}}(\psi)$ is a grid vertex whenever j is greater than zero. Point $q_{\mathrm{split}}(\psi)$ is not a grid vertex only if ψ is a unit cube or if ψ is contained in a unit cube. In all such cases, ψ is never subdivided and $q_{\mathrm{split}}(\psi)$ is irrelevant.

Let q equal $q_{\mathrm{split}}(\psi)$ for mesh element ψ where ψ is not a unit cube and is not contained in a unit cube. If $\Gamma^M(q).b$ is true, then ψ is split. If ψ is a cube, then ψ is split into eight subcubes whenever $\Gamma^M(q).t$ is X and is split into six pyramids whenever $\Gamma^M(q).t$ is Y. If ψ is a pyramid, then ψ is split into four subpyramids whenever $\Gamma^M(q).t$ is X and is split into four tetrahedra whenever $\Gamma^M(q).t$ is Y. If ψ is a tetrahedron and $\Gamma^M(q).b$ is true, then $\Gamma^M(q).t$ is always Y and ψ is split into two tetrahedra. If $\Gamma^M(q).b$ is false, then ψ is not split and $\Gamma^M(q).t$ is undefined.

The cubic subdivisions of a cubic region form a tree. The root of the tree represents the original cubic region. Each node in the tree represents a cubic subregion $\mathbf{c}$ with edge length 2^j. If j is positive, then the node has eight children, one for each subcube of $\mathbf{c}$. If j is zero, then the node is a leaf. The cubic subdivisions of an octree partition form a **cube forest**. The root of each tree in this forest is a cubic region in the octree partition. Each tree represents the cubic subdivisions of its corresponding region.

Input : Γ^M is an $n_1 \times n_2 \times n_3$ array of Boolean values.

`SatisfyMultiresPoly($\Gamma^M$)`

1 $\mathcal{P}_{oct} \leftarrow$ octree partition of Γ^M;
2 `SatisfyCondition1($\Gamma^M$, $\mathcal{P}_{oct}$);`
3 Let k be the smallest integer such that $2^k + 1 \geq \max(n_1, n_2, n_3)$;
4 for $j = 1$ to k do
5   `SatisfyCondition2($\Gamma^M$, $j$);`
6   `SatisfyCondition3($\Gamma^M$, $j$);`
7   `SatisfyCondition4($\Gamma^M$, $j$);`
8 end

Algorithm 10.1. Procedure SATISFYMULTIRESPOLY for multiresolution mesh of cubes, pyramids, and tetrahedra.

For each square **f** and cube **c**, let $q_{center}(\mathbf{f})$ and $q_{center}(\mathbf{c})$ be the centers of **f** and **c**, respectively. For each line segment **e**, let $q_{center}(\mathbf{e})$ be the midpoint of **e**.

Multiresolution Conditions (Convex Polyhedral Mesh). *Let* **c** *be a cubic region at an internal node in the cube forest.*

1. *For every corner q of a region in the octree partition, $\Gamma^M(q).b$ is true and has type X.*

2. *Let* **f** *be a facet of* **c** *and let* **e** *be an edge of* **f**.

   (a) *If $\Gamma^M(q_{center}(\mathbf{c}))$ is true and has type X, then $\Gamma^M(q_{center}(\mathbf{f})).b$ is true and has type X.*

   (b) *If $\Gamma^M(q_{center}(\mathbf{f}))$ is true and has type X, then $\Gamma^M(q_{center}(\mathbf{e})).b$ is true.*

3. *Let* **f** *be a facet of* **c** *and let* **e** *be an edge of* **f**.

   (a) *If $\Gamma^M(q_{center}(\mathbf{e})).b$ is true, then $\Gamma^M(q_{center}(\mathbf{f})).b$ is true.*

   (b) *If $\Gamma^M(q_{center}(\mathbf{f})).b$ is true, then $\Gamma^M(q_{center}(\mathbf{c})).b$ is true.*

4. *If* **c** *is a child of region* **c'** *in the cube forest, and if $\Gamma^M(q_{center}(\mathbf{c})).b$ is true, then $\Gamma^M(q_{center}(\mathbf{c'}))$ is true and has type X.*

We claim that a multiresolution grid that satisfies the Multiresolution Conditions defines a multiresolution convex polyhedral mesh.

Proposition 10.1. *If an $n_1 \times n_2 \times n_3$ multiresolution grid satisfies the Multiresolution Conditions, then the set of cubes, pyramids, and tetrahedra determined by that grid form a convex polyhedral mesh.*

Proof is in Section 10.1.5.

Input : Γ^M is an $n_1 \times n_2 \times n_3$ array of Boolean values.
2^j is the length of the cubic regions begin processed.

SatisfyCondition1(Γ^M, j)

1   foreach region $\mathbf{c}$ in $\mathcal{P}_j(\Gamma^M)$ do
2   foreach corner v of $\mathbf{c}$ do
3       $\Gamma^M(v).b \leftarrow$ true;
4       $\Gamma^M(v).t \leftarrow X$;
5   end
6   end

Algorithm 10.2. SATISFYCONDITION1.

Input : Γ^M is an $n_1 \times n_2 \times n_3$ array of Boolean values.
2^j is the length of the cubic regions begin processed.

SatisfyCondition2(Γ^M, j)

1   foreach region $\mathbf{c}$ in $\mathcal{P}_j(\Gamma^M)$ do
2   if $(\Gamma^M(q_{\text{center}}(\mathbf{c})).b =$ true and $\Gamma^M(q_{\text{center}}(\mathbf{c})).t = X)$ then
3   foreach facet $\mathbf{f}$ of $\mathbf{c}$ do
4         $\Gamma^M(q_{\text{center}}(\mathbf{f})).b \leftarrow$ true;
5         $\Gamma^M(q_{\text{center}}(\mathbf{f})).t \leftarrow X$;
6   end
7   end
8   end
9   foreach facet $\mathbf{f}$ in $\mathcal{P}_j(\Gamma^M)$ do
10   if $(\Gamma^M(q_{\text{center}}(\mathbf{f})).b =$ true and $\Gamma^M(q_{\text{center}}(\mathbf{f})).t = X)$ then
11   foreach edge $\mathbf{e}$ of $\mathbf{f}$ do
12         $\Gamma^M(q_{\text{center}}(\mathbf{e})).b \leftarrow$ true;
13   end
14   end
15   end

Algorithm 10.3. SATISFYCONDITION2.

As in Section 9.2.3, we wish to turn a grid of Boolean values into a multiresolution grid satisfying the Multiresolution Conditions for convex polyhedral meshes. We present procedure SATISFYMULTIRESPOLY for doing so. (See Algorithm 10.1.)

Procedure SATISFYMULTIRESPOLY processes cubic regions in the cube forest of Γ^M, modifying $\Gamma^M(v)$ for vertices v at the center of region faces to satisfy each

Input   : Γ^M is an $n_1 \times n_2 \times n_3$ array of Boolean values.
          2^j is the length of the cubic regions begin processed.

`SatisfyCondition3`(Γ^M, j)

1 foreach facet **f** in $\mathcal{P}_j(\Gamma^M)$ do
2 | foreach edge **e** of **f** do
3 | | if $(\Gamma^M(q_{\text{center}}(\mathbf{e})).b = \text{true}$ and $\Gamma^M(q_{\text{center}}(\mathbf{f})).b = \text{false})$ then
4 | | | $\Gamma^M(q_{\text{center}}(\mathbf{f})).b \leftarrow \text{true};$
5 | | | $\Gamma^M(q_{\text{center}}(\mathbf{f})).t \leftarrow Y;$
6 | | end
7 | end
8 end
9 foreach region **c** in $\mathcal{P}_j(\Gamma^M)$ do
10 | foreach facet **f** of **c** do
11 | | if $(\Gamma^M(q_{\text{center}}(\mathbf{f})).b = \text{true}$ and $\Gamma^M(q_{\text{center}}(\mathbf{c})).b = \text{false})$ then
12 | | | $\Gamma^M(q_{\text{center}}(\mathbf{c})).b \leftarrow \text{true};$
13 | | | $\Gamma^M(q_{\text{center}}(\mathbf{c})).t \leftarrow Y;$
14 | | end
15 | end
16 end

Algorithm 10.4. SATISFYCONDITION3.

of the Multiresolution Conditions. As in Section 9.2.2, let Γ_j^M be the largest subgrid of Γ^M containing $(0,0,0)$ that can be partitioned into cubic regions of edge length 2^j. Let $\mathcal{P}_j(\Gamma^M)$ be the partition of subgrid Γ_j^M into cubic regions of edge length 2^j.

For each corner v of a region in the octree partition, set $\Gamma^M(v)$ to true and type X satisfying Multiresolution Condition 1 (Algorithm 10.2). For each cubic region **c** in $\mathcal{P}_j(\Gamma^M)$, if $\Gamma^M(q_{\text{center}}(\mathbf{c}))$ is true and has type X, then set $\Gamma^M(q_{\text{center}}(\mathbf{f}))$ to true and type X for each facet **f** of **c** satisfying Multiresolution Condition 2a (Algorithm 10.3). For each facet **f** in $\mathcal{P}_j(\Gamma^M)$, if $\Gamma^M(q_{\text{center}}(\mathbf{f}))$ is true and has type X, then set $\Gamma^M(q_{\text{center}}(\mathbf{e}))$ to true for each edge **e** of **f** satisfying Multiresolution Condition 2b.

For each facet **f** in $\mathcal{P}_j(\Gamma^M)$ and each edge **e** of **f**, if $\Gamma^M(q_{\text{center}}(\mathbf{e})).b$ is true and $\Gamma^M(q_{\text{center}}(\mathbf{f})).b$ is false, then set $\Gamma^M(q_{\text{center}}(\mathbf{f}))$ to true and type Y satisfying Multiresolution Condition 3a (Algorithm 10.4). For each cubic region **c** in $\mathcal{P}_j(\Gamma^M)$ and each facet **f** of **c**, if $\Gamma^M(q_{\text{center}}(\mathbf{f})).b$ is true and $\Gamma^M(q_{\text{center}}(\mathbf{c})).b$ is false, then set $\Gamma^M(q_{\text{center}}(\mathbf{c}))$ to true and type Y satisfying Multiresolution Condition 3b.

Input : Γ^M is an $n_1 \times n_2 \times n_3$ array of Boolean values.
2^j is the length of the cubic regions begin processed.

`SatisfyCondition4`(Γ^M, j)

1 foreach region **c** in $\mathcal{P}_j(\Gamma^M)$ do
2   if $\Gamma^M(q_{\text{center}}(\mathbf{c})).b = \text{true}$ then
3   if **c** is one of the 8 subcubes of a cubic region $\mathbf{c}' \in \mathcal{P}_{j+1}(\Gamma^M)$ then
4        $\Gamma^M(q_{\text{center}}(\mathbf{c}')).b \leftarrow \text{true}$;
5        $\Gamma^M(q_{\text{center}}(\mathbf{c}')).t \leftarrow X$;
6   end
7   end
8 end

Algorithm 10.5. SATISFYCONDITION4.

Input : Γ^M is an $n_1 \times n_2 \times n_3$ array of Boolean values satisfying the Multiresolution Conditions.

`IdentifyX`(Γ^M)

1 Let k be the smallest integer such that $2^k + 1 \geq \max(n_1, n_2, n_3)$;
2 for $j = 1$ to k do
3   foreach facet **f** in $\mathcal{P}_j(\Gamma^M)$ do
4   if $\Gamma^M(q_{\text{center}}(\mathbf{e}).b$ is true for every edge **e** of **f** then
5        $\Gamma^M(q_{\text{center}}(\mathbf{f})).t \leftarrow X$;
6   end
7   end
8   foreach cubic region **c** in $\mathcal{P}_j(\Gamma^M)$ do
9   if $\Gamma^M(q_{\text{center}}(\mathbf{f})).b$ is true for every facet **f** of **c** then
10       $\Gamma^M(q_{\text{center}}(\mathbf{c})).t \leftarrow X$;
11   end
12   end
13 end

Algorithm 10.6. IDENTIFYX.

Finally, for each cubic region **c** in $\mathcal{P}_j(\Gamma^M)$, if **c** is one of the eight subcubes of a cubic region $\mathbf{c}'$ in $\mathcal{P}_{j+1}(\Gamma^M)$ and $\Gamma^M(q_{\text{center}}(\mathbf{c})).b$ is true, then set $\Gamma^M(q_{\text{center}}(\mathbf{c}'))$ to true and type X satisfying Multiresolution Condition 4 (Algorithm 10.5).

The order in which these operations are applied is important. Smaller cubic regions should be processed before larger ones. The operations to satisfy con-

---

Input   : Γ^S is an $n_1 \times n_2 \times n_3$ grid.
         RegionList is a list of regions in Γ^S.
Output   : Γ^M is an $n_1 \times n_2 \times n_3$ array of Boolean values satisfying the
         Multiresolution Condition.

`CreateMultiresPolyGrid`(Γ^S, **RegionList**, Γ^M)

1 Initialize $\Gamma^M(v)$ to false for all vertices v;
2 foreach region **R** in **RegionList** do
3   foreach vertex v in **R** do
4        $\Gamma^M(v) \leftarrow$ true;
5   end
6 end
7 `SatisfyMultiresPoly` (Γ^M);
8 `IdentifyX` (Γ^M);

Algorithm 10.7. CREATEMULTIRESPOLYGRID.

ditions 2a and 2b should be applied to all cubic regions of a given size before operations to satisfy conditions 3a and 3b.

While procedure SATISFYMULTIRESPOLY (Algorithm 10.1) solves the problem of satisfying the Multiresolution Conditions, we present an additional procedure, IDENTIFYX, that increases the number of type X vertices in a multiresolution grid. (See Algorithm 10.6.) For each facet $\mathbf{f}$ of $\mathcal{P}_j(\Gamma^M)$, if $\Gamma^M(q_{\mathrm{center}}(\mathbf{e})).b$ is true for each edge $\mathbf{e}$ of $\mathbf{f}$, set $\Gamma^M(q_{\mathrm{center}}(\mathbf{f})).t$ to type X. For each cubic region $\mathbf{c}$ of $\mathcal{P}_j(\Gamma^M)$, if $\Gamma^M(q_{\mathrm{center}}(\mathbf{f})).b$ is true for each facet $\mathbf{f}$ of $\mathbf{c}$, then set $\Gamma^M(q_{\mathrm{center}}(\mathbf{c})).t$ to type X. After these two operations are applied to all facets and all cubic regions, Multiresolution Conditions 2a and 2b are still satisfied. Procedure IDENTIFYX causes subdivisions of cubes into subcubes to be preferred to subdivisions into pyramids and subdivisions of pyramids into subpyramids to be preferred to subdivisions into tetrahedra.

A cube $\mathbf{c}$ divides into only six pyramids. However, if $\Gamma^M(q_{\mathrm{center}}(\mathbf{f})).b$ is true for each facet of $\mathbf{c}$, then each pyramid will be subdivided into four subpyramids or tetrahedra. Similarly, a pyramid $\mathbf{p}$ with square base $\mathbf{f}$ divides into only four tetrahedra, but, if $\Gamma^M(q_{\mathrm{center}}(\mathbf{e})).b$ is true for each edge of $\mathbf{f}$, then each tetrahedron will be split into two. Thus, procedure IDENTIFYX reduces the number of elements in the polyhedral subdivision.

10.1.3 Isosurface Extraction

The algorithm for extracting an isosurface from the multiresolution mesh of cubes, pyramids, and tetrahedra is called MULTIRESOLUTION MARCHING CPT (Algorithm 10.8). CPT stands for cubes, pyramids, and tetrahedra. The

---

Input : Γ^S is an $n_1 \times n_2 \times n_3$ array of scalar values.
Coord is a 3D array of (x, y, z) coordinates.
RegionList is a list of regions in Γ^S.
σ is an isovalue.

MultiresMarchingCPT(Γ^S, Coord, RegionList, σ)

1 CreateMultiresPolyGrid(Γ^S, RegionList, Γ^M);
2 Read cube lookup table into Table$_C$;
3 Read pyramid lookup table into Table$_P$;
4 Read tetrahedron lookup table into Table$_T$;
5 T $\leftarrow \emptyset$;
6 $\mathcal{P}_{oct} \leftarrow$ octree partition of Γ^S;
7 foreach cubic region $\mathbf{c} \in \mathcal{P}_{oct}$ do
8 | MultiresCube(Γ^S, Γ^M, σ, $\mathbf{c}$, T, Table$_C$, Table$_P$, Table$_T$);
9 end
   /* *Compute isosurface vertex coordinates using linear interpolation* */
10 E $\leftarrow \bigcup_{(\mathbf{e}_1, \mathbf{e}_2, \mathbf{e}_3) \in \mathsf{T}} \{\mathbf{e}_1, \mathbf{e}_2, \mathbf{e}_3\}$;
11 foreach edge $e \in$ E with endpoints (i_1, j_1, k_1) and (i_2, j_2, k_2) do
12 | $u_e \leftarrow$ LinearInterpolation
13 | (Coord$[i_1, j_1, k_1]$, $\Gamma^S[i_1, j_1, k_1]$, Coord$[i_2, j_2, k_2]$, $\Gamma^S[i_2, j_2, k_2]$, σ);
14 end
   /* *Convert T to set of triangles* */
15 $\Upsilon \leftarrow \emptyset$;
16 foreach triple of edges $(\mathbf{e}_1, \mathbf{e}_2, \mathbf{e}_3) \in$ T do
17 | $\Upsilon \leftarrow \Upsilon \cup \{u_{\mathbf{e}_1}, u_{\mathbf{e}_2}, u_{\mathbf{e}_3}\}$;
18 end

Algorithm 10.8. MULTIRESOLUTION MARCHING CPT.

algorithm is similar to MULTIRESOLUTION MARCHING TETRAHEDRA (Algorithm 9.7).

Input to MULTIRESOLUTION MARCHING CPT is a scalar grid, Γ^S; an array, Coord, of grid coordinates; a set, RegionList, of high-resolution regions; and a scalar value σ. Start by converting the high-resolution regions to a multiresolution grid, Γ^M (Algorithm 10.7). Set every vertex within the high-resolution regions to true. Set all other vertices to false. Apply procedure SATISFYMULTIRESPOLY (Algorithm 10.1) to the multiresolution grid so that it satisfies the Multiresolution Conditions for multiresolution meshes of cubes, pyramids, and tetrahedra. Apply procedure IDENTIFYX (Algorithm 10.6) to reduce the number of polyhedral elements in the multiresolution mesh.

MULTIRESOLUTION MARCHING CPT requires three isosurface lookup tables: one for cubes, one for pyramids, and one for tetrahedra. Read the three isosurface lookup tables from preconstructed data files.

Input　：Γ^S is an $n_1 \times n_2 \times n_3$ array of scalar values.
　　　　　Γ^M is an $n_1 \times n_2 \times n_3$ array of Boolean values
　　　　　　　satisfying the Multiresolution Condition.
　　　　　σ is an isovalue.
　　　　　$\mathbf{c}$ is a cubic region whose edges lengths are a power of 2.
　　　　　T is a set of edge triples representing isosurface triangles.
　　　　　Tbl_C is the isosurface lookup table for cubes.
　　　　　Tbl_P is the isosurface lookup table for pyramids.
　　　　　Tbl_T is the isosurface lookup table for tetrahedra.

`MultiresCube`(Γ^S, Γ^M, σ, $\mathbf{c}$, T, Tbl_C, Tbl_P, Tbl_T)

1　$L \leftarrow$ length of the edges of $\mathbf{c}$;
2　if $(L = 1$ or $\Gamma^M(q_{\mathrm{center}}(\mathbf{c})).b =$ false$)$ then
3　　│　`ExtractIsoFromCube`(Γ^S, σ, $\mathbf{c}$, T, Tbl_C);
4　else
5　　│　if $\Gamma^M(q_{\mathrm{center}}(\mathbf{c})).t = X)$ then
6　　│　　│　foreach vertex v of cubic region $\mathbf{c}$ do
7　　│　　│　　│　$\mathbf{c}' \leftarrow$ cubic region with opposite vertices v and $q_{\mathrm{center}}(\mathbf{c})$;
8　　│　　│　　│　`MultiresCube`(Γ^S, Γ^M, σ, $\mathbf{c}'$, T, Tbl_C, Tbl_P, Tbl_T) ;
9　　│　　│　end
10　│　else
　　│　　│　/* $\Gamma^M(q_{\mathrm{center}}(\mathbf{c})).t = Y$　　　　　　　　　　　　　*/
11　│　　│　foreach facet $\mathbf{f}$ of cubic region $\mathbf{c}$ do
12　│　　│　　│　$p \leftarrow$ pyramid with base $\mathbf{f}$ and apex $q_{\mathrm{center}}(\mathbf{c})$;
13　│　　│　　│　`MultiresPyramid`(Γ^S, Γ^M, σ, $\mathbf{p}$, T, Tbl_P, Tbl_T);
14　│　　│　end
15　│　end
16　end

Algorithm 10.9. MULTIRES CUBE.

To extract the isosurface, create an octree partition $\mathcal{P}_{oct}$ of Γ^S. For each cubic region $\mathbf{c}$ in $\mathcal{P}_{oct}$, call procedure MULTIRES CUBE (Algorithm 10.9).

Input to procedure MULTIRES CUBE is the scalar grid, Γ^S; the multiresolution grid Γ^M; the isovalue σ; a cubic region $\mathbf{c}$; and the three isosurface lookup tables for cubes, pyramids, and tetrahedra. If $\mathbf{c}$ is a $1 \times 1 \times 1$ unit cube or $\Gamma^M(q_{\mathrm{center}}(\mathbf{c})).b$ is false, then call procedure EXTRACTISOFROMCUBE (Algorithm 10.12) to extract an isosurface patch from $\mathbf{c}$. Otherwise, divide $\mathbf{c}$ into subcubes or pyramids, depending on the value of $\Gamma^M(q_{\mathrm{center}}(\mathbf{c})).t$.

If $\Gamma^M(q_{\mathrm{center}}(\mathbf{c})).t$ equals X, then divide $\mathbf{c}$ into eight subcubes and recursively call MULTIRESCUBE (Algorithm 10.9) on each subcube. If $\Gamma^M(q_{\mathrm{center}}(\mathbf{c})).t$ equals Y, then divide $\mathbf{c}$ into six pyramids with apex $q_{\mathrm{center}}(\mathbf{c})$ and call procedure MULTIRESPYRAMID (Algorithm 10.10) on each pyramid.

Input : Γ^S is an $n_1 \times n_2 \times n_3$ array of scalar values.
Γ^M is an $n_1 \times n_2 \times n_3$ array of Boolean values
satisfying the Multiresolution Condition.
σ is an isovalue.
$\mathbf{p}$ is a pyramid.
T is a set of edge triples representing isosurface triangles.
Tbl_P is the isosurface lookup table for pyramids.
Tbl_T is the isosurface lookup table for tetrahedra.

MultiresPyramid(Γ^S, Γ^M, σ, $\mathbf{p}$, T, Tbl_P, Tbl_T)

```
 1  f ← square base of pyramid p;
 2  q ← apex of pyramid p;
 3  if (Γ^M(q_center(f)).b = false) then
 4  │   ExtractIsoFromPyramid(Γ^S, σ, p,T,Tbl_P);
 5  else
 6  │   if (Γ^M(q_center(f).t = X) then
 7  │   │   foreach vertex v of square f do
 8  │   │   │   f' ← square with opposite vertices v and q_center(f);
 9  │   │   │   p' ← pyramid with apex q and base f';
10  │   │   │   ExtractIsoFromPyramid(Γ^S, σ, p',T,Tbl_P);
11  │   │   end
12  │   else
       │   │   /* (Γ^M(q_center(f).t = Y )                          */
13  │   │   foreach edge e of square f do
14  │   │   │   (v1, v2) ← endpoints of edge e;
       │   │   │   /* Construct isosurface in tetrahedron (v1, v2, q_center(f), q)  */
15  │   │   │   MultiresTet(Γ^S, Γ^M, σ, v1, v2, q_center(f), q, T, Tbl_T);
16  │   │   end
17  │   end
18  end
```

Algorithm 10.10. MULTIRES PYRAMID.

Input to procedure MULTIRESPYRAMID is the scalar grid Γ^S; the multiresolution grid Γ^M; the isovalue σ; a pyramid $\mathbf{p}$; and the isosurface lookup tables for pyramids and tetrahedra. Let q be the apex and $\mathbf{f}$ the square base of pyramid $\mathbf{p}$. If $\Gamma^M(q_{\text{center}}(\mathbf{f})).b$ is false, then call procedure EXTRACTISOFROMPYRAMID (Algorithm 10.13) to extract an isosurface patch from pyramid $\mathbf{p}$. Otherwise, divide $\mathbf{p}$ into subpyramids or tetrahedra, depending on the value of $\Gamma^M(q_{\text{center}}(\mathbf{p})).t$.

If $\Gamma^M(q_{\text{center}}(\mathbf{p})).t$ equals X, then divide $\mathbf{f}'$ into four subsquares and divide $\mathbf{p}$ into four subpyramids whose bases are the four subsquares of $\mathbf{f}$. Call procedure EXTRACTISOFROMPYRAMID to extract an isosurface patch from each subpyra-

Input : Γ^S is an $n_1 \times n_2 \times n_3$ array of scalar values.
Γ^M is an $n_1 \times n_2 \times n_3$ array of Boolean values
satisfying the Multiresolution Condition.
σ is an isovalue.
v_1, v_2, v_3, v_4 are the tetrahedron vertices.
v_1 and v_2 are also vertices of the cubic region containing
the tetrahedron.
T is a set of edge triples representing isosurface triangles.
Tbl_T is the isosurface lookup table for tetrahedra.

$\text{MultiresTet}(\Gamma^S, \Gamma^M, \sigma, v_1, v_2, v_3, v_4, \text{T}, \text{Tbl}_T)$

1 $q \leftarrow$ midpoint of edge (v_1, v_2);
2 if $(\Gamma^M(q).b = \text{false})$ then
3 | $\text{ExtractIsoFromTet}(\Gamma^S, \sigma, v_1, v_2, v_3, v_4)$;
4 else
5 | $\text{ExtractIsoFromTet}(\Gamma^S, \sigma, v_1, q, v_3, v_4)$;
6 | $\text{ExtractIsoFromTet}(\Gamma^S, \sigma, q, v_2, v_3, v_4)$;
7 end

Algorithm 10.11. MULTIRES TET.

mid. If $\Gamma^M(q_{\text{center}}(\mathbf{p})).t$ equals Y, then divide $\mathbf{p}$ into four tetrahedra, one for each edge of square $\mathbf{f}$. The four vertices of a tetrahedra are the endpoints of an edge of $\mathbf{f}$, the point $q_{\text{center}}(\mathbf{f})$ at the center of $\mathbf{f}$, and the apex q of $\mathbf{p}$. Call Procedure MULTIRESTET (Algorithm 10.11) on each tetrahedron.

Input to procedure MULTIRESTET is the scalar grid Γ^S; the multiresolution grid Γ^M; the isovalue σ; the four vertices, v_1, v_2, v_3 and v_4 of a tetrahedron $\mathbf{t}$; and the isosurface lookup table for tetrahedra. Vertices v_1 and v_2 are also vertices of the cubic region containing the tetrahedron. Let q be the midpoint of line segment (v_1, v_2). If $\Gamma^M(q).b$ is false, then call procedure EXTRACTISOFROMTET (Algorithm 10.14) to extract an isosurface patch from the tetrahedron. If $\Gamma^M(q).b$ is false, then split $\mathbf{t}$ into two congruent tetrahedra, $\mathbf{t}'$ and $\mathbf{t}''$, at point q and call EXTRACTISOFROMTET on $\mathbf{t}'$ and $\mathbf{t}''$.

The final step of MULTIRESOLUTION MARCHING CPT is assigning geometric locations to the isosurface vertices.

10.1.4 Isosurface Properties

By Proposition 10.1, the set of cubes, pyramids, and tetrahedra represented by the multiresolution grid form a convex polyhedral mesh. Algorithm MULTIRES-OLUTION MARCHING CPT is equivalent to applying MARCHING POLYHEDRA

Input : Γ^S is an $n_1 \times n_2 \times n_3$ array of scalar values.
σ is an isovalue.
c is a cubic region.
T is a set of edge triples representing isosurface triangles.
Tbl_C is the isosurface lookup table for cubic regions.

$\mathsf{ExtractIsoFromCube}(\Gamma^S, \sigma, \mathbf{c}, \mathsf{T}, \mathsf{Tbl}_T)$

1   $(v_1, v_2, \ldots, v_8) \leftarrow$ vertices of **c** in lexicographic order;
  /* Assign "+" or "−" signs to each vertex   */
2   for $i = 1$ to 8 do
3   if $\Gamma^S(q_i) < \sigma$ then $\mathsf{Sign}[v_i] \leftarrow$ "−";
4   else $\mathsf{Sign}[q_i] \leftarrow$ "+";   /* $\Gamma^S(v_i) \geq \sigma$ */
5   end
6   $\kappa \leftarrow (\mathsf{Sign}[v_1], \mathsf{Sign}[v_2], \ldots, \mathsf{Sign}[v_8])$;
7   foreach edge triple $(\mathbf{e}_1, \mathbf{e}_2, \mathbf{e}_3) \in \mathsf{Table}_C[\kappa]$ do
8   for $j = 1$ to 3 do
9       $\mathbf{e}'_i \leftarrow$ edge of **c** corresponding to $\mathbf{e}_i$;
10   end
11   Insert edge triple $(\mathbf{e}'_1, \mathbf{e}'_2, \mathbf{e}'_3)$ into T;
12   end

Algorithm 10.12. EXTRACTISOFROMCUBE.

(Section 5.4) to that polyhedral mesh. Thus, the properties of the MULTIRES-OLUTION MARCHING CPT isosurface are exactly the properties listed in Section 5.4.1 for the MARCHING POLYHEDRA isosurface.

MULTIRESOLUTION MARCHING CPT constructs an isosurface from a tetrahedral mesh covering a rectangular region. Thus, the condition in Section 5.4.1 about the tetrahedral mesh covering a 3-manifold with boundary is implicit and can be omitted. The condition about the mesh elements being affine transformations of cubes, tetrahedra, and pyramids is also automatically fulfilled.

10.1.5   Triangulation Proof

In this section, we prove Proposition 10.1. To prove this proposition, we must show that the intersection of every two polyhedra of the polyhedral subdivision is either the empty set or a vertex, edge, or facet of each polyhedron. (See Appendix B.5.)

Lemma 10.2. *Let Γ^M be an $n_1 \times n_2 \times n_3$ multiresolution grid satisfying the Multiresolution Conditions. If q is a vertex of the polyhedral subdivision defined by Γ^M, then $\Gamma^M(q).b$ is true.*

Input : Γ^S is an $n_1 \times n_2 \times n_3$ array of scalar values.
σ is an isovalue.
p is a pyramid.
T is a set of edge triples representing isosurface triangles.
Tbl_P is the isosurface lookup table for pyramids.

ExtractIsoFromPyramid(Γ^S, σ, **p**, T, Tbl_T)

1 $q_0 \leftarrow$ apex of pyramid **p**;
2 **f** $\leftarrow$ square base of pyramid **p**;
3 $q_1, q_2, q_3, q_4 \leftarrow$ vertices of **f**;
   /* Assign "+" or "−" signs to each vertex   */
4 **for** $i = 0$ to 4 **do**
5 | **if** $\Gamma^S(q_i) < \sigma$ **then** $\mathsf{Sign}[q_i] \leftarrow$ "−";
6 | **else** $\mathsf{Sign}[q_i] \leftarrow$ "+";   /* $\Gamma^S(q_i) \geq \sigma$ */
7 **end**
8 $\kappa \leftarrow (\mathsf{Sign}[q_0], \mathsf{Sign}[q_1], \mathsf{Sign}[q_2], \mathsf{Sign}[q_3], \mathsf{Sign}[q_4])$;
9 **foreach** edge triple $(\mathbf{e}_1, \mathbf{e}_2, \mathbf{e}_3) \in \mathsf{Tbl}_P[\kappa]$ **do**
10 | **for** $j = 1$ to 3 **do**
11 | | $\mathbf{e}'_i \leftarrow$ edge of pyramid **p** corresponding to $\mathbf{e}_i$;
12 | **end**
13 | Insert edge triple $(\mathbf{e}'_1, \mathbf{e}'_2, \mathbf{e}'_3)$ into T;
14 **end**

Algorithm 10.13. EXTRACTISOFROMPYRAMID.

Proof: If q is an initial vertex of the octree subdivision of Γ^M, then $\Gamma^M(q).b$ is true by Multiresolution Condition 4.

If q was created by dividing a cubic region ψ into six pyramids or a pyramid ψ into four tetrahedra or a tetrahedron ψ into two tetrahedra, then q equals $q_{\mathrm{split}}(\psi)$. Since ψ is subdivided, $\Gamma^M(q_{\mathrm{split}}(\psi)) = \Gamma^M(q)$ must be true.

Assume q was created by dividing a cubic region **c** into eight subcubes. Since **c** is divided into subcubes, $\Gamma^M(q_{\mathrm{center}}(\mathbf{c}))$ is true and of type X. By Multiresolution Condition 2a, $\Gamma^M(q_{\mathrm{center}}(\mathbf{f}))$ is true and has type X for every facet **f** of ψ. By Multiresolution Condition 2b, $\Gamma^M(q_{\mathrm{center}}(\mathbf{e})).b$ is true for every edge **e** of **c**. Thus $\Gamma^M(q)$ is true for every vertex q created by dividing a cubic region **c** into eight subcubes.

Assume q was created by dividing a pyramid **p** into four subpyramids. Let **f** be the square face of **c**. Point $q_{\mathrm{split}}(\mathbf{p})$ equals $q_{\mathrm{center}}(\mathbf{f})$. Since **p** is divided into subpyramids, $\Gamma^M(q_{\mathrm{center}}(\mathbf{f}))$ is true and of type X. By Multiresolution Condition 2b, $\Gamma^M(q_{\mathrm{center}}(\mathbf{e})).b$ is true for every edge **e** of **c**. Thus $\Gamma^M(q)$ is true for every vertex q created by dividing a pyramid **p** into four subpyramids. $\square$

Input : Γ^S is an $n_1 \times n_2 \times n_3$ array of scalar values.

σ is an isovalue.

v_1, v_2, v_3, v_4 are the tetrahedron vertices.

T is a set of edge triples representing isosurface triangles.

Tbl_T is the isosurface lookup table for tetrahedra.

ExtractIsoFromTet(Γ^S, σ, v_1, v_2, v_3, v_4, T, Tbl_T)

/* Assign "+" or "−" signs to each vertex   */
1 for $i = 1$ to 4 do
2   if $\Gamma^S(v_i) < \sigma$ then $\mathsf{Sign}[v_i] \leftarrow$ "−";
3   else $\mathsf{Sign}[v_i] \leftarrow$ "+";   /* $\Gamma^S(v_i) \geq \sigma$ */
4 end
5 $\kappa \leftarrow (\mathsf{Sign}[v_1], \mathsf{Sign}[v_2], \mathsf{Sign}[v_3], \mathsf{Sign}[v_4])$;
6 foreach edge triple $(\mathbf{e}_1, \mathbf{e}_2, \mathbf{e}_3) \in \mathsf{Tbl}_T[\kappa]$ do
7   for $j = 1$ to 3 do
8         $\mathbf{e}'_i \leftarrow$ edge of tetrahedron $\mathbf{t}$ corresponding to $\mathbf{e}_i$;
9   end
10   Insert edge triple $(\mathbf{e}'_1, \mathbf{e}'_2, \mathbf{e}'_3)$ into T;
11 end

Algorithm 10.14. EXTRACTISOFROMTET.

Lemma 10.3. *Let Γ^M be an $n_1 \times n_2 \times n_3$ multiresolution grid satisfying the Multiresolution Conditions. Let F be the set of all cubic regions in the cube forest on Γ^M and all edges and facets of those cubic regions. If q is a vertex of the polyhedral subdivision defined by Γ^M and q lies in the interior of $\mathbf{g} \in F$, then $\Gamma^M(q_{\mathrm{center}}(\mathbf{g})).b$ is true.*

Proof: By Lemma 10.2, $\Gamma^M(q).b$ is true. Point q equals $q_{\mathrm{center}}(\mathbf{g}')$ for some $\mathbf{g}' \in F$. If $\mathbf{g}$ equals $\mathbf{g}'$, then the lemma is proven.

Assume $\mathbf{g}$ does not equal $\mathbf{g}'$. Let $\mathbf{c}$ and $\mathbf{c}'$ be the smallest cubic regions in F that contain $\mathbf{g}$ and $\mathbf{g}'$, respectively. If $\mathbf{c}'$ equals $\mathbf{c}$, then $\mathbf{g}'$ is an edge or facet of $\mathbf{g}$. By Multiresolution Conditions 3a and 3b, $\Gamma^M(q_{\mathrm{center}}(\mathbf{g})).b$ is true, proving the lemma.

Assume $\mathbf{c}'$ does not equal $\mathbf{c}$. Since q equals $q_{\mathrm{center}}(\mathbf{g}')$ and lies on the interior of $\mathbf{g}$, region $\mathbf{c}'$ must be a proper subset of $\mathbf{c}$. Since $\mathbf{c}'$ is a subset of $\mathbf{c}$, region $\mathbf{c}'$ is a descendant of region $\mathbf{c}$ in the cube forest. By Multiresolution Condition 4, $\Gamma^M(q_{\mathrm{center}}(\mathbf{c}))$ is true and has type X. By Multiresolution Conditions 2a and 2b, $\Gamma^M(q_{\mathrm{center}}(\mathbf{g})).b$ is true. $\qquad \square$

For each pyramid $\mathbf{p}$ created by subdividing a cubic region $\mathbf{c}$ into six pyramids, let $C(\mathbf{p})$ equal $\mathbf{c}$. For each cubic region $\mathbf{c}$, let $C(\mathbf{c})$ equal $\mathbf{c}$. For every other polyhedron ψ created by subdivisions of a pyramid $\mathbf{p}$, let $C(\psi)$ equal $C(\mathbf{p})$.

Polyhedron ψ is created by subdivisions of cubic region $C(\psi)$. Note that if ψ is the subpyramid of a pyramid or a type A tetrahedron, then $C(\psi)$ is not the smallest cubic region containing ψ. Rather, if ψ is the subpyramid of a pyramid or a type A tetrahedron, then the smallest cubic region containing ψ is one of the eight octants of $C(\psi)$.

Lemma 10.4. *Let Γ^M be an $n_1 \times n_2 \times n_3$ multiresolution grid satisfying the Multiresolution Conditions. If v is a vertex and ψ is a polyhedron in the polyhedral subdivision defined by Γ^M, then v does not lie in the interior of any edge or facet of ψ.*

Proof: Assume that v was in the interior of an edge or facet $\mathbf{g}$ of ψ. We consider three cases, based on ψ.

Case I: Polyhedron ψ is a cubic region $\mathbf{c}$.

By Lemma 10.3, $\Gamma^M(q_{\mathrm{center}}(\mathbf{g})).b$ is true. By Multiresolution Conditions 3a and 3b, $\Gamma^M(q_{\mathrm{center}}(\mathbf{c})).b$ is true and $\mathbf{c}$ is divided into smaller polyhedra, a contradiction.

Case II: Polyhedron ψ is a pyramid $\mathbf{p}$.

Since pyramid $\mathbf{p}$ is contained in $C(\mathbf{p})$, region $C(\mathbf{p})$ is divided into six pyramids. Because any further subdivisions of these pyramids only adds vertices on the boundary of $C(\mathbf{p})$, $\mathbf{g}$ must either be the square base $\mathbf{f}$ of $\mathbf{p}$ or an edge of that base.

If $\mathbf{p}$ is one of the six pyramids in the subdivision of $C(\mathbf{p})$, then $\mathbf{g}$ is an edge or facet of $C(\mathbf{p})$. By Lemma 10.3, $\Gamma^M(q_{\mathrm{center}}(\mathbf{g})).b$ is true. By Multiresolution Condition 1a, $\Gamma^M(q_{\mathrm{center}}(\mathbf{f})).b$ is true and $\mathbf{p}$ is divided into smaller polyhedra, a contradiction.

If $\mathbf{p}$ is a subpyramid of a pyramid, then $\mathbf{g}$ is an edge or facet of some cubic region $\mathbf{c}'$ properly contained in $C(\mathbf{p})$. By Lemma 10.3, $\Gamma^M(q_{\mathrm{center}}(\mathbf{g})).b$ is true. By Multiresolution Conditions 3a and 3b, $\Gamma^M(q_{\mathrm{center}}(\mathbf{c}')).b$ is true. By Multiresolution Condition 2, $\Gamma^M(q_{\mathrm{center}}(C(\mathbf{p})))$ is true and has type X. Thus, $C(\mathbf{p})$ is divided into eight subcubes, a contradiction.

Case III: Polyhedron ψ is a tetrahedron $\mathbf{t}$.

Since tetrahedron $\mathbf{t}$ is contained in $C(\mathbf{t})$, region $C(\mathbf{t})$ is divided into six pyramids. Since any further subdivisions of these six pyramids only adds vertices on the boundary of $C(\mathbf{t})$, v must lie on the boundary of $C(\mathbf{t})$.

If $\mathbf{t}$ is a type A tetrahedron and v equals $q_{\mathrm{split}}(\mathbf{t})$, then $\mathbf{t}$ is split smaller tetrahedra, a contradiction. If $\mathbf{t}$ is not a type A tetrahedron or v does not equal $q_{\mathrm{split}}(\mathbf{t})$, then v is not at the center of any edge or facet of $C(\mathbf{t})$. Let $\mathbf{c}'$ be a subcube of $C(\mathbf{t})$ containing v. Since v is not at the center of any edge or facet of $C(\mathbf{t})$, vertex v must lie in the interior of an edge or facet $\mathbf{g}'$ of $\mathbf{c}'$. By Lemma 10.3, $\Gamma^M(q_{\mathrm{center}}(\mathbf{g})).b$ is true. By Multiresolution Conditions 3a and 3b, $\Gamma^M(q_{\mathrm{center}}(\mathbf{c}')).b$ is true. By Multiresolution Condition 4, $\Gamma^M(q_{\mathrm{center}}(C(\mathbf{t})))$ is true and has type X. Thus, $C(\mathbf{t})$ is divided into eight subcubes, a contradiction. $\square$

Corollary 10.5. *Let Γ^M be an $n_1 \times n_2 \times n_3$ multiresolution grid satisfying the Multiresolution Conditions. If ψ and ψ' are distinct polyhedra in the polyhedron subdivision defined by Γ^M, and $C(\psi)$ equals $C(\psi')$, then $\psi \cap \psi'$ is a face of ψ and of ψ'.*

Proof: In any subdivision of $C(\psi) = C(\psi')$ into pyramids and tetrahedra, either the intersection of two polyhedra is a face of both or the vertex of one polyhedron lies in the interior of an edge or facet of the other. By Lemma 10.4, no vertex of ψ lies on the interior of an edge or facet of ψ' and no vertex of ψ' lies on the interior of an edge or facet of ψ. Thus, the intersection of ψ and ψ' is a face of ψ and of ψ'. $\qquad\square$

Lemma 10.6. *Let Γ^M be an $n_1 \times n_2 \times n_3$ multiresolution grid satisfying the Multiresolution Conditions. Let ψ be a polyhedron in the polyhedron subdivision defined by Γ^M. If v is in the interior of a facet of $C(\psi)$ and $\Gamma^M(v)$ is true and has type X, then ψ is a subpyramid of a pyramid.*

Proof: Let $\mathbf{f}$ be the facet of $\mathbf{c} = C(\psi)$ containing v. We claim that v is the center of $\mathbf{f}$. If not, then v lies in the interior of an edge or facet of one of the eight congruent subcubes of $\mathbf{c}$. Let $\mathbf{c}'$ be one of the subcubes containing v. By Multiresolution Conditions 3a and 3b, $\Gamma^M(q_{\text{center}}(\mathbf{c}')).b$ is true. By Multiresolution Condition 4, $\Gamma^M(q_{\text{center}}(\mathbf{c}))$ is true and has type X. Thus, $\mathbf{c}$ is divided into eight subcubes and ψ cannot be a polyhedron in the polyhedron subdivision, a contradiction. We conclude that v must be at the center of $\mathbf{f}$ and $\Gamma^M(q_{\text{center}}(\mathbf{f}))$ is true and has type X.

Since $\Gamma^M(q_{\text{center}}(\mathbf{f})).b$ is true, $\Gamma^M(q_{\text{center}}(\mathbf{c})).b$ is also true (Multiresolution Condition 3b). If $\Gamma^M(q_{\text{center}}(\mathbf{c}))$ had type X, then $\mathbf{c}$ would be divided into eight subcubes and again ψ would not be a polyhedron in the polyhedron subdivision. Thus, $\Gamma^M(q_{\text{center}}(\mathbf{c}))$ is true and has type Y.

Cubic region $\mathbf{c}$ is divided into six pyramids. One of these six pyramids contains $\mathbf{f}$ as its base. Let $\mathbf{p}$ be that pyramid. Since $\Gamma^M(q_{\text{center}}(\mathbf{f})).b$ is true, $\mathbf{p}$ is divided into four subpyramids sharing the vertex v. No other polyhedra in $\mathbf{c}$ contain v. Thus ψ must be a subpyramid of a pyramid. $\qquad\square$

For convenience, Proposition 10.1 is restated here.

Proposition 10.1. *If an $n_1 \times n_2 \times n_3$ multiresolution grid satisfies the Multiresolution Conditions, then the set of cubes, pyramids, and tetrahedra determined by that grid form a convex polyhedral mesh.*

Proof: Let ψ and ψ' be polyhedra in the subdivision. Assume $\psi \cap \psi'$ was not a facet of ψ. By Lemma 10.4, no vertex of ψ' lies in the interior of an edge or facet of ψ.

If ψ was a tetrahedron, then some vertex of ψ' would be in the interior of an edge or facet of ψ, a contradiction. The same holds if ψ' were a cube or pyramid. The remaining cases are ψ is a cube or pyramid and ψ' is a tetrahedron $\mathbf{t}$.

Assume ψ' is a tetrahedron $\mathbf{t}$. Some vertex v of $\mathbf{t}$ lies at the center of a facet $\mathbf{f}$ of $C(\mathbf{t})$. By Corollary 10.5, $C(\psi)$ does not equal $C(\mathbf{t})$. Since v does not in the interior of an edge or facet of ψ and no vertex of ψ lies in the interior of an edge or facet of $\mathbf{t}$ (Lemma 10.4), v must be a vertex of ψ. To conclude our proof, we will show that $\Gamma^M(v)$ is true and has type X. By Lemma 10.6, ψ' is a subpyramid of a pyramid, contradicting the assumption that ψ' is a tetrahedron.

We consider three cases, one where ψ is a cube, one where ψ is one of the six pyramids in the subdivision of $C(\psi)$, and one where ψ is a subpyramid of a pyramid.

Case I: Polyhedron ψ is a cube $\mathbf{c}$ and ψ' is a tetrahedron $\mathbf{t}$.

Some vertex v of $\mathbf{c}$ lies in the interior of $\mathbf{f}$. If $\mathbf{c}$ is a region in the octree partition, then, by Multiresolution Condition 1, $\Gamma^M(v)$ is true and has type X. If $\mathbf{c}$ is not a region in the octree partition, then $\mathbf{c}$ is created by the division of some cubic region $\mathbf{c}'$ into cubes. Square $\mathbf{f}$ is a facet of $\mathbf{c}'$ and vertex v is the center of $\mathbf{f}$. By Multiresolution Conditions 2a and 4, $\Gamma^M(v)$ is again true and has type X. By Lemma 10.6, ψ' is a subpyramid of a pyramid and cannot be a tetrahedron.

Case IIa: Polyhedron ψ is a pyramid $\mathbf{p}$ in the division of $C(\mathbf{p})$ into six pyramids and ψ' is a tetrahedron $\mathbf{t}$.

If $C(\mathbf{p})$ is a region in the octree partition, then, by Multiresolution Condition 1, $\Gamma^M(v)$ is true and has type X. If $C(\mathbf{p})$ is not a region in the octree partition, then $C(\mathbf{p})$ is created by the division of some cubic region $\mathbf{c}'$ into cubes. Square $\mathbf{f}$ is a facet of $\mathbf{c}'$ and vertex v is the center of $\mathbf{f}$. By Multiresolution Conditions 2a and 4, $\Gamma^M(v)$ is true and has type X. By Lemma 10.6, ψ' is a subpyramid of a pyramid and cannot be a tetrahedron.

Case IIb: Polyhedron ψ is the subpyramid $\mathbf{p}$ of a pyramid $\mathbf{p}'$ and ψ' is a tetrahedron $\mathbf{t}$.

Square $\mathbf{f}$ is the base of $\mathbf{p}'$ and vertex v is the center of $\mathbf{f}$. Since $\mathbf{p}'$ is split into subpyramids, $\Gamma^M(v)$ is true and has type X. By Lemma 10.6, ψ' is a subpyramid of a pyramid and cannot be a tetrahedron. $\square$

10.2   Multiresolution Surface Nets

The dual contouring algorithm SURFACE NETS (Section 3.2) can be easily modified to produce multiresolution isosurfaces. We call this algorithm MULTIRESOLUTION SURFACE NETS.

10.2.1   Definitions

Vertex, edge, and face properties. Many of the definitions of vertex, edge, and face properties are the same as in Section 3.1. For review, we summarize them here.

A grid vertex is positive if its scalar value is greater than or equal to σ and negative if its scalar value is less than σ. A positive vertex is strictly positive if its scalar value does not equal the isovalue.

A grid face (vertex, edge, facet, or cube) is positive if all its vertices are positive and and negative if all its vertices are negative. A positive grid face is strictly positive if all its vertices are strictly positive.

A grid edge is bipolar if one endpoint is positive and one endpoint is negative. A grid square or cube or any face of a square or cube is active if it has at least one negative and one positive vertex.

The centroid of a set of points $P = \{p_1, p_2, \ldots, p_k\}$ is

$$\frac{p_1 + p_2 + \ldots p_k}{k}.$$

Vertices and edges of cubic regions. A convex polyhedral mesh has the property that the intersection of every two mesh elements $\mathbf{c}$ and $\mathbf{c}'$ is a face of $\mathbf{c}$ *and* a face of $\mathbf{c}'$. A partition of the regular grid into cubic regions of different sizes does not have this property and does not form a convex polyhedral mesh. Nevertheless, a partition of a set of cubic regions of different sizes may have the following weaker property.

Definition 10.7. A set of cubic regions is aligned if the intersection of every two cubic regions $\mathbf{c}$ and $\mathbf{c}'$ is a face of either $\mathbf{c}$ *or* a face of $\mathbf{c}'$.

A set of cubic regions that form a convex polyhedral mesh is aligned but the converse may not be true. If a set of cubic regions form a convex polyhedral mesh, then the intersection of every two regions $\mathbf{c}$ and $\mathbf{c}'$ is a face of $\mathbf{c}$ *and* a face of $\mathbf{c}'$.

In a set $\mathbb{C}$ of cubic regions, there are some edges that are not split into smaller edges by any other vertices or edges of $\mathbb{C}$. We call such edges primary edges.

Definition 10.8. An edge $\mathbf{e}$ in a set of cubic regions is a primary edge if the intersection of $\mathbf{e}$ and every cubic region is the empty set or an endpoint of $\mathbf{e}$ or $\mathbf{e}$ itself.

If $\mathbf{e}$ is an edge in an aligned set of cubic regions, then the set of primary edges contained in $\mathbf{e}$ cover $\mathbf{e}$ and partition $\mathbf{e}$.

We characterize the vertices of a set of cubic regions as follows.

Definition 10.9. Let V be the set of vertices of a set $\mathbb{C}$ of cubic regions.

1. A vertex $v \in V$ is hanging if it lies in the interior of some edge or facet of some $\mathbf{c} \in \mathbb{C}$.

2. A vertex $v \in V$ is conforming if it is not hanging.

Interior and boundary faces and regions. In Section 3.2, we identified interior edges of a regular grid as edges that were contained in four grid cubes. This definition is not appropriate for a set of cubic regions of different sizes. Instead, we give a topological definition.

If $\mathbb{C}$ is a set of cubic regions, then $|\mathbb{C}|$ is the union of all the cubic regions in $\mathbb{C}$.

Definition 10.10. Let $\mathbb{C}$ be a set of cubic regions.

- An edge $\mathbf{e}$ of $\mathbb{C}$ is an interior edge if the interior of $\mathbf{e}$ is contained in the interior of $|\mathbb{C}|$.

- An edge of $\mathbb{C}$ is a boundary edge if the interior of $\mathbf{e}$ is not contained in the interior of $|\mathbb{C}|$.

- A facet $\mathbf{f}$ of $\mathbb{C}$ is an interior facet if the interior of $\mathbf{f}$ is contained in the interior of $|\mathbb{C}|$.

- A facet of $\mathbb{C}$ is a boundary facet if the interior of $\mathbf{f}$ is not contained in the interior of $|\mathbb{C}|$.

Note that a primary edge may be contained in only three cubic regions and still be an interior edge.

The definition of interior and boundary cubes is based on interior and boundary edges and facets.

Definition 10.11. Let $\mathbb{C}$ be a set of cubic regions.

- A cubic region $\mathbf{c}$ of $\mathbb{C}$ is an interior region if every edge and facet of $\mathbf{c}$ is an interior edge or facet.

- A cubic region $\mathbf{c}$ of $\mathbb{C}$ is a boundary region if some edge or facet of $\mathbf{c}$ is a boundary edge or facet.

Let $\mathbb{C}$ be a a set of cubic regions. We use the notation $\mathbb{C}_{\text{inner}}$ to denote the set of inner cubic regions of $\mathbb{C}$.

10.2.2 Algorithm

Input to MULTIRESOLUTION SURFACE NETS is an isovalue, an aligned set $\mathbb{C}$ of cubic regions partitioning a regular grid, and a scalar value assigned to each vertex of $\mathbb{C}$. Add an isosurface vertex $w_{\mathbf{c}}$ to each grid cube $\mathbf{c}$ with at least one positive and at least one negative vertex. A primary interior edge $\mathbf{e}$ is contained in three or four cubes. For each primary interior edge $\mathbf{e}$, if $\mathbf{e}$ is bipolar and contained in three cubes, $\mathbf{c}_1$, $\mathbf{c}_2$, and $\mathbf{c}_3$, add an isosurface triangle with vertices $w_{\mathbf{c}_1}$, $w_{\mathbf{c}_2}$, and $w_{\mathbf{c}_3}$. If $\mathbf{e}$ is bipolar and contained in four cubes, $\mathbf{c}_1$, $\mathbf{c}_2$, $\mathbf{c}_3$ and

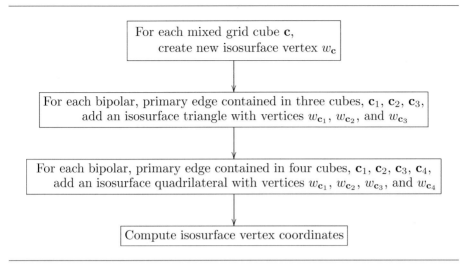

Figure 10.1. MULTIRESOLUTION SURFACE NETS (3D).

$\mathbf{c}_4$, add an isosurface quadrilateral with vertices $w_{\mathbf{c}_1}$, $w_{\mathbf{c}_2}$, $w_{\mathbf{c}_3}$, and $w_{\mathbf{c}_4}$, which is the dual of the four cubes. (See Figure 10.1.)

To position the isosurface vertex within each grid cube, use linear interpolation (Section 1.7.2) to approximate the intersection of the isosurface and all the bipolar edges. Take the centroid of all the approximation points as the location of the isosurface vertex.

More specifically, let σ be the isovalue and s_v be the scalar value at a grid vertex v. For each bipolar edge $e = [u, v]$, let p_e be the point $\alpha u + (1 - \alpha)v$ where $\alpha = (s_v - \sigma)/(s_v - s_u)$. Note that since u and v have different sign, scalar s_u does not equal s_v and the denominator $(s_v - s_u)$ is never zero. If $p_1, \ldots, p_k$ are the interpolation points in the grid cube $\mathbf{c}$, then locate isosurface vertex $w_{\mathbf{c}}$ at $(p_1 + \cdots + p_k)/k$, the centroid of $p_1, \ldots, p_k$.

MULTIRESOLUTION SURFACE NETS creates a mesh of tetrahedra and quadrilaterals but the vertices of each quadrilateral do not necessarily lie in a plane. To create a piecewise linear surface, triangulate each quadrilateral, subdividing it into triangles.

10.2.3 Isosurface Properties

MULTIRESOLUTION SURFACE NETS builds the isosurface from an aligned, multiresolution set of cubic regions. Because the cubic regions do not form a convex polyhedral mesh, MULTIRESOLUTION SURFACE NETS isosurfaces do not have many of the properties of SURFACE NETS isosurfaces.

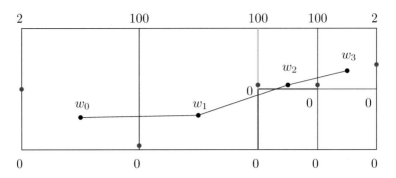

Figure 10.2. Example of an isocontour with isovalue 1 that intersects two negative primary edges (blue) and does not intersect a bipolar primary edge (green). Isocontour vertices w_0, w_1, w_2, and w_3 are at the centroid of the interpolated intersection points (red) of the isocontour and the primary square edges.

As shown in a 2D illustration in Figure 10.2, the MULTIRESOLUTION SUR-FACE NETS isosurface may not intersect a bipolar primary edge. The isosurface may intersect a primary, negative edge or a primary, strictly positive edge. However, the isosurface intersects all bipolar, primary edges whose endpoints are both conforming. The isosurface does not intersect any primary, negative edges or primary, strictly positive edges whose endpoints are both conforming.

MULTIRESOLUTION SURFACE NETS returns a finite set, Υ, of triangles. The isosurface is the union of these triangles. The vertices of the isosurface are the triangle vertices.

The following properties apply to all isosurfaces produced by the SURFACE NETS algorithm.

Property 1. *The isosurface is piecewise linear.*

Property 2. *Set Υ contains at most $2m$ triangles where m is the number of bipolar primary edges.*

Property 3. *The isosurface does not contain any conforming vertices whose scalar values do not equal the isovalue.*

Property 4. *The isosurface intersects at exactly one point every interior, bipolar, primary edge whose positive endpoint is strictly positive and whose endpoints are both conforming.*

Property 5. *The isosurface does not intersect any primary, negative edges or any primary, strictly positive edges whose endpoints are both conforming.*

Property 6. *The isosurface strictly separates the strictly positive conforming vertices of $\mathbb{C}_{\text{inner}}$ from the negative conforming vertices of $\mathbb{C}_{\text{inner}}$.*

The following property applies to the MULTIRESOLUTION SURFACE NETS isosurfaces whose isovalues do not equal the scalar value of any vertex.

Property 7. *Set Υ does not contain any zero-area triangles or duplicate triangles.*

The MULTIRESOLUTION SURFACE NETS isosurface is typically not a manifold, even if the isovalue does not equal the scalar value of any vertex.

Proofs of all the MULTIRESOLUTION SURFACE NETS properties are similar to proofs for SURFACE NETS (Section 3.2.5, 3.2.6, and 3.2.7) and are omitted.

Algorithm MULTIRESOLUTION SURFACE NETS positions isosurface vertices at the centroids of the intersections of the isosurface and primary grid edges. There are numerous possible variations on this step of the algorithm.

10.3   Multiresolution in 4D

Sections 10.3.1 and 10.3.2 contain an extension of MULTIRESOLUTION MARCHING TETRAHEDRA to four dimensions. Sections 10.3.3 and 10.3.4 contain the four-dimensional extension of MULTIRESOLUTION MARCHING CPT.

In Section 6.7, we extended Algorithm SURFACE NETS to four dimensions. For each bipolar, 4D edge $\mathbf{e}$, we generated an isosurface hexahedron. This hexahedron is dual to the eight grid hypercubes containing $\mathbf{e}$. In a multiresolution grid of hypercubes, a "primary" edge can be contained in four, five, six, seven, or eight hypercubes. The dual of these hypercubes may not be realizable as a polytope. Because of this difficulty, we could not find a satisfactory generalization of MULTIRESOLUTION SURFACE NETS to four dimensions.

10.3.1   Multiresolution Triangulations

Let Γ^S be a scalar grid with vertex dimensions $n_1 \times n_2 \times n_3 \times n_4$. Let $|\Gamma|$ be the region covered by grid Γ^S. We describe how to construct a multiresolution triangulation of $|\Gamma|$.

As in Section 9.2.3 for 3D, we first partition $|\Gamma|$ into four-dimensional cubic regions. Let k be the smallest integer such that $2^k + 1 \geq \max(n_1, n_2, n_3, n_4)$. Let $\mathbf{c}^*$ be a hypercube with edge length 2^k and with opposing vertices at $(0, 0, 0, 0)$ and $(2^k, 2^k, 2^k, 2^k)$. Region $\mathbf{c}^*$ contains $|\Gamma|$. If $|\Gamma|$ does not equal $\mathbf{c}^*$, divide $\mathbf{c}^*$ into sixteen hypercubes. Recursively subdivide each of the hypercubes until each hypercube is either contained in $|\Gamma|$ or is disjoint from the interior of $|\Gamma|$. After k partitions, each hypercube has edge length one. Thus the recursive subdivision stops after at most k steps. Let $\mathcal{P}_{oct}$ be the set of hypercubes that are contained in $|\Gamma|$. Set $\mathcal{P}_{oct}$ forms a partition of $|\Gamma|$.

Region $\mathbf{c}^*$ and its subdivisions form an octree containing $|\Gamma|$. (See Section 8.2.1 for a definition of octrees. As previously noted, the term *octree* is misleading in four dimensions. Each hypercube is partitioned into sixteen, not eight, hypercubes.) Set $\mathcal{P}_{oct}$ is the leaves of this octree that represent regions contained in $|\Gamma|$. We call $\mathcal{P}_{oct}$ the octree partition of $|\Gamma|$.

We triangulate each hypercube in the octree partition as follows. Consider the hypercube $\mathbf{c} \in \mathcal{P}_{oct}$ with opposing vertices v_0 and v_{15}. Vertex v_0 lies in four facets $\mathbf{f}_1, \mathbf{f}_2, \mathbf{f}_3$, and $\mathbf{f}_4$ of $\mathbf{c}$. Each facet has a diagonal d_i connecting v_0 to the facet vertex opposite v_0. For instance, if $v_0, v_1, \ldots, v_7$ are the vertices of $\mathbf{f}_1$ and v_7 is opposite v_0 in $\mathbf{f}_1$, then d_i connects v_0 to v_7.

Each facet $\mathbf{f}_i$ has a triangulation into six tetrahedra sharing edge d_i. The convex hulls of each of these tetrahedra and point v_{15} form polyhedra in four dimensions. This polyhedra each has five vertices and are four-dimensional simplices. Since $\mathbf{c}$ has four facets containing v_0 and each facet has six tetrahedra, this construction creates twenty-four simplices. These twenty-four simplices triangulate $\mathbf{c}$. Each simplex contains the diagonal (v_0, v_{15}) of $\mathbf{c}$ as an edge.

Let $\mathbf{t}$ be a four-dimensional simplex and let q be some point at the interior of some edge $\mathbf{e}$ of $\mathbf{t}$. Let p_0 and p_1 be the endpoints of $\mathbf{e}$ and let p_2, p_3, and p_4 be the other vertices of $\mathbf{t}$. To split simplex $\mathbf{t}$ at point q, we replace $\mathbf{t}$ with simplices $\mathbf{t}_0'$ and $\mathbf{t}_1'$ where $\mathbf{t}_0'$ has vertices p_0, q, p_2, p_3, p_4 and $\mathbf{t}_1'$ has vertices q, p_1, p_2, p_3, p_4. Simplices $\mathbf{t}_0'$ and $\mathbf{t}_1'$ partition $\mathbf{t}$. Simplex $\mathbf{t}$ is bisected at point q, if $\mathbf{t}_0'$ and $\mathbf{t}_1'$ are congruent.

Consider a tetrahedron $\mathbf{t}$ lying in $\mathbf{f}_1$ and the simplex $\mathbf{t}$ that is the convex hull of $\mathbf{t}$ and v_{15}. Point $q_{center}(\mathbf{c})$, the center of $\mathbf{c}$, is also the midpoint of edge (v_0, v_{15}) of $\mathbf{t}$. Split $\mathbf{t}$ at $q_{center}(\mathbf{c})$ into two congruent simplices, $\mathbf{t}_1'$ and $\mathbf{t}_2'$, one containing vertex v_0 and the other containing vertex v_{15}.

A four-dimensional simplex has type A if it is similar to $\mathbf{t}$, i.e., it is congruent to some scalar multiple of $\mathbf{t}$. A simplex has type B if it is similar to $\mathbf{t}_1'$ and $\mathbf{t}_2'$.

Point $q_{center}(\mathbf{f}_1)$, the center of facet $\mathbf{f}_1$, is also the midpoint of an edge of $\mathbf{t}_0'$ and $\mathbf{t}_1'$. Split $\mathbf{t}_0'$ and $\mathbf{t}_1'$ at $q_{center}(\mathbf{f}_1)$ into four congruent simplices.

Let $\mathbf{t}''$ be a subsimplex of $\mathbf{t}_0'$ or $\mathbf{t}_1'$. A simplex has type C if it is similar to $\mathbf{t}''$. Simplex $\mathbf{t}''$ contains $q_{center}(\mathbf{g})$, the center of $\mathbf{g}$, for some two-dimensional face $\mathbf{g}$ of $\mathbf{c}$. Split $\mathbf{t}''$ at $q_{center}(\mathbf{g})$.

Let $\mathbf{t}'''$ be a subsimplex of $\mathbf{t}''$. A simplex has type D if it is similar to $\mathbf{t}''$. Simplex $\mathbf{t}'''$ shares one edge $\mathbf{e}$ with $\mathbf{c}$. Split $\mathbf{t}'''$ at point $q_{center}(\mathbf{e})$, the midpoint of edge $\mathbf{e}$. Splitting $\mathbf{t}'''$ at $q_{center}(\mathbf{e})$ creates two simplices that are similar to simplex $\mathbf{t}$, i.e., they are congruent to the simplex formed by shrinking $\mathbf{t}$ by a factor of one-half.

The original twenty-four type A simplices sharing diagonal (v_0, v_{15}) divide into forty-eight congruent type B simplices, each containing the center of some facet of $\mathbf{c}$. These forty-eight simplices divide into ninety-six congruent type C simplices, each containing the center of some two-dimensional face of $\mathbf{c}$. The ninety-six simplices divide into 192 type D simplices, each sharing exactly one edge with $\mathbf{c}$. The 192 simplices divide into 384 type A simplices. Since these 384 simplices are similar to the original 24 simplices, the four bisection steps can be repeated on the 384 simplices, creating simplices whose edges are one-fourth the original size. Repeated application of the bisection steps can lead to simplices of arbitrarily small size.

We note a few things about this recursive bisection scheme. First, the angle between a simplicial edge and a coordinate axis is zero, forty-five, or ninety degrees. Second, all simplices are divided into four types of similar simplices. Third, after four bisection steps, each of the resulting 384 simplices has an edge from one of the original hypercube vertices v_i to $q_{center}(\mathbf{c})$, the center of the hypercube. Conversely, for each cube vertex v_i, exactly forty-eight of the 384 simplices contain the edge $(v_i, q_{center}(\mathbf{c}))$, and these forty-eight simplices form a triangulation of the hypercube with opposite vertices v_i and $q_{center}(\mathbf{c})$. Thus the four bisection steps convert a triangulation of a hypercube into a triangulation of the sixteen subcubes of that hypercube. Repeating the bisection steps another four times gives a triangulation of 16^2 subcubes of the original cube. If all the edges of $\mathbf{c}$ have length $2^k + 1$, then repeating the bisection steps $4k$ times gives a triangulation of the 2^{4k} unit hypercubes in $\mathbf{c}$.

A hypercube $\mathbf{c}$ in the octree partition $\mathcal{P}_{oct}$ has eight diagonals. Each diagonal generates a different triangulation of $\mathbf{c}$ composed of simplices sharing that diagonal as an edge. If $\mathbf{c}$ does not equal $\mathbf{c}^*$, then $\mathbf{c}$ was created by subdividing a hypercube $\mathbf{c}'$ into sixteen octants. The center $q_{center}(\mathbf{c}')$ of $\mathbf{c}'$ is a vertex of $\mathbf{c}$. Let $d_{split}(\mathbf{c})$ be the diagonal of $\mathbf{c}$ connecting $q_{center}(\mathbf{c}')$ to the opposite vertex in $\mathbf{c}$. If $\mathbf{c}$ equals $\mathbf{c}^*$, then let $d_{split}(\mathbf{c})$ be the diagonal from $(0,0,0,0)$ to $(2^k, 2^k, 2^k, 2^k)$. (As in 3D, $\mathbf{c} \in \mathcal{P}_{oct}$ can only equal $\mathbf{c}^*$ if $\mathbf{c}^*$ equals $|\Gamma|$. In that case, $\mathcal{P}_{oct}$ contains only $\mathbf{c}^*$.) Choose the triangulation of $\mathbf{c}$ whose simplices share edge $d_{split}(\mathbf{c})$. Label this triangulation $\tau^{\mathbf{c}}$.

Hypercube $\mathbf{c}$ and its subdivisions form a tree called the full-resolution tree on $\mathbf{c}$. The root of the tree represents the entire region $\mathbf{c}$. The root has forty-eight children, one for each simplex in $\tau^{\mathbf{c}}$. Each tree node other than the root represents a four-dimensional simplex. The two children of a node representing simplex $\mathbf{t}$ are the two simplices created by bisecting $\mathbf{t}$ into two congruent simplices. All the simplices represented at a given level in the tree are congruent. The height of the tree is $3j + 1$ where 2^j is the length of the edges of $\mathbf{c}$. Every simplex produced by $3j$ or fewer bisections of $\tau^{\mathbf{c}}$ is represented by some node in the tree. The union of all the full-resolution trees over all hypercubes in the octree subdivision is called the full-resolution forest.

The union of the triangulations $\tau^{\mathbf{c}}$ form a subdivision of $|\Gamma|$ into simplices, although not necessarily a triangulation of $|\Gamma|$. Bisecting simplices increases the resolution of the subdivision. By choosing to bisect some simplices but not others, we can increase the resolution in areas of interest.

As in 3D, we use an $n_1 \times n_2 \times n_3 \times n_4$ multiresolution grid Γ^M to determine which simplices are bisected. For each simplex $\mathbf{t}$, let $q_{split}(\mathbf{t})$ be the point used to split $\mathbf{t}$ into two congruent tetrahedra. Point $q_{split}(\mathbf{t})$ is the midpoint of some edge of $\mathbf{t}$. For instance, if $\mathbf{t}$ is a type A tetrahedron containing a diagonal of hypercube $\mathbf{c}$, then $q_{split}(\mathbf{t})$ is the center, $q_{center}(\mathbf{c})$, of $\mathbf{c}$.

If simplex $\mathbf{t}$ is not a leaf of the full-resolution forest, then the coordinates of $q_{\mathrm{split}}(\mathbf{t})$ are all integers and $q_{\mathrm{split}}(\mathbf{t})$ is represented by a vertex in grid Γ^M. Simplex $\mathbf{t}$ is bisected if and only if $\mathbf{t}$ is not at a leaf of the full-resolution forest and $\Gamma^M(q_{\mathrm{split}}(\mathbf{t}))$ is true. If simplex $\mathbf{t}$ is a leaf of the full-resolution forest, then $q_{\mathrm{split}}(\mathbf{t})$ is not represented by any vertex in Γ^M but $\mathbf{t}$ is never split into smaller simplices.

The multiresolution grid Γ^M represents a multiresolution subdivision of $|\Gamma|$ as follows. Starting from the triangulations $\tau^{\mathbf{c}}$ for all $\mathbf{c} \in \mathcal{P}_{oct}$, bisect simplex $\mathbf{t} \in \tau^{\mathbf{c}}$ if $q_{\mathrm{split}}(\mathbf{t})$ is true. Repeat this process on the resulting simplices until either $\mathbf{t}$ is either a leaf of the full-resolution forest or $q_{\mathrm{split}}(\mathbf{t})$ is false. The resulting simplices form a multiresolution subdivision of $|\Gamma|$.

To ensure that this multiresolution subdivision is a triangulation, we need additional conditions on the multiresolution grid.

Multiresolution Conditions (4D, Forest). *Let simplex $\mathbf{t}$ be an internal node and let hypercube $\mathbf{c}$ be represented by a root node in the full-resolution forest.*

1. *If $\mathbf{t}$ is a child of tetrahedron $\mathbf{t}'$ in the full-resolution forest and $q_{\mathrm{split}}(\mathbf{t})$ is true, then $q_{\mathrm{split}}(\mathbf{t}')$ is true.*

2. *Every corner v of region $\mathbf{c}$ is true.*

We claim that a multiresolution grid that satisfies the Multiresolution Conditions defines a multiresolution triangulation.

Proposition 10.12. *If an $n_1 \times n_2 \times n_3 \times n_4$ multiresolution grid Γ^M satisfies the Multiresolution Conditions, then the set of simplices defined by Γ^M is a triangulation of the region covered by Γ^M.*

The proof is similar to the proof of 9.7 and is omitted.

The 4D algorithm for converting an array of Boolean values into a multiresolution grid satisfying the Multiresolution Conditions parallels Algorithm 9.6 for 3D multiresolution grids. The only difference is that we must consider 2D face centers in addition to grid edges and facets.

Let k be the smallest integer such that $2^k + 1 \geq \max(n_1, n_2, n_3, n_4)$. Let Γ_j^M be the largest subgrid of Γ^M containing $(0,0,0,0)$ that can be partitioned into hypercubes of edge length 2^j. Let $\mathcal{P}_j(\Gamma^M)$ be the partition of Γ_j^M into hypercubes with edge length 2^j. For j from one to k, if $\Gamma^M(q_{\mathrm{center}}(\mathbf{e}))$ is true for some edge $\mathbf{e}$ in $\mathcal{P}_j(\Gamma^M)$, then set $\Gamma^M(q_{\mathrm{center}}(\mathbf{g}))$ to true for all two-dimensional faces $g \in \mathcal{P}_j(\Gamma^M)$ containing $\mathbf{e}$. If $\Gamma^M(q_{\mathrm{center}}(\mathbf{g}))$ is true for some two-dimensional face $\mathbf{g}$ in $\mathcal{P}_j(\Gamma^M)$, then set $\Gamma^M(q_{\mathrm{center}}(\mathbf{f}))$ to true for all three-dimensional facets $f \in \mathcal{P}_j(\Gamma^M)$ containing $\mathbf{g}$. If $\Gamma^M(q_{\mathrm{center}}(\mathbf{f}))$ is true for some three-dimensional facet $\mathbf{f}$ in $\mathcal{P}_j(\Gamma^M)$, then set $\Gamma^M(q_{\mathrm{center}}(\mathbf{c}))$ to true for all four-dimensional hypercubes $\mathbf{c} \in \mathcal{P}_j(\Gamma^M)$ containing $\mathbf{f}$. As in three dimensions, d-dimensional faces must be processed before $(d+1)$-dimensional faces, and $\mathcal{P}_j(\Gamma^M)$ must be processed before $\mathcal{P}_{j+1}(\Gamma^M)$. Pseudocode for Procedure SATISFYMULTIRES4D is presented in Algorithm 10.15.

Input : Γ^M is an $n_1 \times n_2 \times n_3 \times n_4$ array of Boolean values.

`SatisfyMultires4D(`Γ^M`)`

1 $\mathcal{P}_{oct} \leftarrow$ octree partition of Γ^M;
2 foreach hypercube $\mathbf{c} \in \mathcal{P}_{oct}$ do
3   foreach corner v of $\mathbf{c}$ do
4        $\Gamma^M(v) \leftarrow$ true;
5   end
6 end
7 Let k be the smallest integer such that $2^k + 1 \geq \max(n_1, n_2, n_3, n_4)$;
8 for $j = 1$ to k do
      /* $\mathcal{P}_j(\Gamma^M)$ is the partition of subgrid Γ_j^M of Γ^M into hypercubes of
      length 2^j where Γ_j^M is the largest subgrid of Γ^M containing $(0,0,0)$
      that can be partitioned into hypercubes of length 2^j.   */
9   for $d \leftarrow 1$ to 3 do
10   foreach d-dimensional face $\mathbf{f}$ in $\mathcal{P}_j(\Gamma^M)$ do
         /* $q_{\text{center}}(\mathbf{f})$ is the center of face $\mathbf{f}$ */
11   if $(\Gamma^M(q_{\text{center}}(\mathbf{f})) =$ true$)$ then
12   foreach $(d+1)$-dimensional face $g \in \mathcal{P}_j(\Gamma^M)$ containing $\mathbf{f}$ do
13             $\Gamma^M(q_{\text{center}}(\mathbf{g})) \leftarrow$ true;
14   end
15   end
16   end
17   end
18   foreach hypercube $\mathbf{c}$ in $\mathcal{P}_j(\Gamma^M)$ do
19   if $(\Gamma^M(q_{\text{center}}(\mathbf{c})) =$ true$)$ then
20   foreach corner v of $\mathbf{c}$ do
21          $\Gamma^M(v) \leftarrow$ true;
22   end
23   end
24   end
25 end

Algorithm 10.15. Algorithm SATISFYMULTIRES4D for four-dimensional grids.

10.3.2 Multiresolution Marching Simplices (4D)

Algorithm MULTIRESOLUTION MARCHING SIMPLICES (Figure 10.3) is the application of the 4D MARCHING SIMPLICES from Section 6.4.5 to a multiresolu-

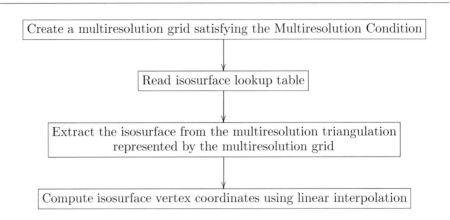

Figure 10.3. MULTIRESOLUTION MARCHING SIMPLICES (4D) for $n_1 \times n_2 \times n_3 \times n_4$ grids.

tion simplicial mesh. However, instead of constructing the multiresolution mesh and then applying MARCHING SIMPLICES, MULTIRESOLUTION MARCHING SIMPLICES constructs each element of the multiresolution mesh and directly extracts the isosurface patch from that element.

Input to MULTIRESOLUTION MARCHING SIMPLICES is a scalar grid, Γ^S; an array, **Coord**, of grid coordinates; a set, **RegionList**, of high-resolution regions; and a scalar value σ. Start by converting the high-resolution regions to a multiresolution grid, Γ^M. (See Algorithm 10.16, CREATEMULTIRESGRID4D.) Set every vertex within the high-resolution regions to true. Set all other vertices to false. Apply procedure SATISFYMULTIRES (Algorithm 10.15) to the multiresolution grid so that it satisfies the Multiresolution Conditions.

Read the isosurface lookup table from a preconstructed data file. To extract the isosurface, create an octree partition $\mathcal{P}_{oct}$ of Γ^S. For each hypercube **c** in $\mathcal{P}_{oct}$, construct the triangulation $\tau^{\mathbf{c}}$ consisting of forty-eight simplices sharing the edge $d_{\text{split}}(\mathbf{c})$. Recursively descend the full-resolution tree on **c**, identifying and processing simplices in the tree. Upon reaching a simplex **t**, if **t** is a leaf of the tree or if $\Gamma^M(q_{\text{split}}(\mathbf{t}))$ is false, then extract an isosurface patch from **t**. Otherwise, **t** is split into two simplices, $\mathbf{t}'$ and $\mathbf{t}''$, and we repeat the procedure on $\mathbf{t}'$ and $\mathbf{t}''$.

In the final step, assign geometric locations to the isosurface vertices based on the scalar values at the endpoints of these bipolar edges.

To extract an isosurface patch from **t**, determine the entry in the isosurface lookup table matching the configuration of positive and negative vertex labels on **t**. Retrieve from this entry the set of isosurface tetrahedra representing the isosurface patch.

Input : Γ^S is an $n_1 \times n_2 \times n_3 \times n_4$ grid.
         RegionList is a list of regions in Γ^S.

Output : Γ^M is an $n_1 \times n_2 \times n_3 \times n_4$ array of Boolean values satisfying the Multiresolution Condition.

`CreateMultiresGrid4D(`Γ^S`, RegionList, `Γ^M`)`

1   Initialize $\Gamma^M(v)$ to false for all vertices v;
2   foreach region R in RegionList do
3   foreach vertex v in R do
4         $\Gamma^M(v) \leftarrow$ true;
5   end
6   end
7   `SatisfyMultires4D`(Γ^M);

Algorithm 10.16. Procedure CREATEMULTIRESGRID4D for four-dimensional grids.

Let $\mathbf{t}_S$ be the "generic" simplex used in building the lookup table. As with MARCHING SIMPLICES, it is important to map the ith vertex of $\mathbf{t}$ in lexicographic order to the ith vertex of $\mathbf{t}_S$. If this order is not preserved, then the isosurface could have "cracks" or self-intersections as discussed in Section 6.4.

Instead of recursively constructing simplices in the multiresolution mesh, we could iteratively enumerate and process all tetrahedra in the full-resolution forest. The iterative algorithm is a 4D version of Algorithm 9.10 in Section 9.2.7. Details are left to the reader.

10.3.3 Multiresolution Convex Polyhedral Mesh

To extend MULTIRESOLUTION MARCHING CPT to four dimensions, we need to build multiresolution convex polyhedral meshes in four dimensions. In 3D, the mesh is built from three convex polyhedra, cubes, square pyramids, and tetrahedra. In 4D, the mesh is built from four convex polytopes, hypercubes, two different types of pyramids, and four-dimensional simplices. The two types of pyramids have different convex polyhedra as bases. If the 4D pyramid has a 3D cube as a base, then we call it a pyramid over a cube. If the 4D pyramid has a 3D square pyramid as a base, then we call it a pyramid over a (square) pyramid.

A hypercube is divided in one of two ways. A hypercube is sometimes divided into sixteen congruent subcubes sharing a vertex at the hypercube center. A hypercube can also be divided into eight congruent pyramids over cubes whose apices are the hypercube center and whose bases are the eight hypercube facets.

A pyramid over a cube is divided in two ways. A pyramid over a cube is sometimes divided into eight congruent pyramids over cubes by splitting its base into eight subcubes and connecting its apex to each of the eight subcubes. A pyramid over a cube can also be divided into six pyramids over square pyramids. Divide the cube forming the base of the pyramid into six square pyramids by connecting the center of the cube to the six square cube facets. Connect each of the square pyramids to the apex of the original pyramid over a cube, forming six pyramids over pyramids.

A pyramid over a square pyramid can also be divided in two ways. A pyramid over a square pyramid can be divided into four pyramids over square pyramids as follows. Let $\mathbf{p}$ be a 3D square pyramid and let $\mathbf{p}'$ be a 4D pyramid whose base is $\mathbf{p}$. Split $\mathbf{p}$ into four square pyramids by splitting its square face into four subsquares and joining each subsquare with the apex of $\mathbf{p}$. Split $\mathbf{p}'$ into four pyramids over a pyramid by joining each of the four square subpyramids of $\mathbf{p}$ with the apex of $\mathbf{p}'$.

A pyramid over a square pyramid can also be divided into four simplices. Again let $\mathbf{p}$ be a 3D square pyramid and let $\mathbf{p}'$ be a 4D pyramid whose base is $\mathbf{p}$. Split $\mathbf{p}$ into four tetrahedra by splitting its square base into four triangles and connecting each triangle with the apex of $\mathbf{p}$. Split $\mathbf{p}'$ into four simplices by connecting each of the four tetrahedra to the apex of $\mathbf{p}'$. These simplices have type D as defined in Section 10.3.1.

A type D simplex $\mathbf{t}$ is created by splitting a hypercube $\mathbf{c}$ into pyramids over cubes, splitting one of these pyramids over cubes into pyramids over pyramids, and splitting one of the pyramids over pyramids into simplices. Simplex $\mathbf{t}$ shares one edge $\mathbf{e}$ with hypercube $\mathbf{c}$. Bisecting $\mathbf{t}$ at the midpoint of $\mathbf{e}$ splits it into two type A simplices. Type A simplices are never subdivided. Similarly, if a pyramid over a cube is subdivided into eight pyramids over cubes, then the eight subpyramids are never subdivided. If a pyramid over a pyramid is subdivided into four pyramids over pyramids, then the four subpyramids are never subdivided.

As in 3D, we use a multiresolution grid, Γ^M, to specify the multiresolution convex polyhedral mesh. Each grid element $\Gamma^M(q)$ has a Boolean value, $\Gamma^M(q).b$, and a split type, $\Gamma^M(q).t$. We say that $\Gamma^M(q)$ is true or false or has type X or Y to mean that $\Gamma^M(q).b$ is true or false or $\Gamma^M(q).t$ equals X or Y.

For each polytope ψ, we define a point $q_{\text{split}}(\psi)$ that determines if and how ψ is split into smaller polyhedra. For a hypercube $\mathbf{c}$, point $q_{\text{split}}(\mathbf{c})$ is the center of $\mathbf{c}$. For a pyramid, $\mathbf{p}$, over a cube, point $q_{\text{split}}(\mathbf{p})$ is the center of the base of $\mathbf{p}$. For a pyramid, $\mathbf{p}$, over a square pyramid, $\mathbf{p}'$, point $q_{\text{split}}(\mathbf{p})$ is the center of the base of $\mathbf{p}'$. For a tetrahedron $\mathbf{t}_D$ of type D, point $q_{\text{split}}(\mathbf{t}_D)$ is the midpoint of the edge $\mathbf{t}_D$ shares with the cubic region containing $\mathbf{t}_D$. A tetrahedron $\mathbf{t}_A$ of type A is never split and $q_{\text{split}}(\mathbf{t}_A)$ is undefined.

Point $q_{\text{split}}(\psi)$ is always the center of some hypercube or the center of some face of a hypercube. Because these cubic regions have edge length 2^j, point $q_{\text{split}}(\psi)$ is a grid vertex whenever j is greater than zero. Point $q_{\text{split}}(\psi)$ is not a grid vertex only if ψ is a unit hypercube or if ψ is contained in a unit hypercube. In all such cases, ψ is never subdivided and $q_{\text{split}}(\psi)$ is irrelevant.

Let q equal $q_{\text{split}}(\psi)$ for mesh element ψ where ψ is not a unit hypercube and is not contained in a unit hypercube. If $\Gamma^M(q).b$ is true, then ψ is split. If ψ is a hypercube, then ψ is split into sixteen hypercubes whenever $\Gamma^M(q).t$ is X and is split into eight pyramids over cubes whenever $\Gamma^M(q).t$ is Y. If ψ is a pyramid over a cube, then ψ is split into eight pyramids over cubes whenever $\Gamma^M(q).t$ is X and is split into six pyramids over pyramids whenever $\Gamma^M(q).t$ is Y. If ψ is a pyramid over a square pyramid, then ψ is split into four pyramids over square pyramids whenever $\Gamma^M(q).t$ is X and is split into four simplices whenever $\Gamma^M(q).t$ is Y. If ψ is a simplex and $\Gamma^M(q).b$ is true, then $\Gamma^M(q).t$ is always Y and ψ is split into two simplices. If $\Gamma^M(q).b$ is false, then ψ is not split and $\Gamma^M(q).t$ is undefined.

The cubic subdivisions of a hypercube form a tree. The root of the tree represents the original hypercube. Each node in the tree represents a hypercube $\mathbf{c}$ with edge length 2^j. If j is positive, then the node has sixteen children, one for each subcube of $\mathbf{c}$. If j is zero, then the node is a leaf. The cubic subdivisions of an octree partition form a hypercube forest. The root of each tree in this forest is a hypercube in the octree partition. Each tree represents the cubic subdivisions of its corresponding region.

For each face $\mathbf{f}$ of a hypercube $\mathbf{c}$, let $q_{\text{center}}(\mathbf{f})$ be the center of $\mathbf{f}$. When $\mathbf{f}$ is a vertex of $\mathbf{c}$, $q_{\text{center}}(\mathbf{f})$ equals $\mathbf{f}$. When $\mathbf{f}$ is an edge of $\mathbf{c}$, $q_{\text{center}}(\mathbf{f})$ is the midpoint of $\mathbf{f}$. Face $\mathbf{f}$ can equal the entire hypercube $\mathbf{c}$, in which case, $q_{\text{center}}(\mathbf{f})$ equals the center of $\mathbf{c} = \mathbf{f}$.

Multiresolution Conditions (4D Convex Polyhedral Mesh). *Let d equal 1, 2, or 3. Let $\mathbf{c}$ be a hypercube at an internal node in the hypercube forest, let $\mathbf{f}$ be a $(d+1)$-dimensional face of $\mathbf{c}$, and let $\mathbf{g}$ be a d-dimensional face of $\mathbf{f}$.*

1. *For every corner q of a region in the octree partition, $\Gamma^M(q).b$ is true and has type X.*

2. *If $\Gamma^M(q_{\text{center}}(\mathbf{f}))$ is true and has type X, then $\Gamma^M(q_{\text{center}}(\mathbf{g}))$ is true and has type X.*

3. *If $\Gamma^M(q_{\text{center}}(\mathbf{g}))$ is true, then $\Gamma^M(q_{\text{center}}(\mathbf{f}))$ is true.*

4. *If $\mathbf{c}$ is a child of region $\mathbf{c}'$ in the hypercube forest, and $\Gamma^M(q_{\text{center}}(\mathbf{c})).b$ is true, then $\Gamma^M(q_{\text{center}}(\mathbf{c}'))$ is true and has type X.*

We sometimes use the term d-face in place of d-dimensional face. Note that a hypercube $\mathbf{c}$ is a four-dimensional face of itself. Thus Multiresolution Conditions 2 and 3 apply to each hypercube $\mathbf{c}$ and each facet $\mathbf{g}$ of $\mathbf{c}$. A multiresolution grid that satisfies the Multiresolution Conditions defines a multiresolution convex polyhedral mesh.

Proposition 10.13. *If an $n_1 \times n_2 \times n_3 \times n_4$ resolution grid satisfies the Multiresolution Conditions, then the set of hypercubes, pyramids over cubes, pyramids over square pyramids, and simplices determined by that grid form a convex polyhedral mesh.*

Input : Γ^M is an $n_1 \times n_2 \times n_3 \times n_4$ array of Boolean values.

SatisfyMultiresPoly4D(Γ^M)

1 $\mathcal{P}_{oct} \leftarrow$ octree partition of Γ^M;
2 Satisfy4DCondition1(Γ^M, $\mathcal{P}_{oct}$);
3 Let k be the smallest integer such that $2^k + 1 \geq \max(n_1, n_2, n_3, n_4)$;
4 for $j = 1$ to k do
5 $\quad$ Satisfy4DCondition2(Γ^M, j);
6 $\quad$ Satisfy4DCondition3(Γ^M, j);
7 $\quad$ Satisfy4DCondition4(Γ^M, j);
8 end

Algorithm 10.17. SATISFYMULTIRESPOLY4D.

Input : Γ^M is an $n_1 \times n_2 \times n_3 \times n_4$ array of Boolean values.
$\qquad$ 2^j is the length of the cubic regions begin processed.

Satisfy4DCondition1(Γ^M, j)

1 foreach hypercube $\mathbf{c}$ in $\mathcal{P}_j(\Gamma^M)$ do
2 $\quad$ foreach corner v of $\mathbf{c}$ do
3 $\quad\quad$ $\Gamma^M(v).b \leftarrow$ true;
4 $\quad\quad$ $\Gamma^M(v).t \leftarrow X$;
5 $\quad$ end
6 end

Algorithm 10.18. SATISFY4DCONDITION1.

The proof is similar to the proof of Proposition 10.1 and is omitted.

Given a grid of Boolean values, we modify it to satisfy the Multiresolution Conditions by setting $\Gamma^M(q_{center}(\mathbf{f})).b$ to true as appropriate (Algorithm 10.17, SATISFYMULTIRESPOLY4D). Let Γ_j^M be the largest subgrid of Γ^M containing $(0, 0, 0)$ that can be partitioned into cubic regions of edge length 2^j. Let $\mathcal{P}_j(\Gamma^M)$ be the partition of subgrid Γ_j^M into cubic regions of edge length 2^j. Process the edges, 2-faces, and facets of $\mathcal{P}_j(\Gamma^M)$ to satisfy each of the four Multiresolution Conditions (Algorithms 10.18, 10.19, 10.20, and 10.21). The order in which faces are processed is important, since setting $\Gamma^M(q_{center}(\mathbf{f})).b$ to true or $\Gamma^M(q_{center}(\mathbf{f})).t$ to X can affect $\Gamma^M(q_{center}(\mathbf{g}))$ for faces $\mathbf{g}$ containing or contained in $\mathbf{f}$.

Procedure SATISFYMULTIRESPOLY4D (Algorithm 10.17) satisfies the Multiresolution Condition but an additional procedure, IDENTIFYX4D, increases the number of vertices of type X (Algorithm 10.22). For each face $\mathbf{f}$ of $\mathcal{P}_j(\Gamma^M)$, if

Input : Γ^M is an $n_1 \times n_2 \times n_3 \times n_4$ array of Boolean values.
  2^j is the length of the cubic regions begin processed.

Satisfy4DCondition2(Γ^M, j)

```
1  for d = 3, 2, 1 do
2      foreach (d+1)-dimensional face f in Pⱼ(Γᴹ) do
3          if (Γᴹ(q_center(f)).b = true and Γᴹ(q_center(f)).t = X) then
4              foreach d-dimensional face g of f do
5                  Γᴹ(q_center(g)).b ← true;
6                  Γᴹ(q_center(g)).t ← X;
7              end
8          end
9      end
10 end
```

Algorithm 10.19. SATISFY4DCONDITION2.

Input : Γ^M is an $n_1 \times n_2 \times n_3 \times n_4$ array of Boolean values.
  2^j is the length of the cubic regions begin processed.

Satisfy4DCondition3(Γ^M, j)

```
1  for d = 1, 2, 3 do
2      foreach (d+1)-dimensional face f in Pⱼ(Γᴹ) do
3          foreach d-dimensional face g of f do
4              if (Γᴹ(q_center(g)).b = true and Γᴹ(q_center(f)).b = false) then
5                  Γᴹ(q_center(f)).b ← true;
6                  Γᴹ(q_center(f)).t ← Y;
7              end
8          end
9      end
10 end
```

Algorithm 10.20. SATISFY4DCONDITION3.

$\Gamma^M(q_{\text{center}}(\mathbf{g})).b$ is true for all the subfaces of $\mathbf{f}$, then set $\Gamma^M(q_{\text{center}}(\mathbf{g})).t$ to type X. The Boolean $(\Gamma^M(q_{\text{center}}(\mathbf{g})).b$ must already be true, since Γ^M satisfies the Multiresolution Conditions.) Increasing the number of type X vertices decreases the number of elements in the multiresolution mesh and the number of elements in the isosurface.

Input : Γ^M is an $n_1 \times n_2 \times n_3 \times n_4$ array of Boolean values.
         2^j is the length of the cubic regions begin processed.

`Satisfy4DCondition4(`Γ^M, j`)`

1 foreach hypercube **c** in $\mathcal{P}_j(\Gamma^M)$ do
2 | if $\Gamma^M(q_{\text{center}}(\mathbf{c})).b = $ true then
3 | | if **c** is one of the 16 subcubes of a hypercube $\mathbf{c}' \in \mathcal{P}_{j+1}(\Gamma^M)$ then
4 | | | $\Gamma^M(q_{\text{center}}(\mathbf{c}')).b \leftarrow$ true;
5 | | | $\Gamma^M(q_{\text{center}}(\mathbf{c}')).t \leftarrow X$;
6 | | end
7 | end
8 end

Algorithm 10.21. SATISFY4DCONDITION4.

Input : Γ^M is an $n_1 \times n_2 \times n_3 \times n_4$ array of Boolean values satisfying the
         Multiresolution Conditions.

`IdentifyX4D(`Γ^M`)`

1 Let k be the smallest integer such that $2^k + 1 \geq \max(n_1, n_2, n_3, n_4)$;
2 for $j = 1$ to k do
3 | for $d = 1, 2, 3$ do
4 | | foreach face **f** of dimension $(d+1)$ in $\mathcal{P}_j(\Gamma^M)$ do
5 | | | if $\Gamma^M(q_{\text{center}}(\mathbf{g}).b$ is true for every d-face **g** of **f** then
6 | | | | $\Gamma^M(q_{\text{center}}(\mathbf{f})).t \leftarrow X$;
7 | | | end
8 | | end
9 | end
10 end

Algorithm 10.22. IDENTIFYX4D.

10.3.4 Multiresolution Marching HPS

The four-dimensional version of MULTIRESOLUTION MARCHING CPT is called
MULTIRESOLUTION MARCHING HPS (Figure 10.4). HPS stands for hypercubes,
pyramids, and simplices.

Input to MULTIRESOLUTION MARCHING HPS is a scalar grid, Γ^S; an array,
Coord, of grid coordinates; a set, **RegionList**, of high-resolution regions; and a
scalar value σ. In its first step, the algorithm converts the high-resolution regions

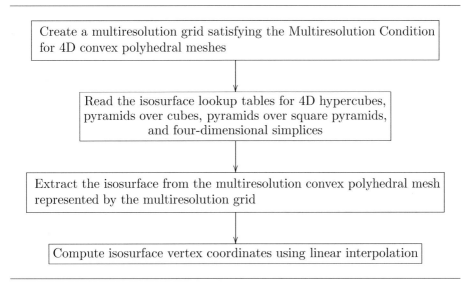

Figure 10.4. MULTIRESOLUTION MARCHING HPS for four-dimensional grids.

to a multiresolution grid, Γ^M. Set every vertex within a high-resolution region to true. Set all other vertices to false. Apply procedure SATISFYMULTIRESPOLY4D to the multiresolution grid so that is satisfies the Multiresolution Conditions. Apply procedure IDENTIFYX4D to reduce the number of polyhedral elements in the multiresolution grid.

In the second step, the algorithm reads four isosurface lookup tables, one for each polytope type. In the third step, MULTIRESOLUTION MARCHING HPS extracts the isosurface from the multiresolution mesh. Create a 4D octree partition, $\mathcal{P}_{oct}(\Gamma^M)$, of Γ^M.

For each cubic region $\mathbf{c} \in \mathcal{P}_{oct}(\Gamma^M)$, if $\Gamma^M(q_{center}(\mathbf{c})).b$ is false, then extract an isosurface patch from cube $\mathbf{c}$ using the isosurface lookup table for cubes. If $\Gamma^M(q_{center}(\mathbf{c}))$ is true and has type X, then divide $\mathbf{c}$ into sixteen hypercubes and recursively process each hypercube. If $\Gamma^M(q_{center}(\mathbf{c}))$ is true and has type Y, then divide $\mathbf{c}$ into eight pyramids over cubes.

For each pyramid $\mathbf{p}$ over a cube, if $\Gamma^M(q_{center}(\mathbf{p}))$ is false, then extract an isosurface patch from pyramid $\mathbf{p}$ using the isosurface lookup table for pyramids over cubes. If $\Gamma^M(q_{center}(\mathbf{p}))$ is true and has type X, then divide $\mathbf{p}$ into eight pyramids over cubes. Extract an isosurface patch from each of the eight subpyramids using the isosurface lookup table for pyramids over cubes. If $\Gamma^M(q_{center}(\mathbf{p}))$ is true and has type Y, then divide $\mathbf{p}$ into six pyramids over square pyramids.

For each pyramid $\mathbf{p}$ over a square pyramid, if $\Gamma^M(q_{center}(\mathbf{p}))$ is false, then extract an isosurface patch from pyramid $\mathbf{p}$ using the isosurface lookup table for pyramids over square pyramids. If $\Gamma^M(q_{center}(\mathbf{p}))$ is true and has type X, then divide **p** into four pyramids over square pyramids. Extract an isosurface patch from each of the four subpyramids using the isosurface lookup table for pyramids over pyramids. If $\Gamma^M(q_{\text{center}}(\mathbf{p}))$ is true and has type Y, then divide **p** into four type D simplices.

For each type D simplex **t**, if $\Gamma^M(q_{\text{center}}(\mathbf{t}))$ is false, then extract an isosurface patch from simplex **t** using the isosurface lookup table for four-dimensional simplices. If $\Gamma^M(q_{\text{center}}(\mathbf{t}))$ is true, then split **t** into two congruent type A simplices. Type A simplices are never subdivided. Extract the isosurface patch from each of the type A simplices using the isosurface lookup table for four-dimensional simplices.

In the final step, assign geometric locations to the isosurface vertices using linear interpolation.

As discussed in Section 6.3.1, it is crucial that isosurface patch triangulations match along their common boundary. The isosurface lookup tables for the different polytopes must be compatible and the mesh vertices must be properly mapped to the isosurface table vertices. As discussed in Section 6.5.3, the apex of a 4D pyramid over a cube should be mapped to the apex in the isosurface lookup table while the base vertices should be mapped in lexicographic order. Similarly, to map a 4D pyramid **p** over a 3D square pyramid **p**′ to a 4D pyramid ψ over a 3D square pyramid ψ' in the isosurface lookup table, the apex of **p** should map to the apex of ψ, the apex of **p**′ should map to the apex of ψ', and the base vertices of **p**′ should map in lexicographic order to the base vertices of ψ'.

10.4 Notes and Comments

Weber et al. [Weber et al., 2001] give a multiresolution polyhedral mesh with grids of cubes connected by pyramids and tetrahedra. They analyze 64 configurations of pyramids and tetrahedra to form these connections, exploiting some symmetry to simplify the analysis.

The multiresolution CPT (cube, pyramid, and tetrahedra) mesh presented in this chapter is a special case of diamond hierarchies discussed by Pascucci in [Pascucci, 2002]. Weiss and De Floriani in [Weiss and De Floriani, 2011] discuss variations and applications of these diamond hierarchies.

Other algorithms that construct multiresolution isosurfaces are described in [Ho et al., 2005, Schaefer and Warren, 2004, Kazhdan and Hoppe, 2007, Manson and Schaefer, 2010]. In general, the isosurfaces constructed by these algorithms are manifolds.

Multiresolution dual contouring algorithms are presented in [Perry and Frisken, 2001, Ju et al., 2002, Schaefer and Warren, 2002, Varadhan et al., 2003, Zhang et al., 2004]. Multiresolution dual contouring algorithms that produce manifolds are described in [Ashida and Badler, 2003, Greß and Klein, 2004, Schaefer et al., 2007].

Barry and Wood [Barry and Wood, 2007] give a dual contouring algorithm for constructing a multiresolution isosurface with a detailed normal map. The normal map contains information about the surface normals in low resolution regions.

CHAPTER 11

ISOVALUES

The previous chapters presented algorithms and data structures for building an isosurface for a given isovalue. In this chapter, we discuss algorithms and techniques for choosing this isovalue.

Choice of isovalue is application dependent. Different imaging devices, different materials, and different areas of interest dictate different choices for the isovalue. Nevertheless, there are some generic algorithms and techniques that can assist in finding the isovalue that determines the desired isosurface.

Different materials are often represented by different isovalues within a volumetric data set. (See Figures 11.1 and 11.2.) Each of the isosurfaces corresponding to each of the isovalues represents a different material surface. Thus, the goal is to determine a set of isovalues, not necessarily a single one, that may be significant.

Often a range of isovalues generates similar isosurfaces with small, insignificant differences between them. Isovalues within the range generate "equivalent" isosurfaces. Thus, the goal may be restated as determining a set of potentially significant isovalue ranges. Sampling one isovalue from each range and constructing the corresponding isosurface gives a set of potentially significant isosurfaces.

The simplest technique for choosing isovalues is to plot the frequency of each scalar value among the set of grid vertices. Using this plot, one selects a sample set of isovalues representing significantly different isosurfaces. Frequency plots of vertex scalar values are discussed in Section 11.1.

If the data set consists of scalar integers, use isovalues $(x + 0.5)$, where x is an integer. For data sets of integers and noninteger isovalues, the MARCHING CUBES and MANIFOLD DUAL MARCHING CUBES isosurfaces are always manifolds, and the returned sets of triangles contain no zero-area or duplicate triangles (Sections 2.3.3 and 3.3.6).

Another approach to representing scalar frequencies is to plot the number of grid edges or grid cubes containing each scalar value. As defined in Chapter 8, the span of a grid cube is the interval from the minimum to the maximum

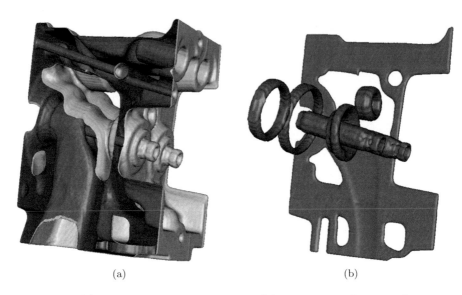

(a)   (b)

Figure 11.1. (a) Engine isosurface, isovalue 80. (b) Engine isosurface, isovalue 170.

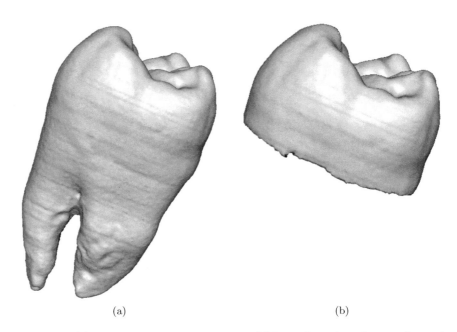

(a)   (b)

Figure 11.2. (a) Tooth isosurface, isovalue 640. (b) Isosurface of tooth enamel, isovalue 920.

scalar value of any cube vertex. The span of a grid edge is the interval from the minimum to the maximum scalar value of the two edge vertices. The plot represents the number of grid edges or grid cubes whose spans contain each scalar value. Frequency plots based on spans of grid edges and cubes are discussed in Section 11.2.1.

As discussed in Section 11.1, frequency plots of vertex scalar values represent interval volumes, the volume between two isosurfaces. In contrast, frequency plots of grid edges or grid cubes represent isosurface area. Section 11.2.2 describes how grid edges and grid cubes can be used to represent interval volumes.

The boundary between materials in a scalar field is often marked by a sharp change in the scalar values at the boundary. This change is reflected in a gradient with a large magnitude. Representations of the gradient are discussed in Section 11.3.

11.1   Counting Grid Vertices

The frequency of scalar value σ in a scalar grid is the number of grid vertices with scalar value σ. Let $\mathcal{F}_v(\sigma)$ denote the frequency of scalar value σ. When the scalar values are integers in a small range, the frequency of scalar values can be represented by a histogram, a bar graph representing the frequencies. The height of the bar corresponding to scalar value σ is $\mathcal{F}_v(\sigma)$, the frequency of σ. Figure 11.3 contains an example of a histogram of scalar frequencies. Note the the y-axis is represented on a logarithmic scale.

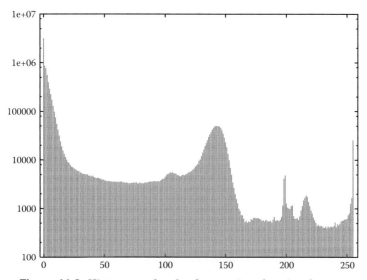

Figure 11.3. Histogram of scalar frequencies of engine data set.

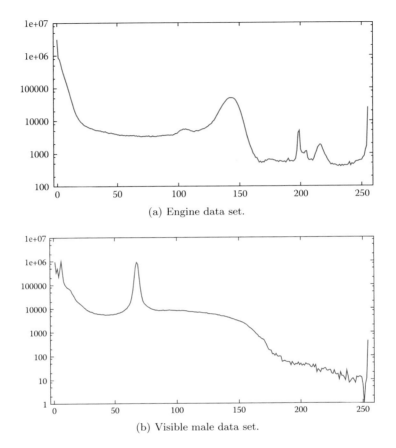

(a) Engine data set.

(b) Visible male data set.

Figure 11.4. Frequency polygons representing scalar frequencies.

When the scalar values are integers in a large range or the image resolution is small, the bars in a histogram become very dense. It is then better to represent the frequencies by the polygonal line with vertices $(\sigma, \mathcal{F}_v(\sigma))$. (See Figure 11.4.) Such a polygonal line is called a frequency polygon.

Scalar values may also be floating point numbers. In that case, the range $[s_{\min}, s_{\max}]$ of scalar values is partitioned into some fixed number, k, of uniform size bins. Bin $i < k$ has a subrange $[s_{i-1}, s_i)$ or, equivalently, $\{\sigma : s_{i-1} \leq \sigma < s_i\}$. Note that this subrange is closed at s_{i-1} and open at s_i. The value of s_0 and s_k are $s_{\min}$ and $s_{\max}$, respectively. We count the number of scalar values in each subrange.

The frequency of scalar values in the range $[\sigma_0, \sigma_1)$ (i.e., $\{\sigma : \sigma_0 \leq \sigma < \sigma_1\}$) is the number of grid vertices with scalar values in the range $[\sigma_0, \sigma_1)$. This frequency is denoted $\mathcal{F}_v(\sigma_0, \sigma_1)$.

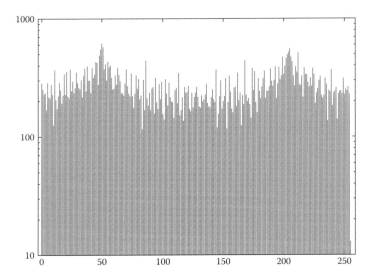

Figure 11.5. Histogram of scalar frequencies of Marschner-Lobb data set. The Marschner-Lobb data set is a sampling of the Marschner-Lobb function in the region $[-1, 1] \times [-1, 1] \times [-1, 1]$ on a regular $41 \times 41 \times 41$ grid. See Figure 11.9 for the definition of the Marschner-Lobb function.

The range $[\sigma_0, \sigma_1)$ is open at σ_1. It is useful to have a slightly different version of $\mathcal{F}_v$ on the closed range $[\sigma_0, \sigma_1]$. Let $\overline{\mathcal{F}}_v(\sigma_0, \sigma_1)$ denote the number of grid vertices with scalar values in the closed range $[\sigma_0, \sigma_1]$.

The frequency of scalar values in bin $i < k$ is the frequency, $\mathcal{F}_v(s_{i-1}, s_i)$, of scalar values in $[s_{i-1}, s_i)$. The frequency of scalar values in bin k is the frequency, $\overline{\mathcal{F}}_v(s_{k-1}, s_k)$, of scalar values in $[s_{k-1}, s_k]$. The frequency of scalar values is plotted either as a histogram or as a frequency polygon.

Bins are also useful on small grids of integer scalar values. On small grids with relatively few vertices, sampling frequencies of consecutive integers can vary greatly, creating "noisy" plots. (See Figure 11.5.) Partitioning the range into a small number of bins and counting the frequency of scalar values within each bin effectively smooths these plots, creating a much more useful representation of the data. (See Figure 11.6.)

The plot of scalar frequencies typically has some "peaks" of high-frequency scalar values separating regions of scalar values with approximately uniform frequencies. The high-frequency scalar values are often not good choices for isovalues. Zero typically has the highest frequency, often an order of magnitude larger than the frequency of any other scalar value. Isovalues at or near zero do not usually generate interesting or significant isosurfaces.

High-frequency scalar values other than zero often represent transitions between two surfaces representing different materials. (See Figures 11.7 and 11.8.)

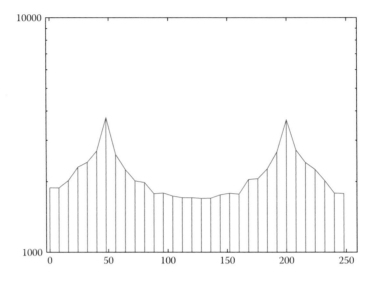

Figure 11.6. Plot of scalar frequencies of Marschner-Lobb data set using bins of size eight. Note that the range of frequencies has changed by about a factor of eight since each bin represents eight values instead of one.

Scalar values from the regions between "peaks" represent the materials themselves and are a better choice for isovalues than scalar values at the peaks.

To compare plots of frequencies from different data sets, it is useful to have a normalized version of the frequency. To normalize the frequencies, $\mathcal{F}_v(\sigma_0, \sigma_1)$ we divide by the total number, N_Γ, of the grid vertices of Γ, and by the size $\sigma_1 - \sigma_0$ of the bin. (Note that N_Γ is the total number of vertices, not the number of vertices along a grid axis. For an $n \times n \times n$ grid, N_Γ equals n^3.)

Define the **normalized frequency** of scalar values as

$$\mathcal{F}_v^*(\sigma_0, \sigma_1) = \frac{\mathcal{F}_v(\sigma_0, \sigma_1)}{N_\Gamma \times (\sigma_1 - \sigma_0)},$$

$$\overline{\mathcal{F}}_v^*(\sigma_0, \sigma_1) = \frac{\overline{\mathcal{F}}_v(\sigma_0, \sigma_1)}{N_\Gamma \times (\sigma_1 - \sigma_0)}.$$

When the range of scalar values is partitioned into some fixed number of bins of uniform size α, the normalized frequencies are simply the bin frequencies divided by $N_\Gamma \times \alpha$.

As either the number of grid vertices, N_Γ, or the bin size, $(\sigma_1 - \sigma_0)$, increase, the values $\mathcal{F}_v(\sigma_0, \sigma_1)$ and $\overline{\mathcal{F}}_v(\sigma_0, \sigma_1)$ increase proportionally, so $\mathcal{F}_v^*(\sigma_0, \sigma_1)$ and $\overline{\mathcal{F}}_v^*(\sigma_0, \sigma_1)$ remain relatively the same. Thus the normalized frequency measures depend primarily on the function ϕ, not on the grid or bin sizes.

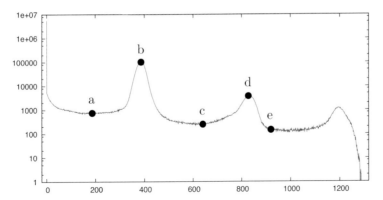

Figure 11.7. Scalar frequencies of tooth data set. Labeled points correspond to isovalues of images in Figure 11.8.

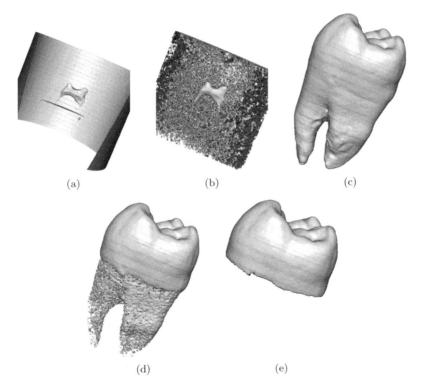

(a) (b) (c)

(d) (e)

Figure 11.8. Tooth data set. (a) Isovalue 187. Isosurface represents container around tooth and root of tooth. Isosurface is clipped to show root of tooth. (b) High-frequency isovalue 386. Isosurface represents the transition between the container surface (isovalue 187) and the tooth surface (isovalue 640). (c) Isovalue 640. Isosurface represents tooth surface. (d) High-frequency isovalue 828. Isosurface represents the transition between the tooth surface (isovalue 640) and the tooth enamel surface (isovalue 920). (e) Isovalue 920. Isosurface represents tooth enamel.

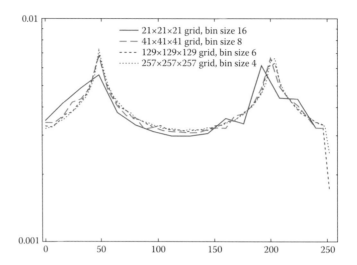

Figure 11.9. Normalized scalar frequencies for the Marschner-Lobb function $\phi(x, y, z) = (1 - \sin(\pi z / 2)) + (1 + \cos(12\pi \cos(\pi r / 2))) / 10$ where $r = \sqrt{x^2 + y^2}$ and $-1 \leq x, y, z \leq 1$. The function was sampled on four different regular grids. Plots of normalized frequencies on a $21 \times 21 \times 21$ grid with bin size 16, a $41 \times 41 \times 41$ grid with bin size 8, a $127 \times 127 \times 127$ grid with bin size 6, and a $257 \times 257 \times 257$ grid with bin size 4.

Figure 11.9 contains an example of normalized scalar frequency plots of the Marschner-Lobb function using different grid and bin sizes. Figure 11.10 contains normalized plots for the engine, visible male, and Marschner-Lobb data sets.

If the range $[s_{\min}, s_{\max}]$ of scalar values are partitioned into k bins $[s_{i-1}, s_i)$, then

$$\sum_{i=1}^{k-1} (\mathcal{F}_v^*(s_{i-1}, s_i) \times (s_i - s_{i-1})) + \overline{\mathcal{F}}_v^*(s_{k-1}, s_k) \times (s_k - s_{k-1}) = 1.$$

What property of the data is being measured by the frequency of scalar values? Let $f : \mathcal{D} \to \mathbb{R}$ define a scalar field on a rectangular region $\mathcal{D} \subset \mathbb{R}^3$ and let Γ be a regular grid sampling of ϕ. To make the dependence of $\mathcal{F}_v$ and $\overline{\mathcal{F}}_v$ on Γ explicit, we use the notation $\mathcal{F}_v(\sigma_0, \sigma_1, \Gamma)$ and $\overline{\mathcal{F}}_v(\sigma_0, \sigma_1, \Gamma)$ in place of $\mathcal{F}_v(\sigma_0, \sigma_1)$ and $\overline{\mathcal{F}}_v(\sigma_0, \sigma_1)$, respectively.

The value $\overline{\mathcal{F}}_v(\sigma_0, \sigma_1, \Gamma)$ is the number of vertices of Γ in the region $\{x \in \mathbb{R}^3 : \sigma_0 \leq \phi(x) \leq \sigma_1\}$. This region is $\mathcal{I}_\phi(\sigma_0, \sigma_1)$, the interval volume defined in Chapter 7. The ratios $\mathcal{F}_v(\sigma_0, \sigma_1, \Gamma)/N_\Gamma$ and $\overline{\mathcal{F}}_v(\sigma_0, \sigma_1, \Gamma)/N_\Gamma$ approximate $|\mathcal{I}_\phi(\sigma_0, \sigma_1)|/|\mathcal{D}|$ where $|\mathcal{I}_\phi(\sigma_0, \sigma_1)|$ is the volume of $\mathcal{I}_\phi(\sigma_0, \sigma_1)$ and $|\mathcal{D}|$ is the volume of the domain $\mathcal{D}$.

By repeatedly partitioning each cube in a grid into eight subcubes, we get a sequence of grids, each one with eight times the number of cubes and approx-

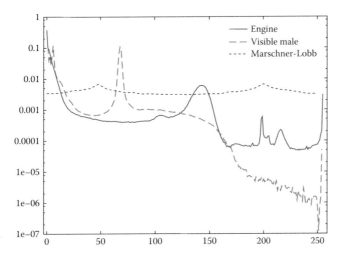

Figure 11.10. Normalized scalar frequencies of engine, visible male, and Marschner-Lobb data sets. The Marschner-Lobb data set samples the Marschner-Lobb function on a $41 \times 41 \times 41$ regular grid. Plot of Marschner-Lobb data set uses size eight bins.

imately eight times the number of vertices of its predecessor. As the number of grid vertices, N_Γ, goes to infinity, $\mathcal{F}_v(\sigma_0, \sigma_1, \Gamma)/N_\Gamma$ and $\overline{\mathcal{F}}_v(\sigma_0, \sigma_1, \Gamma)/N_\Gamma$ approach $|\mathcal{I}_\phi(\sigma_0, \sigma_1)|/|\mathcal{D}|$.

11.2 Counting Grid Edges and Grid Cubes

Grid edges and grid cubes can be used to measure isosurface area and to measure the interval volume between isosurfaces.

11.2.1 Measuring Isosurface Area

As previously defined, the span of a grid cube is the interval from the minimum to the maximum scalar value of any cube vertex. The span of a grid edge is the interval from the minimum to the maximum scalar value of the two edge vertices.

When the scalar values are integers, we plot the number of grid edges or grid cubes whose span contains each scalar value. When the scalar values are floating point numbers, the range $[s_{\min}, s_{\max}]$ of scalar values is sampled at regular intervals by a fixed number of scalar values. We plot the number of grid edges or grid cubes whose span intersects each scalar value. Plots can either be histograms or polygonal lines. Figure 11.11 contains examples of plots.

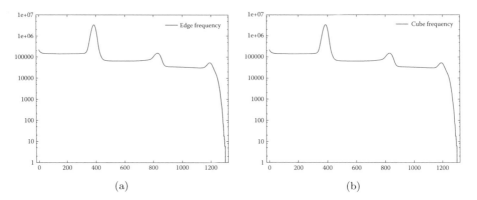

Figure 11.11. Tooth data set. (a) Number of grid edges whose spans contain the scalar value. (b) Number of grid cubes whose spans contain the scalar value.

As with plots of scalar frequencies, we define a normalized plot of frequencies of edge or cube spans. Let $\mathcal{F}_e(\sigma)$ and $\mathcal{F}_c(\sigma)$ be the number of edges or cubes, respectively, whose spans contain the scalar value σ.

To normalize the plot of edge and cube frequencies, we divide by the sum of $\mathcal{F}_e(\sigma)$ or $\mathcal{F}_c(\sigma)$ and by the average bin size. Let $\{s_i : i = 0, \ldots, k\}$ be the $k + 1$ scalar values of the range $[s_{\min}, s_{\max}]$ sampled at regular intervals where $s_0 = s_{\min}$ and $s_k = s_{\max}$. Let $\alpha = (s_{\max} - s_{\min})/k$ be the average bin size. Define normalized values $\mathcal{F}_e^*(\sigma)$ and $\mathcal{F}_c^*(\sigma)$ as

$$\mathcal{F}_e^*(\sigma) = \frac{\mathcal{F}_e(\sigma)}{\alpha \sum_{i=0}^{k} \mathcal{F}_e(s_i)},$$

$$\mathcal{F}_c^*(\sigma) = \frac{\mathcal{F}_c(\sigma)}{\alpha \sum_{i=0}^{k} \mathcal{F}_c(s_i)}.$$

Note that as the number k of samples increases, the decrease in the average bin size, α, is proportional to the increase in $\sum_{i=0}^{k} \mathcal{F}_c(s_i)$, so $\mathcal{F}_e^*(\sigma)$ and $\mathcal{F}_c^*(\sigma)$ remain approximately the same.

The normalized frequencies sum to $1/\alpha$ so that

$$\alpha \sum_{i=0}^{k} \mathcal{F}_e^*(s_i) = 1,$$

$$\alpha \sum_{i=0}^{k} \mathcal{F}_c^*(s_i) = 1.$$

As illustrated in Figure 11.12, normalized edge and cube frequencies depend primarily on the function ϕ, not the grid or bin sizes.

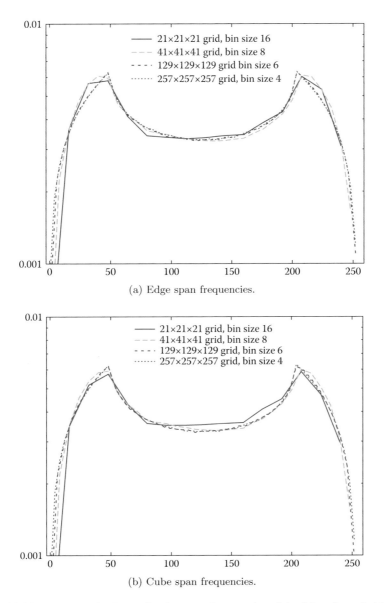

(a) Edge span frequencies.

(b) Cube span frequencies.

Figure 11.12. Normalized edge and cube span frequencies of the Marschner-Lobb function for a $21 \times 21 \times 21$ grid with bin size 16, a $41 \times 41 \times 41$ grid with bin size 8, a $129 \times 129 \times 129$ grid with bin size 6, and a $257 \times 257 \times 257$ grid with bin size 4.

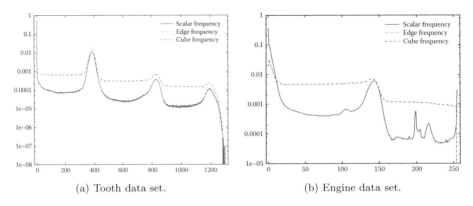

(a) Tooth data set. (b) Engine data set.

Figure 11.13. Graphs of $\mathcal{F}_v^*$, $\mathcal{F}_e^*$, and $\mathcal{F}_c^*$.

The functions $\mathcal{F}_e(\sigma)$ and $\mathcal{F}_c(\sigma)$ approximate the number of edges and cubes, respectively, intersecting the level set $\phi^{-1}(\sigma)$. This number can be used as a relative measure of the area $|\phi^{-1}(\sigma)|$ of $\phi^{-1}(\sigma)$.

More specifically, let $f : \mathcal{D} \to \mathbb{R}$ define a scalar field on a rectangular region $\mathcal{D} \subset \mathbb{R}^3$ and let Γ be a regular grid sampling of ϕ. To make the dependence of $\mathcal{F}_e$ and $\mathcal{F}_c$ on Γ explicit, we use the notation $\mathcal{F}_e(\sigma, \Gamma)$ and $\mathcal{F}_c(\sigma, \Gamma)$ in place of $\mathcal{F}_e(\sigma)$ and $\mathcal{F}_c(\sigma)$, respectively. The ratios $\mathcal{F}_e(\sigma, \Gamma)/\mathcal{F}_e(\sigma', \Gamma)$ and $\mathcal{F}_c(\sigma, \Gamma)/\mathcal{F}_c(\sigma', \Gamma)$ approximate the ratio of areas $|\phi^{-1}(\sigma)|/|\phi^{-1}(\sigma')|$ where σ and σ' are scalar values in the range of ϕ.

By repeatedly partitioning each cube in a grid into eight subcubes, we get a sequence of grids, each one with eight times the number of cubes and approximately eight times the number of vertices of its predecessor. As the number of grid vertices goes to infinity, $\mathcal{F}_e(\sigma, \Gamma)/\mathcal{F}_e(\sigma', \Gamma)$ and $\mathcal{F}_c(\sigma, \Gamma)/\mathcal{F}_c(\sigma', \Gamma)$ approach $|\phi^{-1}(\sigma)|/|\phi^{-1}(\sigma')|$.

11.2.2 Measuring Interval Volumes

Functions $\mathcal{F}_e$ and $\mathcal{F}_c$ approximate isosurface area while $\mathcal{F}_v$ approximates the interval volume between two isosurfaces. The two measures are not necessarily the same, even after normalization. (See Figure 11.13.) The measure of the interval volume between two isosurfaces with isovalues σ_0 and σ_1 depends upon the "distance" between the isosurfaces. In contrast, the area of the isosurface with isovalue σ_0 is independent of its "distance" to any other isosurface.

The interval volume can be measured using grid edges or grid cubes. Consider an edge whose endpoints p and q have scalar values s_p and s_q, respectively. If the interval $[\sigma_0, \sigma_1]$ intersects the edge span $[s_p, s_q]$, then the interval volume $\mathcal{I}_\phi(\sigma_0, \sigma_1)$ intersects edge (p, q). However, if $|s_q - s_p|$ is much larger than $|\sigma_1 - \sigma_0|$,

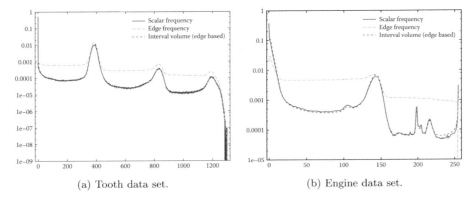

(a) Tooth data set.   (b) Engine data set.

Figure 11.14. Graphs of $\mathcal{F}_v^*$, $\mathcal{F}_e^*$, and Vol_e^*.

then $\mathcal{I}_\phi(\sigma_0, \sigma_1)$ will intersect only a fraction of edge (p, q). This fraction is $|\sigma_1 - \sigma_0|/|s_q - s_p|$.[1]

Let $E(\sigma_0, \sigma_1)$ be the set of grid edges (p, q) whose span, $[s_p, s_q]$, intersects $[\sigma_0, \sigma_1]$. Define $\mathrm{Vol}_e(\sigma_0, \sigma_1)$ as

$$\mathrm{Vol}_e(\sigma_0, \sigma_1) = \sum_{\substack{(p,q) \in E(\sigma_0, \sigma_1) \\ |s_p - s_q| \le |\sigma_1 - \sigma_0|}} \frac{1}{3} + \sum_{\substack{(p,q) \in E(\sigma_0, \sigma_1) \\ |s_q - s_p| > |\sigma_1 - \sigma_0|}} \frac{|\sigma_1 - \sigma_0|}{3|s_q - s_p|}.$$

The division by 3 is to compensate for the number of grid edges being about three times the number of grid vertices. Note that the values in the second sum are always less than or equal to one. The value $\mathrm{Vol}_e(\sigma_0, \sigma_1)$ approximates the size of the interval volume $\mathcal{I}_\phi(\sigma_0, \sigma_1)$.

To normalize Vol_e, we divide by the sum of Vol_e and by the bin size, $\sigma_1 - \sigma_0$. Let $\{s_i : i = 0, \ldots, k\}$ be the $k+1$ scalar values of the range $[s_{\min}, s_{\max}]$ sampled at regular intervals where $s_0 = s_{\min}$ and $s_k = s_{\max}$. Define normalized values $\mathrm{Vol}_e^*(\sigma_0, \sigma_1)$ as

$$\mathrm{Vol}_e^*(\sigma_0, \sigma_1) = \frac{\mathrm{Vol}_e(\sigma_0, \sigma_1)}{(\sigma_1 - \sigma_0) \sum_{i=1}^{k} \mathrm{Vol}_e(s_{i-1}, s_i)}.$$

Figure 11.14 contains examples comparing $\mathcal{F}_e^*$ and Vol_e^*.

To measure the interval volume using grid cubes, let $\mathrm{Span}(c)$ be the span of cube c, i.e., the interval $[\sigma, \sigma']$ where σ is the lowest scalar value of any vertex in c and σ' is the greatest. The notation $|\mathrm{Span}(c)| = |\sigma' - \sigma|$ represents the size of the span of c. Let $C(\sigma_0, \sigma_1)$ be the set of grid cubes whose span intersects

---

[1]Alternate derivations ([Bachthaler and Weiskopf, 2008] and [Scheidegger et al., 2008]) are based on the coarea formula. See notes and comments in Section 11.4.

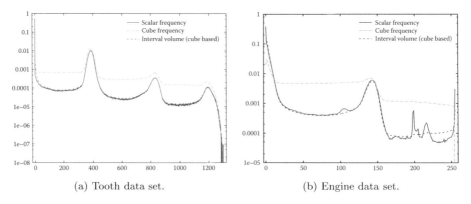

(a) Tooth data set.   (b) Engine data set.

Figure 11.15. Graphs of $\mathcal{F}_v^*$, $\mathcal{F}_c^*$, and Vol_c^*.

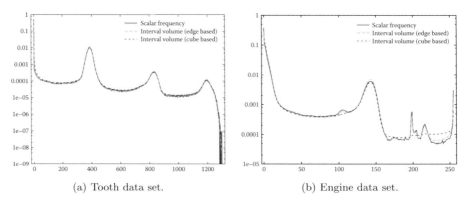

(a) Tooth data set.   (b) Engine data set.

Figure 11.16. Graphs of $\mathcal{F}_v^*$, Vol_e^*, and Vol_c^*.

$[\sigma_0, \sigma_1]$. Define $\mathrm{Vol}_c(\sigma_0, \sigma_1)$ as

$$\mathrm{Vol}_c(\sigma_0, \sigma_1) = \sum_{\substack{c \in C(\sigma_0, \sigma_1) \\ |\mathrm{Span}(c)| \le |\sigma_1 - \sigma_0|}} 1 + \sum_{\substack{(p,q) \in E(\sigma_0, \sigma_1) \\ |\mathrm{Span}(c)| > |\sigma_1 - \sigma_0|}} \frac{|\sigma_1 - \sigma_0|}{|\mathrm{Span}(c)|}.$$

The value $\mathrm{Vol}_c(\sigma_0, \sigma_1)$ approximates the size of the interval volume $\mathcal{I}_\phi(\sigma_0, \sigma_1)$.

Define $\mathrm{Vol}_c^*(\sigma_0, \sigma_1)$, the normalized version of $\mathrm{Vol}_c(\sigma_0, \sigma_1)$, as

$$\mathrm{Vol}_c^*(\sigma_0, \sigma_1) = \frac{\mathrm{Vol}_c(\sigma_0, \sigma_1)}{(\sigma_1 - \sigma_0) \sum_{i=1}^{k} \mathrm{Vol}_c(s_{i-1}, s_i)}.$$

Since the number of grid cubes is approximately the number of grid vertices, there is no division by 3.

Figure 11.15 contains examples comparing $\mathcal{F}_c^*$ and Vol_c^*. Figure 11.16 contains examples comparing $\mathcal{F}_v^*$, Vol_e^*, and Vol_c^*.

11.3 Measuring Gradients

Given a scalar field $\phi : \mathbb{R}^3 \to \mathbb{R}$, the gradient of ϕ is

$$\nabla \phi = \left(\frac{\partial \phi}{\partial x}, \frac{\partial \phi}{\partial y}, \frac{\partial \phi}{\partial z} \right).$$

The magnitude of the gradient is

$$|\nabla \phi| = \sqrt{\left(\frac{\partial \phi}{\partial x} \right)^2 + \left(\frac{\partial \phi}{\partial y} \right)^2 + \left(\frac{\partial \phi}{\partial z} \right)^2}.$$

High gradient magnitudes often indicate changes in the materials represented by the data set. High gradient magnitudes are often better indicators of significant isosurfaces than scalar frequencies.

There are two measures of gradients over a level set. The total gradient of ϕ over a level set $\sigma = \phi^{-1}(\sigma)$ is

$$\int_{S=\phi^{-1}(\sigma)} |\nabla \phi| dS.$$

The total gradient integrates the gradient magnitude over the entire level set. Thus level sets with large gradients but a small area can have the same total gradient as level sets with small gradients but a large area. The mean gradient of ϕ over a level set $\sigma = \phi^{-1}(\sigma)$ is

$$\frac{\int_{S=\phi^{-1}(\sigma)} |\nabla \phi| dS}{|\phi^{-1}(\sigma)|},$$

where $|\phi^{-1}(\sigma)|$ is the area of $\phi^{-1}(\sigma)$. It is sometimes a better indicator of significant isovalues than the total gradient.

Edge and cube spans can be used to approximate the relative magnitude of gradients (Section 11.3.1). They can also be computed using the Laplacian (Section 11.3.2).

11.3.1 Gradients Based on Grid Edges and Grid Cubes

The simplest approximation to the gradient is the difference between the scalar values of grid edge endpoints. Summing $|s_q - s_p|$ over all edges (p, q) that intersect an isosurface gives a relative measure of the total gradient.

Let $E(\sigma)$ be the set of grid edges (p, q) that intersect the isosurface with isovalue σ. Define $\text{TotalGrad}_e(\sigma)$ as

$$\text{TotalGrad}_e(\sigma) = \sum_{(p,q) \in E(\sigma)} \frac{|s_q - s_p|}{3}.$$

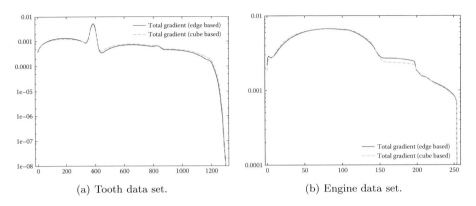

(a) Tooth data set.   (b) Engine data set.

Figure 11.17. Graphs of TotalGrad_e^* and TotalGrad_c^*.

Again, the division by 3 compensates for the number of grid edges being about three times the number of grid vertices.

The ratio $\text{TotalGrad}_e(\sigma)/\text{TotalGrad}_e(\sigma')$ approximates the ratio of total gradients,

$$\frac{\int_{S=\phi^{-1}(\sigma)} |\nabla\phi| dS}{\int_{S'=\phi^{-1}(\sigma')} |\nabla\phi| dS'}.$$

Cube spans can also be used for a relative measure of the total gradient. As in Section 11.2.2, let $\text{Span}(c)$ be the interval $[\sigma, \sigma']$ where σ is the lowest scalar value of any vertex in c and σ' is the greatest. Let $C(\sigma)$ be the set of grid cubes that intersect the isosurface with isovalue σ. Define $\text{TotalGrad}_c(\sigma)$ as

$$\text{TotalGrad}_c(\sigma) = \sum_{c \in C(\sigma)} |\text{Span}(c)|.$$

The ratio $\text{TotalGrad}_c(\sigma)/\text{TotalGrad}_c(\sigma')$ also approximates the corresponding ratio of total gradients.

As with all other measures, normalized versions are useful for comparing different data sets. Let $\{s_i : i = 0, \ldots, k\}$ be the $k+1$ scalar values of the range $[s_{\min}, s_{\max}]$ sampled at regular intervals where $s_0 = s_{\min}$ and $s_k = s_{\max}$. Let $\alpha = (s_{\max} - s_{\min})/k$ be the average bin size. The normalized versions of $\text{TotalGrad}_e(\sigma)$ and $\text{TotalGrad}_c(\sigma)$ are

$$\text{TotalGrad}_e^*(\sigma) = \frac{\text{TotalGrad}_e(\sigma)}{\alpha \sum_{i=0}^k \text{TotalGrad}_e(s_i)},$$

$$\text{TotalGrad}_c^*(\sigma) = \frac{\text{TotalGrad}_c(\sigma)}{\alpha \sum_{i=0}^k \text{TotalGrad}_c(s_i)}.$$

As before, α equals the average bin size, $(s_{\max} - s_{\min})/k$.

Figure 11.17 contains examples of plots of TotalGrad_e^* and TotalGrad_c^*.

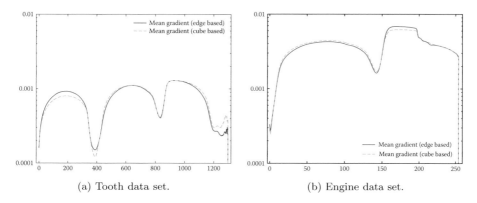

(a) Tooth data set.   (b) Engine data set.

Figure 11.18. Graphs of MeanGrad$_e^*$ and MeanGrad$_c^*$.

The edge frequency, $\mathcal{F}_e(\sigma)$, approximates the isosurface area (up to a constant factor.) The mean gradient is computed by dividing by $\mathcal{F}_e(\sigma)$. Define MeanGrad$_e(\sigma)$ and MeanGrad$_c(\sigma)$ as

$$\text{MeanGrad}_e(\sigma) \;=\; \frac{\text{TotalGrad}_e(\sigma)}{\mathcal{F}_e(\sigma)},$$

$$\text{MeanGrad}_c(\sigma) \;=\; \frac{\text{TotalGrad}_c(\sigma)}{\mathcal{F}_e(\sigma)}.$$

We use $\mathcal{F}_e(\sigma)$, not $\mathcal{F}_c(\sigma)$, in the definition of MeanGrad$_c$. We could have used $\mathcal{F}_c(\sigma)$, but $\mathcal{F}_e(\sigma)$ seems to be a slightly more accurate measure of relative isosurface area.

Normalized versions of MeanGrad$_e(\sigma)$ and MeanGrad$_c(\sigma)$ are

$$\text{MeanGrad}_e^*(\sigma) \;=\; \frac{\text{MeanGrad}_e(\sigma)}{\alpha \sum_{i=0}^{k} \text{MeanGrad}_e(s_i)},$$

$$\text{MeanGrad}_c^*(\sigma) \;=\; \frac{\text{MeanGrad}_c(\sigma)}{\alpha \sum_{i=0}^{k} \text{MeanGrad}_c(s_i)}.$$

Figure 11.18 contains examples of plots of MeanGrad$_e^*$ and MeanGrad$_c^*$.

Figure 11.19 compares plots of the scalar frequency, isosurface area, total gradient, and mean gradient. The plots of the scalar frequency and isosurface area are similar but not quite the same. Peaks in those plots correspond to isosurfaces representing transitions between surfaces. These isosurfaces have large amounts of noise. The significant isosurfaces lie between the peaks. Some of the peaks in the plot of total gradient correspond to significant isosurfaces, but others correspond to isosurfaces with large amounts of noise. Some significant isovalues also fail to correspond to any peak.

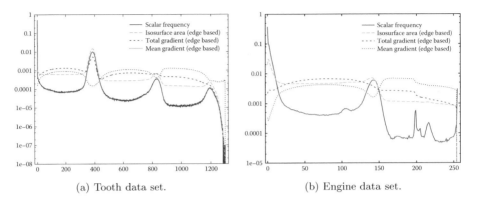

(a) Tooth data set.   (b) Engine data set.

Figure 11.19. Comparisons of scalar frequency, isosurface area, total gradient, and mean gradient.

The most useful plot in Figure 11.19 is the plot of the mean gradient. The three distinct peaks in the tooth data set correspond to three distinct, significant isovalues. Similarly, the two distinct peaks in the engine data set correspond to two distinct, significant isovalues.

11.3.2 Gradients Based on the Laplacian

Let $\Psi : \mathbb{R}^3 \to \mathbb{R}^3$ be a continuously differentiable vector field defined as

$$\Psi(x, y, z) = (\Psi_x(x, y, z), \Psi_y(x, y, z), \Psi_z(x, y, z)).$$

The divergence of Ψ is

$$\mathrm{div}(\Psi) = \frac{\partial \Psi_x(x, y, z)}{\partial x} + \frac{\partial \Psi_y(x, y, z)}{\partial y} + \frac{\partial \Psi_z(x, y, z)}{\partial z}.$$

Let $\mathbb{X}$ be a compact volume with piecewise-smooth boundary $\partial \mathbb{X}$. The divergence theorem is

$$\int_{\mathbb{X}} \mathrm{div}(\Psi) d\mathbb{X} = \int_{S = \partial \mathbb{X}} \Psi \cdot \mathbf{n} \, dS,$$

where $\mathbf{n}$ is the unit normal vector to S that points away from $\mathbb{X}$. The divergence theorem is a special case of Stoke's theorem.

When Ψ is the gradient vector field $\nabla \phi = \left(\frac{\partial \phi}{\partial x}, \frac{\partial \phi}{\partial y}, \frac{\partial \phi}{\partial z} \right)$ of a continuously differentiable scalar field $\phi : \mathbb{R}^3 \to \mathbb{R}$ and $\mathbb{X}$ equals $\{x : \phi(x) \geq \sigma\}$, the divergence theorem becomes

$$\int_{\mathbb{X} = \phi^{-1}([\sigma, \infty])} \mathrm{div}(\nabla \phi) d\mathbb{X} = \int_{S = \phi^{-1}(\sigma)} |\nabla \phi| \, dS.$$

The divergence of the gradient of ϕ is called the Laplacian of ϕ and is denoted $\Delta\phi$. The Laplacian can also be defined directly as

$$\Delta\phi = \frac{\partial^2\phi}{\partial x^2} + \frac{\partial^2\phi}{\partial y^2} + \frac{\partial^2\phi}{\partial z^2}.$$

Using the Laplacian, the divergence theorem becomes

$$\int_{\mathbb{X}=\phi^{-1}([\sigma,\infty])} \Delta\phi \, d\mathbb{X} = \int_{S=\phi^{-1}(\sigma)} |\nabla\phi| \, dS. \tag{11.1}$$

Pekar, Wiemker, and Hempel [Pekar et al., 2001] used the Laplacian and Equation (11.1) to compute total and mean gradients over a level set. Let V_Γ be the set of grid vertices of grid Γ. Define $\text{TotalGrad}_{\mathcal{L}}(\sigma)$ as

$$\text{TotalGrad}_{\mathcal{L}}(\sigma) = \sum_{\substack{p \in V_\Gamma \\ \phi(p) \geq \sigma}} \Delta\phi(p).$$

If $\{s_i : i = 0, \ldots, k\}$ are $k+1$ scalar values in increasing order, then

$$\text{TotalGrad}_{\mathcal{L}}(s_i) = \text{TotalGrad}_{\mathcal{L}}(s_{i+1}) + \sum_{\substack{p \in V_\Gamma \\ s_i \leq \phi(p) < s_{i+1}}} \Delta\phi(p). \tag{11.2}$$

Algorithm 11.1 applies Equation (11.2) to compute $\text{TotalGrad}_{\mathcal{L}}$.

Computing $\text{TotalGrad}_{\mathcal{L}}$ requires computing the Laplacian at grid points (i_x, i_y, i_z). The partial second derivatives can be computed by the formulas [Cheney and Kincaid, 2007]

$$\frac{\partial^2\phi}{\partial x^2}(i_x, i_y, i_z) = \phi(i_x - 1, i_y, i_z) - 2\phi(i_x, i_y, i_z) + \phi(i_x + 1, i_y, i_z),$$

$$\frac{\partial^2\phi}{\partial y^2}(i_x, i_y, i_z) = \phi(i_x, i_y - 1, i_z) - 2\phi(i_x, i_y, i_z) + \phi(i_x, i_y + 1, i_z),$$

$$\frac{\partial^2\phi}{\partial z^2}(i_x, i_y, i_z) = \phi(i_x, i_y, i_z - 1) - 2\phi(i_x, i_y, i_z) + \phi(i_x, i_y, i_z + 1).$$

Note that these formulas only work for points at the grid interior. The Laplacian is computed by summing the three partial derivatives. (See Algorithm 11.2.)

The running time of Algorithm 11.1 depends upon the time to divide the interior grid vertices into bins (Step 1). If the scalar values in S are spaced at regular intervals, then the bin containing a grid vertex can be determined in constant time and Step 1 takes time proportional to the number of grid vertices. The rest of the algorithm takes time proportional to the number of grid vertices.

Input   : F is a 3D array of scalar values.

S is an array of scalar values in increasing order.

k is the index of the last element of array S.

Result   : An array G where G[i] is total gradient of the isosurface with isovalue S[i].

LaplacianTotalGradient(F,Coord,S)

1 Divide the interior grid vertices (i_x, i_y, i_z) into k bins where:
   - $(i_x, i_y, i_z) \in$ Bin[j] for $j < k$ if S[j] $\leq F[i_x, i_y, i_z] <$ S[$j+1$], and
   - $(i_x, i_y, i_z) \in$ Bin[k] if S[k] $\leq F[i_x, i_y, i_z]$;
2 G[k] $\leftarrow$ 0;
3 foreach interior grid vertex $(i_x, i_y, i_z) \in$ Bin[k] do
4 | G[k] $\leftarrow$ G[k]+Laplace(F, i_x, i_y, i_z);
5 end
6 for $j \leftarrow k-1$ downto 0 do
7 | G[j] $\leftarrow$ G[$j+1$];
8 | foreach interior grid vertex $(i_x, i_y, i_z) \in$ Bin[j] do
9 | | G[j] $\leftarrow$ G[j]+Laplace(F, i_x, i_y, i_z);
10 | end
11 end

Algorithm 11.1. LAPLACIANTOTALGRADIENT.

Input   : F is a 3D array of scalar values.

Coordinates i_x, i_y and i_z of an interior grid point.

Output   : The Laplacian of F at (i_x, i_y, i_z).

Laplace(F, i_x, i_y, i_z)

1 ddx $\leftarrow F[i_x - 1, i_y, i_z] - 2F[i_x, i_y, i_z] + F[i_x + 1, i_y, i_z]$;
2 ddy $\leftarrow F[i_x, i_y - 1, i_z] - 2F[i_x, i_y, i_z] + F[i_x, i_y + 1, i_z]$;
3 ddz $\leftarrow F[i_x, i_y, i_z - 1] - 2F[i_x, i_y, i_z] + F[i_x, i_y, i_z + 1]$;
4 L $\leftarrow$ ddx + ddy + ddz;
5 return L;

Algorithm 11.2. LAPLACE.

Algorithm 11.1 ignores the Laplacian of vertices on the grid boundary. A modified formula could be used to compute the Laplacian of boundary vertices, but in practice the Laplacian of the boundary vertices has little effect on the sum.

TotalGrad$_\mathcal{L}(\sigma)$ approximates the total gradient $\int_{S=\phi^{-1}(\sigma)} |\nabla\phi| dS$. Note that this contrasts with TotalGrad$_e(\sigma)$ and TotalGrad$_c(\sigma)$, which approximate the total gradient only up to some constant factor. (See Figure 11.20.)

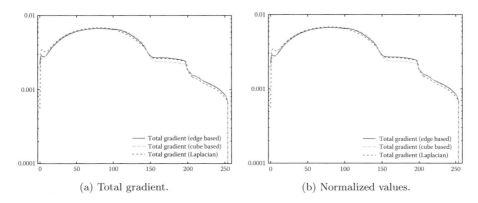

(a) Total gradient.   (b) Normalized values.

Figure 11.20. Engine data set. (a) Functions TotalGrad$_e$, TotalGrad$_c$, and TotalGrad$_\mathcal{L}$. Function TotalGrad$_\mathcal{L}$ is the Laplacian-based approximation for the total gradient. (b) Normalized versions: TotalGrad$_e^*$, TotalGrad$_c^*$, and TotalGrad$_\mathcal{L}^*$.

The normalized version of TotalGrad$_\mathcal{L}$, the Laplacian-based computation of the mean gradient, and its normalized version are

$$\mathrm{TotalGrad}_\mathcal{L}^*(\sigma) = \frac{\mathrm{TotalGrad}_\mathcal{L}(\sigma)}{\alpha \sum_{i=0}^{k} \mathrm{TotalGrad}_\mathcal{L}(s_i)},$$

$$\mathrm{MeanGrad}_\mathcal{L}(\sigma) = \frac{\mathrm{TotalGrad}_\mathcal{L}(\sigma)}{\mathcal{F}_e(\sigma)},$$

$$\mathrm{MeanGrad}_\mathcal{L}^*(\sigma) = \frac{\mathrm{MeanGrad}_\mathcal{L}(\sigma)}{\alpha \sum_{i=0}^{k} \mathrm{MeanGrad}_\mathcal{L}(s_i)}.$$

Equation 11.1 still holds if ϕ is defined on some subset $\mathcal{D}$ of $\mathbb{R}^3$ as long as $\mathbb{X} = \{x \in \mathcal{D} : \phi(x) \geq \sigma\}$ is bounded by $\phi^{-1}(\sigma)$. However, if the level set $\phi^{-1}(\sigma)$ has a boundary, then it no longer bounds $\mathbb{X}$. Thus, TotalGrad$_\mathcal{L}(\sigma)$ is not a good measure of the total gradient of isosurfaces with boundary. (It is also not a good measure if the points $\{p \in V_\Gamma : \phi(p) \geq \sigma\}$ lie between the isosurface and the grid boundary.)

For example, Figure 11.21(a) contains a plot of TotalGrad$_e$ and TotalGrad$_\mathcal{L}$ for the Gaussian function $\phi(x,y,z) = e^{-(x^2+y^2+z^2)/2}$ over the domain $\mathcal{D} = \{(x,y,z) : -1 \leq x,y,z \leq 1\}$. Note that TotalGrad$_e(\sigma)$ and TotalGrad$_\mathcal{L}(\sigma)$ agree for $\sigma > 0.6$ but diverge drastically when $\sigma < 0.6$. (The actual point of divergence is $\sigma = e^{-1/2} \approx 0.6065$.) The reason that TotalGrad$_\mathcal{L}$ diverges from the total gradient for $\sigma < 0.6$ is that isosurfaces with isovalue σ intersect the grid boundary and no longer "contain" the set $\{p \in V_\Gamma : \phi(p) \geq \sigma\}$.

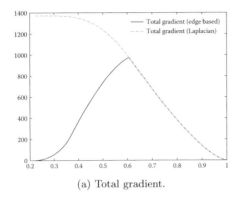

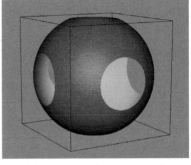

(a) Total gradient.   (b) Isosurface (isovalue 0.55).

Figure 11.21. Gaussian function $\phi(x,y,z) = e^{-(x^2+y^2+z^2)/2}$ for $-1 \leq x, y, z \leq 1$. (a) TotalGrad$_e$ and TotalGrad$_\mathcal{L}$. (b) Isosurface for isovalue 0.55. The boundaries of the isosurface lie on the bounding box of the grid.

11.4   Notes and Comments

Bajaj, Pascucci, and Schikore [Bajaj et al., 1997] suggested plotting isosurface metrics as a function of isovalue and using those metrics to select isovalues. They call the set of computed metrics the contour spectrum of the scalar field. Pekar et al. [Pekar et al., 2001] gave the algorithm in Section 11.3.2 for computing the total and mean gradients using the Laplacian.

Carr, Duffy, and Denby [Carr et al., 2006a] and Scheidegger et al. [Scheidegger et al., 2008] analyze the relationship between histograms of scalar values and isosurface area. Duffy, Carr, and Moller [Duffy et al., 2012] discuss the effects of quantization of the scalar values associated with grid or mesh vertices on computing isosurface statistics.

Scheidegger et al. [Scheidegger et al., 2008] and Bachthaler and Weiskopf [Bachthaler and Weiskopf, 2008] relate the isosurface area to the interval volume based on the coarea formula by Federer. (See [Morgan, 2009] for a thorough explanation of the formula.) The formula relates the integral over an interval volume to an integral over a range of level sets. The coarea formula is

$$\int_{x \in \mathcal{I}_\phi(\sigma_0, \sigma_1)} \psi(x)|\nabla\phi(x)|dV = \int_{\sigma_0 \leq h \leq \sigma_1} \left(\int_{x \in \phi^{-1}(h)} \psi(x)dS \right) dh,$$

where ϕ and ψ are functions from $\mathbb{R}^3$ to R and $\mathcal{I}_\phi(\sigma_0, \sigma_1)$ is the interval volume between σ_0 and σ_1 and $|\nabla\phi(x)|$ is the magnitude of the gradient of ϕ. The term on the left of the equal sign integrates ψ directly over the entire volume while the term on the right integrates ψ over the isosurfaces $\phi^{-1}(h)$. The formula indicates that a scaling factor of $|\nabla \phi(x)|$ must be used in transforming from an integral over the volume to an integral over isosurfaces.

Instead of representing attributes as a function of scalar value, one could compute two (or more) attributes for each grid vertex or cube and plot these attributes as points. For instance, plot the point $(s_{\min}(\mathbf{c}), s_{\max}(\mathbf{c}))$ for each grid cube $\mathbf{c}$. For another example, let s_v be the scalar value and let g_v be the gradient at grid vertex v. Plot (s_v, g_v) for each grid vertex v. Such plots are called scatter plots. For large data sets, it is necessary to subsample the data set before creating the scatter plot.

Scatter plots are simply collection of points. It is often hard to see structure in such collections of points. Bachthaler and Weisskopf [Bachthaler and Weiskopf, 2008, Bachthaler and Weiskopf, 2009] describe a continuous version of scatter plots for scalar data sets that often gives better insight into the data set.

Tenginakai et al. [Tenginakai et al., 2001, Tenginakai and Machiraju, 2002] compute moments of scalar values in windows around grid vertices. They use such moments to determine scalar values near boundary regions in the data set. Such scalar values are good candidates for isovalues.

Pekar et al. in [Pekar et al., 2001] suggest comparing isovalues by computing the fractal dimension of the correspoinding isosurfaces. Khoury and Wenger [Khoury and Wenger, 2010] give a simple way for computing the isosurface fractal dimension. They show that the isosurface fractal dimension is correlated with noise in the isosurface and inversely correlated with the gradient magnitude. Plotting the fractal dimension as a function of the isovalue gives another graph of isovalue properties and another way to look for significant isovalues.

Bruckner and Möller [Bruckner and Möller, 2010] compute the similarity between isosurfaces for different isovalues. This computation can be used to identify stable isovalues where small perturbations in the isovalue will not substantially change the isosurface. It also identifies ranges of isovalues that generate similar isosurfaces.

Marschner and Lobb [Marschner and Lobb, 1994] suggested the function in Figure 11.9 for evaluating surface reconstruction algorithms. The Marschner-Lobb data set is a sampling of this function.

CHAPTER 12

CONTOUR TREES

Let $\phi : \mathcal{D} \to \mathbb{R}$ be a continuous scalar field where $\mathcal{D}$ is a rectangular region in $\mathbb{R}^d$. More generally, $\mathcal{D}$ is homeomorphic to a closed ball in $\mathbb{R}^d$. A connected component of a level set $\phi^{-1}(\sigma)$ is a maximally connected subset of $\phi^{-1}(\sigma)$, i.e., a connected subset of $\phi^{-1}(\sigma)$ that is not contained in any other connected subset of $\phi^{-1}(\sigma)$. As σ changes, the connected components of $\phi^{-1}(\sigma)$ change. New components appear, existing components disappear, multiple components join together to form one component, or a single component splits apart into multiple components. A contour tree is a graph that represents these components and their changes.

In this chapter, we present an elegant algorithm by Carr, Snoeyink, and Axen [Carr et al., 2003] for constructing the contour tree of a piecewise linear scalar field defined by a tetrahedral mesh. The algorithm constructs a "join" tree representing the joining of contours and a "split" tree representing the splitting of contours, and then the algorithm combines the two trees into a "merge" tree. This merge tree is equivalent to the contour tree.

Section 12.2 contains a formal definition of contour trees and some contour tree properties. Section 12.3 gives the definition of join, split, and merge trees. Section 12.4 presents algorithms for constructing the join, split, and merge trees. Section 12.5 presents algorithms for constructing contour trees from tetrahedral meshes and from regular grids. Proofs of all propositions are contained in Section 12.6.

12.1   Examples of Contour Trees

The standard, intuitive definition of a contour tree for a scalar field $\phi : \mathcal{D} \to \mathbb{R}$ is as follows. Collapse each connected component of each level set $\phi^{-1}(\sigma)$ into a single point. The resulting "graph" is a contour tree.

379

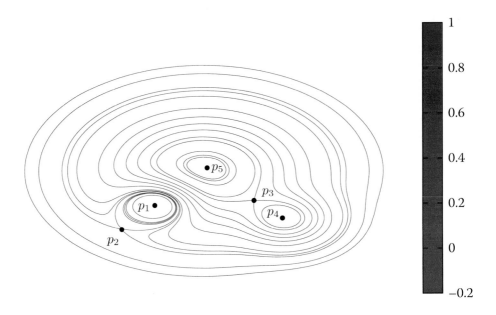

Figure 12.1. Contour lines for a scalar field $\phi : \mathcal{D} \to \mathbb{R}$. Domain $\mathcal{D}$ is the region bounded by the outermost contour. Point p_1 is a local minimum. Points p_4 and p_5 are local maxima. Connected components of contours are joined at point p_2 and split at point p_3. The outermost contour is also a local minimum.

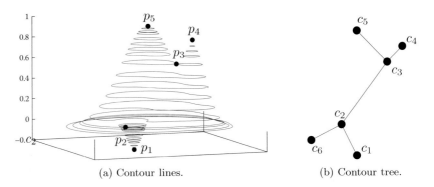

(a) Contour lines.   (b) Contour tree.

Figure 12.2. (a) Contour lines lifted into 3D. (b) Vertices c_1, c_4, and c_5 correspond to points p_1, p_4, and p_5, respectively. Vertices c_2 and c_3 correspond to the contour lines through p_2 and p_3, respectively. Vertex c_6 corresponds to the outermost contour, which is the boundary of $\mathcal{D}$.

For instance, Figure 12.1 contains contours of the function

$$\phi(x,y) = e^{-(x^2+y^2)} + (2/3)e^{-5((x-1)^2+(y+1)^2)} - (2/3)e^{-5((x+0.5)^2+(y+0.6)^2)}.$$
$$(12.1)$$

Set $\mathcal{D}$ is the region bounded by the outermost contour. The scalar function ϕ forms a surface $\{(x,y,\phi(x,y))\}$ in $\mathbb{R}^3$. Figure 12.2(a) displays a plot of the contours on that surface.

Collapsing the connected components of the contours of ϕ into level sets gives the contour tree in Figure 12.2(b). The collapse is perhaps best visualized by comparing Figure 12.2(a) with Figure 12.2(b). The connected components on the horizontal plane $z = \sigma$ map to points at height σ in the contour tree.

For isovalues in the range [-0.8,0.06], the isocontour has two components, one around the point p_1 and one near the boundary of the domain $\mathcal{D}$. Contour tree edge (c_1, c_2) corresponds to the isocontours around p_1. Contour tree edge (c_6, c_2) corresponds to the isocontours near $\partial \mathcal{D}$. These two components join at scalar value 0.06. The isocontour at scalar value 0.06 passes through point p_2. Edges (c_1, c_2) and (c_6, c_2) join at contour tree vertex c_2. Contour tree vertex c_2 corresponds to the isocontour passing through point p_2.

For isovalues in the range [0.06,0.6], the isocontour has one component. Contour tree edge (c_2, c_3) corresponds to this set of isocontours. At scalar value 0.6, this component splits into two components, one around point p_4 and one around point p_5. The isocontour at scalar value 0.6 passes through point p_3. Similarly, contour tree edge (c_2, c_3) splits at c_3 into edge (c_3, c_4) and edge (c_3, c_5). Contour tree vertex c_3 corresponds to the isocontour through point p_3. Contour tree edge (c_3, c_4) corresponds to the isocontours around p_4 and contour tree edge (c_3, c_5) corresponds to the isocontours around p_5.

For a 3D example, Figure 12.3(a) contains the contour tree for the hydrogen atom data set.[1] Each node corresponds to a connected component in some isosurface. The values inside the nodes correspond to the scalar value of the connected components. In this data set, connected components only split apart, they never join together.

For isovalues in the range [0,12], the isosurface consists of two components (Figure 12.3(b)). One of these components is a torus, while the other is two spheres connected by a tube through the torus. At isovalue 12, the component of two spheres connected by a tube splits into three parts. For isovalues in the range [12,36], the isosurface has four components (Figure 12.3(c)). At isovalue 50, the torus disappears, leaving three connected components (Figure 12.3(d)).

---

[1] The hydrogen atom data is "a simulation of the probability distribution of the electron in a hydrogen atom residing in a strong magnetic field" (www.volvis.org).

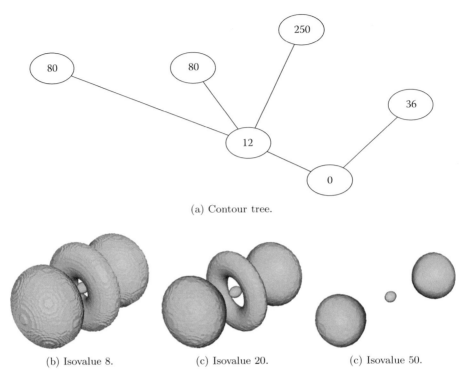

(a) Contour tree.

(b) Isovalue 8.   (c) Isovalue 20.   (c) Isovalue 50.

Figure 12.3. (a) Contour tree (simplified) of the hydrogen atom data set. (Two tree nodes representing artifacts in the data set were removed.) (b) Isosurface for isovalue 8, which has two connected components. (The two spheres at either end are connected by a tube throught the center of the torus and form a single component.) (c) Isosurface for isovalue 20, which has four connected components. (d) Isosurface for isovalue 50, which has three connected components.

As another 3D example, Figure 12.4(a) contains a simplified version of the contour tree for the nucleon data set.[2] In this data set, connected components only join together; they never split apart.

For isovalues in the range [13,103], the isosurface consists of three components, a sphere containing two other spheres (Figure 12.4(b)). At isovalue 103, one of the inner spheres joins the outer sphere. For isovalues in the range [103,161], the isosurface has two components (Figure 12.4(c)). At isovalue 161, the two components merge into a single component (Figure 12.4(d)).

---

[2]The nucleon data is a "simulation of the two-body distribution probability of a nucleon in the atomic nucleus 160 if a second nucleon is known to be positioned at $r' = (2 \text{ fm}, 0.0)$" (www.volvis.org).

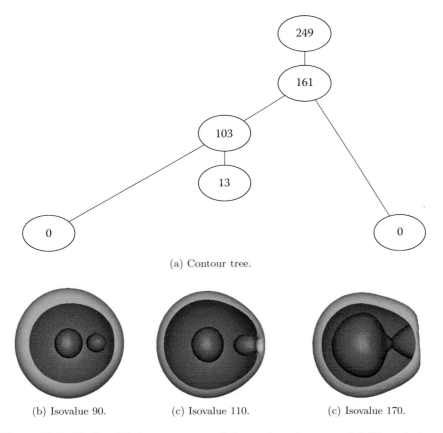

(a) Contour tree.

(b) Isovalue 90.   (c) Isovalue 110.   (c) Isovalue 170.

Figure 12.4. (a) Simplified contour tree of the nucleon data set. (b) Clipped view of isosurface for isovalue 90. Isosurface has three connected components. (c) Clipped view of isosurface for isovalue 110. Isosurface has two connected components. (d) Clipped view of isosurface for isovalue 170. Isosurface has one connected component.

12.2   Definition of Contour Tree

While the description of contour trees based on collapsing connected components to points is quite intuitive, the formal definition requires using equivalence classes and topological quotients. Moreover, while the resulting set has the appearance of a graph, formally defining the vertices and edges of that graph can be tricky.

A different way of viewing contour trees is as a tree whose vertices are a set of connected components of contours. Tree edges represent the containment or nesting of one contour component inside another. Such a tree is called a nesting diagram. We will give a formal definition of contour trees based on this approach. This definition will be used throughout this chapter.

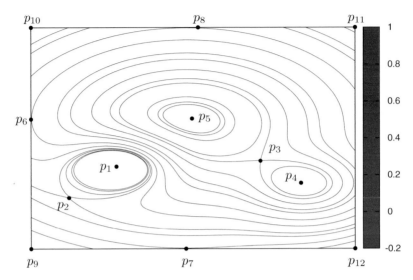

Figure 12.5. Contour lines for a scalar field $\phi : \mathcal{D}' \to \mathbb{R}$ where domain $\mathcal{D}'$ is the rectangle region $[-1.3 : 1.4] \times [-2 : 1.5]$. Points p_1, p_9, p_{10}, p_{11}, and p_{12} are local minima. Points p_4 and p_5 are local maxima. Connected components of contours are joined at points p_2, p_6, and p_8 and split at point p_3.

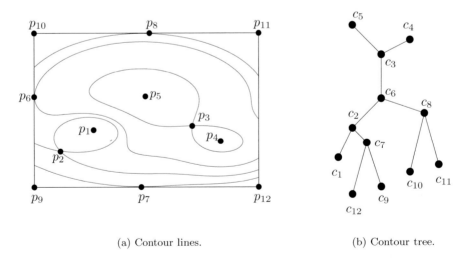

(a) Contour lines.   (b) Contour tree.

Figure 12.6. (a) Critical connected components of the scalar field in Figure 12.5. (b) Vertices c_1, c_4, c_5, c_9, c_{10}, c_{11}, and c_{12} correspond to points p_1, p_4, p_5, p_9, p_{10}, p_{11}, and p_{12}, respectively. Vertices c_2, c_3, c_6, c_7, and c_8 correspond to the contour lines through p_2, p_3, p_6, p_7, and p_8, respectively.

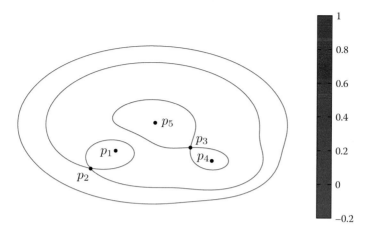

Figure 12.7. Critical connected components of the scalar field in Figure 12.1.

To represent changes in level set components, we need to identify components where changes occur. Let $\phi : \mathcal{D} \to \mathbb{R}$ be a continuous scalar field. A connected component λ of a level set of ϕ is a local minimum if $\phi(p) \geq \phi(\lambda)$ for every point p in a sufficiently small neighborhood of λ. Component λ is a local maximum if $\phi(p) \leq \phi(\lambda)$ for every point p in a sufficiently small neighborhood of λ. Typically, local minima and local maxima are single points, although they could be arbitrary connected sets.

For example, point p_1 in Figure 12.1 is a local minimum. Points p_4 and p_5 are local maxima. The boundary of the domain $\mathcal{D}$ is also a local minimum since $\phi : \mathcal{D} \to \mathbb{R}$ is locally minimal on its boundary. A different domain with a different boundary would have a different local minimum.

A connected component λ of a level set is regular if λ is not a local minimum or local maximum and $\mathcal{D} - \lambda$ has exactly two connected components. Component λ is critical if it is not regular.[3] Figure 12.7 contains the critical connected components of the level sets of $\phi : \mathcal{D} \to \mathbb{R}$.

The critical components depend upon the domain $\mathcal{D}$. If the domain $\mathcal{D}$ changes, then the critical connected components of $\phi : \mathcal{D} \to \mathbb{R}$ change. In Figure 12.1, the domain $\mathcal{D}$ is bounded by a contour. In Figure 12.5, the function ϕ is the same, but the domain $\mathcal{D}'$ is the rectangular region $[-1.3 : 1.4] \times [-2 : 1.5]$. Each of the corners of $\mathcal{D}'$ are local minima. For instance, p_9 is the lower-left corner of $\mathcal{D}'$. There is a small neighborhood $\mathbb{N}_{p_9} \subset \mathcal{D}'$ of p_9, such that $\phi(p) \geq \phi(p_9)$ for every point $p \in \mathbb{N}_{p_9}$. Note that $\mathbb{N}_{p_9}$ is restricted to $\mathcal{D}'$. Set $\mathbb{N}_{p_9}$ is not a neighborhood of p_9 within $\mathbb{R}^2$, only within rectangle $\mathcal{D}'$.

In Figure 12.5, the components of level sets through points p_6, p_7, and p_8 partition $\mathcal{D}'$ into three separate pieces. Thus these level set components are also

---

[3]This usage of the terms *regular* and *critical* is different from the usage in differential geometry and Morse theory, although the regular level sets defined here play a similar role to the regular and critical sets in differential geometry.

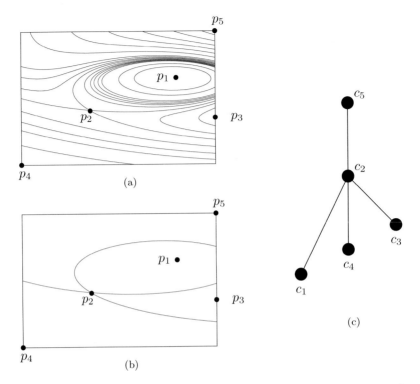

(a)

(b)

(c)

Figure 12.8. Contour lines for a scalar field $\phi : \mathcal{D}'' \to \mathbb{R}$ where domain $\mathcal{D}''$ is the rectangle region $[-1.3 : -0.4] \times [-2 : 0.0]$. Function ϕ is given by equation 12.1. Points p_1, p_3, and p_4 are local minima. Point p_4 is a local maximum. Connected components of contours are joined at point p_2.

critical. However, when ϕ is defined over $\mathcal{D}$, these components separate $\mathcal{D}$ into two pieces and are regular.

Another example is provided in Figure 12.8. The function ϕ is the same as before but the domain $\mathcal{D}''$ is the rectangle $[-1.3 : -0.4] \times [-2 : 0.0]$. Corner point p_4 is a local minimum and corner point p_5 is a local maximum. The two other corner points are not minima or maxima. Point p_3 on the boundary of $\mathcal{D}$ is also a local minima.

A set of connected components induces a graph whose vertices correspond to the connected components.

Definition 12.1. Let Σ be a closed subset of $\mathcal{D}$ and let Λ be the set of connected components of Σ. Set Λ induces graph $\mathcal{G}$ if

1. for each connected component $\lambda \in \Lambda$, graph $\mathcal{G}$ has a vertex v_λ;

2. graph $\mathcal{G}$ has edge $(v_\lambda, v_{\lambda'})$ if λ and λ' are both on the boundary of some connected component μ of $\mathcal{D} - \Sigma$.

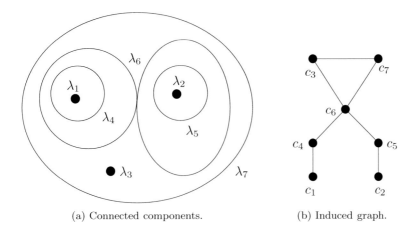

(a) Connected components.   (b) Induced graph.

Figure 12.9. (a) Connected components, $\lambda_1, \lambda_2, \ldots, \lambda_7$. Component λ_6 is the union of a circle and an ellipse. (b) Graph induced by set $\Lambda = \{\lambda_1, \ldots, \lambda_7\}$. Note that the graph is not a tree.

For instance, in Figure 12.9(a) set Λ is the set of points and curves $\{\lambda_1, \ldots, \lambda_7\}$. Set Σ is the union, $\cup_i \lambda_i$, of the elements of Λ. If $\mathcal{D}$ is the region bounded by λ_7, then $\mathcal{D} - \Sigma$ is the white space between the points and curves in Figure 12.9(a). Set Λ induces the graph in Figure 12.9(b).

We will let Λ be some connected components of level sets of some function $\phi : \mathcal{D} \to \mathbb{R}$. In general, the graph $\mathcal{G}$ induced by Λ is not a tree. However, under appropriate conditions, $\mathcal{G}$ is a tree.

Proposition 12.2. *Let $\phi : \mathcal{D} \to \mathbb{R}$ be a continuous scalar field and let Λ be some of the connected components of level sets of ϕ. If*

1. $\mathcal{D}$ is homeomorphic to an open or closed ball in $\mathbb{R}^d$,

2. Λ contains all the critical connected components of level sets of ϕ,

then the graph $\mathcal{G}$ induced by Λ is a tree.

The proof is in Section 12.6.1.

If Λ equals the set of critical connected components of the level set of ϕ and contains no other components, then we call the graph induced by Λ the contour tree of ϕ.

Definition 12.3. Let $\mathcal{D} \subset \mathbb{R}^d$ be a region homeomorphic to an open or closed ball in $\mathbb{R}^d$. The contour tree of a continuous function $\phi : \mathcal{D} \to \mathbb{R}$ is the graph induced by the set of critical connected components of the level sets of ϕ.

Figure 12.6(b) is an example of a contour tree. Each tree vertex corresponds to a critical connected component in Figure 12.6(a). The critical connected

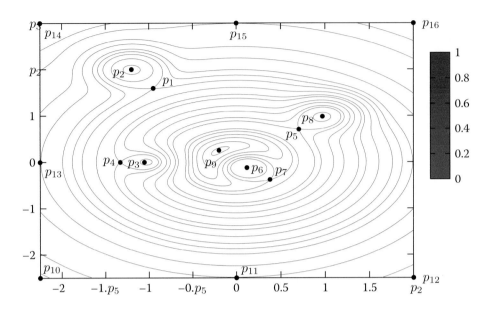

Figure 12.10. Contour lines for the scalar field $\phi' : \mathcal{D} \to \mathbb{R}$ where the domain $\mathcal{D}$ is $[-2.25, 2] \times [-2.5, 3]$. Function ϕ' is given by equation 12.2. Points p_3, p_6, p_{10}, p_{12}, p_{14}, and p_{16} are local minimum. Points p_2, p_8, and p_9 are local maxima. Contour lines join together or split apart at points p_1, p_4, p_5, p_7, p_{11}, p_{13}, and p_{15}.

components lie on the boundary of white regions in Figure 12.6(a). Two vertices in the contour tree are connected by an edge if the corresponding critical connected components lie on the boundary of the same white region.

Another example of a contour tree is in Figure 12.8(c). Contour tree vertex c_2 corresponds to a critical connected component through point p_2, which bounds four separate white regions. (See Figure 12.8(b).) Therefore, vertex c_2 is connected to four other vertices in the contour tree.

The contour trees constructed so far have been fairly simple. Figures 12.10 and 12.11 contain an example of a slightly more complicated contour tree. The function to generate this contour tree is

$$\phi'(x,y) \;=\; e^{-(x^2+y^2)} + \left(\frac{1}{2}\right) e^{-10((x-1)^2+(y-1)^2)} + \left(\frac{1}{8}\right) e^{-10((x+1.2)^2+(y-2)^2)}$$

$$- \left(\frac{1}{3}\right) e^{-10((x+0.9)^2+y^2))} - \left(\frac{1}{2}\right) e^{-10((x-0.1)^2+(y+0.1)^2)}. \quad (12.2)$$

Let Λ be some connected components of level sets of $\phi : \mathcal{D} \to \mathbb{R}$. When Λ properly contains the set of critical connected components of ϕ, the graph induced by Λ is called the **augmented contour tree** of ϕ.

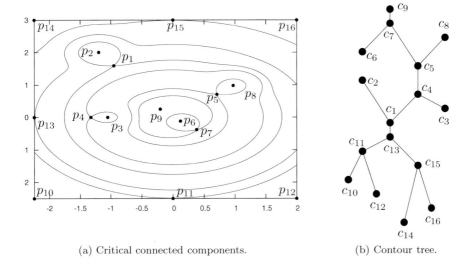

(a) Critical connected components.   (b) Contour tree.

Figure 12.11. (a) Critical connected components of level sets of $\phi' : \mathcal{D} \to \mathbb{R}$ where the domain $\mathcal{D}$ is $[-2.25, 2] \times [-2.5, 3]$. (b) Contour tree of $\phi' : \mathcal{D} \to \mathbb{R}$.

Definition 12.4. Let $\mathcal{D} \subset \mathbb{R}^d$ be a region homeomorphic to an open or closed ball in $\mathbb{R}^d$. The augmented contour tree of a continuous function $\phi : \mathcal{D} \to \mathbb{R}$ is the graph induced by a set of connected components of the level sets of ϕ that properly contains the set of critical connected components of the level sets of ϕ.

Since regular connected components of the level sets of ϕ split $\mathcal{D}$ exactly into two components, an augmented contour tree is a subdivision of a contour tree of ϕ.

The following sections describe how to build a contour tree for a piecewise linear function determined by a simplicial mesh.

12.3 Join, Split, and Merge Trees

A contour tree can be "decomposed" into two trees, a "join tree" representing contours that join together and a "split tree" representing contours that split apart as the scalar value increases. Conversely, the join and split trees of a contour tree can be "merged" to reconstruct the contour tree.

Join and split trees are defined for any graph with scalar values at the vertices, not just for contour trees. This section contains definitions and properties of join and split trees. In Section 12.4.1, we describe how to construct join and split trees and in Section 12.4.2 we present an algorithm for merging join and split trees of a contour tree.

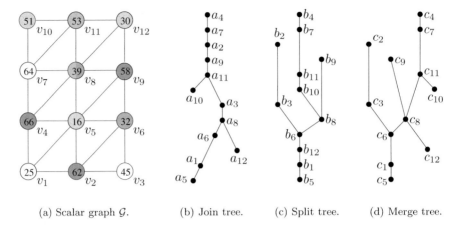

| (a) Scalar graph $\mathcal{G}$. | (b) Join tree. | (c) Split tree. | (d) Merge tree. |

Figure 12.12. (a) Scalar values are written inside the circles representing the vertices. Blue vertices v_5, v_{10}, and v_{12} are local minima. Magenta vertices v_2, v_4, and v_9 are local maxima. Light green vertex v_{11} joins two sets of vertices. Dark green vertex v_6 splits two sets of vertices. Green vertex v_8 both joins and splits sets of vertices. White vertices are neither local minima nor local maxima and do not join or split any sets of vertices. (b) Join tree $\mathcal{T}_J$ of $\mathcal{G}$. (c) Split tree $\mathcal{T}_S$ of $\mathcal{G}$. Note that the root b_5 of the split tree is drawn at the bottom and the leaves are drawn at the top. (d) Merge tree of $\mathcal{T}_J$ and $\mathcal{T}_S$ and of the scalar graph $\mathcal{G}$.

12.3.1   Join and Split Trees

A scalar graph is a graph where each node has a scalar value.[4] (See Figure 12.12.) A scalar tree is a scalar graph which is a tree. Note that a contour tree is a scalar tree.

A distinct valued graph is a scalar graph where no two nodes have the same scalar value.

We define join and split trees on distinct valued graphs.

Let $\mathcal{G}$ be a connected, distinct valued graph on n vertices, $\{v_1, v_2, \ldots, v_n\}$. Let $\mathcal{G}^-(s)$ be the subgraph of $\mathcal{G}$ induced by vertices whose scalar values are less than s. A connected component of $\mathcal{G}^-(s)$ is a maximally connected induced subgraph of $\mathcal{G}^-(s)$. (See Appendix C.) As s increases, new connected components of $\mathcal{G}^-(s)$ appear and multiple components join together to form one component. Components of $\mathcal{G}^-(s)$ do not disappear and do not split apart. The join tree of $\mathcal{G}$ represents the connected components of $\mathcal{G}^-(s)$ as a function of s.

For notational convenience, let $\mathcal{G}^-(v_i)$ be $\mathcal{G}^-(v_i.\mathsf{scalar})$ where v_i is a vertex of $\mathcal{G}$ and $v_i.\mathsf{scalar}$ is the scalar value of v_i. Vertex v_i is connected to a connected component of $\mathcal{G}^-(v_i)$ if there is an edge between v_i and some vertex of $\mathcal{G}^-(v_i)$.

---

[4] Scalar graphs are called "height graphs" by Carr et al. in [Carr et al., 2003].

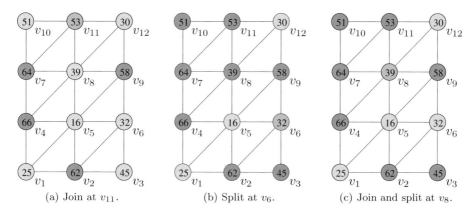

(a) Join at v_{11}. (b) Split at v_6. (c) Join and split at v_8.

Figure 12.13. Examples of joins and splits. The join or split vertex is colored green. Blue vertices have scalar value below the join or split vertex. Magenta vertices have scalar values above the join or split vertex. (a) Vertex v_{11} joins vertex v_{10} to vertices $\{v_1, v_3, v_5, v_6, v_8, v_{12}\}$. (b) Vertex v_6 splits vertices $\{v_2, v_3\}$ from vertices $\{v_4, v_7, v_8, v_9, v_{10}, v_{11}\}$. (c) Vertex v_8 joins vertices $\{v_1, v_5, v_6\}$ to vertex v_{12} and splits vertex v_9 from $\{v_4, v_7, v_{10}, v_{11}\}$.

Definition 12.5. The join tree of $\mathcal{G}$ is a tree with a vertex a_i that corresponds to each vertex v_i of $\mathcal{G}$. Vertex a_i is a child of vertex a_j in the join tree of $\mathcal{G}$ if

1. vertex v_j is connected to some connected component λ of $\mathcal{G}^-(v_j)$,

2. vertex v_i is the vertex of λ with maximum scalar value.

For instance, in Figure 12.13(a), vertices v_1, v_3, v_5, v_6, v_8, v_{10}, and v_{12} have scalar values less than v_{11}. Vertices v_1, v_3, v_5, v_6, v_8, and v_{12} form one component while vertex v_{10} forms another. Vertex v_3 has the highest scalar value in its component so a_3 is a child of a_{11}. Similarly, vertex v_{10} has the highest scalar value in its component and so a_{10} is a child of a_{11}. The join tree is given in Figure 12.13(b).

The number of children of vertex a_j corresponds to the number of connected components of $\mathcal{G}^-(v_j)$ that join together at v_j. The root a_r of the join tree corresponds to the vertex v_r of $\mathcal{G}$ with the highest scalar value. Since $\mathcal{G}$ is connected, vertex v_r is connected to every connected component of $\mathcal{G}^-(v_r)$.

Let $\mathcal{G}^+(s)$ be the subgraph of $\mathcal{G}$ induced by vertices whose scalar values are greater than s. As s increases, connected components of $\mathcal{G}^+(s)$ disappear and single components split apart to form multiple components. Components of $\mathcal{G}^+(s)$ do not appear and do not split apart. The split tree of $\mathcal{G}$ represents the connected components of $\mathcal{G}^+(s)$ as a function of s.

Let $\mathcal{G}^+(v_i)$ be $\mathcal{G}^+(v_i.\text{scalar})$ where v_i is a vertex of $\mathcal{G}$ and $v_i.\text{scalar}$ is the scalar value of v_i. Vertex v_i is connected to a connected component of $\mathcal{G}^+(v_i)$ if there is an edge between v_i and some vertex of $\mathcal{G}^+(v_i)$.

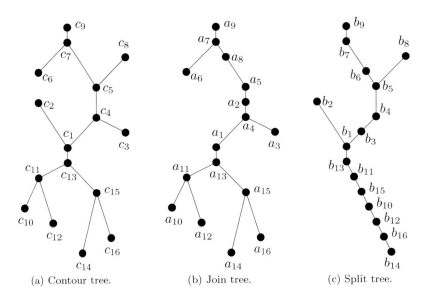

(a) Contour tree. (b) Join tree. (c) Split tree.

Figure 12.14. The contour tree from Figure 12.11 and its join and split trees. Vertex a_9 is the root of the join tree. Vertex b_{14} is the root of the split tree.

Definition 12.6. The split tree of $\mathcal{G}$ is a tree with a vertex b_i that corresponds to each vertex v_i of $\mathcal{G}$. Vertex b_i is a child of vertex b_j in the split tree of $\mathcal{G}$ if

1. vertex v_j is connected to some connected component λ of $\mathcal{G}^+(v_j)$,

2. vertex v_i is the vertex of λ with minimum scalar value.

In Figure 12.13(b), vertices v_2, v_3, v_4, v_7, v_8, v_9, v_{10}, and v_{11} have scalar values greater than v_6. Vertices v_2 and v_3 form one connected component while vertices $v_4, v_7, v_8, v_9, v_{10}, v_{11}$ form another. Since vertex v_3 has the lowest scalar value in its component, vertex b_3 is a child of b_6 in the split tree. Similarly, v_8 has the lowest scalar value in its component, so b_8 is also a child of b_6. The split tree is given in Figure 12.12(c).

The number of children of vertex b_j corresponds to the number of connected components of $\mathcal{G}^+(v_j)$ that are connected to v_j. The root $b_{r'}$ of the split tree corresponds to the vertex $v_{r'}$ of $\mathcal{G}$ with the lowest scalar value. Since $\mathcal{G}$ is connected, vertex $v_{r'}$ is connected to every connected component of $\mathcal{G}^+(v_{r'})$.

A vertex can both split and join sets of vertices. For instance, in Figure 12.13(c), vertex v_8 joins vertices v_1, v_5, and v_6 to vertex v_{12}. Vertex v_8 also splits vertices v_4, v_7, v_{10}, and v_{11} from vertex v_9. Thus, in Figure 12.12, vertex a_8 has two children in the join tree and vertex b_8 has two children in the split tree.

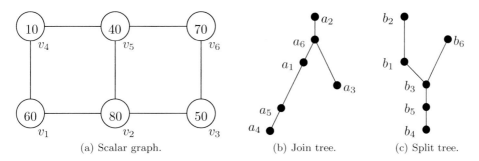

(a) Scalar graph. (b) Join tree. (c) Split tree.

Figure 12.15. (a) A scalar graph $\mathcal{G}$ with no merge tree. Scalar values are written inside the circles representing the vertices. (b) Join tree of $\mathcal{G}$. (c) Split tree of $\mathcal{G}$.

As previously noted, a contour tree $\mathcal{T}_C$ is a scalar graph. If $\mathcal{T}_C$ is distinct valued, then it has join and split trees. For instance, Figure 12.14 presents the join and split trees of the contour tree from Figure 12.11.

The join tree of a contour tree represents connected components of level sets that join together as the scalar value increases. The split tree represents connected components of level sets that split apart.

12.3.2 Merge Trees

Let $\mathcal{T}_J$ and $\mathcal{T}_S$ be the join and split trees of some connected, distinct valued graph $\mathcal{G}$.

Definition 12.7. The merge tree of $\mathcal{T}_J$ and $\mathcal{T}_S$ is a scalar tree $\mathcal{T}_M$ such that $\mathcal{T}_J$ and $\mathcal{T}_S$ are join and split trees of $\mathcal{T}_M$. The merge tree of a scalar graph $\mathcal{G}$ is the merge tree of its join and split trees.

Figure 12.12 contains an example of a merge tree.

Not every graph $\mathcal{G}$ has a merge tree. For instance, Figure 12.15 contains a graph $\mathcal{G}$ and its join tree $\mathcal{T}_J$ and split tree $\mathcal{T}_S$. There is no tree $\mathcal{T}$ such that $\mathcal{T}_J$ and $\mathcal{T}_S$ are the join and split trees of $\mathcal{T}$. (See Section 12.6.3 for an explanation of why there is no such $\mathcal{T}$ for this graph.) Note that graph $\mathcal{G}$ in Figure 12.15 is the vertices and edges of a regular grid.

If a pair of join and split trees have a merge tree, then that merge tree is unique.

Proposition 12.8. If $\mathcal{T}_J$ and $\mathcal{T}_S$ are the join and split trees, respectively, of two trees, $\mathcal{T}$ and $\mathcal{T}'$, then $\mathcal{T}$ is isomorphic to $\mathcal{T}'$.

(See Appendix C for a definition of graph isomorphism.) The proof is in Section 12.6.

A scalar simplicial mesh is a simplicial mesh with a scalar value associated to each vertex. Simplicial mesh Γ determines a piecewise linear function ϕ that is linear in each simplex of the mesh. (See Appendix B.7.) Let λ be the connected components of levels sets of ϕ that contain the vertices of Γ. Since level sets only merge or split at mesh vertices, set λ contains all the critical connected components of level sets of ϕ. Let $\mathcal{T}_C$ be the augmented contour tree of ϕ induced by λ. We call $\mathcal{T}_C$ the augmented contour tree of scalar simplicial mesh Γ.

A scalar simplicial mesh, Γ, is distinct valued if no two mesh vertices have the same scalar value. The 1-skeleton of a scalar simplicial mesh is the scalar graph consisting of the vertices and edges of Γ. The augmented contour tree of Γ is the merge tree of its 1-skeleton.

Proposition 12.9. *Let Γ be a distinct valued simplicial mesh covering a rectangular region in $\mathbb{R}^d$. If $\mathcal{G}$ is the 1-skeleton of Γ, then the augmented contour tree of Γ is the merge tree of $\mathcal{G}$.*

In the next section, we describe an algorithm for constructing the merge tree of $\mathcal{T}_S$ and $\mathcal{T}_J$. By Proposition 12.9, this merge tree is the augmented contour of $\mathcal{G}$.

12.4 Constructing Join, Split, and Merge Trees

In this section, we present algorithms by Carr, Snoeyink, and Axen [Carr et al., 2003] for constructing the join, split, and merge trees of a graph $\mathcal{G}$. To construct the join tree of a graph $\mathcal{G}$, sort the vertices of $\mathcal{G}$ by increasing scalar value, "grow" $\mathcal{G}$ by adding vertices one at a time in order, and add edges to the join tree whenever two connected components in $\mathcal{G}$ are joined together. Do the same to construct the split tree of $\mathcal{G}$, except that the vertices are processed by decreasing scalar value. Combine the join and split trees of $\mathcal{G}$ into the merge tree by pruning leaves from the join or split trees and adding corresponding leaves to the merge tree.

Section 12.4.1 contains the algorithms for constructing the join and split trees. Section 12.4.2 contains the algorithm for constructing the merge tree and some properties from [Carr et al., 2003] that form the basis of the algorithm.

12.4.1 Constructing Join and Split Trees

To construct a join tree, we must be able to detect when two sets of mesh vertices are joined by a vertex v. Fortunately, there is an efficient union-find data structure for detecting the union of disjoint sets. A union-find data structure supports an operation union, which forms the union of two sets, and the operation find, which finds a representative element of a set. Efficient implementations and analysis of the union-find data structure can be found in [Cormen et al., 2001].

```
Input    : Scalar graph 𝒢.
           Sequence (v₁, v₂, ..., vₙ) of vertices of 𝒢 sorted
                by increasing scalar value.
Result   : A join tree.

ConstructJoin(𝒢, (v₁, v₂, ..., vₙ))
```

1 for $i = 1$ to n do
2   CreateSet(i);
3   Create join node a_i;
4   a_i.parent ← NULL;
5   v_i.jNode ← a_i;
6   foreach vertex $v_j \in V(\mathcal{G})$ adjacent to v_i do
7   if $(j < i)$ and $(\text{Set}(i) \neq \text{Set}(j))$ then
8   k ← MaxSet(j);
9   a ← v_k.jNode;
10   a.parent ← v_i.jNode;
11   Union(i,j);
           /* Note: i now equals MaxSet(j)   */
12   end
13   end
14 end

Algorithm 12.1. CONSTRUCT JOIN.

In addition to the union and find operations, we will also need to retrieve the maximum element of a set. We can do this efficiently by storing this maximum element for each set. When applying the union operation to two sets, A and B, compare their maxima, m_A and m_B, respectively, and set the maximum of the new set to be the maximum of m_A and m_B.

We construct a join tree of a scalar graph by first sorting vertices in increasing order of scalar value. We process the vertices by increasing order as follows. For vertex v_i, we create a set containing index i. Create a join tree node a_i. For each graph vertex v_j adjacent to v_i, if v_j.scalar $< v_i$.scalar and the set containing i is different from the set containing j, get the maximum element k of the set containing j and make a_i the parent of a_k in the join tree. Union the set containing i with the set containing j. The algorithm, CONSTRUCT JOIN, is presented in Algorithm 12.1.

Construction of a split tree is similar to construction of a join tree. Instead of storing and retrieving maximum elements of sets, we store and retrieve minimum elements of sets. When applying the union operation to two sets, A and B, compare their minima, m_A and m_B, respectively, and set the minima of the new set to be the minima of m_A and m_B.

Input   : Scalar graph $\mathcal{G}$.

               Sequence $(v_1, v_2, \ldots, v_n)$ of vertices of $\mathcal{G}$ sorted
                   by increasing scalar value.

Result : A split tree.

ConstructSplit$(\mathcal{G}, (v_1, v_2, \ldots, v_n))$

1 for $i = n$ downto 1 do
2   CreateSet(i);
3   Create split node b;
4   b_i.parent $\leftarrow$ NULL;
5   v_i.sNode $\leftarrow b_i$;
6   foreach vertex $v_j \in V(\mathcal{G})$ adjacent to v_i do
7   if $(j > i)$ and $(\mathtt{Set}(i) \neq \mathtt{Set}(j))$ then
8              $k \leftarrow$ MinSet(j);
9              $b \leftarrow v_k$.sNode;
10   b.parent $\leftarrow v_i$.sNode;
11   Union(i,j);
             /* Note: i now equals MinSet(j) */
12   end
13   end
14 end

Algorithm 12.2. CONSTRUCT SPLIT.

As in construction of a join tree, sort vertices in order of scalar value. However, instead of processing vertices in increasing order, process them in decreasing order. For vertex v_i, create a set containing index i. Create a split tree node b_i. For each graph vertex v_j adjacent to v_i, if v_j.scalar $> v_i$.scalar and the set containing i is different from the set containing j, get the minimum element k of the set containing j and make b_i the parent of b_k in the split tree. Union the set containing i with the set containing j. The algorithm, CONSTRUCT SPLIT, is presented in Algorithm 12.2.

12.4.2   Merging Join and Split Trees

Before describing the merge tree algorithm, we present some propositions from [Carr et al., 2003] that form the basis of the algorithm. The proofs of all these propositions are in Section 12.6.

Both join and split trees have a root vertex. The root of the join tree is the vertex with highest scalar value while the root of the split tree is the vertex with the lowest scalar value. For each edge (v_i, v_j) of a rooted tree, either v_i is a child of v_j or v_j is a child of v_i. (See Appendix C for properties of rooted trees.)

For a vertex a_i of a join tree, let a_i.numChildren be the number of children of a_i in the join tree. Similarly, let b_i.numChildren be the number of children of a vertex b_i in a corresponding split tree.

The following proposition relates the number of children of a_i and b_i to the degree of c_i.

Proposition 12.10. *Let $\mathcal{T}$ be a tree whose join and split trees are $\mathcal{T}_J$ and $\mathcal{T}_S$. For each vertex $c_i \in V(\mathcal{T})$ corresponding to vertices $a_i \in V(\mathcal{T}_J)$ and $b_i \in V(\mathcal{T}_S)$, the degree of c_i equals $(a_i$.numChildren $+ b_i$.numChildren$)$.*

The following proposition allows us to identify some of the merge tree edges.

Proposition 12.11. *Let $\mathcal{T}_J$ and $\mathcal{T}_S$ be the join and split trees, respectively, of scalar tree $\mathcal{T}$, where $a_i \in V(\mathcal{T}_J)$ and $b_i \in V(\mathcal{T}_S)$ correspond to $c_i \in V(\mathcal{T})$.*

1. *If a_i has no children and $a_j \in V(\mathcal{T}_J)$ is the parent of a_i, then (c_i, c_j) is an edge of $\mathcal{T}$.*

2. *If b_i has no children and $b_j \in V(\mathcal{T}_S)$ is the parent of b_i, then (c_i, c_j) is an edge of $\mathcal{T}$.*

The deletion of a vertex v_i from a graph $\mathcal{G}$, denoted $\mathcal{G} - v_i$, is the removal of v_i and all its incident edges from the graph. Formally,

$$
\begin{aligned}
V(\mathcal{G} - v_i) &= V(\mathcal{G}) - \{v_i\}, \\
E(\mathcal{G} - v_i) &= E(\mathcal{G}) - \{(v_i, v_j) : (v_i, v_j) \in E(\mathcal{G})\}.
\end{aligned}
$$

The reduction of a rooted tree $\mathcal{T}$ by a vertex v_i, denoted $\mathcal{T} \ominus v_i$, is the deletion of v_i from $\mathcal{T}$ and the connection of each child of v_i in $\mathcal{T}$ to the parent of v_i in $\mathcal{T}$. Formally,

$$
V(\mathcal{T} \ominus v_i) = V(\mathcal{T} - v_i),
$$
$$
E(\mathcal{T} \ominus v_i) = E(\mathcal{T} - v_i) \cup \{(v_j, v_k) : v_j \text{ is a child and } v_k \text{ is a parent of } v_i \text{ in } \mathcal{T}\}.
$$

If v_i is the root of $\mathcal{T}$, then the reduction of $\mathcal{T}$ by v_i is equivalent to the deletion of v_i from $\mathcal{T}$. Each child of v_i becomes a root.

Let a_i and b_i be corresponding nodes of join and split trees, $\mathcal{T}_J$ and $\mathcal{T}_S$, of the merge tree $\mathcal{T}_M$. If a_j.numChildren $+ b_j$.numChildren equals one, then a_j, b_j, and c_j can be removed from $\mathcal{T}_M$, $\mathcal{T}_J$, and $\mathcal{T}_S$ as shown in Proposition 12.12.

Proposition 12.12. *Let $\mathcal{T}_J$ and $\mathcal{T}_S$ be the join and split trees, respectively, of scalar tree $\mathcal{T}$, where $a_i \in V(\mathcal{T}_J)$ and $b_i \in V(\mathcal{T}_S)$ correspond to $c_i \in V(\mathcal{T})$.*

If a_j.numChildren $+ b_j$.numChildren equals one, then

1. *the join tree of $\mathcal{T} - c_j$ is $\mathcal{T}_J \ominus a_j$,*

2. *the split tree of $\mathcal{T} - c_j$ is $\mathcal{T}_S \ominus b_j$.*

The next proposition guarantees that for at least some $c_k \in V(\mathcal{T}_M)$, Proposition 12.12 applies.

Proposition 12.13. *Let $\mathcal{T}_J$ and $\mathcal{T}_S$ be the join and split trees, respectively, of a tree $\mathcal{T}$ with $n \geq 2$ vertices, where $a_i \in V(\mathcal{T}_J)$ and $b_i \in V(\mathcal{T}_S)$ correspond to $v_i \in V(\mathcal{G})$. For at least one $k \in [1, \ldots, n]$,*

$$a_k.\text{numChildren} + b_k.\text{numChildren} = 1.$$

Propositions 12.10–12.13 can be applied to an algorithm to construct a tree $\mathcal{T}_M$ whose join and split trees are $\mathcal{T}_J$ and $\mathcal{T}_S$.

By Proposition 12.13, $(a_k.\text{numChildren} + b_k.\text{numChildren})$ equals one for some $k \in [1, \ldots, n]$. By Proposition 12.12, the join and split trees of $\mathcal{T}_M - c_k$ are $\mathcal{T}_J \ominus a_k$ and $\mathcal{T}_S \ominus b_k$. Thus, $\mathcal{T}_M - c_k$ is the merge tree of $\mathcal{T}_J \ominus a_k$ and $\mathcal{T}_S \ominus b_k$. Recursively, apply the algorithm to construct $\mathcal{T}_M - c_k$. Applying Proposition 12.11 gives an edge $\mathbf{e}$ incident on c_k in $\mathcal{T}_M$. By Proposition 12.10, vertex c_k has degree one in $\mathcal{T}_M$, so only edge $\mathbf{e}$ is incident on c_k. Construct $\mathcal{T}_M$ by adding vertex c_k and edge $\mathbf{e}$ to $\mathcal{T}_M - c_k$.

More specifically, algorithm MERGE JOIN SPLIT creates a merge tree as follows. Start by creating a queue containing all $i \in [1, \ldots, n]$ such that the sum $a_i.\text{numChildren} + b_i.\text{numChildren}$ is one. While the queue has two or more elements, remove one element i from the queue and do the following.

If $a_i \in \mathcal{T}_J$ has no children, then let a_k be the parent of a_i in the join tree. (Note that a_i cannot be the root of $\mathcal{T}_J$. If a_i were the root of $\mathcal{T}_J$, then $\mathcal{T}_J$ would have only one vertex and the queue would have only one element.) Add edge (c_i, c_k) to the merge tree. Delete a_i from $\mathcal{T}_J$. Replace b_i in $\mathcal{T}_S$ with an edge connecting its child and its parent. (If b_i is the root of $\mathcal{T}_S$, then make the child of b_i the root of $\mathcal{T}_S$.) If $a_k.\text{numChildren} + b_k.\text{numChildren}$ is one, add k to the queue.

If $a_i \in \mathcal{T}_J$ has one or more children, then $b_i \in \mathcal{T}_S$ has no children. Let b_k be the parent of b_i in the split tree. (Note that b_i cannot be the root of $\mathcal{T}_S$.) Add edge (c_i, c_k) to the contour tree. Delete b_i from $\mathcal{T}_S$. Replace a_i in $\mathcal{T}_J$ with an edge connecting its child and its parent. (If a_i is the root of $\mathcal{T}_J$, then make the child of a_i the root of $\mathcal{T}_J$.) If $a_k.\text{numChildren} + b_k.\text{numChildren}$ is one, add k to the queue. Pseudocode is presented in Algorithm 12.3.

12.4.3 Running Time Analysis

The running times of CONSTRUCT JOIN (Algorithm 12.1) and CONSTRUCT SPLIT (Algorithm 12.2) are dominated by the time to perform the Union and

Input   : A join tree $\mathcal{T}_J$ with vertices $\{a_1, \ldots, a_n\}$.
              A split tree $\mathcal{T}_S$ with vertices $\{b_1, \ldots, b_n\}$.
              Vertex $a_i \in \mathcal{T}_J$ corresponds to vertex $b_i \in \mathcal{T}_S$.
Result : A merge tree $\mathcal{T}_M$ with vertices $\{c_1, \ldots, c_n\}$.
              Vertex c_i is a join (minimum) vertex in $\mathcal{T}_M$ if and only if
                    a_i is a join (minimum) vertex in $\mathcal{T}_J$.
              Vertex c_i is a split (maximum) vertex in $\mathcal{T}_C$ if and only if
                    b_i is a split (maximum) vertex in $\mathcal{T}_S$.

`MergeJoinSplit`$(\mathcal{T}_J, \mathcal{T}_S)$

1 foreach vertex pair (a_i, b_i) where $a_i \in \mathcal{T}_J$ and $b_i \in \mathcal{T}_S$ do
2   Create contour tree vertex c_i;
3   if $(a_i.\mathsf{numChildren} + b_i.\mathsf{numChildren} = 1)$ then $\mathsf{Q.Enqueue}(i)$;
4 end
5 while $(\mathsf{Q.Size}() > 1)$ do
6       $i \leftarrow \mathsf{Q.Dequeue}()$;
7   if $(a_i.\mathsf{numChildren} = 0)$ then
8            $k \leftarrow$ index such that $a_k = a_i.\mathsf{parent}$;
9            $c_i.\mathsf{parent} \leftarrow c_k$;
10   else
11          $k \leftarrow$ index such that $b_k = b_i.\mathsf{parent}$;
12          $c_i.\mathsf{parent} \leftarrow c_k$;
13   end
14     $\mathcal{T}_J \leftarrow \mathcal{T}_J \ominus a_i$;
15     $\mathcal{T}_S \leftarrow \mathcal{T}_S \ominus b_i$;
16   if $(a_k.\mathsf{numChildren} + b_k.\mathsf{numChildren} = 1)$ then $\mathsf{Q.Enqueue}(k)$;
17 end

Algorithm 12.3. MERGE JOIN SPLIT.

Set operations. If implemented using a union-find data structure as described in [Cormen et al., 2001], the total running time is $\Theta(m\alpha(n))$ where n is the number of graph vertices and m is the number of graph edges. Function $\alpha(n)$ is the inverse of the Ackermann function and grows very slowly. (See [Cormen et al., 2001] for the analysis of the union-find data structure.)

Both CONSTRUCT JOIN and CONSTRUCT SPLIT assume that the input sequence of vertices is already sorted by increasing scalar value. Sorting the input vertices would require additional time. General purpose sorting algorithms require $\Theta(n \log n)$ time. However, if the input scalar values are integers in a fixed range, say 0 to 4095 $(2^{12} - 1)$, then counting sort can be used to sort the vertices in $\Theta(n)$ time. (See [Cormen et al., 2001] for a description and analysis of counting sort.)

Algorithm MERGE JOIN SPLIT (Algorithm 12.3) contains one for (Steps 1–4) loop and one while loop (Steps 5–17). The for loop executes in $\Theta(n)$ time. At each iteration of the while loop, one node is removed from $\mathcal{T}_J$ and one node is removed from $\mathcal{T}_S$. Thus the while loop also takes $\Theta(n)$ time and the algorithm takes $\Theta(n)$ time.

The running times are summarized in the following proposition.

Proposition 12.14. *Let n be the number of graph vertices and m the number of graph edges.*

- *Algorithm* CONSTRUCT JOIN *creates a join tree from a sequence of graph vertices sorted by increasing scalar value in $\Theta(m\alpha(n))$ time.*

- *Algorithm* CONSTRUCT SPLIT *creates a split tree from a sequence of graph vertices sorted by increasing scalar value in $\Theta(m\alpha(n))$ time.*

- *Algorithm* MERGE JOIN SPLIT *merges a join and split tree in $\Theta(n)$ time.*

12.5   Constructing Contour Trees

To construct the contour tree for a scalar simplicial mesh Γ, we simply construct the merge tree of the 1-skeleton of Γ. We handle vertices with identical scalar values by symbolically perturbing the scalar values.

Constructing a contour tree from a regular grid is more difficult. In Proposition 12.17, we show that the isosurfaces constructed by MARCHING CUBES (using an appropriate lookup table) are homeomorphic to level sets of a particular triangulation of the regular grid. We construct the contour tree of the regular grid by constructing the merge tree of the 1-skeleton of this triangulation.

12.5.1   Simplicial Meshes

Let Γ be a distinct valued scalar simplicial mesh and let $\phi : \mathbb{R}^3 \to \mathbb{R}$ be the piecewise linear function determined by Γ. Let $\mathcal{G}$ be the 1-skeleton of Γ. Let $\mathcal{T}_J$ and $\mathcal{T}_S$ be the join and split trees, respectively, of $\mathcal{G}$. By Proposition 12.9, the merge tree of $\mathcal{T}_J$ and $\mathcal{T}_S$ is an augmented contour tree of ϕ.

Algorithm CONSTRUCT CTREE is as follows. (See Figure 12.16.) Sort the vertices of $\mathcal{G}$ in increasing order. Apply algorithms CONSTRUCT JOIN and CONSTRUCT SPLIT (Algorithms 12.1 and 12.2) to construct the join and split trees of $\mathcal{G}$. Apply algorithm MERGE JOIN SPLIT (Algorithm 12.3) to construct the merge tree of $\mathcal{G}$ from the join and split trees.

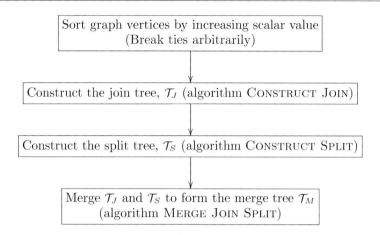

Figure 12.16. Algorithm CONSTRUCT CTREE. Input is a regular scalar graph. If $\mathcal{G}$ represents a continuous scalar field $\phi : \mathbb{R}^d \to \mathbb{R}$, then $\mathcal{T}_M$ is an augmented contour tree of ϕ.

Starting with Section 12.3, we have assumed that all vertices of the scalar graph $\mathcal{G}$ have different scalar values. We can drop this assumption by symbolically perturbing each scalar value by a small amount so that no two vertices have the same scalar value. To simulate this perturbation, simply break ties arbitrarily when sorting the vertices of $\mathcal{G}$ by scalar value. If vertex v_i comes before v_j in sorted order, then the perturbed value $v_i.\mathsf{scalar} + \epsilon_i$ of v_i is strictly less than the perturbed value $v_j.\mathsf{scalar} + \epsilon_j$ of v_j. Note that algorithm CONSTRUCT CTREE never uses the scalar values after sorting the vertices and so the rest of the algorithm is unchanged.

Let n be the number of vertices and m the number of edges of a scalar simplicial mesh. By Proposition 12.14 (Section 12.4.3), the running time of CONSTRUCT JOIN and CONSTRUCT SPLIT is $\Theta(m\alpha(n))$. The running time of MERGE JOIN SPLIT is $\Theta(n)$. We can always use a general purpose sorting algorithm to sort the mesh vertices in $\Theta(n \log n)$ time. Thus, algorithm CONSTRUCT CTREE runs in $\Theta(n \log n + m\alpha(n))$ time.

Proposition 12.15. *Let n be the number of mesh vertices and m the number of mesh edges. Algorithm CONSTRUCT CTREE creates a contour tree in $\Theta(n \log n + m\alpha(n))$ time.*

If $\mathcal{G}$ has numerous vertices with the same scalar value, then its contour tree may have large connected clusters of vertices with the same scalar value. The join and merge vertices in these clusters are a result of the arbitrary way ties were broken in the sorting by scalar value. Thus, it is desirable to collapse

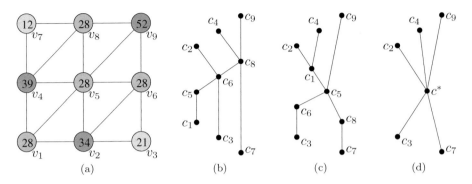

Figure 12.17. Different contour trees produced by different ordering of vertices with identical scalar values. (a) Scalar graph $\mathcal{G}$. Scalar values of vertices v_1, v_5, v_6, and v_8 are identical. Scalar values are written inside the circles representing the vertices. (b) Augmented contour tree for vertex order (v_1, v_5, v_6, v_8) of identical vertices. (c) Augmented contour tree for vertex order (v_8, v_6, v_5, v_1) of identical vertices. (d) Tree produced by collapsing vertices c_1, c_5, c_6, and c_8 into a single vertex c^*.

Input   : A tree $\mathcal{T}$ with vertices $\{c_1, \ldots, c_n\}$.

Result : Modify $\mathcal{T}$ by merging adjacent vertices with the same scalar value.

MergeIdenticalScalar($\mathcal{T}$)

1   foreach tree edge (c_i, c_j) do
2   if $(c_i.\mathsf{scalar} = c_j.\mathsf{scalar})$ then
3   foreach vertex c_k adjacent to c_j do
4   if $(c_k \neq c_i)$ then Add edge (c_i, c_k) to $\mathcal{T}$;
5   end
6   Delete c_j from $\mathcal{T}$;
7   end
8 end

Algorithm 12.4. MERGE IDENTICAL SCALAR.

---

such clusters into single vertices. (See Figure 12.17.) For each contour tree edge (v_i, v_j) where $v_i.\mathsf{scalar}$ equals $v_j.\mathsf{scalar}$, delete vertex v_j and connect all its incident edges to v_i. Pseudocode for algorithm MERGE IDENTICAL SCALAR is presented in Algorithm 12.4.

Algorithm CONSTRUCT CTREE constructs an augmented contour tree, $\mathcal{T}_C$. To reduce $\mathcal{T}_C$ to a contour tree, we must remove the vertices that are not join, split, local minima, or local maxima. A vertex of an augmented contour tree is regular if it is not a join, split, local minima, or local maxima vertex. A vertex is regular if and only if it is adjacent to exactly one vertex v_j with

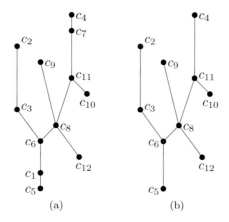

Figure 12.18. (a) Augmented contour tree. (b) Contour tree. Regular vertices c_1 and c_7 are in the augmented contour tree but not in the contour tree.

Input   : A tree $\mathcal{T}$ with vertices $\{c_1, \ldots, c_n\}$.
Result   : Modify $\mathcal{T}$ by removing each regular vertex and connecting its
           adjacent vertices by an edge.

`RemoveRegularVertices(`$\mathcal{T}$`)`

1   foreach tree vertex c_i do
2   if (c_i is adjacent to exactly one c_j with c_i.scalar $> c_j$.scalar) and
3   (c_i is adjacent to exactly one c_k with c_i.scalar $< c_k$.scalar) then
4   Add edge (c_j, c_k) to $\mathcal{T}$;
5   Delete c_i from $\mathcal{T}$;
6   end
7   end

Algorithm 12.5. REMOVE REGULAR VERTICES.

v_i.scalar $> v_j$.scalar and exactly one vertex v_k with v_i.scalar $< v_k$.scalar. We delete each such v_i from $\mathcal{T}_C$ and connect its two adjacent vertices by an edge. (See Figure 12.18.). Pseudocode for algorithm REMOVE REGULAR VERTICES is presented in Algorithm 12.5.

12.5.2   Regular Scalar Grids

To construct a contour tree from a regular scalar grid, we could uniformly tetrahedralize the grid cubes into six tetrahedra as described in Section 2.4.3 and construct a contour tree on the resulting tetrahedral mesh. This contour tree will represent the topology of the isosurfaces constructed by applying the MARCH-ING SIMPLICES algorithm to the tetrahedral mesh. However, the topology of the MARCHING CUBES isosurfaces does not necessarily match the topology of the isosurfaces created by applying MARCHING SIMPLICES to the tetrahedralized grid. Different connected components in the MARCHING CUBES isosurface may form a single component in the MARCHING SIMPLICES isosurface and vice versa.

The uniform tetrahedralization into six tetrahedra is not the only way to tetrahedralize a cube. In Section 6.4.1, we defined the incremental triangulation of $\text{conv}(p_1, p_2, \ldots, p_n)$ induced by the sequence $(p_1, p_2, \ldots, p_n)$. The incremental triangulation is the triangulation of $\text{conv}(p_1, p_2, \ldots, p_n)$ induced by adding points to the convex hull in the order $p_1, p_2, \ldots, p_n$.

Let Γ be a regular grid and let $(v_1, v_2, \ldots, v_n)$ be a sequence of vertices of Γ. The sequence determines an incremental triangulation of each cube of Γ. Moreover, the triangulations of any two adjacent grid cubes match on their common face so the union of all the tetrahedra form a triangulation of the entire grid.

Definition 12.16. The triangulation of regular grid Γ induced by the sequence $(v_1, v_2, \ldots, v_n)$ of vertices of Γ is the union of the incremental triangulations of each cube of Γ induced by the sequence $(v_1, v_2, \ldots, v_n)$.

In Section 5.2, we described an algorithm, IsoMID3D, to automatically generate an isosurface lookup table for 3D cubes. The table generated by this algorithm is exactly the same as the isosurface lookup table for MARCHING CUBES (Figure 2.16) except for ambiguous configuration 2C. In configuration 2C, Iso-MID3D generates the isosurface patch 2C-I in Figure 2.22.

A scalar grid is distinct valued if no two vertices of the grid have the same value.

Proposition 12.17. *Let Γ be a distinct valued regular scalar grid. Let τ be the triangulation of Γ induced by sorting the vertices of Γ in decreasing order of scalar value. Let ϕ be the piecewise linear function determined by τ. For each scalar value σ, the level set $\phi^{-1}(\sigma)$ is homeomorphic to the isosurfaces constructed by* MARCHING CUBES *using the* IsoMID3D *lookup table.*

Given a distinct valued regular grid, Γ, we construct the triangulation τ induced by sorting the vertices of Γ in decreasing order of scalar value. We compute the contour tree of τ. The contour tree of τ represents the isosurface generated by the MARCHING CUBES algorithm using the IsoMID3D lookup table.

Constructing a triangulation τ of Γ is time- and space-consuming for any reasonably sized grid Γ. However, we can construct the contour tree of τ directly from Γ as described below.

The 1-skeleton of a regular scalar grid Γ is the scalar graph consisting of its vertices and edges. The 26-connectivity graph of Γ is the scalar graph for every

---

Input   : Regular grid Γ.
          Sequence $(v_1, v_2, \ldots, v_n)$ of vertices of Γ sorted
             by increasing scalar value.
Result : A join tree.

`GridConstructJoin(`Γ`, (`$v_1, v_2, \ldots, v_n$`))`

```
 1 for i = 1 to n do
 2  │  CreateSet(i);
 3  │  Create join node a_i;
 4  │  a_i.parent ← NULL;
 5  │  v_i.jNode ← a_i;
 6  │  foreach edge (v_i, v_j) of Γ do
 7  │  │  if (j < i) and (Set(i) ≠ Set(j)) then
 8  │  │  │  k ← MaxSet(j);
 9  │  │  │  a ← v_k.jNode;
10  │  │  │  a.parent ← v_i.jNode;
11  │  │  │  Union(i,j);
    │  │  │  /* Note: i now equals MaxSet(j)              */
12  │  │  end
13  │  end
14 end
```

Algorithm 12.6. GRID CONSTRUCT JOIN.

cube $\mathbf{c} \in \Gamma$; every pair of vertices $v, v' \in \mathbf{c}$ are connected by an edge. (A vertex in the interior of Γ is connected to 26 vertices in the 26-connectivity graph.) The following proposition relates the 1-skeleton of τ to the 1-skeleton and 26-connectivity graph of Γ.

Proposition 12.18. *Let Γ be a distinct valued regular scalar grid. Let τ be the triangulation of Γ induced by sorting the vertices of Γ in decreasing order of scalar value. Let $\mathcal{G}_\tau$ be the 1-skeleton of τ, let $\mathcal{G}$ be the 1-skeleton of Γ, and let $\widetilde{\mathcal{G}}$ be the 26-connectivity graph of Γ.*

1. The join tree of $\mathcal{G}_\tau$ is the join tree of $\mathcal{G}$.

2. The split tree of $\mathcal{G}_\tau$ is the split tree of $\widetilde{\mathcal{G}}$.

Algorithm 12.6, GRID CONSTRUCT JOIN, constructs the join tree of the 1-skeleton $\mathcal{G}$ of Γ. By Proposition 12.18, this join tree is also the join tree of the 1-skeleton of τ. The edges of $\mathcal{G}$ are simply the edges of Γ, so there is no need to explicitly construct $\mathcal{G}$.

Algorithm 12.7, GRID CONSTRUCT SPLIT, constructs the split tree of the 26-connectivity graph $\widetilde{\mathcal{G}}$ of Γ. By Proposition 12.18, this split tree is also the

---

Input   : Regular grid Γ.
        Sequence $(v_1, v_2, \ldots, v_n)$ of vertices of Γ sorted
            by increasing scalar value.
Result : A split tree.

GridConstructSplit$(\Gamma, (v_1, v_2, \ldots, v_n))$

```
 1  for i = n downto 1 do
 2      CreateSet(i);
 3      Create split node b_i;
 4      b_i.parent ← NULL;
 5      v_i.sNode ← b_i;
 6      foreach vertex v_j, which shares a grid cube with v_i do
 7          if (j > i) and (Set(i) ≠ Set(j)) then
 8              k ← MaxSet(j);
 9              b ← v_k.sNode;
10              b.parent ← v_i.sNode;
11              Union(i,j);
                /* Note: i now equals MaxSet(j)                    */
12          end
13      end
14  end
```

Algorithm 12.7. GRID CONSTRUCT SPLIT.

split tree of the 1-skeleton of τ. The edges of $\widetilde{\mathcal{G}}$ are determined by the cubes of Γ, so again there is no need to explicitly construct $\widetilde{\mathcal{G}}$.

Algorithm GRID CONSTRUCT CTREE constructs a contour tree from a regular scalar grid as follows. (See Figure 12.19.) Sort the grid vertices by increasing scalar value. Symbolically perturb the scalar values to break any ties. Call GRID CONSTRUCT JOIN and GRID CONSTRUCT SPLIT to construct the join and split trees. Call MERGE JOIN SPLIT to merge the join and split trees into a contour tree.

The running times of GRID CONSTRUCT JOIN and GRID CONSTRUCT SPLIT are the same as the running times of CONSTRUCT JOIN and CONSTRUCT SPLIT. Thus the running time of GRID CONSTRUCT CTREE is the same as the running time of CONSTRUCT CTREE. However, the number of edges in the regular grid is about three times the number N of grid vertices, so the running time can be expressed just in terms of N. Since $(N\alpha(N))$ is less than $(N \log N)$ for large N, the $(N\alpha(N))$ term is absorbed into the $(N \log N)$ term.

Proposition 12.19. *Let N be the number of vertices of a regular scalar grid Γ. Algorithm* GRID CONSTRUCT CTREE *creates a contour tree from grid Γ in* $\Theta(N \log N)$ *time.*

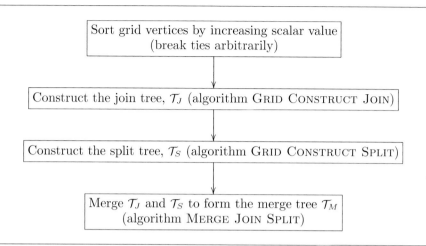

Figure 12.19. Algorithm GRID CONSTRUCT CTREE. Input is a scalar grid Γ.

Algorithms MERGE IDENTICAL SCALAR and REMOVE REGULAR VERTICES can be used to merge vertices with identical scalar values and remove regular vertices from a contour tree produced from a regular scalar grid.

12.6 Theory and Proofs

12.6.1 Contour Tree Proofs

We prove that if $\mathcal{D}$ is homeomorphic to an open or closed ball and λ contains the critical connected components of level sets of ϕ, then the graph induced by λ is a tree.

Proposition 12.2. *Let $\phi : \mathcal{D} \to \mathbb{R}$ be a continuous scalar field and let Λ be some of the connected components of level sets of ϕ. If*

1. *$\mathcal{D}$ is homeomorphic to an open or closed ball in $\mathbb{R}^d$,*

2. *Λ contains all the critical connected components of level sets of ϕ,*

then the graph $\mathcal{G}$ induced by Λ is a tree.

Proof: Let Λ be a set of connected components of the level sets of ϕ that contains all the critical connected components of the level sets of ϕ. Let Σ be the union of all the elements of Λ. Let $\mathcal{G}$ be the graph induced by Λ.

Since Λ contains all the critical connected components of level sets, each connected component μ of $\mathcal{D} - \Sigma$ is bounded by at most two elements, λ_1 and λ_2, of Λ. Thus, each edge $\mathbf{e} = (v_{\lambda_1}, v_{\lambda_2})$ of $\mathcal{G}$ corresponds to a connected component μ of $\mathcal{D} - \Sigma$. Region μ separates λ_1 from λ_2 so deleting $\mathbf{e}$ from $\mathcal{G}$ disconnects v_{λ_1} from v_{λ_2}. Thus $\mathcal{G}$ has no cycles.

There is a path in $\mathcal{D}$ between any two elements of $\Lambda_{\mathbf{f}}$, so there is a path in $\mathcal{G}$ between the corresponding vertices in $\mathcal{G}$. Thus $\mathcal{G}$ is connected.

Since $\mathcal{G}$ is a connected graph with no cycles, $\mathcal{G}$ is a tree. $\square$

12.6.2 Merge Tree Proofs

This section contains all the proofs of Propositions 12.8–12.13. The proof of Proposition 12.8 requires Propositions 12.10, 12.12, and 12.13 and is delayed until the end of this section.

As previously defined, $\mathcal{G}^-(v_i)$ is the subgraph of a scalar graph $\mathcal{G}$ induced by vertices whose scalar values are less than v_i.scalar, the scalar value of v_i. Graph $\mathcal{G}^+(v_i)$ is the subgraph of a scalar graph $\mathcal{G}$ induced by vertices whose scalar values are greater than v_i.scalar. Similarly, $\mathcal{T}^-(c_i)$ is the subgraph of the scalar tree $\mathcal{T}$ induced by vertices whose scalar values are less than v_i.scalar and $\mathcal{T}^+(c_i)$ is the subgraph of $\mathcal{T}$ induced by vertices whose scalar values are greater than v_i.scalar.

Proposition 12.9 relates the augmented contour tree of a scalar simplicial mesh to the scalar graph formed by its vertices and edges.

Proposition 12.9. *Let Γ be a distinct valued simplicial mesh covering a rectangular region in $\mathbb{R}^d$. If $\mathcal{G}$ is the 1-skeleton of Γ, then the augmented contour tree of Γ is the merge tree of $\mathcal{G}$.*

Proof: Let Γ be a distinct valued scalar simplicial mesh and let $\mathcal{G}$ be the scalar graph formed by the vertices and edges of Γ. Let $\phi : \mathcal{D} \to \mathbb{R}$ be the piecewise linear function defined by Γ.

Consider a level set $\phi^{-1}(v_i.\text{scalar})$ for some mesh vertex v_i. This level set separates $\mathcal{D}$ into connected components and partitions the vertices of Γ into subsets where two vertices are in the same subset if they are in the same component. (See an example in Figure 12.20.) The vertices of each subset either all have scalar value less than v_i.scalar or all have scalar value greater than v_i.scalar. The subsets whose vertices have scalar value less than v_i.scalar are exactly the sets of vertices of the connected components of $\mathcal{G}^-(v_i)$. The subsets whose vertices have scalar value greater than v_i.scalar are exactly the sets of vertices of the components of $\mathcal{G}^+(v_i)$.

Let $\mathcal{T}_C$ be the augmented contour tree of Γ. Let $\mathcal{T}_J$ and $\mathcal{T}_J'$ be the join trees of $\mathcal{T}_C$ and $\mathcal{G}$, respectively. Let a_i, a_i', and c_i be the vertices of $\mathcal{T}_J$, $\mathcal{T}_J'$, and $\mathcal{T}_C$ that correspond to mesh vertex v_i.

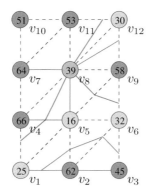

Figure 12.20. Level set $\phi^{-1}(v_8.\text{scalar})$ where ϕ is the piecewise linear function defined by the scalar mesh.

Let λ_i be the connected component of the level set $\phi^{-1}(v_i.\text{scalar})$ that contains v_i. The connected components of $\mathcal{D} - \phi^{-1}(v_i.\text{scalar})$ that have λ_i on its boundary correspond to the connected components of $\mathcal{G}^-(v_i)$ and $\mathcal{G}^+(v_i)$, which are connected to v_i. Thus U is the set of vertices of some connected component λ of $\mathcal{T}_C^-(c_i)$ if and only if $\{v_j : c_j \in U\}$ is the set of vertices of some connected component λ' of $\mathcal{G}^-(v_i)$. Moreover, c_i is connected to λ if and only if v_i is connected to λ'. Thus, $a_j \in \mathcal{T}_J$ is a child of $a_i \in \mathcal{T}_J$ if and only if $a_j' \in \mathcal{T}_J'$ is a child of $a_i' \in \mathcal{T}_J'$ and the join tree of $\mathcal{T}_C$ is isomorphic to the join tree of $\mathcal{G}$. Similarly, the split tree of $\mathcal{T}_C$ is isomorphic to the split tree of $\mathcal{G}$.

Since the join and split trees of $\mathcal{T}_C$ are isomorphic to $\mathcal{T}_J'$ and $\mathcal{T}_S'$, respectively, tree $\mathcal{T}_C$ is the merge tree of $\mathcal{T}_J'$ and $\mathcal{T}_S'$. By definition, $\mathcal{T}_C$ is the merge tree of $\mathcal{G}$.□

The degree of a vertex $v_i \in \mathcal{G}$ is the number of edges incident on v_i. For a scalar graph $\mathcal{G}$, define the **down degree** of a vertex v_i as the number of edges (v_i, v_j) incident on v_i where $v_i.\text{scalar} > v_j.\text{scalar}$. Define the **up degree** of a vertex v_i as the number of edges (v_i, v_j) incident on v_i where $v_i.\text{scalar} < v_j.\text{scalar}$.

The following lemma relates up and down degrees of merge tree vertices to the number of children in the join and split trees.

Lemma 12.20. *Let $\mathcal{T}$ be a scalar tree whose fully augmented join and split trees are $\mathcal{T}_J$ and $\mathcal{T}_S$. For each vertex $c_i \in V(\mathcal{T})$ corresponding to vertices $a_i \in V(\mathcal{T}_J)$ and $b_i \in V(\mathcal{T}_S)$,*

1. the down degree of c_i equals a_i.numChildren.

2. the up degree of c_i equals b_i.numChildren.

Proof: The number of children of a_i is the number of connected components of $\mathcal{T}^-(c_i)$. Since $\mathcal{T}$ is a tree, the number of connected components of $\mathcal{T}^-(c_i)$ is the down degree of c_i. Thus, the number of children of a_i is the down degree of c_i. By a similar argument, the number of children of b_i is the up degree of c_i.   □

Proposition 12.10.   *Let $\mathcal{T}$ be a tree whose join and split trees are $\mathcal{T}_J$ and $\mathcal{T}_S$. For each vertex $c_i \in V(\mathcal{T})$ corresponding to vertices $a_i \in V(\mathcal{T}_J)$ and $b_i \in V(\mathcal{T}_S)$, the degree of c_i equals $(a_i.\mathsf{numChildren} + b_i.\mathsf{numChildren})$.*

Proof: The degree of c_i equals the down degree plus the up degree of c_i. By Lemma 12.20, the down degree plus the up degree equals

$$(a_i.\mathsf{numChildren} + b_i.\mathsf{numChildren}). \qquad \qquad \square$$

Proposition 12.11.   *Let $\mathcal{T}_J$ and $\mathcal{T}_S$ be the join and split trees, respectively, of scalar tree $\mathcal{T}$, where $a_i \in V(\mathcal{T}_J)$ and $b_i \in V(\mathcal{T}_S)$ correspond to $c_i \in V(\mathcal{T})$.*

1. *If a_i has no children and $a_j \in V(\mathcal{T}_J)$ is the parent of a_i, then (c_i, c_j) is an edge of $\mathcal{T}$.*

2. *If b_i has no children and $b_j \in V(\mathcal{T}_S)$ is the parent of b_i, then (c_i, c_j) is an edge of $\mathcal{T}$.*

Proof: We prove only Statement 1 of the proposition. The proof of Statement 2 is similar.

Let a_i be a vertex of $\mathcal{T}_J$ with no children and let a_j be the parent of a_i. By definition of the join tree (Definition 12.5), vertex c_j is connected to some component λ of $\mathcal{T}^-(c_j)$ where c_i is the element of λ with the maximum scalar value. By Lemma 12.20, the down degree of c_i is zero. Thus, c_i is not connected to any vertices $c_k \in \lambda$ with scalar value less than $c_i.\mathsf{scalar}$. On the other hand, c_i has maximum scalar value in λ so c_i is not connected to any vertices $c_k \in \lambda$ with scalar value greater than $c_i.\mathsf{scalar}$. Thus c_i is not connected to any other vertices in λ. Since c_i is in λ and λ is connected, vertex c_i is only the vertex in λ. Since c_j is connected to λ, (c_i, c_j) is an edge of $\mathcal{T}$. $\qquad \square$

Proposition 12.21. *Let $\mathcal{T}_J$ and $\mathcal{T}_S$ be the join and split trees, respectively, of scalar tree $\mathcal{T}$, where $a_i \in V(\mathcal{T}_J)$ and $b_i \in V(\mathcal{T}_S)$ correspond to $c_i \in V(\mathcal{T})$. If $a_j.\mathsf{numChildren} + b_j.\mathsf{numChildren}$ equals one, then*

1. *the join tree of $\mathcal{T} - c_j$ is $\mathcal{T}_J \ominus a_j$,*

2. *the split tree of $\mathcal{T} - c_j$ is $\mathcal{T}_S \ominus b_j$.*

Proof: Let $\mathcal{T}'$ equal $\mathcal{T} - \{c_j\}$. By Proposition 12.10, vertex c_j has degree one and so $\mathcal{T}'$ is a tree. Let $\mathcal{T}'_J$ be the join tree of $\mathcal{T}'$ and a'_i be the vertex of $\mathcal{T}'_J$ corresponding to $c_i \in \mathcal{T}'$.

We show that every edge in $\mathcal{T}_J \ominus a_j$ corresponds to an edge of $\mathcal{T}'_J$. We consider two cases based on the number of children of a_j.

Case I: Vertex a_j has zero children.

Consider an edge $(a_i, a_k) \in \mathcal{T}_J \ominus a_j$ where a_i is a child of a_k. Since a_j has zero children, a_i is also a child of a_k in $\mathcal{T}_J$. By definition of the join tree (Definition 12.5), vertex c_i is the vertex with maximum scalar value in some connected component λ of $\mathcal{T}^-(c_k)$ and c_k is connected to λ. If c_j is not an element of λ, then λ is also a connected component of $\mathcal{T}'^-(c_k)$ and c_i is a vertex with maximum scalar value in λ. Thus, a_i' is a child of a_k' in $\mathcal{T}_J'$. If c_j is an element of λ, then $\lambda - c_j$ is a connected component of $\mathcal{T}'^-(c_k)$ (since c_j has degree one). Vertex c_i is still the vertex with maximum scalar value in $\lambda - c_j$. Again a_i' is a child of a_k' in $\mathcal{T}_J'$.

Case II: Vertex a_j has one child.

Consider an edge $(a_i, a_k) \in \mathcal{T}_J \ominus a_j$ where a_i is a child of a_k. Assume a_i is not the child of a_j in $\mathcal{T}_J$. In this case, a_i is a child of a_k in $\mathcal{T}_J$. By definition of the join tree (Definition 12.5), vertex c_i is the vertex with maximum scalar value in some connected component λ of $\mathcal{T}^-(c_k)$ and c_k is connected to λ. If c_j is not an element of λ, then $\lambda' = \lambda$ is a connected component of $\mathcal{T}'^-(c_k)$. If c_j is an element of λ, then $\lambda' = \lambda - c_j$ is a connected component of $\mathcal{T}'^-(c_k)$. In either case, c_i is a vertex with maximum scalar value in λ', so a_i' is a child of a_k' in $\mathcal{T}_J'$.

Assume a_i is the child of a_j in $\mathcal{T}_J$. Since a_i is the child of a_k in $\mathcal{T}_J \ominus a_j$, vertex a_j is the child of a_k in $\mathcal{T}_J$. Let λ be the connected component of $\mathcal{T}^-(c_k)$ containing c_j. Since c_j has degree one, $\lambda' = \lambda - c_j$ is connected. Since c_k is connected to λ, vertex c_k is also connected to λ'. Thus λ' is a connected component of $\mathcal{T}_k'^-$. Since a_i is the child of a_j in $\mathcal{T}_J$ vertex c_i is in λ and λ' is the vertex with maximum scalar value in λ'. Thus, a_i' is a child of a_k' in $\mathcal{T}_J'$.

Since a_j has at most one child in $\mathcal{T}_J$, graph $\mathcal{T}_J \ominus a_j$ is a tree. Since every edge (a_i, a_k) in $\mathcal{T}_J \ominus a_j$ corresponds to an edge (a_i', a_k') in $\mathcal{T}_J'$ and trees $\mathcal{T}_J \ominus a_j$ and $\mathcal{T}_J'$ have the same number of vertices and edges, $\mathcal{T}_J'$ is isomorphic to $\mathcal{T}_J \ominus a_j$.

A similar argument shows that the split tree of $\mathcal{T}'$ is $\mathcal{T}_S' \ominus b_j$. $\qquad \square$

Proposition 12.22. *Let* $\mathcal{T}_J$ *and* $\mathcal{T}_S$ *be the join and split trees, respectively, of a tree* $\mathcal{T}$ *with* $n \geq 2$ *vertices, where* $a_i \in V(\mathcal{T}_J)$ *and* $b_i \in V(\mathcal{T}_S)$ *correspond to* $v_i \in V(\mathcal{G})$. *For at least one* $k \in [1, \ldots, n]$,

$$a_k.\text{numChildren} + b_k.\text{numChildren} = 1.$$

Proof: Let n be the number of vertices of $\mathcal{T}_J$ and $\mathcal{T}_S$. Since a tree on n vertices has $n - 1$ edges,

$$\sum_{i=1}^{n} a_i.\text{numChildren} = n - 1,$$

$$\sum_{i=1}^{n} b_i.\text{numChildren} = n - 1.$$

Thus,

$$\sum_{i=1}^{n} (a_i.\text{numChildren} + b_i.\text{numChildren}) = 2(n-1),$$

and $(a_k.\text{numChildren} + b_k.\text{numChildren}) \leq 1$ for some $k \in [1, \dots, n]$.

Let v_k be the vertex of $\mathcal{G}$ corresponding to a_k and b_k. Since $\mathcal{G}$ is connected and has at least two vertices, v_k is adjacent to some vertex v_j of $\mathcal{G}$. If $v_j.\text{scalar}$ is less than $v_k.\text{scalar}$, then v_k is connected to some component of $\mathcal{G}_k^-$ and a_k is at least one. If $v_j.\text{scalar}$ is greater than $v_k.\text{scalar}$, then v_k is connected to some component of $\mathcal{G}_k^+$ and b_k is at least one. Thus $(a_k.\text{numChildren} + b_k.\text{numChildren}) \geq 1$ and so $(a_k.\text{numChildren} + b_k.\text{numChildren})$ equals 1. $\square$

Finally, we prove that the merge tree is unique.

Proposition 12.8.   *If $\mathcal{T}_J$ and $\mathcal{T}_S$ are the join and split trees, respectively, of two trees, $\mathcal{T}$ and $\mathcal{T}'$, then $\mathcal{T}$ is isomorphic to $\mathcal{T}'$.*

Proof: Let c_i and c_i' be the vertices of $\mathcal{T}$ and $\mathcal{T}'$, respectively, corresponding to vertex a_i of $\mathcal{T}_J$ and b_i of $\mathcal{T}_S$. We prove there is an isomorphism from $\mathcal{T}$ to $\mathcal{T}'$ that maps c_i to c_i'. The proof is by induction. If $\mathcal{T}_J$ and $\mathcal{T}_S$ each contain one or two vertices, then the claim is trivially true.

Assume that the claim holds when each tree has n vertices. We show that the claim holds when each tree has $n+1$ vertices.

By Proposition 12.13, there is some $k \in [1, \dots, n+1]$ such that $(a_k + b_k)$ equals one. Assume that a_k has zero children and b_k has one child. Let a_j be the parent of a_k in the join tree. By Proposition 12.10, vertices c_k and c_k' have degree one. Thus edge (c_k, c_j) is the sole edge of $\mathcal{T}$ incident on c_k and (c_k', c_j') is the sole edge of $\mathcal{T}$ incident on c_k'. By Proposition 12.12, the join tree of $\mathcal{T} - c_k$ and of $\mathcal{T}' - c_k'$ is $\mathcal{T}_J \ominus a_k$ and the split tree of $\mathcal{T} - c_k$ and of $\mathcal{T}' - c_k'$ is $\mathcal{T}_S \ominus b_k$. By the induction assumption, there is an isomorphism from $\mathcal{T} - c_k$ to $\mathcal{T}' - c_k'$ that maps $c_i \in V(\mathcal{T} - c_k)$ to $c_i' \in V(\mathcal{T}' - c_k')$. Extending this by mapping c_k to c_k' gives an isomorphism from $\mathcal{T}$ to $\mathcal{T}'$, which maps $c_i \in V(\mathcal{T})$ to $c_i' \in V(\mathcal{T}')$. $\square$

12.6.3   No Merge Tree

In Section 12.3.2, Figure 12.15, we gave an example of a scalar graph formed from the edges of a regular grid that we claimed had no merge tree. In this section, we prove that claim.

Proposition 12.23. *There is no merge tree for the scalar graph $\mathcal{G}$ in Figure 12.15.*

Proof: The join tree, $\mathcal{T}_J$, and the split tree, $\mathcal{T}_S$, of $\mathcal{G}$ are drawn in Figures 12.21(a) and 12.21(b), respectively. Assume that there was some tree $\mathcal{T}_M$ such that the

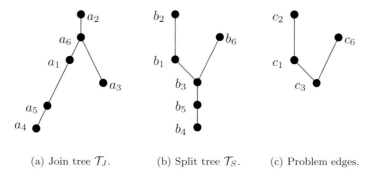

(a) Join tree $\mathcal{T}_J$.   (b) Split tree $\mathcal{T}_S$.   (c) Problem edges.

Figure 12.21. A pair of join and split trees that have no merge tree. (a) Join tree $\mathcal{T}_J$ of the scalar graph in Figure 12.15. (b) Split tree $\mathcal{T}_S$ of the scalar graph in Figure 12.15. (c) A merge tree of $\mathcal{T}_J$ and $\mathcal{T}_S$ would have to contain edges (c_1, c_2), (c_3, c_1), and (c_3, c_6). However, $\mathcal{T}_J$ cannot be the join tree of any graph with those edges.

join tree of $\mathcal{T}_M$ was $\mathcal{T}_J$ and the split tree of $\mathcal{T}_M$ was $\mathcal{T}_S$. Let c_i be the vertex of $\mathcal{T}_M$ corresponding to vertex a_i of $\mathcal{T}_J$ and vertex b_i of $\mathcal{T}_S$.

Vertex b_2 has no children in $\mathcal{T}_S$. By Proposition 12.11, (c_1, c_2) is an edge of $\mathcal{T}_M$. By Proposition 12.10, there is no other edge incident on c_2 in $\mathcal{T}_M$. Vertex b_3 has two children in $\mathcal{T}_S$. By Lemma 12.20, the up degree of c_3 is two. There are only three vertices, c_1, c_2, and c_6, whose scalar values are greater than c_3. Since (c_1, c_2) is the only edge incident on c_2, vertex c_3 must be connected to c_1 and c_6. (See Figure 12.21(c).)

Since c_3 is connected to c_1 and c_6, vertices c_1 and c_3 lie in the same connected component of $\mathcal{T}_M^-(c_6)$. By definition of the join tree, vertex a_3 is not a child of a_6 in the join tree of $\mathcal{T}_M$. Thus, $\mathcal{T}_J$ is not the join tree of $\mathcal{T}_M$, contradicting the assumption that $\mathcal{T}_M$ is the merge tree of $\mathcal{T}_J$ and $\mathcal{T}_S$. We conclude that there is no tree $\mathcal{T}_M$ whose join and split trees are $\mathcal{T}_J$ and $\mathcal{T}_S$. $\square$

12.6.4 Regular Grid Proofs

This section contains the proofs of Propositions 12.17 and 12.18 from Section 12.5.2.

As in Chapter 2, a vertex is positive, "+" if its scalar value is greater than or equal to the isovalue, and negative, "−", if its scalar value is below the isovalue. An edge is positive if both its endpoints are positive and is negative if both endpoints are negative.

Proposition 12.17. *Let Γ be a distinct valued regular scalar grid. Let τ be the triangulation of Γ induced by sorting the vertices of Γ in decreasing order of scalar value. Let ϕ be the piecewise linear function determined by τ. For each scalar* *value* σ, *the level set* $\phi^{-1}(\sigma)$ *is homeomorphic to the isosurfaces constructed by* MARCHING CUBES *using the* ISOMID3D *lookup table.*

Proof: Assign a positive, "+", label to grid vertices whose scalar values are greater than or equal to σ and a negative, "−", label to vertices whose scalar values are less than σ. Let $\mathbf{c}$ be some cube of the grid. Since the triangulation τ is induced by sorting vertices in decreasing scalar order, the positive vertices in $\mathbf{c}$ are connected by positive edges of τ. The level set $\phi^{-1}(\sigma)$ does not intersect any positive edges. Therefore, the level set $\phi^{-1}(\sigma)$ does not separate any of the positive vertices in $\mathbf{c}$. Similarly, $\phi^{-1}(\sigma)$ does not separate any positive vertices on any face $\mathbf{f}$ of $\mathbf{c}$.

Algorithm ISOMID3D constructs an isosurface patch in $\mathbf{c}$ that also does not separate any of the positive vertices in $\mathbf{c}$ or in any face $\mathbf{f}$ of $\mathbf{c}$. Let S_σ be the isosurface with isovalue σ constructed by the Marching Cubes algorithm and let F be the union of all the grid faces of Γ. Since both $\phi^{-1}(\sigma)$ and S_σ do not separate positive vertices in face $\mathbf{f}$, some curve of $\phi^{-1}(\sigma) \cap \mathbf{f}$ connects grid edges $\mathbf{e}_1$ and $\mathbf{e}_2$ if and only if some curve in $S_\sigma \cap \mathbf{f}$ connects $\mathbf{e}_1$ and $\mathbf{e}_2$. Thus, there is a homeomorphism that maps $\phi^{-1}(\sigma) \cap F$ to $S_\sigma \cap F$.

For all isosurface lookup table configurations, except 2C, the isosurface patch is a union of disks. For all such configurations, $\phi^{-1}(\sigma) \cap \mathbf{c}$ is a union of corresponding disks. For configuration 2C, the isosurface patch is a cylinder containing two diagonally opposite positive vertices. These two vertices have greatest scalar value in $\mathbf{c}$, so τ contains the diagonal edge connecting these vertices and $\phi^{-1}(\sigma) \cap \mathbf{c}$ is a cylinder containing the two vertices. Thus, for every grid cube $\mathbf{c}$, the homeomorphism from $\phi^{-1}(\sigma) \cap F$ to $S_\sigma \cap F$ can be extended to the level set and isosurface in the interior of each grid cube. Thus the homeomorphism can be extended to the entire level set and isosurface. $\qquad\square$

The following lemma is used in proving Proposition 12.18.

Lemma 12.24. *Let* Γ *be a distinct valued regular scalar grid. Let* τ *be the triangulation of* Γ *induced by sorting the vertices of* Γ *in decreasing order of scalar value. Let* $\mathcal{G}_\tau$ *be the 1-skeleton of* τ, *let* $\mathcal{G}$ *be the 1-skeleton of* Γ, *and let* $\widetilde{\mathcal{G}}$ *be the 26-connectivity graph of* Γ.

1. *If* (v, v') *is an edge of* $\mathcal{G}_\tau$, *then there is some path from* v *to* v' *in* $\mathcal{G}$ *whose vertices all have scalar values at most* $\max(v.\mathsf{scalar}, v'.\mathsf{scalar})$.

2. *If* (v, v') *is an edge of* $\widetilde{\mathcal{G}}$, *then there is some path from* v *to* v' *in* $\mathcal{G}_\tau$ *whose vertices all have scalar values at least* $\min(v.\mathsf{scalar}, v'.\mathsf{scalar})$.

Proof of 1: Edge (v, v') is either an edge of $\mathcal{G}$ or the diagonal of some grid face $\mathbf{f}$ or the diagonal of some grid cube $\mathbf{c}$. We consider each of the three cases.

Case I: Edge (v, v') is an edge of $\mathcal{G}$.
    Statement 1 is trivially true since (v, v') is a path from v to v'.

Case II: Edge (v, v') is the diagonal of some grid face **f**.

Triangulation τ restricted to **f** is a triangulation of **f** containing edge (v, v'). Since τ is induced by adding vertices in decreasing scalar order, some third vertex v'' of **f** must have scalar value less than v.scalar and v'.scalar. Thus (v, v'', v') is a path from v to v' whose vertices all have scalar value at most $\max(v$.scalar$, v'$.scalar$)$.

Case III: Edge (v, v') is the diagonal of some grid cube **c**.

Let s equal $\max(v$.scalar$, v'$.scalar$)$. Let W be the set of three vertices in cube **c** adjacent to v and let W' be the set of three vertices in cube **c** adjacent to v'.

Triangulation τ restricted to **c** is a triangulation of **c** containing edge (v, v'). Diagonal (v, v') passes through the interior of triangle formed by the three vertices of W. Since τ is induced by adding vertices in decreasing scalar order, at least one of the scalar values of the vertices in W. must be less than s. Similarly, at least one of the scalar values of the vertices in W' must be less than s.

Let w and w' be the vertices in W and W', respectively, where w.scalar and w'.scalar are less than s. If (w, w') is an edge of cube **c**, then (v, w, w', v') is a path from v to v' whose vertices all have scalar value at most $\max(v$.scalar$, v'$.scalar$)$. If (w, w') is not an edge of cube **c**, then (w, w') is a diagonal of the cube. Let W'' be the four vertices of **c** that are not v, v', w, or w'. Diagonal (v, v') passes thought the interior of the convex hull of W''. Thus w''.scalar is less than s for at least one vertex $w'' \in W''$. If vertex w'' is adjacent to w, then (v, w, w'', v') is a path from v to v' whose vertices all have scalar value at most $\max(v$.scalar$, v'$.scalar$)$. Otherwise, vertex w'' is adjacent to w' and (v, w'', w', v') is a path from v to v' whose vertices all have scalar value at most $\max(v$.scalar$, v'$.scalar$)$. $\square$

Proof of 2: Since (v, v') is an edge of $\widetilde{\mathcal{G}}$, some cube **c** of Γ contains both v and v'. Let W be the vertices of **c** whose scalar values are at least $\min(v$.scalar$, v'$.scalar$)$. Since triangulation τ is induced by adding vertices in decreasing scalar order, triangulation τ restricted to the vertices of W form a triangulation of the convex hull of W. Since the convex hull is connected, there is a path from v to v' in W. Thus, there is a path from v to v' whose vertices all have scalar value at least $\min(v$.scalar$, v'$.scalar$)$. $\square$

Proposition 12.18.   *Let Γ be a distinct valued regular scalar grid. Let τ be the triangulation of Γ induced by sorting the vertices of Γ in decreasing order of scalar value. Let $\mathcal{G}_\tau$ be the 1-skeleton of τ, let $\mathcal{G}$ be the 1-skeleton of Γ, and let $\widetilde{\mathcal{G}}$ be the 26-connectivity graph of Γ.*

1. *The join tree of $\mathcal{G}_\tau$ is the join tree of $\mathcal{G}$.*

2. *The split tree of $\mathcal{G}_\tau$ is the split tree of $\widetilde{\mathcal{G}}$.*

Proof of 1: As in Section 12.3, let $\mathcal{G}^-(v_i)$ and $\mathcal{G}_\tau^-(v_i)$ be the set of vertices of $\mathcal{G}$ and $\mathcal{G}_\tau$ with scalar value strictly less than v_i.scalar. Since $\mathcal{G}$ is a subset of $\mathcal{G}_\tau$, every connected component of $\mathcal{G}^-(v_i)$ is contained in a connected component of $\mathcal{G}_\tau^-(v_i)$. By Lemma 12.24, Statement 1, if (v, v') are an edge of $\mathcal{G}_\tau^-(v_i)$, then v and v' are connected by some path in $\mathcal{G}^-(v_i)$. Thus, two vertices are in the same connected component of $\mathcal{G}_\tau^-(v_i)$ if and only if they are in the same component of $\mathcal{G}^-(v_i)$. Similarly, vertex v_i is connected to the component containing v_j in $\mathcal{G}_\tau^-(v_i)$ if and only if v_j is connected to the component containing v_j in $\mathcal{G}_\tau^-(v_i)$. Since the definition of the join tree is based on the connected components of $\mathcal{G}^-(v_i)$ and $\mathcal{G}_\tau^-(v_i)$ that are connected to v_i, the join tree of $\mathcal{G}$ is the same as the join tree of $\mathcal{G}_\tau$. $\qquad\square$

Proof of 2: As in Section 12.3, let $\widetilde{\mathcal{G}}^+(v_i)$ and $\mathcal{G}_\tau^+(v_i)$ be the set of vertices of $\widetilde{\mathcal{G}}$ and $\mathcal{G}_\tau$ with scalar value strictly greater than v_i.scalar. Since $\mathcal{G}_\tau$ is a subset of $\widetilde{\mathcal{G}}$, every connected component of $\mathcal{G}_\tau^+(v_i)$ is contained in a connected component of $\widetilde{\mathcal{G}}^+(v_i)$. By Lemma 12.24, Statement 2, if (v, v') are an edge of $\widetilde{\mathcal{G}}^+(v_i)$, then v and v' are connected by some path in $\mathcal{G}_\tau^+(v_i)$. Thus, two vertices are in the same connected component of $\widetilde{\mathcal{G}}^+(v_i)$ if and only if they are in the same component of $\mathcal{G}_\tau^-(v_i)$. Similarly, vertex v_i is connected to the component containing v_j in $\widetilde{\mathcal{G}}^+(v_i)$ if and only if v_j is connected to the component containing v_j in $\mathcal{G}_\tau^+(v_i)$. Since the definition of the split tree is based on the connected components of $\mathcal{G}^+(v_i)$ and $\mathcal{G}_\tau^+(v_i)$ that are connected to v_i, the split tree of $\widetilde{\mathcal{G}}$ is the same as the split tree of $\mathcal{G}_\tau$. $\qquad\square$

12.7   Simplification of Contour Trees

Even isosurfaces in the simplest, computer-generated data sets may contain many small connected components that are artifacts of the grid sampling or of the data set boundary. The small connected components create numerous nodes in the contour tree, making it difficult to visualize. For instance, the full contour tree of the nucleon data (Figure 12.4) contains 41 nodes. Figure 12.4(a) is a simplified version of the contour tree with many nodes removed.

Data sets created by imaging devices contain numerous connected components caused by noise in the data. The full contour tree of the engine data set in Figure 11.1 contains 123,864 nodes. The full contour tree of the tooth data set in Figure 11.2 contains 576,118 nodes.

Numerous metrics can be used to remove nodes from contour trees. One simple metric is the scalar value associated with the nodes. Consider Figure 12.22 containing part of the full contour tree. The tree nodes contain the scalar values of the associated contour trees. All the tree nodes in this portion of the contour tree contain scalar values of 247, 248, or 249. Since the scalar values

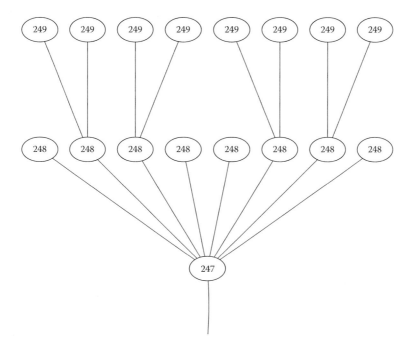

Figure 12.22. Part of the full contour tree of the nucleon data set. Tree nodes contain the scalar values of the associated isosurface component.

in the data set vary from 0 to 249, the difference between 247, 248, and 249 is insignificant. Thus these nodes can all be collapsed into a single node, as is done in Figure 12.4(a).

Another metric is the size of the region bounded by the isosurface component. The number of grid vertices in a region is a good approximation to its size. Each contour tree node is associated with a region in the mesh or grid and a set of mesh or grid vertices contained in that region. Remove leaves whose associated vertices are fewer than a certain threshold.

Figure 12.23 shows a full contour tree of the hydrogen atom data set. Each node represents some connected component in an isosurface. It contains the number of grid vertices in a region bounded by that component. Two leaves have fewer than 200 vertices in their associated regions. These leaves represent connected components near the boundary that separate a small number of grid vertices from the rest of the grid. These leaves were eliminated from the hydrogen atom contour tree previously shown in Figure 12.3(a).

Connected components of isosurfaces that pass through grid vertices correspond to nodes of the fully augmented join tree and the fully augmented split tree. The number of grid vertices in the region bounded by such a component is the number of nodes in the subtree rooted at the corresponding node in the join

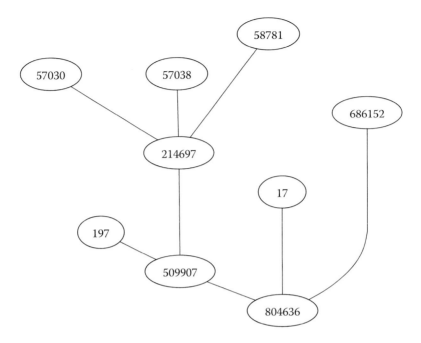

Figure 12.23. The full contour tree of the hydrogen atom data set. Each node contains the number of grid vertices in an associated region.

or split tree. The number of nodes in all such subtrees can be computed in linear time. Thus, the number of grid vertices in the region bounded by a component can be computed in linear time.

Carr, Snoeyink, and van de Panne in [Carr et al., 2004] discuss other measures that can be used for contour tree simplication. They also give an explicit algorithm for simplifying the contour trees under these metrics.

12.8 Applications

Contour trees contain a wealth of information about scalar fields. As shown in Figures 12.3 and 12.4, they show how isosurfaces change as the isovalue changes. They also show the relationship between isosurfaces of different scalar values.

Contour trees can be used to select isovalues. Isovalues at joins or splits in the contour tree produce joins or splits of connected components in the isosurface. These isovalues may warrant further investigation. On the other hand, there are no joins or splits of the connected components represented by a contour tree edge. The isosurface component from a single isovalue chosen along that contour edge may suffice to represent all the components represented by that edge.

Contour trees do not model geometric changes in the isosurface. They also do not model the creation of tunnels in the isosurface, although such information can be added to the contour tree. (See [Pascucci and Cole-McLaughlin, 2002].) Thus, there are limits on the information provided by contour trees and on their ability to indicate all significant isovalues.

In Section 8.4, we discussed constructing an isosurface by starting from a seed set of grid cubes, determining which elements of the seed set intersect the isosurface, and propagating the isosurface from those cube. A key problem is to find a small or the smallest complete seed set of grid cubes that intersects every connected component of every isovalue.

A grid vertex v with scalar value s_v is contained in some connected component of the isosurface with isovalue s_v. Grid vertex v maps to the point $\chi(v)$ in the contour tree corresponding to this connected component. A grid edge $\mathbf{e} = (v_0, v_1)$ maps to the path $\zeta_{\mathbf{e}}$ from $\chi(v_0)$ to $\chi(v_1)$ in the contour tree. A grid cube $\mathbf{c}$ has a vertex v_0 with minimum scalar value and a vertex v_1 with maximum scalar value. Cube $\mathbf{c}$ maps to the path $\zeta_{\mathbf{c}}$ from $\chi(v_0)$ to $\chi(v_1)$ in the contour tree. A complete seed set of grid cubes corresponds to a set $\mathbb{C}$ of grid cubes such that $\{\zeta_{\mathbf{c}} : \mathbf{c} \in \mathbb{C}\}$ covers the entire contour tree. A minimal complete seed set is a minimal covering of the contour tree by the $\zeta_{\mathbf{c}}$. Van Kreveld et al. [van Kreveld et al., 1997] solve the minimal covering problem to construct a minimal complete seed set.

The seed set constructed by van Kreveld et al. [van Kreveld et al., 1997] may contain a large number of seeds scattered across the grid. In a different approach, Carr and Snoeyink [Carr and Snoeyink, 2003] show that seed sets for connected components corresponding to a contour tree edge can be quickly constructed from the seed set for the connected components corresponding to the edge endpoints. The seed set is the set of cubes intersected by a path in the grid from grid vertex v_0 to grid vertex v_1 where $\chi(v_0)$ is one endpoint of the contour tree edge and $\chi(v_1)$ is another. To construct an isosurface for a given isovalue, determine all the contour tree edges whose span contains the isovalue, construct the seed sets corresponding to those edges, and propagate the isosurface from those seed sets.

Contour trees can also be used to select and reconstruct individual connnected components of an isosurface [Carr and Snoeyink, 2003]. Each contour tree edge corresponds to a seed set. A point on the contour tree edge corresponds to an isosurface component intersecting a grid cube in that seed set. Select a point with scalar value σ on some contour tree edge, construct the seed set corresponding the contour tree edge, find the grid cube whose span contains σ, and propagate the isosurface from that grid cube.

The correspondence between contour tree edges and connected components allows one to select or deselect individual isosurface components for display. Moreover, one can also simultaneously select and display connected components from isosurfaces with different isovalues.

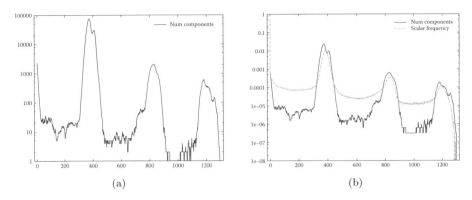

(a)   (b)

Figure 12.24. Tooth data set. (a) Number of connected components. (b) Number of connected components (normalized) and scalar frequencies (normalized).

In the previous section, we discussed contour tree simplification. Contour tree simplification can be used to simplify isosurfaces. Removing leaves and their incident edges from the contour tree corresponds to removing the corresponding connected components from the isosurface. Each point in the simplified contour tree corresponds to a connected component of some isosurface. Extending the approach in [Carr and Snoeyink, 2003], Carr et al. [Carr et al., 2004] show how to find a grid cube that intersects that component from the simplified contour tree.

Finally, contour trees can be used to count the number of components even in complex or noisy data set. Figure 12.24(a) contains a plot of the number of components as a function of the scalar value for the tooth data set. The plot was produced by first constructing the contour tree for the data set and then computing the number of contour tree edges intersecting scalar values from 0 to 1,300. Scalar values with large numbers of components correspond to noisy isosurfaces. The peaks correspond to the peaks in the graph of the scalar frequencies. (See Figure 12.24(b).)

12.9   Notes and Comments

Boyell and Ruston described contour trees in [Boyell and Ruston, 1963] and used them to represent contour lines on 2D terrains. Van Kreveld et al. [van Kreveld et al., 1997] gave an algorithm to compute contour trees of 2D scalar meshes in $\Theta(N \log N)$ time algorithm where N is the number of mesh elements. Tarasov and Vyalyi [Tarasov and Vyalyi, 1998] showed how to compute contour trees in 3D scalar meshes.

The algorithms in Sections 12.4 and 12.5 are from [Carr et al., 2003] by Carr, Snoeyink, and Axen. The algorithm computes contour trees for scalar meshes in any dimension and runs in $\Theta(n \log n + N\alpha(N))$ time where n is the number of mesh vertices and N is the number of mesh simplices. The function $\alpha(n)$ is the inverse of the Ackermann function and grows very slowly. Carr and Snoeyink [Carr and Snoeyink, 2009] show how the contour tree algorithm can be modified to model the variation in isosurface topology produced by different isosurface construction algorithms. Pascucci and Cole-McLaughlin [Pascucci and Cole-McLaughlin, 2002] add the computation of Betti numbers to the algorithm in [Carr et al., 2003].

Pascucci and Cole-McLaughlin [Pascucci and Cole-McLaughlin, 2002] present an algorithm for constructing contour trees that runs in $\Theta(n+k \log k)$ time where k is the number of "critical" points. Chiang et al. [Chiang et al., 2005] give an algorithm that runs in $\Theta(n+k' \log k')$ time where k' is the number of "component critical" mesh vertices. (See the references for the definitions of "critical" and "component critical.") Both papers improved the running time of [Carr et al., 2003] by reducing the time to construct join and split trees.

Van Kreveld et al. [van Kreveld et al., 1997] and Carr and Snoeyink [Carr and Snoeyink, 2003] use contour trees to construct small seed sets. Carr et al. in [Carr et al., 2004] discuss techniques for simplifying contour trees. Cox et al. [Cox et al., 2003] use an augmented split tree to partition the scalar data into "topological zones" and reduce I/O and disk seek times in constructing isosurface on large data sets. (They call the split tree a "criticality tree".)

Contour trees are special cases of Reeb graphs defined by and named after Georges Reeb [Reeb, 1946]. Algorithms for computing Reeb graphs are in [Shinagawa and Kunii, 1991, Cole-McLaughlin et al., 2003, Doraiswamy and Natarajan, 2008, Doraiswamy and Natarajan, 2009, Pascucci et al., 2007, Tierny et al., 2009, Harvey et al., 2010, Parsa, 2012].

APPENDIX A

GEOMETRY

A.1 Affine Hull

A set of points $\mathbb{X} \subseteq \mathbb{R}^d$ is an affine subspace of $\mathbb{R}^d$ if, for every pair of points $p_1, p_2 \in \mathbb{X}$, the line through p_1 and p_2 is in $\mathbb{X}$. Equivalently, $\mathbb{X}$ is an affine subspace if, for every finite subset $\{p_1, \ldots, p_k\} \subseteq \mathbb{X}$, set $\mathbb{X}$ contains the set

$$\{\alpha_1 p_1 + \alpha_2 p_2 + \ldots + \alpha_k p_k : \alpha_1, \alpha_2, \ldots, \alpha_k \in \mathbb{R} \text{ and } \alpha_1 + \alpha_2 + \ldots + \alpha_k = 1\}.$$

Examples of affine subspaces are a point, a line, a plane, or a hyperplane.

Each affine subspace $\mathbb{X} \subseteq \mathbb{R}^d$ is congruent to $\mathbb{R}^k$ for some k, i.e., there is a one-to-one and onto mapping of $\mathbb{X}$ to $\mathbb{R}^k$ that preserves distances between points. The dimension of $\mathbb{X}$ is k where $\mathbb{X}$ is congruent to $\mathbb{R}^k$.

The affine hull of a set $\mathbb{X} \subseteq \mathbb{R}^d$ is the smallest affine subspace containing $\mathbb{X}$. Equivalently, the affine hull of a set $\mathbb{X} \subset \mathbb{R}^d$ is the set

$$\{\alpha_1 p_1 + \alpha_2 p_2 + \ldots + \alpha_k p_k : \{p_1, p_2, \ldots, p_k\} \text{ is a finite subset of } \mathbb{X} \text{ and}$$
$$\alpha_1, \alpha_2, \ldots, \alpha_k \in \mathbb{R} \text{ and } \alpha_1 + \alpha_2 + \ldots + \alpha_k = 1\}.$$

A finite set P of points is in general position if, for every $p \in P$, the affine hull of $P - \{p\}$ does not contain p. Equivalently, a set P of $k + 1 \le d + 1$ points is in general position in $\mathbb{R}^d$ if no k-dimensional affine subspace of $\mathbb{R}^d$ contains P.

A.2 Convexity

A set of points $\mathbb{X} \subseteq \mathbb{R}^d$ is convex if, for every pair of points $p_1, p_2 \in \mathbb{X}$, the line segment from p_1 to p_2 is in $\mathbb{X}$. Equivalently, $\mathbb{X}$ is convex if, for every finite subset

$\{p_1, \ldots, p_k\} \subseteq \mathbb{X}$, set $\mathbb{X}$ contains the set

$$\{\alpha_1 p_1 + \alpha_2 p_2 + \ldots + \alpha_k p_k : \alpha_1, \alpha_2, \ldots, \alpha_k \in \mathbb{R} \text{ and } \alpha_1 + \alpha_2 + \ldots + \alpha_k = 1 \text{ and}$$
$$\alpha_1, \alpha_2, \ldots, \alpha_k \geq 0\}.$$

Examples of convex sets are a line segment, a triangle, a tetrahedron, or a cube. Other examples of convex sets are the area bounded by a circle or ellipse in $\mathbb{R}^2$ or the volume bounded by a sphere or ellipsoid in $\mathbb{R}^3$. Note that every affine subspace is convex.

The convex hull of a set $\mathbb{X} \subseteq \mathbb{R}^d$ is the smallest convex set containing $\mathbb{X}$. Equivalently, the convex hull of a set $\mathbb{X} \subset \mathbb{R}^d$ is the set

$$\text{conv}(\mathbb{X}) = \{\alpha_1 p_1 + \alpha_2 p_2 + \ldots + \alpha_k p_k : \tag{A.1}$$
$$\{p_1, p_2, \ldots, p_k\} \text{ is a finite subset of } \mathbb{X} \text{ and}$$
$$\alpha_1, \alpha_2, \ldots, \alpha_k \in \mathbb{R} \text{ and } \alpha_1 + \alpha_2 + \ldots + \alpha_k = 1 \text{ and}$$
$$\alpha_1, \alpha_2, \ldots, \alpha_k \geq 0\}.$$

The convex hull of $\mathbb{X}$ is contained in the affine hull of $\mathbb{X}$.

A.3   Convex Polytope

A convex polytope $\mathbf{c}$ is the convex hull of a finite set of points P in $\mathbb{R}^d$. Examples of convex polytopes are a triangle, a tetrahedron, or a cube. The dimension of a convex polytope $\mathbf{c}$ is the dimension of the affine hull of $\mathbf{c}$. A three-dimensional convex polytope is called a convex polyhedron.

A vertex of a convex polytope $\mathbf{c}$ is a point $p \in \mathbf{c}$ such that $\text{conv}(\mathbf{c} - \{p\})$ does not equal $\mathbf{c}$. If P is a finite point set and V is the set of vertices of the convex polytope $\text{conv}(P)$, then V is a subset of P. Thus, a convex polytope has a finite set of vertices.

Let $\mathbf{c}$ be a convex polytope and let V be its set of vertices. The (polytope) interior of $\mathbf{c}$ is the set

$$\text{int}(\mathbf{c}) = \{\alpha_1 p_1 + \alpha_2 p_2 + \ldots + \alpha_k p_k : \tag{A.2}$$
$$\{p_1, p_2, \ldots, p_k\} \text{ is a finite subset of } P \text{ and}$$
$$\alpha_1, \alpha_2, \ldots, \alpha_k \in \mathbb{R} \text{ and } \alpha_1 + \alpha_2 + \ldots + \alpha_k = 1 \text{ and}$$
$$\alpha_1, \alpha_2, \ldots, \alpha_k > 0\}.$$

Note the difference from Equation A.1. In Equation A.1, each α_i is greater than or equal to zero. In Equation A.2, each α_i is strictly greater than zero.

The (polytope) boundary of $\mathbf{c}$, denoted $\partial \mathbf{c}$, is all the points not on the interior of $\mathbf{c}$, i.e., $\mathbf{c} - \text{int}(\mathbf{c})$.

Let $\mathbf{c}$ be a convex polytope and let V be its set of vertices. A face of $\mathbf{c}$ is $\mathrm{conv}(V')$ where V' is a subset of V and the affine hull of V' does not intersect $\mathrm{conv}(V - V')$. A d-face of c is a d-dimensional face of c. The convex polytope $\mathbf{c}$ is a face of itself. The proper faces of $\mathbf{c}$ are the faces of $\mathbf{c}$ except for $\mathbf{c}$.

A hyperplane $\mathbf{h}$ supports a convex polytope $\mathbf{c}$ if $\mathbf{h}$ intersects $\mathbf{c}$ and one of the closed half-spaces bounded by $\mathbf{h}$ contains $\mathbf{c}$. Set $\mathbf{f}$ is a face of $\mathbf{c}$ if and only if $\mathbf{f}$ equals $\mathbf{h} \cap \mathbf{c}$ for some supporting hyperplane $\mathbf{h}$ of $\mathbf{c}$.

A.4   Simplex

A simplex is the simplest type of convex polytope. A k-simplex in $\mathbb{R}^d$ is the convex hull of a set of $k + 1 \leq d + 1$ points in general position in $\mathbb{R}^d$. The points are called the simplex vertices. A triangle is a 2-simplex. A tetrahedron is a 3-simplex.

The faces of a k-simplex $\mathbf{t}$ are the convex hulls of all subsets of the vertices of $\mathbf{t}$. Simplex $\mathbf{t}$ is a face of itself. The proper faces of a k-simplex $\mathbf{t}$ are the convex hulls of all proper subsets of the vertices of $\mathbf{t}$. Equivalently, the proper faces of $\mathbf{t}$ are the faces of $\mathbf{t}$ except for $\mathbf{t}$.

The boundary of a simplex $\mathbf{t}$, denoted $\partial\mathbf{t}$, is the union of all its proper faces. The interior of a simplex $\mathbf{t}$, denoted $\mathrm{int}(\mathbf{t})$, is all the points not in the boundary of $\mathbf{t}$, i.e., $\mathbf{t} - \partial\mathbf{t}$. Note that the simplex vertices, simplex boundary, and simplex interior of a simplex $\mathbf{t}$ are the same as the polytope vertices, polytope boundary, and polytope interior of $\mathbf{t}$.

A.5   Barycentric Coordinates

Let $\{p_1, \ldots, p_{d+1}\}$ be a set of $d+1$ points in general position in $\mathbb{R}^d$. For each point $q \in \mathbb{R}^d$ and scalar values $\alpha_1, \ldots, \alpha_{d+1} \in \mathbb{R}$ consider the set $d + 1$ of equations:

$$\sum_{i=1}^{d+1} \alpha_i p_i = q, \tag{A.3}$$

$$\sum_{i=1}^{d+1} \alpha_i = 1. \tag{A.4}$$

Note that Equation A.3 represents d separate equations. Equations A.3 and A.4 are $d + 1$ equations in $d + 1$ unknowns $\alpha_i \in \mathbb{R}$. Since the points p_i are in general position, these equations are independent and have a single unique solution. The set $\alpha_1, \ldots, \alpha_{d+1}$ satisfying these equations are called the barycentric coordinates of point q.

A.6   Linear Function

A function $\mu : \mathbb{R}^d \to \mathbb{R}$ is linear if it can be written as

$$\mu(x_1, x_2, \ldots, x_d) = a_1 x_1 + a_2 x_2 + \ldots + a_d x_d$$

for some scalar constants $a_1, a_2, \ldots, a_d \in \mathbb{R}^d$.

Let $\{p_1, \ldots, p_{d+1}\}$ be a set of $d + 1$ points in general position in $\mathbb{R}^d$ where each p_i is associated with a scalar s_i. The scalar values s_i determine a linear function $\mu : \mathbb{R}^d \to \mathbb{R}$ defined as

$$\mu(q) = \sum_{i=1}^{d+1} \alpha_i s_i,$$

where $(\alpha_1, \ldots, \alpha_{d+1})$ are the barycentric coordinates of q. Note that $\mu(p_i)$ equals s_i.

A function $\mu : \mathbb{R}^d \to \mathbb{R}$ is affine if it can be written as $\mu(x_1, x_2, \ldots, x_d) = a_0 + a_1 x_1 + a_2 x_2 + \ldots + a_d x_d$ for some scalar constants $a_0, a_1, a_2, \ldots, a_d \in \mathbb{R}^d$.

Let $\mathbb{X}$ and $\mathbb{Y}$ be subsets of $\mathbb{R}^d$. An affine transformation is an affine function μ such that $\mu(\mathbb{X})$ equals $\mathbb{Y}$.

A.7   Congruent and Similar

Two convex polytopes in $\mathbb{R}^d$ are congruent if they have the same shape and size. More specifically, convex polytope $\mathbf{c}_1$ is congruent to convex polytope $\mathbf{c}_2$ if there is a sequence of translations, rotations, and reflections that transforms $\mathbf{c}_1$ into $\mathbf{c}_2$.

Two convex polytopes in $\mathbb{R}^d$ are similar if they have the same shape. More specifically, convex polytope $\mathbf{c}_1$ is similar to convex polytope $\mathbf{c}_2$ if $\mathbf{c}_1$ can be scaled uniformly in all directions into a polytope $\mathbf{c}_1'$ such that $\mathbf{c}_1'$ is congruent to $\mathbf{c}_2$.

APPENDIX B

TOPOLOGY

B.1 Interiors and Boundaries

The (topological) interior of a set of points $\mathbb{X} \subseteq \mathbb{R}^d$, denoted $\text{int}(\mathbb{X})$, is the points of $\mathbb{X}$ that have neighborhoods contained in $\mathbb{X}$. The (topological) boundary of a set $\mathbb{X} \subseteq \mathbb{R}^d$, denoted $\partial \mathbb{X}$, is the set of points of $\text{cl}(\mathbb{X})$, the closure of $\mathbb{X}$, that do not have neighborhoods contained in $\mathbb{X}$. Formally,

$$
\begin{aligned}
\text{int}(\mathbb{X}) &= \{p \in \mathbb{X} : \mathbb{N}_p \subseteq \mathbb{X} \text{ for some neighborhood } \mathbb{N}_p \text{ of } p\}, \\
\partial \mathbb{X} &= \text{cl}(\mathbb{X}) - \text{int}(\mathbb{X}) \\
&= \{p \in \text{cl}(\mathbb{X}) : \mathbb{N}_p \not\subseteq \mathbb{X} \text{ for any neighborhood } \mathbb{N}_p \text{ of } p\}.
\end{aligned}
$$

Note that by definition $\text{int}(\mathbb{X})$ is a subset of $\mathbb{X}$.

The topological boundaries and interiors of set P depends on the ambient space $\mathbb{R}^d$ containing $\mathbb{X}$. If $\mathbb{X}$ is a two-dimensional disk in $\mathbb{R}^2$, then $\partial \mathbb{X}$ is a circle and $\text{int}(\mathbb{X})$ is the open two-dimensional disk bounded by this circle. However, if $\mathbb{X}$ is a two-dimensional disk in $\mathbb{R}^3$, then all points in $\mathbb{X}$ are on the boundary of $\mathbb{X}$ and $\text{int}(\mathbb{X})$ is the empty set.

To define a variation of interior and boundary that is independent of the ambient space $\mathbb{R}^d$, we use the affine hull. As defined in Appendix A, Section A.1, the affine hull of a set $\mathbb{X} \subseteq \mathbb{R}^d$ of points is the smallest affine subspace containing $\mathbb{X}$. The relative interior of a set $\mathbb{X} \subseteq \mathbb{R}^d$, denoted $\text{relint}(\mathbb{X})$, is the set of all points $p \in \mathbb{X}$ with some neighborhood $\mathbb{N}_p$ such that $\mathbb{N}_p \cap \text{aff}(\mathbb{X}) \subseteq \mathbb{X}$. The relative boundary of $\mathbb{X}$, denoted $\text{relbnd}(\mathbb{X})$, is the set of points in $\text{cl}(\mathbb{X})$ that are not in $\text{relint}(\mathbb{X})$. Formally,

$$
\text{relint}(\mathbb{X}) = \{p \in \mathbb{X} : \mathbb{N}_p \cap \text{aff}(\mathbb{X}) \subseteq \mathbb{X} \text{ for some neighborhood } \mathbb{N}_p \text{ of } p\},
$$
$$
\text{relbnd}(\mathbb{X}) = \text{cl}(\mathbb{X}) - \text{relint}(\mathbb{X}).
$$

If $\mathbf{c}$ is a convex polytope, then the polytope boundary and polytope interior of $\mathbf{c}$ as defined in Appendix A, Section A.3, are the relative boundary and relative interior, respectively, of $\mathbf{c}$.

The following two lemmas relate the closure and interior of a point set $\mathbb{X}$.

Lemma B.1. *For any closed set* $\mathbb{X} \subseteq \mathbb{R}^d$,

$$\mathrm{cl}(\mathrm{int}(\mathbb{X})) \subseteq \mathbb{X}.$$

Proof: Let p be a point in $\mathrm{cl}(\mathrm{int}(\mathbb{X}))$. Since p is in $\mathrm{cl}(\mathrm{int}(\mathbb{X}))$, there is an infinite sequence $q_1, q_2, \ldots$ of points $q_i \in \mathrm{int}(\mathbb{X})$ such that

$$\lim_{n \to \infty} q_i = p.$$

Since $\mathrm{int}(\mathbb{X})$ is a subset of $\mathbb{X}$, each q_i is in $\mathbb{X}$. Since $\mathbb{X}$ is closed, point p is in $\mathbb{X}$. Thus, $\mathrm{cl}(\mathrm{int}(\mathbb{X}))$ is a subset of $\mathbb{X}$.   □

Lemma B.2. *Let* $\mathbb{Y}_1$ *be a closed subset of* $\mathbb{X} \subseteq \mathbb{R}^d$. *If* $\mathbb{Y}_2$ *equals* $\mathrm{cl}(\mathrm{int}(\mathbb{Y}_1))$, *then* $\mathrm{cl}(\mathbb{X} - \mathbb{Y}_1)$ *equals* $\mathrm{cl}(\mathbb{X} - \mathbb{Y}_2)$.

Proof: By Lemma B.1, set $\mathbb{Y}_2$ is a subset of $\mathbb{Y}_1$. Thus $\mathbb{X} - \mathbb{Y}_1$ is a subset of $\mathbb{Y}_2$ and so $\mathrm{cl}(\mathbb{X} - \mathbb{Y}_1)$ is a subset of $\mathrm{cl}(\mathbb{X} - \mathbb{Y}_2)$.

Let p be a point in $\mathrm{cl}(\mathbb{X} - \mathbb{Y}_2)$. Since p is in $\mathrm{cl}(\mathbb{X} - \mathbb{Y}_2)$, for every $\epsilon > 0$, there is a point $q \in \mathbb{X} - \mathbb{Y}_2 = \mathbb{X} - \mathrm{cl}(\mathrm{int}(\mathbb{Y}_1))$ that is distance at most $\epsilon/2$ from p. Let N_q be the set of all points in $\mathbb{X}$ at distance at most $\epsilon/2$ from q. Since q is not an element of $\mathrm{int}(\mathbb{Y}_1)$, set N_q is not a subset of $\mathbb{Y}_1$. Thus there exists a $q' \in N_q$ such that q' is not in $\mathbb{Y}_1$. Since the distance from q' to q is at most $\epsilon/2$ and the distance from q to p is at most $\epsilon/2$, the distance from q' to p is at most ϵ. For every ϵ there is a point $q' \in \mathbb{X} - \mathbb{Y}_1$ at distance at most ϵ from p. Thus point p is in $\mathrm{cl}(\mathbb{X} - \mathbb{Y}_1)$ and $\mathrm{cl}(\mathbb{X} - \mathbb{Y}_2)$ is a subset of $\mathrm{cl}(\mathbb{X} - \mathbb{Y}_1)$.

Since $\mathrm{cl}(\mathbb{X} - \mathbb{Y}_1)$ is a subset of $\mathrm{cl}(\mathbb{X} - \mathbb{Y}_2)$ and $\mathrm{cl}(\mathbb{X} - \mathbb{Y}_2)$ is a subset of $\mathrm{cl}(\mathbb{X} - \mathbb{Y}_1)$, set $\mathrm{cl}(\mathbb{X} - \mathbb{Y}_1)$ equals $\mathrm{cl}(\mathbb{X} - \mathbb{Y}_2)$.   □

B.2   Homeomorphism

Let $\mathbb{X}$ and $\mathbb{Y}$ be subsets of $\mathbb{R}^d$. Function $\mu : \mathbb{X} \to \mathbb{Y}$ is a homeomorphism from $\mathbb{X}$ to $\mathbb{Y}$ if μ is continuous, one-to-one, and onto and has continuous inverse μ^{-1}. Two spaces, $\mathbb{X}$ and $\mathbb{Y}$, are homeomorphic if there is a homeomorphism from $\mathbb{X}$ to $\mathbb{Y}$.

A (closed) topological ball is a set $\mathbb{X}$ that is homeomorphic to a closed ball $\mathbb{B}^d$. The dimension of the topological ball is d. A simple curve is a set $\mathbb{X}$ that is homeomorphic to the unit interval $[0, 1]$. A simple closed curve is a set $\mathbb{X}$ that is homeomorphic to the unit circle $\{(x, y) : x^2 + y^2 = 1\}$.

If $\mathbb{X}$ is closed and bounded and μ is continuous, one-to-one, and onto, then μ^{-1} is continuous and μ is a homeomorphism.

Lemma B.3. *Let $\mathbb{X}$ and $\mathbb{Y}$ be subsets of $\mathbb{R}^d$. If $\mathbb{X}$ is closed and bounded, and $\mu : \mathbb{X} \to \mathbb{Y}$ is continuous, one-to-one, and onto, then μ is a homeomorphism from $\mathbb{X}$ to $\mathbb{Y}$.*

Proof: Let $(y_1, y_2, y_3, \ldots)$ be a sequence of points in $\mathbb{Y}$ that approach some limit point $y \in \mathbb{R}^d$. Since μ is one-to-one, it has an inverse function $\mu^{-1} : \mathbb{Y} \to \mathbb{X}$. Let x_i equal $\mu^{-1}(y_i)$. Since $\mathbb{X}$ is closed and bounded, some subsequence of $(x_1, x_2, x_3, \ldots)$ approaches a limit point $x \in \mathbb{X}$. Since μ is continuous, the corresponding subsequence of $(\mu(x_1), \mu(x_2), \mu(x_3), \ldots)$ approaches $\mu(x)$. However, $(\mu(x_1), \mu(x_2), \mu(x_3), \ldots)$ equals $(y_1, y_2, y_3, \ldots)$, which approaches y. Thus $\mu(x)$ equals y. Since μ is onto, point y is in $\mathbb{Y}$. Since the subsequence $(y_1, y_2, y_3, \ldots)$ approaches point $y \in \mathbb{Y}$ and $(\mu^{-1}(y_1), \mu^{-1}(y_2), \mu^{-1}(y_3), \ldots)$ approaches $x = \mu^{-1}(y)$, function μ^{-1} is continuous. Since μ is continuous, one-to-one, and onto and μ^{-1} is continuous, function μ is a homeomorphism. $\square$

B.3 Manifolds

A k-dimensional manifold (k-manifold) is a set of points whose local topology is the same as $\mathbb{R}^k$. A simple closed curve is a 1-manifold while a two-dimensional surface such as a sphere is a 2-manifold. Formally, a k-manifold $\mathbb{M}$ is a topological space such that for every point $p \in \mathbb{M}$, some open neighborhood of p is homeomorphic to $\mathbb{R}^k$.

A k-manifold with boundary is a set of points whose local topology is either the same as $\mathbb{R}^k$ or as some half-space in $\mathbb{R}^k$. A half-space in $\mathbb{R}^3$ or a ball in $\mathbb{R}^3$ are examples of manifolds with boundary. Let $\mathbb{R}^{k+}$ be the half-space consisting of all points whose last coordinate is greater than or equal to zero. A k-manifold with boundary, labeled $\mathbb{M}$, is a topological space such that for every point $p \in \mathbb{M}$, some open neighborhood of p is either homeomorphic to $\mathbb{R}^k$ or homeomorphic to $\mathbb{R}^{k+}$.

The (manifold) interior of $\mathbb{M}$, denoted int($\mathbb{M}$), is the set of points $p \in \mathbb{M}$ such that some open neighborhood of p is homeomorphic to $\mathbb{R}^k$. The (manifold) boundary of $\mathbb{M}$, denoted $\partial\mathbb{M}$, is the set of points $p \in \mathbb{M}$ such that a homeomorphism maps p to the origin and some neighborhood of p to $\mathbb{R}^{k+}$.

The topological interior and boundary of a manifold $\mathbb{M}$ can differ from the manifold interior and boundary of $\mathbb{M}$. For instance, if $\mathbb{M}$ is the unit disk $\{(x, y, 0) : x^2 + y^2 \leq 1\}$ lying in the xy-plane in $\mathbb{R}^3$, then the topological interior of $\mathbb{M}$ is the empty set, while the manifold interior is the open disk

$\{(x, y, 0) : x^2 + y^2 < 1\}$. The topological boundary of $\mathbb{M}$ is all of $\mathbb{M}$ while the manifold boundary interior is the circle $\{(x, y, 0) : x^2 + y^2 = 1\}$. If $\mathbb{M}$ is a k-manifold whose affine hull has dimension k, then the relative interior of $\mathbb{M}$ equals the manifold interior of $\mathbb{M}$ and the relative boundary of $\mathbb{M}$ equals the manifold boundary of $\mathbb{M}$.

Every convex polytope $\mathbf{c}$ is a manifold with boundary. The polytope dimension of convex polytope $\mathbf{c}$ as defined in Appendix A, Section A.3, equals the manifold dimension of $\mathbf{c}$. The polytope interior and boundary of $\mathbf{c}$ equal the manifold interior and boundary of $\mathbf{c}$.

Every k-simplex $\mathbf{t}$ is a k-manifold with boundary. The simplex interior and boundary of $\mathbf{t}$ equal the manifold interior and boundary of $\mathbf{t}$.

B.4   Triangulations

As defined in Appendix A, Section A.4, a k-simplex in $\mathbb{R}^d$ is the convex hull of a set of $k + 1 \leq d + 1$ points in general position in $\mathbb{R}^d$.

Definition B.4. A triangulation τ is a set of simplices such that for every pair of simplices $\mathbf{t}, \mathbf{t}' \in \tau$, the intersection, $\mathbf{t} \cap \mathbf{t}'$, is either empty or a face of each simplex.

Mathematics texts usually add a formal requirement to Definition B.4 that if simplex $\mathbf{t}$ is in τ, then every face of $\mathbf{t}$ is in τ. Since we represent the triangulation of a d-dimensional region by a set of d-dimensional simplices, it is more convenient not to include this requirement in the definition. A triangulation τ can always be made to conform to this requirement by adding to τ all faces of all simplices in τ.

The set $\bigcup_{\mathbf{t} \in \tau} \mathbf{t}$ is denoted by $|\tau|$. It is the set of all points in all simplices of τ. We refer to τ as a triangulation of a space $\mathbb{X} \subseteq \mathbb{R}^d$ if $\mathbb{X}$ equals $|\tau|$. A tetrahedralization of a space $\mathbb{X} \subseteq \mathbb{R}^d$ is a triangulation of $\mathbb{X}$ whose simplices are a set of tetrahedra (3-simplices) and their faces.

A set $\mathbb{X} \subseteq \mathbb{R}^d$ is piecewise linear if $\mathbb{X}$ has some triangulation.

B.5   Convex Polytopal Meshes

As defined in Appendix A, Section A.3, a convex polyhedron is the convex hull of a finite set P of points in $\mathbb{R}^3$, and a convex polytope is the convex hull of a finite set P of points in $\mathbb{R}^d$.

Definition B.5. A convex polytopal mesh Γ is a set of convex polytopes in $\mathbb{R}^d$ such that for every pair of convex polytopes $\mathbf{c}, \mathbf{c}' \in \Gamma$, the intersection, $\mathbf{c} \cap \mathbf{c}'$, is either empty or a face of each convex polytope.

Mathematics texts usually add a formal requirement that if convex polytope $\mathbf{c}$ is in Γ, then every face of $\mathbf{c}$ is in Γ.

The set $\bigcup_{\mathbf{c} \in \Gamma} \mathbf{c}$ is denoted by $|\Gamma|$. It is the set of all points in all polyhedra in Γ.

A convex polytopal mesh in $\mathbb{R}^3$ is called a convex polyhedral mesh. A tetrahedral mesh is a convex polytopal mesh where every mesh element is a tetrahedron. By definition, a convex polytopal mesh where every mesh element is a simplex is a triangulation. It is also sometimes called a simplicial mesh.

B.6   Orientation

Let $v_1, v_2, \ldots, v_{k+1}$ be the vertices of a k-simplex $\mathbf{t}$. There are $(k+1)!$ permutations of the vertices of $\mathbf{t}$. A transposition of a permutation is the exchange of two elements of the permutation. For instance, given permutation (v_1, v_2, v_3, v_4), the transposition of v_2 and v_4 gives the permutation (v_1, v_4, v_3, v_2). Given any permutation, we can generate any other permutation by a sequence of transpositions.

Let permutations Z_1 and Z_2 of the vertices of $\mathbf{t}$ be equivalent if the number of transpositions needed to transform Z_1 into Z_2 is even. This equivalence determines two equivalence classes on the set of all permutations of the vertices of $\mathbf{t}$. These two equivalence classes are called the two orientations of $\mathbf{t}$. A permutation Z of the vertices of $\mathbf{t}$ determines an orientation of $\mathbf{t}$, namely the equivalence class containing Z. An oriented simplex is a simplex with a given orientation. The simplex and orientation is defined by giving some permuation of the vertices of the simplex.

The facets of a k-simplex are $(k-1)$-simplices. The orientation of a k-simplex $\mathbf{t}$ induces an orientation of its facets as follows. Without loss of generality, assume that the orientation of $\mathbf{t}$ is given by the permutation $(v_1, v_2, \ldots, v_{k+1})$. (If not, relabel the vertices so that the orientation of $\mathbf{t}$ is given by this permutation.) A facet of $\mathbf{t}$ has all the same vertices as $\mathbf{t}$ except for one, v_j. If j is odd, then the induced orientation of the facet is the orientation that contains the permutation $(v_1, v_2, \ldots, v_{j-1}, v_{j+1}, \ldots, v_{k+1})$. If j is even, then the induced orientation of the facet is the orientation that *does not* contain the permutation $(v_1, v_2, \ldots, v_{j-1}, v_{j+1}, \ldots, v_{k+1})$.

Let $\mathbf{t}_1$ and $\mathbf{t}_2$ be two oriented k-simplices that share a common facet $\mathbf{f}$. The orientations of $\mathbf{t}_1$ and $\mathbf{t}_2$ are consistent if the orientation of $\mathbf{f}$ induced by $\mathbf{t}_1$ is the opposite of the orientation of $\mathbf{f}$ induced by $\mathbf{t}_2$.

The orientation of an oriented d-simplex $\mathbf{t} = (v_1, \ldots, v_{d+1})$ in $\mathbb{R}^d$ can be calculated using determinants. Let $(v_i^1, v_i^2, \ldots, v_i^d)$ be the coordinates of vertex v_i. Define the determinant of $(v_1, \ldots, v_{d+1})$ as

$$\det(v_1, \ldots, v_{d+1}) = \det \begin{pmatrix} v_1^1 & v_1^2 & \cdots & v_1^d & 1 \\ v_2^1 & v_2^2 & \cdots & v_2^d & 1 \\ \vdots & \vdots & \vdots & \vdots \\ v_{d+1}^1 & v_{d+1}^2 & \cdots & v_{d+1}^d & 1 \end{pmatrix}.$$

Since the vertices of $\mathbf{t}$ are in general position, this determinant is nonzero. Permutations of the vertices of $\mathbf{t}$ that have the same orientation produce the same values for this determinant. Permutations of the vertices of $\mathbf{t}$ that have the opposite orientation produce the value $(-\det(v_1, \ldots, v_{d+1}))$ for this determinant. Thus, the sign of the determinant indicates the orientation of the simplex.

If the sign of $\det(v_1, \ldots, v_{d+1})$ is positive, then we say that $\mathbf{t}$ has positive orientation. If the sign of $\det(v_1, \ldots, v_{d+1})$ is negative, then $\mathbf{t}$ has negative orientation.

The orientation, $(v_1, \ldots, v_d)$, of a $(d-1)$-simplex $\mathbf{t}$ in $\mathbb{R}^d$ determines a vector orthogonal to $\mathbf{t}$. In $\mathbb{R}^3$, this vector is $u = (v_2 - v_1) \times (v_3 - v_1)$ where $\times$ denotes the cross product. Note that other permutations of (v_1, v_2, v_3) that represent the same orientation generate the same vector u. Permutations that represent the opposite orientation generate the vector $-u$.

In $\mathbb{R}^d$, construct the $(d-1) \times d$ matrix $A = \{a_{ij}\}$ whose ith row is $v_i - v_1$. Let C_{ij} be the cofactor of a_{ij} in A. The orientation $(v_1, \ldots, v_d)$ determines the vector $u = (C_{11}, C_{22}, \ldots, C_{dd})$ that is orthogonal to $\mathbf{t}$. Vector u is called the generalized cross product of $((v_2 - v_1), (v_3 - v_1), \ldots, (v_d - v_1))$. Again, any other permutation of $(v_1, \ldots, v_d)$ that represents the same orientation generate the same vector u. Permutations that represent the opposite orientation generate the vector $-u$.

B.7 Piecewise Linear Functions

Let $\{p_1, \ldots, p_{d+1}\}$ be a set of $d+1$ points in general position in $\mathbb{R}^d$ where each p_i is associated with a scalar s_i. As defined in Appendix A, Section A.6, the scalar values s_i determine a linear function $\mu : \mathbb{R}^d \to \mathbb{R}$ defined as

$$\mu(q) = \sum_{i=1}^{d+1} \alpha_i s_i,$$

where $(\alpha_1, \ldots, \alpha_{d+1})$ are the barycentric coordinates of q.

A scalar simplicial mesh is a simplicial mesh where each simplex vertex v_i is associated with a scalar values s_i. Let Γ be a scalar simplicial mesh in $\mathbb{R}^d$. Within each simplex of $\mathbf{t}_j \in \Gamma$, the scalar values at the vertices of $\mathbf{t}_j$ determine a linear function μ_j on $\mathbf{t}_j$. When two simplices $\mathbf{t}_j$ and $\mathbf{t}_{j'}$ share a common face $\mathbf{f}$, the piecewise linear functions μ_j and $\mu_{j'}$ agree on $\mathbf{f}$. Putting all the linear functions μ_j together gives a function ϕ where

$$\phi(q) = \mu_j(q) \text{ if } q \in \mathbf{t}_j.$$

Function ϕ is the piecewise linear function determined by the scalar simplicial mesh Γ.

B.8 Paths and Loops

A path ζ connecting $p \in \mathbb{R}^d$ to $q \in \mathbb{R}^d$ is a continuous map ζ from the unit interval $[0, 1]$ to $\mathbb{R}^d$ such that $\zeta(0)$ equals p and $\zeta(1)$ equals q. If ζ is a homeomorphism, then the image of ζ is a simple curve in $\mathbb{R}^d$ and thus a 1-manifold with boundary.

The interior of a path ζ is the set $\{\zeta(\alpha) : 0 < \alpha < 1\}$, i.e., the image of ζ restricted to the open interval $(0, 1)$. When ζ is a homeomorphism, the interior of path ζ is the same as the manifold interior of $\zeta[0, 1]$.

The interior of path ζ is usually not the same as the topological interior of $\zeta([0, 1]) \subset \mathbb{R}^d$. For instance, if $\zeta : [0, 1] \to \mathbb{R}^2$ is

$$\zeta(\alpha) = (\alpha, \alpha^2),$$

then the topological interior of $\zeta([0, 1])$ is the empty set, while the interior of path ζ is $\{(\alpha, \alpha^2) : 0 < \alpha < 1\}$.

Let $\zeta_1 : [0, 1] \to \mathbb{R}^d$ and $\zeta_2 : [0, 1] \to \mathbb{R}^d$ be two paths where the last point p in ζ_1 is the first point in ζ_2, i.e., $\zeta_1(1)$ equals $\zeta_2(0)$. These two paths can be joined at p to form a new path denoted $\zeta_2 \circ \zeta_1$. Formally, $\zeta_2 \circ \zeta_1$ is defined as a path $\zeta : [0, 1] \to \mathbb{R}^d$ where

$$\zeta(\alpha) = \begin{cases} \zeta_1(2 * \alpha) & \text{if } \alpha \in [0, 1/2], \\ \zeta_2(2 * \alpha - 1) & \text{if } \alpha \in (1/2, 1]. \end{cases}$$

Note that ζ is continuous since both $\zeta_1(2 * (1/2))$ and $\zeta_2(2 * (1/2) - 1)$ equal p.

A loop ζ in $\mathbb{R}^d$ is a continuous map ζ from the unit circle $\mathbb{S}^1$ to $\mathbb{R}^d$. If $\zeta : \mathbb{S}^1 \to \mathbb{R}^d$ is a homeomorphism, then $\zeta(\mathbb{S}^1)$ is a simple closed curve in $\mathbb{R}^d$ and thus a 1-manifold.

B.9　Separation

Let $\mathbb{X}$ and $\mathbb{Y}$ be subsets of $\mathbb{R}^d$.

Definition B.6.

- Set $\mathbb{X}$ separates point $p \in \mathbb{Y}$ from point $q \in \mathbb{Y}$ if every path in $\mathbb{Y}$ connecting p to q intersects $\mathbb{X}$.

- Set $\mathbb{X}$ strictly separates p from q if $\mathbb{X}$ separates p from q and neither p nor q is in $\mathbb{X}$.

Definition B.7.

- Set $\mathbb{X}$ separates $\mathbb{Y}_2 \subseteq \mathbb{Y}$ from $\mathbb{Y}_3 \subseteq \mathbb{Y}$ if $\mathbb{X}$ separates every $p \in \mathbb{Y}_2$ from every $q \in \mathbb{Y}_3$.

- Set $\mathbb{X}$ strictly separates $\mathbb{Y}_2 \subseteq \mathbb{Y}$ from $\mathbb{Y}_3 \subseteq \mathbb{Y}$ if $\mathbb{X}$ separates $\mathbb{Y}_2$ from $\mathbb{Y}_3$ and does not intersect $\mathbb{Y}_2$ or $\mathbb{Y}_3$.

Lemma B.8. *Let $\mathbb{Y}_1$ be a closed subset of $\mathbb{R}^d$ and let $\mathbb{Y}_2$ be a closed subset of $\mathbb{Y}_1$. Set $\mathbb{X} = \mathbb{Y}_2 \cap \mathrm{cl}(\mathbb{Y}_1 - \mathbb{Y}_2)$ separates $\mathbb{Y}_2$ from $\mathbb{Y}_1 - \mathbb{Y}_2$ and strictly separates $\mathbb{Y}_2 - \mathbb{X}$ from $\mathbb{Y}_1 - \mathbb{Y}_2$.*

Proof: Let $\zeta \subset \mathbb{Y}_1$ be a path from $\mathbb{Y}_1 - \mathbb{Y}_2$ to $\mathbb{Y}_2$. Let p be the first point along ζ that lies in $\mathbb{Y}_2$. Since $\mathbb{Y}_2$ is closed, there is such a first point p. Since all points preceding p are in $\mathbb{Y}_1 - \mathbb{Y}_2$, point p is in $\mathrm{cl}(\mathbb{Y}_1 - \mathbb{Y}_2)$. Thus p is in $\mathbb{Y}_2 \cap \mathrm{cl}(\mathbb{Y}_1 - \mathbb{Y}_2) = \mathbb{X}$ and $\mathbb{X}$ separates $\mathbb{Y}_2$ from $\mathbb{Y}_1 - \mathbb{Y}_2$. Since $\mathbb{X}$ does not intersect $\mathbb{Y}_1 - \mathbb{Y}_2$ or $\mathbb{Y}_2 - \mathbb{X}$, set $\mathbb{X}$ strictly separates $\mathbb{Y}_2 - \mathbb{X}$ from $\mathbb{Y}_1 - \mathbb{Y}_2$. □

Corollary B.9. *Let $\mathbb{Y}$ be a subset of $\mathbb{R}^d$. The boundary $\partial \mathbb{Y}$ of $\mathbb{Y}$ strictly separates* $\mathrm{int}(\mathbb{Y})$ *from* $\mathbb{R}^d - \mathbb{Y}$.

Proof: By Lemma B.8, set $\mathbb{X} = \mathbb{Y} \cap \mathrm{cl}(\mathbb{R}^d - \mathbb{Y})$ strictly separates $\mathbb{Y} - \mathbb{X}$ from $\mathbb{R}^d - \mathbb{Y}$. Set $\mathbb{X}$ is a subset of $\partial \mathbb{Y}$, so $\partial \mathbb{Y}$ strictly separates $\mathbb{Y} - \mathbb{X}$ from $\mathbb{R}^d - \mathbb{Y}$. Since $\mathrm{int}(\mathbb{Y})$ is a subset of $\mathbb{Y} - \mathbb{X}$, the boundary $\partial \mathbb{Y}$ strictly separates $\mathrm{int}(\mathbb{Y})$ from $\mathbb{R}^d - \mathbb{Y}$. □

B.10　Compact

A subset $\mathbb{X}$ of $\mathbb{R}^d$ is compact if $\mathbb{X}$ is closed and bounded. A cover of $\mathbb{X}$ is a collection $\mathcal{C}$ of subsets of $\mathbb{R}^d$ such that $\bigcup_{\mathbb{Y} \in \mathcal{C}} \mathbb{Y}$ contains $\mathbb{X}$. A subcover of $\mathcal{C}$ is a cover of $\mathbb{X}$ that is a subset of $\mathcal{C}$.

A cover $\mathcal{C}$ of $\mathbb{X}$ can contain an infinite and even uncountable number of subsets. A finite subcover of $\mathcal{C}$ is a subcover of $\mathcal{C}$ that has a finite number of elements. Proof of the following theorem can be found in [Armstrong, 1983] or any introductory text in point set topology.

Theorem B.10. *A set $\mathbb{X}$ is compact if and only if every cover of $\mathbb{X}$ by a collection of open sets has a finite subcovering.*

The conclusion of Theorem B.10 is often used as the definition of compact, i.e., a set $\mathbb{X}$ is compact if every cover of $\mathbb{X}$ by a collection of open sets has a finite subcovering. Under this definition of compactness, a related theorem is that a set $\mathbb{X} \subset \mathbb{R}^d$ is compact if and only $\mathbb{X}$ is closed and bounded.

B.11 Connected

Let $\mathbb{X}$ be a subset of $\mathbb{R}^d$. Set $\mathbb{X}$ is connected if, for any pair of points $p, q \in \mathbb{X}$, there is a path from p to q in $\mathbb{X}$. A connected component of $\mathbb{X}$ is a maximal connected subset of $\mathbb{X}$, i.e., a connected subset of $\mathbb{X}$ that is not contained in any other connected subset of $\mathbb{X}$. Every subset of $\mathbb{R}^d$ can be partitioned into connected components.

$\mathbb{B}^2$ is the closed unit disk (ball) in $\mathbb{R}^2$. Let $\mathbb{Y}_1$ and $\mathbb{Y}_2$ be subsets of $\mathbb{B}^2$ and let p and q be points in $\partial\mathbb{B}^2 \cap \mathbb{Y}_1 \cap \mathbb{Y}_2$. Lemma B.11 gives conditions under which points p and q are in the same connected component of $\mathbb{Y}_1 \cap \mathbb{Y}_2$.

Because we make no assumptions about $\mathbb{Y}_1$ and $\mathbb{Y}_2$, the proof of Lemma B.11 is a bit technical.

Lemma B.11. *Let $\mathbb{Y}_1$ and $\mathbb{Y}_2$ be subsets of the unit disk $\mathbb{B}^2$ where*

$$\mathrm{cl}(\mathbb{Y}_1) \cap \mathrm{cl}(\mathbb{B}^2 - \mathbb{Y}_1) \cap \mathrm{cl}(\mathbb{Y}_2) \cap \mathrm{cl}(\mathbb{B}^2 - \mathbb{Y}_2) = \emptyset. \tag{B.1}$$

If p and q are distinct points on $\partial\mathbb{B}^2$, the boundary of $\mathbb{B}^2$ where one of the arcs of $\partial\mathbb{B}^2$ from p to q is contained in $\mathbb{Y}_1$ and the other is contained in $\mathbb{Y}_2$, then there is some path from p to q in $\mathbb{Y}_1 \cap \mathbb{Y}_2$.

Proof: By Equation B.1, each point r in $\mathbb{B}^2$ is not in $\mathrm{cl}(\mathbb{Y}_1)$, $\mathrm{cl}(\mathbb{Y}_2)$, $\mathrm{cl}(\mathbb{B}^2 - \mathbb{Y}_1)$, or $\mathrm{cl}(\mathbb{B}^2 - \mathbb{Y}_2)$. Since $\mathrm{cl}(\mathbb{Y}_1)$ is closed, if point r is not in $\mathrm{cl}(\mathbb{Y}_1)$, then there is an disk around r that does not intersect $\mathrm{cl}(\mathbb{Y}_1)$. The same holds for $\mathrm{cl}(\mathbb{Y}_2)$, $\mathrm{cl}(\mathbb{B}^2 - \mathbb{Y}_1)$, and $\mathrm{cl}(\mathbb{B}^2 - \mathbb{Y}_2)$. Thus each point in $\mathbb{B}^2$ lies in some open disk that does not intersect $\mathrm{cl}(\mathbb{Y}_1)$, $\mathrm{cl}(\mathbb{Y}_2)$, $\mathrm{cl}(\mathbb{B}^2 - \mathbb{Y}_1)$, or $\mathrm{cl}(\mathbb{B}^2 - \mathbb{Y}_2)$.

The open disks around points in $\mathbb{B}^2$ form a covering of $\mathbb{B}^2$. Since $\mathbb{B}^2$ is closed and bounded (compact), there is a finite subset $\mathcal{D}$ of the open disks such that $\mathcal{D}$ covers $\mathbb{B}^2$ and no disk of $\mathcal{D}$ is contained in any other disk of $\mathcal{D}$.

Let $\mathbf{d} \in \mathcal{D}$ be a disk that intersects $\mathbb{Y}_1$ and $\mathbb{Y}_2$. Since $\mathbf{d}$ is in $\mathcal{D}$, either $\mathbf{d}$ does not intersect $\mathrm{cl}(\mathbb{B}^2 - \mathbb{Y}_1)$ or $\mathbf{d}$ does not intersect $\mathrm{cl}(\mathbb{B}^2 - \mathbb{Y}_2)$. If $\mathbf{d}$ does

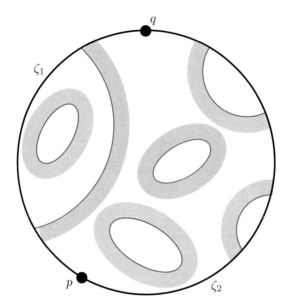

Figure B.1. Arc ζ_1 from p to q is in $\mathbb{Y}_1$. Arc ζ_2 from p to q is in $\mathbb{Y}_2$. Blue curves bound region $\mathbb{Y}_1$ and red curves bound region $\mathbb{Y}_2$. Note that $\mathbb{Y}_1$ is outside the blue ovals and $\mathbb{Y}_2$ is outside the red oval. Blue and red curves do not intersect.

not intersect $\mathrm{cl}(\mathbb{B}^2 - \mathbb{Y}_1)$, then $\mathbf{d}$ is contained in $\mathbb{Y}_1$. If $\mathbf{d}$ does not intersect $\mathrm{cl}(\mathbb{B}^2 - \mathbb{Y}_2)$, then $\mathbf{d}$ is contained in $\mathbb{Y}_2$. Thus either $\mathbf{d}$ is contained in $\mathbb{Y}_1$ or $\mathbf{d}$ is contained in $\mathbb{Y}_2$ (or $\mathbf{d}$ is contained in both.)

Let $\mathcal{D}_1 \subseteq \mathcal{D}$ be the disks in $\mathcal{D}$ that are contained in $\mathbb{Y}_1$. Let $\mathcal{D}_2 \subseteq \mathcal{D}$ be the disks in $\mathcal{D}$ that are contained in $\mathbb{Y}_2$. Every open disk that contains p or q intersects $\mathbb{Y}_1$ and $\mathbb{Y}_2$. By the previous argument, if an open disk $\mathbf{d}$ contains p or q, then $\mathbf{d}$ is either contained in $\mathbb{Y}_1$ or contained in $\mathbb{Y}_2$ (or contained in both). Thus $\mathbf{d}$ is in $\mathcal{D}_1 \cup \mathcal{D}_2$.

Let ζ_1 be the arc of $\partial \mathbb{B}^2$ from p to q that is contained in $\mathbb{Y}_1$ (Figure B.1). Let ζ_2 be the arc of $\partial \mathbb{B}^2$ from p to q that is contained in $\mathbb{Y}_2$.

Assume a disk $\mathbf{d} \in \mathcal{D}_1$ intersects a disk $\mathbf{d}' \in \mathcal{D}_2$. The boundaries of these two disks intersect in two points that may or may not be contained in $\mathbb{B}^2$. Assume some point r is in $\mathbb{B}^2 \cap \partial \mathbf{d} \cap \partial \mathbf{d}'$. Since $\mathbf{d}$ and $\mathbf{d}'$ are open, they do not contain r. Thus some other disk $\mathbf{d}'' \in \mathcal{D}$ contains r. Disk $\mathbf{d}''$ contains a neighborhood of r so $\mathbf{d}''$ intersects both $\mathbf{d}$ and $\mathbf{d}'$. Thus $\mathbf{d}''$ intersects both $\mathbb{Y}_1$ and $\mathbb{Y}_2$. By the previous argument, $\mathbf{d}''$ is either contained in $\mathbb{Y}_1$ or contained in $\mathbb{Y}_2$. Thus $\mathbf{d}''$ is in $\mathcal{D}_1 \cup \mathcal{D}_2$.

Similarly, consider a disk $\mathbf{d} \in \mathcal{D}_1$ that intersects ζ_2 and a point r is in $\mathbb{B}^2 \cap \partial \mathbf{d} \cap \zeta_2$. Since r is not in $\mathbf{d}$, some disk $\mathbf{d}'' \in \mathcal{D}$ contains r. Since $\mathbf{d}''$ intersects $\mathbf{d} \subseteq \mathbb{Y}_1$ and $\zeta_2 \subseteq \mathbb{Y}_2$, disk $\mathbf{d}''$ is either contained in $\mathbb{Y}_1$ or contained in $\mathbb{Y}_2$ (or contained in both) and is in $\mathcal{D}_1 \cup \mathcal{D}_2$.

Finally, consider a disk $\mathbf{d} \in \mathcal{D}_2$ that intersects ζ_1 and a point r in $\mathbb{B}^2 \cap \partial \mathbf{d} \cap \zeta_1$. Some disk $\mathbf{d}'' \in \mathcal{D}$ contains r. Disk $\mathbf{d}''$ is contained in $\mathbb{Y}_1$ or $\mathbb{Y}_2$ and is in $\mathcal{D}_1 \cup \mathcal{D}_2$.

Let $\mathbf{d}$ be a disk in $\mathcal{D}_1$ and $\mathbf{d}'$ be a disk in $\mathcal{D}_2$. The sets $\mathbf{d} \cap \mathbf{d}'$ and $\mathbf{d} \cap \zeta_2$ and $\mathbf{d}' \cap \zeta_1$ are all contained in $\mathbb{Y}_1 \cap \mathbb{Y}_2$. Let $\mathbb{Z}$ be the union of all such sets over all $\mathbf{d} \in \mathcal{D}_1$ and $\mathbf{d}' \in \mathcal{D}_2$. Formally, define $\mathbb{Z}$ as

$$\bigcup_{\mathbf{d} \in \mathcal{D}_1, \mathbf{d}' \in \mathcal{D}_2} \Big((\mathbf{d} \cap \mathbf{d}') \cup (\mathbf{d} \cap \zeta_2) \cup (\mathbf{d}' \cap \zeta_1) \Big).$$

Note that p is in $\mathbb{Z}$ since p is in $(\mathbf{d} \cap \zeta_2)$ for some $\mathbf{d} \in \mathcal{D}_1$ or p is in $(\mathbf{d}' \cap \zeta_1)$ for some $\mathbf{d}' \in \mathcal{D}_2$. Similarly, q is in $\mathbb{Z}$.

We claim that there is a path in $\mathbb{Z}$ connecting p to q. Assume that it is not. Let $\mathbb{Z}_p$ be the connected component of $\mathbb{Z}$ containing p. Since there is no path from p to q in $\mathbb{Z}_p$, some connected component $\mathbb{X}$ of $\mathrm{cl}(\mathbb{Z}_p) \cap \mathrm{cl}(\mathbb{B}^2 - \mathbb{Z}_p)$ separates p from q in $\mathbb{B}^2$. To separate p from q, $\mathbb{X}$ must intersect both ζ_1 and ζ_2. Let r_1 and r_2 be the points on $\mathbb{X} \cap \zeta_1$ and $\mathbb{X} \cap \zeta_2$, respectively, that are closest to q. Let ζ_3 be a path in $\mathbb{X}$ from r_1 to r_2.

Since every point in $\mathbb{Z}_p$ is inside a disk of $\mathcal{D}_1$ or $\mathcal{D}_2$, set $\mathcal{D}_1 \cap \mathcal{D}_2$ covers $\mathbb{Z}_p$. Some subset of $\mathcal{D}_1 \cup \mathcal{D}_2$ covers ζ_3. Let $(\mathbf{d}_1, \mathbf{d}_2, \ldots, \mathbf{d}_k)$ be a sequence of disks from $\mathcal{D}_1 \cup \mathcal{D}_2$ such that $\bigcup \mathbf{d}_i$ covers ζ_3, set $\mathbf{d}_{i-1} \cap \mathbf{d}_i$ intersects ζ_3 for $i = 2, \ldots, k$, set $\mathbf{d}_1$ contains r_1, and set $\mathbf{d}_k$ contains r_k.

If $\mathbf{d}_{i-1}$ was in $\mathcal{D}_1$ and $\mathbf{d}_i$ was in $\mathcal{D}_2$ or vice versa, then $\mathbf{d}_{i-1} \cap \mathbf{d}_i$ would be in $\mathbb{Z}_p$. Since $\mathbb{X}$ is in $\mathrm{cl}(\mathbb{B}^2 - \mathbb{Z}_p)$, set $\mathbb{X}$ and path ζ_3 would not intersect $\mathbf{d}_{i-1} \cap \mathbf{d}_i$. Thus all the $\mathbf{d}_i$ are in $\mathcal{D}_1$ or all the $\mathbf{d}_i$ are in $\mathcal{D}_2$.

If all the $\mathbf{d}_i$ are in $\mathcal{D}_1$, then $\mathbf{d}_k \cap \zeta_2$ is in $\mathbb{Z}_p$. Since r_2 is in $\mathbf{d}_k$ and $\mathbf{d}_k$ is open, there is some point $r \in \mathbf{d}_k \cap \zeta_2$ that is closer to point q than r_2. Since no point of $\mathbb{X} \cap \zeta_2$ is closer to q than r_2, the arc from r to q does not intersect $\mathbb{X}$. Thus, $r \in \mathbb{Z}_p$ is not separated by $\mathbb{X}$ from q, a contradiction. A similar argument gives a contradiction if all the $\mathbf{d}_i$ are in $\mathcal{D}_2$. Thus, $\mathrm{cl}(Z_p) \cap \mathrm{cl}(D - Z_p)$ does not separate p from q and point q is in Z_p. Since Z_p is a subset of $\mathbb{Y}_1 \cap \mathbb{Y}_2$, there is a path from p to q in $\mathbb{Y}_1 \cap \mathbb{Y}_2$. $\qquad \square$

Set $\mathbb{Z} \subseteq \mathbb{R}^d$ is simply connected if $\mathbb{Z}$ is connected and every loop $\zeta : \mathbb{S}^1 \to \mathbb{Z}$ can be extended to a map $\mu : \mathbb{B}^2 \to \mathbb{Z}$ from the unit disk into $\mathbb{Z}$ such that μ agrees with ζ on every point of $\mathbb{S}^1 = \partial \mathbb{B}^2$.

The following lemma gives conditions under which the intersection of two connected subsets of a simply connected set are connected.

Lemma B.12. *Let $\mathbb{Y}_1$ and $\mathbb{Y}_2$ be subsets of a simply connected set $\mathbb{Z} \subseteq \mathbb{R}^d$. If $\mathbb{Y}_1$ is connected and $\mathbb{Y}_2$ is connected and*

$$\mathrm{cl}(\mathbb{Y}_1) \cap \mathrm{cl}(\mathbb{Z} - \mathbb{Y}_1) \cap \mathrm{cl}(\mathbb{Y}_2) \cap \mathrm{cl}(\mathbb{Z} - \mathbb{Y}_2) = \emptyset, \tag{B.2}$$

then $\mathbb{Y}_1 \cap \mathbb{Y}_2$ is connected.

Proof: Let p and q be points in $\mathbb{Y}_1 \cap \mathbb{Y}_2$. Since $\mathbb{Y}_1$ is connected, there is a path $\zeta_1 \subseteq \mathbb{Y}_1$ from p to q. Since $\mathbb{Y}_2$ is connected, there is a path $\zeta_2 \subseteq \mathbb{Y}_2$ from q to p. Combining these two loops gives a path $\zeta = \zeta_2 \circ \zeta_1$ that starts and ends at p.

Since $\mathbb{Z}$ is simply connected, there is a map μ from the unit disk $\mathbb{B}^2$ into $\mathbb{Z}$ such that μ agrees with ζ on every point of $\mathbb{S}^1$. Some point p' on $\mathbb{S}^1$ maps to p and some point q' maps to q. One arc from p' to q' maps to ζ_1 while the other arc maps to ζ_2. Define $\mathbb{Y}_1' \subseteq \mathbb{B}^2$ as $\{r \in \mathbb{B}^2 : \mu(r) \in \mathbb{Y}_1\}$. Similarly, define $\mathbb{Y}_2' \subseteq \mathbb{B}^2$ as $\{r \in \mathbb{B}^2 : \mu(r) \in \mathbb{Y}_2\}$.

Sets $\mathrm{cl}(\mathbb{Y}_1')$ and $\mathrm{cl}(\mathbb{Y}_2')$ are subsets of $\mathrm{cl}(\mathbb{Y}_1)$ and $\mathrm{cl}(\mathbb{Y}_2)$, respectively. Similarly, $\mathrm{cl}(\mathbb{B}^2 - \mathbb{Y}_1')$ and $\mathrm{cl}(\mathbb{B}^2 - \mathbb{Y}_2')$ are subsets of $\mathrm{cl}(\mathbb{Z} - \mathbb{Y}_1)$ and $\mathrm{cl}(\mathbb{Z} - \mathbb{Y}_2')$, respectively. Thus if some point r is in $\mathrm{cl}(\mathbb{Y}_1') \cap \mathrm{cl}(\mathbb{B}^2 - \mathbb{Y}_1') \cap \mathrm{cl}(\mathbb{Y}_2') \cap \mathrm{cl}(\mathbb{B}^2 - \mathbb{Y}_2)$, then $\mu(r)$ would be in $\mathrm{cl}(\mathbb{Y}_1) \cap \mathrm{cl}(\mathbb{Z} - \mathbb{Y}_1) \cap \mathrm{cl}(\mathbb{Y}_2) \cap \mathrm{cl}(\mathbb{Z} - \mathbb{Y}_2)$. By Equation B.2, $\mathrm{cl}(\mathbb{Y}_1) \cap \mathrm{cl}(\mathbb{Z} - \mathbb{Y}_1) \cap \mathrm{cl}(\mathbb{Y}_2) \cap \mathrm{cl}(\mathbb{Z} - \mathbb{Y}_2)$ is the empty set, so

$$\mathrm{cl}(\mathbb{Y}_1') \cap \mathrm{cl}(\mathbb{B}^2 - \mathbb{Y}_1') \cap \mathrm{cl}(\mathbb{Y}_2') \cap \mathrm{cl}(\mathbb{B}^2 - \mathbb{Y}_2) = \emptyset.$$

By Lemma B.11, there is some path ζ' from p' to q' in $\mathbb{Y}_1' \cap \mathbb{Y}_2'$. Mapping path ζ' into $\mathbb{Z}$ using μ gives a path $\mu(\zeta)$ from p to q in $\mathbb{Y}_1 \cap \mathbb{Y}_2$. Since there is a path between every two points in $\mathbb{Y}_1 \cap \mathbb{Y}_2$, set $\mathbb{Y}_1 \cap \mathbb{Y}_2$ is connected. $\qquad\square$

B.12   Homotopy Map

Let $\mathbb{X}$ and $\mathbb{Y}$ be subsets of $\mathbb{R}^d$. A homotopy map is a continuous function $\eta : \mathbb{X} \times [0, 1] \to \mathbb{Y}$. If $\mu_0 : \mathbb{X} \to \mathbb{Y}$ is the function $\mu_0(x) = \eta(x, 0)$ and $\mu_1 : \mathbb{X} \to \mathbb{Y}$ is the function $\mu_1(x) = \eta(x, 1)$, then η is a homotopy map from function μ_0 to function μ_1.

If $\mathbb{X}$ is a subset of $\mathbb{Y}$, then the inclusion map, $\iota : \mathbb{X} \to \mathbb{Y}$, is the map $\iota(x) = x$ for all $x \in \mathbb{X}$. If $\mathbb{X}_0$ is a subset of $\mathbb{Y}$ and map η is a homotopy map from the inclusion map $\iota_0 : \mathbb{X}_0 \to \mathbb{Y}$ to some map $\mu : \mathbb{X}_0 \to \mathbb{Y}$, then we say that η is a homotopy map from set $\mathbb{X}_0$ to $\mathbb{X}_1 = \mu(\mathbb{X}_0)$. Equivalently, η is a homotopy map from set $\mathbb{X}_0$ to $\mathbb{X}_1$ if $\eta(x, 0) = x$ for all $x \in \mathbb{X}_0$ and $\{\eta(x, 1) : x \in \mathbb{X}_0\}$ equals $\mathbb{X}_1$.

A constant map is a map $\mu : \mathbb{X} \to \mathbb{Y}$ where $\mu(x) = y$ for all $x \in \mathbb{X}$ and some point $y \in \mathbb{Y}$. A function $\mu : \mathbb{X} \to \mathbb{Y}$ is null-homotopic if there is a homotopy map from μ_0 to some constant map.

The identity map on $\mathbb{X}$ is $\mu(x) = x$ for all $x \in \mathbb{X}$. A set $\mathbb{X}$ is contractible if the identity map is null-homotopic, i.e., if there is a homotopy map from the identity map on $\mathbb{X}$ to some constant map.

Let $\eta_1 : \mathbb{X} \times [0, 1] \to \mathbb{Y}$ and $\eta_2 : \mathbb{X} \times [0, 1] \to \mathbb{Y}$ be homotopy maps where $\eta_1(x, 1)$ equals $\eta_2(x, 0)$ for all $x \in \mathbb{X}$. Maps η_1 and η_2 can be combined to form a new homotopy map, denoted $\eta_2 \circ \eta_1$, which first applies η_1 and then applies η_2

to $\mathbb{X}$. Formally, $\eta_2 \circ \eta_1$ is defined as a homotopy map $\eta : \mathbb{X} \times [0,1] \to \mathbb{Y}$ where

$$\eta(x, \alpha) = \begin{cases} \eta_1(x, 2 * \alpha) & \text{if } \alpha \in [0, 1/2], \\ \eta_2(x, 2 * \alpha - 1) & \text{if } \alpha \in (1/2, 1]. \end{cases}$$

Note that η is continuous since $\eta_1(x, 2 * (1/2))$ equals $\eta_2(x, 2 * (1/2) - 1)$ for all $x \in \mathbb{X}$.

The following lemma relates the inclusion map ι to another map μ_p.

Lemma B.13. *Let $\mathbb{X}$ be a subset of $\mathbb{R}^d$, let p be a point in $\mathbb{R}^d - \mathbb{X}$, and let $\mu_p : \mathbb{X} \to \mathbb{S}^{d-1}$ be the map*

$$\mu_p(x) = \frac{x - p}{|x - p|}.$$

The inclusion map $\iota : \mathbb{X} \to \mathbb{R}^d - \{p\}$ is null-homotopic if and only if $\mu_p : \mathbb{X} \to \mathbb{S}^{d-1}$ is null-homotopic.

Proof: Assume there is a homotopy $\eta : \mathbb{X} \times [0,1] \to \mathbb{R}^d - \{p\}$ from the inclusion map $\iota : \mathbb{X} \to \mathbb{R}^d - \{p\}$ to a constant map. Define the homotopy $\eta' : \mathbb{X} \times [0,1] \to \mathbb{S}^{d-1}$ as

$$\eta'(x, \alpha) = \mu_p(\eta(x, \alpha)).$$

Since $\eta(x, \alpha)$ is never equal to p, map η' is well-defined. Since $\eta(x, 1)$ is a constant map, map $\eta'(x, 1)$ is a constant map. Map η' is a homotopy map from μ_p to a constant map. Thus, if ι is null-homotopic, then μ_p is null-homotopic.

To prove the converse, assume there is a homotopy $\eta' : \mathbb{X} \times [0,1] \to \mathbb{S}^{d-1}$ from $\mu_p : \mathbb{X} \to \mathbb{S}^{d-1}$ to a constant map. Let $\mu'_p : \mathbb{X} \to \mathbb{R}^d - \{p\}$ be the function $\mu'_p(x) = \mu_p(x) + p$. Define the homotopy maps $\eta'' : \mathbb{X} \times [0,1] \to \mathbb{R}^d - \{p\}$ and $\eta''' : \mathbb{X} \times [0,1] \to \mathbb{R}^d - \{p\}$ as

$$\eta''(x, \alpha) = (1 - \alpha)(x - p) + \alpha \mu_p(x) + p,$$
$$\eta'''(x, \alpha) = \eta'(x, \alpha) + p.$$

Since $(x - p)$ and $\mu_p(x)$ are nonzero vectors with the same direction, $\eta'(x, \alpha)$ is never equal to p. Map η'' is a homotopy map from the inclusion map ι to μ'_p. Since $\eta'(x, 1)$ is a constant map, map η''' is a homotopy map from μ'_p to a constant map. Combining η'' and η''' into a map $\eta = \eta''' \circ \eta''$ gives a homotopy map $\eta : \mathbb{X} \times [0,1] \to \mathbb{R}^d - \{p\}$ from ι to a constant map. Thus, if μ_p is null-homotopic, then ι is null-homotopic. $\qquad \square$

B.13   Embeddings

Let $\mathbb{X}$ be a closed, bounded subset of $\mathbb{R}^d$. Set $\mathbb{R}^d - \mathbb{X}$ has one or more connected components. One of these components is unbounded. The Borsuk separation theorem gives conditions for a point $p \in \mathbb{R}^d - \mathbb{X}$ to be in this unbounded component. A proof can be found in [Hocking and Young, 1961, p. 275].

Theorem B.14 (Borsuk separation theorem). *Let $\mathbb{X}$ be a closed, bounded subset of $\mathbb{R}^d$ and let p be a point in $\mathbb{R}^d - \mathbb{X}$. Let $\mu_p : \mathbb{X} \to \mathbb{S}^{d-1}$ be the map*

$$\mu_p(x) = \frac{x - p}{|x - p|}. \tag{B.3}$$

Map μ_p is null-homotopic if and only if point p is in the unbounded connected component of $\mathbb{R}^d - \mathbb{X}$.

The following corollary replaces the map $\mu_p : \mathbb{X} \to \mathbb{S}^{d-1}$ with the inclusion map $\iota : \mathbb{X} \to \mathbb{R}^d - \{p\}$.

Corollary B.15. *Let $\mathbb{X}$ be a closed, bounded subset of $\mathbb{R}^d$ and let p be a point in $\mathbb{R}^d - \mathbb{X}$. The inclusion map $\iota : \mathbb{X} \to \mathbb{R}^d - \{p\}$ is null-homotopic if and only if point p is in the unbounded connected component of $\mathbb{R}^d - \mathbb{X}$.*

Proof: Let $\mu : \mathbb{X} \to \mathbb{S}^{d-1}$ be the map

$$\mu_p(x) = \frac{x - p}{|x - p|},$$

as in Equation B.3. By Lemma B.13, the inclusion map $\iota : \mathbb{X} \to \mathbb{R}^d - \{p\}$ is null-homotopic if and only if map μ_p is null-homotopic. By Theorem B.14, the Borsuk separation theorem, map μ_p is null-homotopic if and only if point p is in the unbounded connected component of $\mathbb{R}^d - \mathbb{X}$. Thus the inclusion map $\iota : \mathbb{X} \to \mathbb{R}^d - \{p\}$ is null-homotopic if and only if point p is in the unbounded connected component of $\mathbb{R}^d - \mathbb{X}$. $\square$

A second corollary applies the theorem to apply to the separation of two points in $\mathbb{R}^d$. We first define a mapping $\phi_q : \mathbb{R}^d - \{q\} \to \mathbb{R}^d - \{q\}$ that maps the neighborhood of a point $q \in \mathbb{R}^d$ to points near infinity.

Consider $\mathbb{R}^d$ as a subspace of $\mathbb{R}^{d+1}$. Embed the unit sphere $\mathbb{S}^d$ in $\mathbb{R}^{d+1}$ so that $\mathbb{R}^d$ is a tangent hyperplane to $\mathbb{S}^d$ that touches $\mathbb{S}^d$ at q. Let $q' \in \mathbb{S}^d$ be the point antipodal to q. Rays from q' through $\mathbb{S}^d - \{q\}$ give a homeomorphism ϕ' from $\mathbb{S}^d - \{q'\}$ to $\mathbb{R}^d$. Translate $\mathbb{R}^d$ so that it is tangent to $\mathbb{S}^d$ at q', and take rays from $q \in \mathbb{S}^d$ through $\mathbb{S}^d - \{q\}$ for another homeomorphism ϕ'' from $\mathbb{S}^d - \{q\}$ to $\mathbb{R}^d$. Let $\phi_q = \phi'' \circ \phi'$ be the combination of these homeomorphisms. Function ϕ_q maps the neighborhood of q to points near infinity.

Corollary B.16. *Let p and q be points in $\mathbb{R}^d$ and let $\mathbb{X}$ be a closed, bounded subset of $\mathbb{R}^d$ that does not contain p or q. Points p and q are in the same connected component of $\mathbb{R}^d - \mathbb{X}$ if and only if the inclusion map $\iota : \phi_q(\mathbb{X}) \to \mathbb{R}^d - \{\phi_q(p)\}$ is null-homotopic.*

Proof: Let $\mathbb{X}'$ equal $\phi_q(\mathbb{X})$. Since $\mathbb{X}$ is closed and bounded, set $\mathbb{X}'$ is closed and bounded. Points p and q are in the same connected component of $\mathbb{R}^d - \mathbb{X}'$ if and only if point $\phi_q(p)$ is in the unbounded component of $\mathbb{R}^d - \mathbb{X}'$. By Corollary B.15, point $\phi_q(p)$ is in the unbounded component of $\mathbb{R}^d - \mathbb{X}'$ if and only if the inclusion map ι is null-homotopic. $\qquad\square$

A homotopy map $\eta : \mathbb{X} \times [0,1]$ does not *pass through* point p if $p \notin \eta(\mathbb{X}, \alpha)$ for any $\alpha \in [0,1]$.

Let $\mathbb{Y}$ be a closed, bounded subset of $\mathbb{R}^d$ and let $\mathbb{X}_1$ be a closed subset of $\mathbb{Y}$ that strictly separates point $p \in \mathbb{Y}$ from $q \in \mathbb{Y}$. Let $\eta : \mathbb{X}_1 \times [0,1] \to \mathbb{R}^d$ be a homotopy map from $\mathbb{X}_1$ to $\mathbb{X}_2$ that never passes through p or q. The following three lemmas give conditions under which $\mathbb{X}_2$ also strictly separates p from q in $\mathbb{Y}$. The first lemma (Lemma B.17) is the simplest case where $\mathbb{Y}$ equals $\mathbb{R}^d$.

Lemma B.17. *Let p and q be points in $\mathbb{R}^d$ and let $\mathbb{X}_0$ be a closed, bounded subset of $\mathbb{R}^d$ that strictly separates p from q. If $\eta : \mathbb{X}_0 \times [0,1] \to \mathbb{R}^d$ is a homotopy map from $\mathbb{X}_0$ to $\mathbb{X}_1$ that never passes through p or q, then $\mathbb{X}_1$ strictly separates p from q.*

Proof: Points p and q are in different connected components of $\mathbb{R}^d - \mathbb{X}_0$. By Corollary B.16, the inclusion map $\iota_0 : \phi_q(\mathbb{X}_0) \to \mathbb{R}^d - \{\phi_q(p)\}$ is not homotopic to the constant map.

Let $\iota_1 : \phi_q(\mathbb{X}_1) \to \mathbb{R}^d - \{\phi_q(p)\}$ be the inclusion map of $\phi_q(\mathbb{X}_1)$ into $\mathbb{R}^d - \{\phi_q(p)\}$. Define a homotopy map $\eta' : \mathbb{X}_0 \times [0,1] \to \mathbb{R}^d$ from ι_0 to ι_1 as $\eta'(x, \alpha) = \phi_q(\eta(x, \alpha))$. Note that $\phi'(\mathbb{X}_0, 0)$ equals $\phi_q(\mathbb{X}_0)$ and $\eta'(\mathbb{X}_0, 1)$ equals $\phi_q(\mathbb{X}_1)$.

If there was a homotopy map η'' from ι_1 to the constant map, then $\eta'' \circ \eta'$ would be a homotopy map from ι_0 to the constant map. Since ι_0 is not homotopic to the constant map, neither is ι_1. By Corollary B.16, points p and q are in different connected components of $\mathbb{R}^d - \mathbb{X}_1$ and $\mathbb{X}_1$ strictly separates p from q. $\square$

Lemma B.17 assumed that $\mathbb{X}_0$ separated p and q in $\mathbb{R}^d$. The next lemma assumes that $\mathbb{X}_0$ separates p and q in some $\mathbb{Y} \subseteq \mathbb{R}^d$, i.e., that any path in $\mathbb{Y}$ connecting p and q intersects $\mathbb{X}_0$.

Lemma B.18. *Let $\mathbb{Y}$ be a closed, bounded subset of $\mathbb{R}^d$, and let $\mathbb{X}_0$ be a closed subset of $\mathbb{Y}$ that strictly separates point $p \in \mathbb{Y}$ from point $q \in \mathbb{Y}$. Let $\eta : \mathbb{X}_0 \times [0,1] \to \mathbb{Y}$ be a homotopy map from $\mathbb{X}_0$ to $\mathbb{X}_1 = \{\eta(x,1) : x \in \mathbb{X}_0\}$. If η never passes through p or q and $\{\eta(x, \alpha) : x \in \mathbb{X}_0 \cap \partial\mathbb{Y}\}$ is a subset of $\partial\mathbb{Y}$ for all $\alpha \in [0,1]$, then $\mathbb{X}_1$ strictly separates p from q.*

Proof: Let $\mathbb{Y}_p$ be the connected component of $\mathbb{Y} - \mathbb{X}_0$ that contains p. Since $\mathbb{Y}_p$ is bounded, it is possible to construct a closed and bounded set $\mathbb{Y}'_p \subseteq \mathbb{R}^d$ such that $\mathbb{Y} \cap \mathbb{Y}'_p$ equals $\mathrm{cl}(\mathbb{Y}_p)$ and $\mathbb{Y} \cap \partial\mathbb{Y}'_p$ is a subset of $\mathbb{X}_0$. (See Figure B.2.) Since $\mathbb{Y}_p$ is a subset of $\mathbb{Y} - \mathbb{X}_0$ and $\mathbb{Y} \cap \partial\mathbb{Y}'_p$ is a subset of $\mathbb{X}_0$, set $\mathbb{Y}_p$ is a subset of the interior of $\mathbb{Y}'_p$. Thus p is in the interior of $\mathbb{Y}'_p$. On the other hand, q is not in $\mathbb{Y}'_p$. By Corollary B.9, $\partial\mathbb{Y}'_p$ strictly separates p from q in $\mathbb{R}^d$.

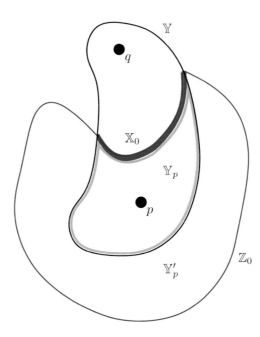

Figure B.2. Red curve $\mathbb{X}_0$ strictly separates p from q in $\mathbb{Y}$ (region bounded by black curve). Region $\mathbb{Y}_p$ (bounded by green curve) is the connected component of $\mathbb{Y} - \mathbb{X}_0$ containing p. Region $\mathbb{Y}'_p$ (bounded by blue curve $\mathbb{Z}_0$) is an extension of $\mathbb{Y}_p$ such that $\mathbb{Y} \cap \mathbb{Y}'_p$ equals $\mathrm{cl}(\mathbb{Y}_p)$ and $\mathbb{Y} \cap \partial \mathbb{Y}'_p$ is a subset of $\mathbb{X}_0$.

Let $\mathbb{Z}_0$ denote $\partial \mathbb{Y}'_p$. Extend η to a homotopy map $\eta' : \mathbb{Z}_0 \times [0,1] \to \mathbb{R}^d$ that agrees with η on $\mathbb{X}_0$ and always maps $\mathbb{Z}_0 - \mathbb{X}_0$ to $\mathbb{R}^d - \mathbb{Y}$. More specifically, $\eta'(x, \alpha)$ equals $\eta(x, \alpha)$ for all $x \in \mathbb{Z}_0 \cap \mathbb{X}_0$ and $\alpha \in [0,1]$, and $\{\eta'(x, \alpha) : x \in \mathbb{Z}_0 - \mathbb{X}_0\}$ is a subset of $\mathbb{R}^d - \mathbb{Y}$ for all $\alpha \in [0,1]$. Let $\mathbb{Z}_1$ equal $\{\eta'(x, 1) : x \in \mathbb{Z}_0\}$. By definition, $\mathbb{Z}_1 \cap \mathbb{Y}$ equals $\mathbb{X}_1$.

Map η' is a homotopy map from $\mathbb{Z}_0$ to $\mathbb{Z}_1$ that does not pass through p or q. Set $\mathbb{Z}_0$ strictly separates p from q in $\mathbb{R}^d$. By Lemma B.17, set $\mathbb{Z}_1$ strictly separates p from q in $\mathbb{R}^d$.

Assume there is path $\zeta \subseteq \mathbb{Y}$ from p to q. Set $\mathbb{Z}_1$ strictly separates p from q so path ζ intersects $\mathbb{Z}_1$. Since ζ is a subset of $\mathbb{Y}$, path ζ does not intersect $\mathbb{R}^d - \mathbb{Y}$ and so must intersect $\mathbb{X}_1 = \mathbb{Z}_1 \cap \mathbb{Y}$. Thus $\mathbb{X}_1$ strictly separates p from q in $\mathbb{Y}$. $\square$

Lemma B.18 assumed that $\eta(x, \alpha)$ was a subset of $\partial \mathbb{Y}$ for all $x \in \mathbb{X}_0 \cap \partial \mathbb{Y}$ and $\alpha \in [0,1]$. The next lemma is based on two sets, $\mathbb{Y}, \widetilde{\mathbb{Y}} \subseteq \mathbb{R}^d$, where $\widetilde{\mathbb{Y}} \subseteq \mathbb{Y}$. The lemma assumes that $\eta(x, \alpha)$ is a subset of $\mathbb{R}^d - \widetilde{\mathbb{Y}}$ for all $x \in \mathbb{X}_0 \cap \partial \mathbb{Y}$.

Lemma B.19. *Let $\mathbb{Y}$ and $\widetilde{\mathbb{Y}}$ be closed, bounded subsets of $\mathbb{R}^d$ where $\widetilde{\mathbb{Y}} \subseteq \mathbb{Y}$, let p, q be distinct points of $\widetilde{\mathbb{Y}}$, and let $\mathbb{X}_0$ be a closed subset of $\mathbb{Y}$ such that $p, q \notin \mathbb{X}_0$*

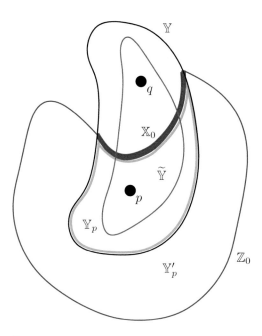

Figure B.3. Region $\widetilde{\mathbb{Y}}$ (bounded by magenta curve) is contained in region $\mathbb{Y}$ (bounded by black curve). Red curve $\mathbb{X}_0$ strictly separates p from q in region $\mathbb{Y}$ (bounded by black curve). Region $\mathbb{Y}_p$ (bounded by green curve) is the connected component of $\mathbb{Y} - \mathbb{X}_0$ containing p. Region $\mathbb{Y}_p'$ (bounded by blue curve $\mathbb{Z}_0$) is an extension of $\mathbb{Y}_p$ such that $\mathbb{Y} \cap \mathbb{Y}_p'$ equals $\mathrm{cl}(\mathbb{Y}_p)$ and $\mathbb{Y} \cap \partial \mathbb{Y}_p'$ is a subset of $\mathbb{X}_0$.

and every path in $\mathbb{Y}$ connecting p to q intersects $\mathbb{X}_0$. Let $\eta : \mathbb{X}_0 \times [0,1] \to \mathbb{Y}$ be a homotopy map from $\mathbb{X}_0$ to $\mathbb{X}_1 = \{\eta(x,1) : x \in \mathbb{X}_0\}$. If η never passes through p or q and $\{\eta(x,\alpha) : x \in \mathbb{X}_0 \cap \partial \mathbb{Y}\}$ is a subset of $\mathrm{cl}(\mathbb{R}^d - \widetilde{\mathbb{Y}})$ for all $\alpha \in [0,1]$, then every path from p to q in $\widetilde{\mathbb{Y}}$ intersects $\mathbb{X}_1 \cap \widetilde{\mathbb{Y}}$.

Proof: Let $\mathbb{Y}_p$ be the connected component of $\mathbb{Y} - \mathbb{X}_0$ that contains p. Since $\mathbb{Y}_p$ is bounded, it is possible to construct a closed and bounded set $\mathbb{Y}_p' \subseteq \mathbb{R}^d$ such that $\mathbb{Y} \cap \mathbb{Y}_p'$ equals $\mathrm{cl}(\mathbb{Y}_p)$ and $\mathbb{Y} \cap \partial \mathbb{Y}_p'$ is a subset of $\mathbb{X}_0$. (See Figure B.3.) Since $\mathbb{Y}_p$ is a subset of $\mathbb{Y} - \mathbb{X}_0$ and $\mathbb{Y} \cap \partial \mathbb{Y}_p'$ is a subset of $\mathbb{X}_0$, set $\mathbb{Y}_p$ is a subset of the interior of $\mathbb{Y}_p'$. Thus p is in the interior of $\mathbb{Y}_p'$. On the other hand, q is not in $\mathbb{Y}_p'$. By Corollary B.9, $\partial \mathbb{Y}_p'$ strictly separates p from q in $\mathbb{R}^d$.

Let $\mathbb{Z}_0$ denote $\partial \mathbb{Y}_p'$. Extend η to a homotopy map $\eta' : \mathbb{Z}_0 \times [0,1] \to \mathbb{R}^d$ that agrees with η on $\mathbb{X}_0$ and always maps $\mathbb{Z}_0 - \mathbb{X}_0$ to $\mathbb{R}^d - \widetilde{\mathbb{Y}}$. More specifically, $\eta'(x,\alpha)$ equals $\eta(x,\alpha)$ for all $x \in \mathbb{Z}_0 \cap \mathbb{X}_0$ and $\alpha \in [0,1]$, and $\{\eta'(x,\alpha) : x \in \mathbb{Z}_0 - \mathbb{X}_0\}$ is a subset of $\mathbb{R}^d - \widetilde{\mathbb{Y}}$ for all $\alpha \in [0,1]$. Let $\mathbb{Z}_1$ equal $\{\eta'(x,1) : x \in \mathbb{Z}_0\}$. By definition, $\mathbb{Z}_1 \cap \widetilde{\mathbb{Y}}$ equals $\mathbb{X}_1 \cap \widetilde{\mathbb{Y}}$.

Map η' is a homotopy map from $\mathbb{Z}_0$ to $\mathbb{Z}_1$ that does not pass through p or q. Set $\mathbb{Z}_0$ strictly separates p from q in $\mathbb{R}^d$. By Lemma B.17, set $\mathbb{Z}_1$ strictly separates p from q in $\mathbb{R}^d$.

Assume there is path $\zeta \subseteq \widetilde{\mathbb{Y}}$ from p to q. Set $\mathbb{Z}_1$ strictly separates p from q so path ζ intersects $\mathbb{Z}_1$. Since ζ is a subset of $\widetilde{\mathbb{Y}}$, path ζ does not intersect $\mathbb{R}^d - \widetilde{\mathbb{Y}}$ and so must intersect $\mathbb{Z}_1 \cap \widetilde{\mathbb{Y}} = \mathbb{X}_1 \cap \widetilde{\mathbb{Y}}$. Thus every path from p to q in $\widetilde{\mathbb{Y}}$ intersects $\mathbb{X}_1 \cap \widetilde{\mathbb{Y}}$. $\qquad\square$

APPENDIX C

GRAPH THEORY

A graph $\mathcal{G}$ is a set $V(\mathcal{G}) = \{v_1, v_2, \ldots\}$ of objects and a set $E(\mathcal{G})$ of pairs (v_i, v_j) of elements of $V(\mathcal{G})$. The set $V(\mathcal{G})$ is called the graph vertices and the set $E(\mathcal{G})$ is called the graph edges. If edge **e** equals (v_i, v_j), then v_i and v_j are called the endpoints of edge **e**. A vertex $v \in V(\mathcal{G})$ is incident on edge $\mathbf{e} \in E(\mathcal{G})$ if v is an endpoint of **e**. Vertex v_i is adjacent to vertex v_j if (v_i, v_j) is an edge of $\mathcal{G}$. The degree of a graph vertex v is the number of edges incident on v. Equivalently, the degree of a graph vertex v is the number of vertices adjacent to v.

A subgraph of a graph $\mathcal{G}$ is a graph whose vertices and edges are a subset of the vertices and edges of $\mathcal{G}$. An induced subgraph of $\mathcal{G}$ is a subset V' of the vertices of $\mathcal{G}$ and a subset E' of the edges of $\mathcal{G}$ containing every edge $\mathbf{e} \in E(\mathcal{G})$ whose endpoints are both in V'.

A path in a graph is a sequence of vertices $(v_{i_1}, v_{i_2}, \ldots, v_{i_k})$ such that $(v_{i_j}, v_{i_{j+1}})$ is an edge of the graph for every i_j. A path is simple if no two vertices in the sequence are the same. A cycle in a graph is a sequence of three or more distinct vertices $(v_{i_1}, v_{i_2}, \ldots, v_{i_k})$ such that $(v_{i_1}, v_{i_2}, \ldots, v_{i_k})$ is a path and (v_{i_1}, v_{i_k}) is an edge. The length of a path or cycle is the number of edges in the path or cycle. The length of a path equals the number of path vertices minus one. The length of a cycle equals the number of cycle vertices.

A vertex $v_i \in V(\mathcal{G})$ is connected to a vertex $v_j \in V(\mathcal{G})$ if there is a path from v_i to v_j in $\mathcal{G}$. A graph $\mathcal{G}$ is connected if every vertex $v_i \in V(\mathcal{G})$ is connected to every other vertex $v_j \in V(\mathcal{G})$. A connected component of $\mathcal{G}$ is a maximally connected, induced subgraph of $\mathcal{G}$, i.e., an induced subgraph $\mathcal{G}'$ of $\mathcal{G}$ that is connected such that any induced subgraph containing $\mathcal{G}'$ is not connected.

A tree is a connected graph with no cycles. A leaf of a tree is a vertex that has a degree of one.

A tree may have a special vertex called the root of the tree. Such trees are sometimes called rooted trees. For every vertex v in a rooted tree, there is a unique simple path from v to the root. Vertex v_i is an ancestor of vertex v_j and v_j is a descendant of v_i if the simple path from v_j to the root contains v_i. Vertex

v_i is a parent of vertex v_j and v_j is a child of v_i if v_i is an ancestor of v_j and (v_i, v_j) is a tree edge. Every vertex other than the root has a single parent.

The depth of a vertex in a rooted tree is the length of the path from the vertex to the root. The height of a tree is the maximum depth of any vertex in the tree. A level of a rooted tree is a set of tree vertices at the same depth in the tree.

A height graph is a graph where each node has a scalar value representing its "height".

A graph $\mathcal{G}$ is isomorphic to graph $\mathcal{G}'$ if there is a one-to-one and onto mapping $\iota : V(\mathcal{G}) \rightarrow V(\mathcal{G}')$ such that $(v_i, v_j) \in E(\mathcal{G})$ if and only if $(\iota(v_i), \iota(v_j)) \in E(\mathcal{G}')$.

APPENDIX D

NOTATION

Greek Letters

α Scalar value.

Δ Triangle.

Γ Regular grid or convex polyhedral mesh.

Γ_{inner} Subgrid of the regular grid Γ composed of interior grid cube cubes of Γ.

Γ_{outer} Set of boundary grid cubes of Γ.

η Homotopy map.

ι Inclusion map $\iota : \mathbb{X} \to \mathbb{Y}$ where $\iota(x) = x$.

κ Configuration of negative and positive vertices. An assignment of "−" and "+" labels to polytope vertices.

λ Connected component.

λ^- Connected component of $\mathcal{G}_\sigma^-$ containing the connected component λ of $\mathcal{G}_\sigma$.

λ^+ Connected component of $\mathcal{G}_\sigma^+$ containing the connected component λ of $\mathcal{G}_\sigma$.

Λ Set of connected components.

π	Orthogonal projection.
μ	Affine transformation, homeomorphism, or continuous map.
σ	Isovalue.
$\sigma_0,\ \sigma_1$	Lower and upper scalar bounds on the interval volume $\{x : \sigma_0 \le \phi(x) \le \sigma_1\}$.
$\Sigma_{\mathbf{c}}^{+}(\sigma)$	Isosurface patch for isovalue σ in polytope $\mathbf{c}$.
$\Sigma_{\Gamma}(\sigma)$	Isosurface for isovalue σ in grid or mesh Γ.
ϕ	Scalar function.
τ	Triangulation.
$\tau^{\mathbf{c}}$	Triangulation of cube $\mathbf{c}$.
$\tau_{\mathrm{cong}}(\mathbf{t})$	Set of tetrahedra or simplices congruent to tetrahedron or simplex $\mathbf{t}$.
$\chi(\mathbf{c}),\ \chi(\Gamma)$	Mapping of polytope $\mathbf{c}$ to $\mathbf{c} \times [0,1]$ and mesh Γ to $\Gamma \times [0,1]$ for constructing interval volumes. Vertices of $\chi(\mathbf{c})$ and $\chi(\Gamma)$ receive "$-$" and "$+$" labels based on "$-$", "$+$", and "$*$" labels of $\mathbf{c}$ and Γ.
ζ	Path.
$\zeta_{\mathbf{c}}$	Path of cubes.
ψ	Polyhedron or polytope.
Ψ	Function from $\mathbb{R}^3 \times [0,1]$ to $\mathbb{R}$ that "lifts" the interval volume $\mathcal{I}_{\phi}(\sigma_0, \sigma_1) \subset \mathbb{R}^3$ to the isosurface in $\Psi^{-1}(0) \subset \mathbb{R}^4$.
Υ	Set of isocontour line segments or isosurface triangles, tetrahedra or simplices, or interval volume tetrahedra.

Roman Letters

$a_{\mathbf{c}},\ a_{\mathbf{f}}$	Center of cube $\mathbf{c}$ or face $\mathbf{f}$.
$\mathbb{B}^d$	Unit ball in $\mathbb{R}^d$.
$\mathbf{c}$	Cube, hypercube, convex polyhedron, or convex polytope.
$\mathbb{C}$	Set of cubic regions.

$\mathbb{C}_{\text{inner}}$	Set of inner cubic regions.
$\mathbb{C}_\lambda$	Set of cubes corresponding to the connected component λ of $\mathcal{G}_\sigma$.
$\mathbb{C}_{\lambda^-}$	Set of cubes corresponding to the connected component λ^- of $\mathcal{G}_\sigma^-$.
$\mathbb{C}_{\lambda^+}$	Set of cubes corresponding to the connected component λ^+ of $\mathcal{G}_\sigma^+$.
$\mathbf{d}$	Disk or diagonal.
$\mathcal{D}$	Function domain.
$\mathbf{e}$	Edge.
$E_\kappa^{+/-}$	Set of bipolar edges of configuration κ.
$\mathbf{f}$	Face or facet of a simplex, polytope, or regular grid.
F	Set of faces or facets.
$\mathcal{F}_c(\sigma)$	Number of cubes whose span contains scalar value σ.
$\mathcal{F}_e(\sigma)$	Number of edges whose span contains scalar value σ.
$\mathcal{F}_v(\sigma)$	Frequency of scalar value σ.
$\mathcal{F}_v(\sigma_0, \sigma_1)$	Frequency of scalar values in the range $[\sigma_0, \sigma_1)$.
$\overline{\mathcal{F}}_v(\sigma_0, \sigma_1)$	Frequency of scalar values in the closed range $[\sigma_0, \sigma_1]$.
$\mathcal{F}_v^*(\sigma_0, \sigma_1)$	Normalized frequency.
$\mathbf{h}$	Plane or hyperplane.
$\mathbf{g}$	Face or facet of a grid or polytope.
$\mathcal{G}$	Graph.
$\mathcal{G}^-(s)$	Subgraph of $\mathcal{G}$ induced by vertices whose scalar values are less than s.
$\mathcal{G}^+(s)$	Subgraph of $\mathcal{G}$ induced by vertices whose scalar values are greater than s.
$\mathcal{G}_\sigma$	Graph representing grid cubes and facets whose spans contain the isovalue σ. Vertices of $\mathcal{G}_\sigma$ are grid cubes $\{\,\mathbf{c} : \text{span of } \mathbf{c} \text{ contains } \sigma\,\}$. Edges of $\mathcal{G}_\sigma$ are $\{\,(\mathbf{c}, \mathbf{c}') : \text{cubes } \mathbf{c} \text{ and } \mathbf{c}' \text{ share a common facet } \mathbf{f} \text{ and the span of } \mathbf{f} \text{ contains } \sigma\}$.

$\mathcal{G}_\sigma^-$ Graph representing grid cubes $\mathbf{c}$ and facets $\mathbf{f}$ where $s_{\min}(\mathbf{c}) \leq \sigma$ and $s_{\min}(\mathbf{f}) \leq \sigma$. Vertices of $\mathcal{G}_\sigma^-$ are cubes $\{\, \mathbf{c} : s_{\min}(\mathbf{c}) \leq \sigma \,\}$. Edges of $\mathcal{G}_\sigma^-$ are $\{\, (\mathbf{c},\mathbf{c}') : $ cubes $\mathbf{c}$ and $\mathbf{c}'$ share a common facet $\mathbf{f}$ and $s_{\min}(\mathbf{f}) \leq \sigma \,\}$.

$\mathcal{G}_\sigma^+$ Graph representing grid cubes $\mathbf{c}$ and facets $\mathbf{f}$ where $s_{\max}(\mathbf{c}) \geq \sigma$ and $s_{\max}(\mathbf{f}) \geq \sigma$. Vertices of $\mathcal{G}_\sigma^+$ are cubes $\{\, \mathbf{c} : s_{\max}(\mathbf{c}) \geq \sigma \,\}$. Edges of $\mathcal{G}_\sigma^+$ are $\{\, (\mathbf{c},\mathbf{c}') : $ cubes $\mathbf{c}$ and $\mathbf{c}'$ share a common facet $\mathbf{f}$ and $s_{\max}(\mathbf{f}) \geq \sigma \,\}$.

$\mathcal{I}_\phi(\sigma_0, \sigma_1)$ Interval volume $\{x : \sigma_0 \leq \phi(x) \leq \sigma_1\}$.

$\mathcal{L}$ Interval volume vertex label "$-$", "$+$", or "$*$".

$L_\mathbf{c}^+(U_\mathbf{c})$ Line segments on the boundary of $R_\mathbf{c}^+(U_\mathbf{c})$ that do not lie on the boundary of the square $\mathbf{c}$. Abbreviated as $L_\mathbf{c}^+$.

$M_\mathbf{c}$ Midpoints of bipolar edges in polytope $\mathbf{c}$.

n_i Number of grid vertices along the ith axis. Vertex dimensions of a 3D grid are $n_1 \times n_2 \times n_3$.

n_x, n_y, n_z, n_w
     Number of grid vertices along axes x, y, z, or w.

$\mathbb{N}_p$ Neighborhood of point p.

N^v, N_Γ^v Number of vertices in a grid Γ.

N^c, N_Γ^c Number of cubes in a grid Γ.

m_i Number of grid edges along the ith axis. Cube dimensions of a 3D grid are $m_1 \times m_2 \times m_3$.

$\mathbb{M}$ Manifold.

$\mathbf{q}$ Quadrilateral.

$r_\mathbf{e}$ Isosurface vertex on edge $\mathbf{e}$.

$\mathbf{r}$ Ray.

$\mathbb{R}$ Real numbers.

$\mathbb{R}^d$ Real d-dimensional coordinate space.

$R_\mathbf{c}^{M+}$ Abbreviation for $R_\mathbf{c}^+(M_\mathbf{c})$.

s_v Scalar value at vertex v.

$R_{\mathbf{c}}^+(U_{\mathbf{c}})$	Positive region in polytope $\mathbf{c}$. Defined as $\mathrm{conv}(U_{\mathbf{c}} \cup V_{\mathbf{c}}^+)$. Abbreviated as $R_{\mathbf{c}}^+$.
$s_{\min}(\mathbf{c})$	Minimum scalar value of any vertex of cube $\mathbf{c}$.
$s_{\min}(\mathbf{f})$	Minimum scalar value of any vertex of facet $\mathbf{f}$.
$s_{\min}(\mathbb{C})$	Minimum of $s_{\min}(\mathbf{c})$ over all cubes $\mathbf{c} \in \mathbb{C}$.
$s_{\max}(\mathbf{c})$	Maximum scalar value of any vertex of cube $\mathbf{c}$.
$s_{\max}(\mathbf{f})$	Maximum scalar value of any vertex of facet $\mathbf{f}$.
$s_{\max}(\mathbb{C})$	Maximum of $s_{\max}(\mathbf{c})$ over all cubes $\mathbf{c} \in \mathbb{C}$.
$S_{\mathbf{c}}^+(U_{\mathbf{c}})$	Surface separating $R_{\mathbf{c}}^+(U_{\mathbf{c}})$ from $\mathbf{c} - R_{\mathbf{c}}^+(U_{\mathbf{c}})$. Defined as $R_{\mathbf{c}}^+(U_{\mathbf{c}}) \cap \mathrm{cl}(\mathbf{c} - R_{\mathbf{c}}^+(U_{\mathbf{c}}))$. Abbreviated as $S_{\mathbf{c}}^+$.
$\mathbb{S}^d$	Unit sphere in $\mathbb{R}^d$.
$\mathbf{t}$	Triangle, tetrahedron, or simplex.
$T_{\mathbf{c}}^+(U_{\mathbf{c}})$	Triangles or simplices in the triangulation of the boundary of $R_{\mathbf{c}}^+(U_{\mathbf{c}})$ that do not lie on the boundary of polytope $\mathbf{c}$. Abbreviated as $T_{\mathbf{c}}^+$.
$T_{\mathbf{c}}^{M+}$	Abbreviation for $T_{\mathbf{c}}^+(M_{\mathbf{c}})$.
$T_{\Gamma}^+(U)$	Union of $T_{\mathbf{c}}^+(U \cap \mathbf{c})$ over all polytopes $\mathbf{c} \in \Gamma$.
$\tilde{T}_{\mathbf{c}}^+(U_{\mathbf{c}})$	Equals $T_{\mathbf{c}}^+(U_{\mathbf{c}})$ when $R_{\mathbf{c}}^+(U_{\mathbf{c}})$ has the same dimension as $\mathbf{c}$. Equals a triangulation of $R_{\mathbf{c}}^+(U_{\mathbf{c}})$ when $R_{\mathbf{c}}^+(U_{\mathbf{c}})$ is contained in a facet of $\mathbf{c}$ and has dimension equal to the dimension of the facet. Abbreviated as $\tilde{T}_{\mathbf{c}}^+$.
$\mathcal{T}_C$	Contour tree.
$\mathcal{T}_J$	Join tree.
$\mathcal{T}_S$	Split tree.
u	Element of an isosurface vertex set.
U	Isosurface vertex set.
$U_{\mathbf{c}}$	Isosurface vertex set for polytope $\mathbf{c}$.
v	Vertex.
$V_{\mathbf{c}}^+$	Positive vertices of polytope $\mathbf{c}$.
$V_{\mathbf{c}}^-$	Negative vertices of polytope $\mathbf{c}$.

$\text{Vol}_c(\sigma_0, \sigma_1)$ Approximate measure of an interval volume based on the number of cubes intersecting the volume.

$\text{Vol}_e(\sigma_0, \sigma_1)$ Approximate measure of an interval volume based on the number of edges intersecting the volume.

w Isosurface vertex.

Z Sequence or permutation.

Z_E Sequence of bipolar edges.

Z_U Sequence of elements of an isosurface vertex set U.

Z_V^+ Sequence of positive vertices.

Operators

$\partial \mathbf{c}$ Boundary of a convex polytope $\mathbf{c}$.

$\partial \mathbb{M}$ Manifold boundary of a manifold $\mathbb{M}$.

∂P Topological boundary of a set of points P.

$\text{aff}(P)$ Affine hull of a set of points P.

$\text{cl}(P)$ Closure of a set of points P.

$\text{conv}(P)$ Convex hull of a set of points P.

$\text{int}(\mathbf{c})$ Interior of a convex polytope $\mathbf{c}$.

$\text{int}(\mathbb{M})$ Manifold interior of a manifold $\mathbb{M}$.

$\text{int}(P)$ Topological interior of a set of points P.

$\text{relbnd}(P)$ Relative boundary of a set of points P.

$\text{relint}(P)$ Relative interior of a set of points P.

BIBLIOGRAPHY

[Alliez et al., 2007] Alliez, P., Attene, M., Gotsman, C., and Ucelli, G. (2007). Recent advances in remeshing of surfaces. In Floriani, L. D. and Spagnuolo, M., editors, *Shape Analysis and Structuring*, pages 53–82. Springer-Verlag, Berlin.

[Armstrong, 1983] Armstrong, M. A. (1983). *Basic topology*. Undergraduate texts in mathematics. Springer-Verlag, New York.

[Ashida and Badler, 2003] Ashida, K. and Badler, N. I. (2003). Feature preserving manifold mesh from an octree. In *Proceedings of the Eighth ACM Symposium on Solid Modeling and Applications*, pages 292–297, New York. ACM Press.

[Attali et al., 2005] Attali, D., Cohen-Steiner, D., and Edelsbrunner, H. (2005). Extraction and simplification of iso-surfaces in tandem. In *Proceedings of the Eurographics Symposium on Geometry Processing*, pages 139–148, Aire-la-Ville, Switzerland. Eurographics Association.

[Bachthaler and Weiskopf, 2008] Bachthaler, S. and Weiskopf, D. (2008). Continuous scatterplots. *IEEE Transactions on Visualization and Computer Graphics*, 14(6):1428–1435.

[Bachthaler and Weiskopf, 2009] Bachthaler, S. and Weiskopf, D. (2009). Efficient and adaptive rendering of 2-D continuous scatterplots. *Computer Graphics Forum*, 28(3):743–750.

[Bai et al., 2006] Bai, Y., Han, X., and Prince, J. (2006). Octree-based topology-preserving isosurface simplification. In *Proceedings of the 2006 Conference on Computer Vision and Pattern Recognition*, page 81, Washington, DC. IEEE Computer Society.

[Bajaj et al., 1997] Bajaj, C. L., Pascucci, V., and Schikore, D. (1997). The contour spectrum. In *Proceedings of IEEE Visualization 1997*, pages 167–173, Los Alamitos, CA. IEEE Computer Society.

[Bajaj et al., 1996] Bajaj, C. L., Pascucci, V., and Schikore, D. R. (1996). Fast isocontouring for improved interactivity. In *Proceedings of the 1996 Symposium on Volume Visualization*, pages 39–46, Los Alamitos, CA. IEEE Computer Society.

[Balmelli et al., 2002] Balmelli, L., Morris, C. J., Taubin, G., and Bernardini, F. (2002). Volume warping for adaptive isosurface extraction. In *Proceedings of IEEE Visualization 2002*, pages 467–474, Los Alamitos, CA. IEEE Computer Society.

[Banks et al., 2004] Banks, D. C., Linton, S. A., and Stockmeyer, P. K. (2004). Counting cases in substitope algorithms. *IEEE Transactions on Visualization and Computer Graphics*, 10:371–384.

[Barry and Wood, 2007] Barry, M. and Wood, Z. (2007). Direct extraction of normal mapped meshes from volume data. In *Proceedings of the International Conference on Advances in Visual Computing*, pages 816–826, Berlin. Springer-Verlag.

[Bhaniramka et al., 2000] Bhaniramka, P., Wenger, R., and Crawfis, R. (2000). Isosurfacing in higher dimensions. In *Proceedings of IEEE Visualization 2000*, pages 267–273, Los Alamitos, CA. IEEE Computer Society.

[Bhaniramka et al., 2004a] Bhaniramka, P., Wenger, R., and Crawfis, R. (2004a). Isosurface construction in any dimension using convex hulls. *IEEE Transactions on Visualization and Computer Graphics*, 10(2):130–141.

[Bhaniramka et al., 2004b] Bhaniramka, P., Zhang, C., Xue, D., Crawfis, R., and Wenger, R. (2004b). Volume interval segmentation and rendering. In *Proceedings of the 2004 IEEE Symposium on Volume Visualization and Graphics*, pages 55–62, Los Alamitos, CA. IEEE Computer Society.

[Bischoff and Kobbelt, 2002] Bischoff, S. and Kobbelt, L. P. (2002). Isosurface reconstruction with topology control. In *Proceedings of the 10th Pacific Conference on Computer Graphics and Applications*, pages 246–255, Los Alamitos, CA. IEEE Computer Society.

[Bischoff et al., 2005] Bischoff, S., Pavic, D., and Kobbelt, L. (2005). Automatic restoration of polygon models. *ACM Transactions on Graphics*, 24(4):1332–1352.

[Bloomenthal, 1988] Bloomenthal, J. (1988). Polygonization of implicit surfaces. *Computer Aided Geometric Design*, 5(4):341–355.

[Bloomenthal, 1994] Bloomenthal, J. (1994). An implicit surface polygonizer. In Heckbert, P., editor, *Graphics Gems IV*, pages 324–349. Academic Press, Boston.

[Bordoloi and Shen, 2003] Bordoloi, U. and Shen, H.-W. (2003). Space efficient fast isosurface extraction for large datasets. In *Proceedings of IEEE Visualization 2003*, pages 201–208, Los Alamitos, CA. IEEE Computer Society.

[Boyell and Ruston, 1963] Boyell, R. L. and Ruston, H. (1963). Hybrid techniques for real-time radar simulation. In *Proceedings of the 1963 Fall Joint Computer Conference*, pages 445–458, New York. ACM Press.

[Bruckner and Möller, 2010] Bruckner, S. and Möller, T. (2010). Isosurface similarity maps. *Computer Graphics Forum*, 29(3):773–782.

[Carneiro et al., 1996] Carneiro, B. P., Silva, C. T., and Kaufman, A. E. (1996). Tetracubes: An algorithm to generate 3D isosurfaces based upon tetrahedra. In *Proceedings of IX Brazilian Symposium on Computer Graphics, Image Processing and Vision*, pages 205–210, Los Alamitos, CA. IEEE Computer Society.

[Carr et al., 2006a] Carr, H., Duffy, B., and Barry, D. (2006a). On histograms and isosurface statistics. *IEEE Transactions on Visualization and Computer Graphics*, 12(5):1259–1266.

[Carr and Max, 2010] Carr, H. and Max, N. (2010). Subdivision analysis of the trilinear interpolant. *IEEE Transactions on Visualization and Computer Graphics*, 16:533–547.

[Carr et al., 2001] Carr, H., Möller, T., and Snoeyink, J. (2001). Simplicial subdivisions and sampling artifacts. In *Proceedings of IEEE Visualization 2001*, pages 99–106, Los Alamitos, CA. IEEE Computer Society.

[Carr et al., 2006b] Carr, H., Möller, T., and Snoeyink, J. (2006b). Artifacts caused by simplicial subdivision. *IEEE Transactions on Visualization and Computer Graphics*, 12:231–242.

[Carr and Snoeyink, 2003] Carr, H. and Snoeyink, J. (2003). Path seeds and flexible isosurfaces using topology for exploratory visualization. In *Proceedings of the Symposium on Data Visualisation*, pages 49–58, Aire-la-Ville, Switzerland. Eurographics Association.

[Carr and Snoeyink, 2009] Carr, H. and Snoeyink, J. (2009). Representing interpolant topology for contour tree computation. In Hege, H.-C., Polthier, K., and Scheuermann, G., editors, *Topology-Based Methods in Visualization II*, Mathematics and Visualization, pages 59–73. Springer-Verlag, Berlin.

[Carr et al., 2003] Carr, H., Snoeyink, J., and Axen, U. (2003). Computing contour trees in all dimensions. *Computational Geometry: Theory and Applications*, 24:75–94.

[Carr et al., 2004] Carr, H., Snoeyink, J., and van de Panne, M. (2004). Simplifying flexible isosurfaces using local geometric measures. In *Proceedings of IEEE Visualization 2004*, pages 497–504, Los Alamitos, CA. IEEE Computer Society.

[Chan and Purisima, 1998] Chan, S. and Purisima, E. (1998). A new tetrahedral tesselation scheme for isosurface generation. *Computers & Graphics*, 22(1):83–90.

[Cheney and Kincaid, 2007] Cheney, E. W. and Kincaid, D. R. (2007). *Numerical Mathematics and Computing*. Brooks/Cole Publishing Co., Pacific Grove, CA.

[Cheng et al., 2004] Cheng, S.-W., Dey, T. K., Ramos, E. A., and Ray, T. (2004). Sampling and meshing a surface with guaranteed topology and geometry. In *Proceedings of the Annual Symposium on Computational Geometry*, pages 280–289, New York. ACM Press.

[Chernyaev, 1995] Chernyaev, E. V. (1995). Marching cubes 33: Construction of topologically correct isosurfaces. Technical Report CN/95-17, CERN, Geneva, Switzerland.

[Chiang et al., 2001] Chiang, Y.-J., Farias, R., Silva, C. T., and Wei, B. (2001). A unified infrastructure for parallel out-of-core isosurface and volume rendering of unstructured grids. In *Symposium on Parallel and Large-Data Visualization and Graphics*, pages 59–66, 151, New York. ACM Press.

[Chiang et al., 2005] Chiang, Y.-J., Lenz, T., Lu, X., and Rote, G. (2005). Simple and optimal output-sensitive construction of contour trees using monotone paths. *Computational Geometry: Theory and Applications*, 30:165–195.

[Chiang and Lu, 2003] Chiang, Y.-J. and Lu, X. (2003). Progressive simplification of tetrahedral meshes preserving all isosurface topologies. *Computer Graphics Forum*, 22(3):493–504.

[Chiang and Silva, 1997] Chiang, Y.-J. and Silva, C. T. (1997). I/O optimal isosurface extraction. In *Proceedings of IEEE Visualization 1997*, pages 293–300, Los Alamitos, CA. IEEE Computer Society.

[Chiang and Silva, 1999] Chiang, Y.-J. and Silva, C. T. (1999). External memory techniques for isosurface extraction in scientific visualization. In Abello, J. M. and Vitter, J. S., editors, *External Memory Algorithms*, pages 247–277. American Mathematical Society, Boston.

[Chiang et al., 1998] Chiang, Y.-J., Silva, C. T., and Schroeder, W. J. (1998). Interactive out-of-core isosurface extraction. In *Proceedings IEEE Visualization 1998*, pages 167–174, Los Alamitos, CA. IEEE Computer Society.

[Cignoni et al., 1994] Cignoni, P., De Floriani, L., Montani, C., Puppo, E., and Scopigno, R. (1994). Multiresolution modeling and visualization of volume data based on simplicial complexes. In *Proceedings of the 1994 Symposium on Volume Visualization*, pages 19–26, New York. ACM Press.

[Cignoni et al., 2000] Cignoni, P., Ganovelli, F., Montani, C., and Scopigno, R. (2000). Reconstruction of topologically correct and adaptive trilinear isosurfaces. *Computers & Graphics*, 24:399–418.

[Cignoni et al., 1997] Cignoni, P., Marino, P., Montani, C., Puppo, E., and Scopigno, R. (1997). Speeding up isosurface extraction using interval trees. *IEEE Transactions on Visualization and Computer Graphics*, 3(2):158–170.

[Cignoni et al., 1998] Cignoni, P., Montani, C., and Scopigno, R. (1998). A comparison of mesh simplification algorithms. *Computers & Graphics*, 22(1):37–54.

[Clarkson et al., 1993] Clarkson, K. L., Mehlhorn, K., and Seidel, R. (1993). Four results on randomized incremental constructions. *Computational Geometry: Theory and Applications*, 3:185–212.

[Cline et al., 1988] Cline, H., Lorensen, W., Ludke, S., Crawford, C., and Teeter, B. (1988). Two algorithms for the three-dimensional reconstruction of tomograms. *Medical Physics*, 15(3):320–327.

[Cole-McLaughlin et al., 2003] Cole-McLaughlin, K., Edelsbrunner, H., Harer, J., Natarajan, V., and Pascucci, V. (2003). Loops in Reeb graphs of 2-manifolds. In *Proceedings of the Annual Symposium on Computational Geometry*, pages 344–350, New York. ACM Press.

[Cormen et al., 2001] Cormen, T. H., Leiserson, C. E., Rivest, R. L., and Stein (2001). *Introduction to Algorithms*. MIT Press, Cambridge, MA.

[Cox et al., 2003] Cox, J., Karron, D. B., and Ferdous, N. (2003). Topological zone organization of scalar volume data. *Journal of Mathematical Imaging and Vision*, 18:95–117.

[Dey and Levine, 2007] Dey, T. K. and Levine, J. A. (2007). Delaunay meshing of isosurfaces. In *International Conference on Shape Modeling and Applications*, pages 241–250, Los Alamitos, CA. IEEE Computer Society.

[Dietrich et al., 2008] Dietrich, C., Scheidegger, C., Comba, J., Nedel, L., and Silva, C. (2008). Edge groups: An approach to understanding the mesh quality of marching methods. *IEEE Transactions on Visualization and Computer Graphics*, 14(6):1651–1666.

[Dietrich et al., 2009a] Dietrich, C., Scheidegger, C., Comba, J., Nedel, L., and Silva, C. (2009a). Marching cubes without skinny triangles. *Computing in Science Engineering*, 11(2):82–87.

[Dietrich et al., 2009b] Dietrich, C., Scheidegger, C., Schreiner, J., Comba, J., Nedel, L., and Silva, C. (2009b). Edge transformations for improving mesh quality of marching cubes. *IEEE Transactions on Visualization and Computer Graphics*, 15(1):150–159.

[Doraiswamy and Natarajan, 2008] Doraiswamy, H. and Natarajan, V. (2008). Efficient output-sensitive construction of Reeb graphs. In *Proceedings of the 19th International Symposium on Algorithms and Computing*, pages 556–567, Berlin. Springer-Verlag.

[Doraiswamy and Natarajan, 2009] Doraiswamy, H. and Natarajan, V. (2009). Efficient algorithms for computing Reeb graphs. *Computational Geometry: Theory and Applications*, 42:606–616.

[Duffy et al., 2012] Duffy, B., Carr, H., and Möller, T. (2012). Integrating isosurface statistics and histograms. *IEEE Transactions on Visualization and Computer Graphics*, 19(2):263–277.

[Dürst, 1988] Dürst, M. J. (1988). Additional reference to Marching Cubes. *Computer Graphics*, 22(2):72–73.

[Edelsbrunner, 1987] Edelsbrunner, H. (1987). *Algorithms in Combinatorial Geometry*. Springer-Verlag, New York.

[Etiene et al., 2012] Etiene, T., Nonato, L. G., Scheidegger, C., Tierny, J., Peters, T. J., Pascucci, V., Kirby, R. M., and Silva, C. T. (2012). Topology verification for isosurface extraction. *IEEE Transactions on Visualization and Computer Graphics*, 18:952–965.

[Etiene et al., 2009] Etiene, T., Scheidegger, C., Nonato, L. G., Kirby, R. M., and Silva, C. (2009). Verifiable visualization for isosurface extraction. *IEEE Transactions on Visualization and Computer Graphics*, 15:1227–1234.

[Fujishiro et al., 1996] Fujishiro, I., Maeda, Y., Sato, H., and Takeshima, Y. (1996). Volumetric data exploration using interval volume. *IEEE Transactions on Visualization and Computer Graphics*, 2(2):144–155.

[Gallagher, 1991] Gallagher, R. S. (1991). Span filtering: an optimization scheme for volume visualization of large finite element models. In *Proceedings of IEEE Visualization 1991*, pages 68–75, Los Alamitos, CA. IEEE Computer Society.

[Gallagher and Nagtegaal, 1989] Gallagher, R. S. and Nagtegaal, J. C. (1989). An efficient 3-D visualization technique for finite element models and other coarse volumes. In *Proceedings of the 16th Annual Conference on Computer Graphics and Interactive Techniques, SIGGRAPH 1989*, pages 185–194, New York. ACM Press.

[Garey and Johnson, 1979] Garey, M. R. and Johnson, D. S. (1979). *Computers and Intractability: A Guide to the Theory of NP-Completeness*. W. H. Freeman, New York.

[Garland, 1999] Garland, M. (1999). Multiresolution modeling: Survey and future opportunities. In *State of the Art Report, Eurographics*, pages 111–131.

[Gelder and Wilhelms, 1990] Gelder, A. V. and Wilhelms, J. (1990). Topological considerations in isosurface generation. *Computer Graphics*, 24(5):79–86.

[Gerstner, 2002] Gerstner, T. (2002). Multiresolution extraction and rendering of transparent isosurfaces. *Computers & Graphics*, 26(2):219–228.

[Gerstner and Pajarola, 2000] Gerstner, T. and Pajarola, R. (2000). Topology preserving and controlled topology simplifying multiresolution isosurface extraction. In *Proceedings of IEEE Visualization 2000*, pages 259–266, Los Alamitos, CA. IEEE Computer Society.

[Gerstner and Rumpf, 2000] Gerstner, T. and Rumpf, M. (2000). Multiresolutional parallel isosurface extraction based on tetrahedral bisection. In Chen, M., Kaufman, A., and Yagel, R., editors, *Volume Graphics*, pages 267–278. Springer-Verlag, London.

[Gibson, 1998a] Gibson, S. F. F. (1998a). Constrained elastic surface nets: Generating smooth surfaces from binary segmented data. In *Proceedings of the First International Conference on Medical Image Computing and Computer-Assisted Intervention, MICCAI 1998*, pages 888–898, Berlin. Springer-Verlag.

[Gibson, 1998b] Gibson, S. F. F. (1998b). Using distance maps for accurate surface representation in sampled volumes. In *Proceedings of the 1998 IEEE Symposium on Volume Visualization*, pages 23–30, Los Alamitos, CA. IEEE Computer Society.

[Giles and Haimes, 1990] Giles, M. and Haimes, R. (1990). Advanced interactive visualization for CFD. *Computing Systems in Engineering*, 1(1):51–62.

[Gregorski et al., 2002] Gregorski, B., Duchaineau, M., Lindstrom, P., Pascucci, V., and Joy, K. I. (2002). Interactive view-dependent rendering of large isosurfaces. In *Proceedings of IEEE Visualization 2002*, pages 475–484, Los Alamitos, CA. IEEE Computer Society.

[Gregorski et al., 2004] Gregorski, B., Senecal, J., Duchaineau, M. A., and Joy, K. I. (2004). Adaptive extraction of time-varying isosurfaces. *IEEE Transactions on Visualization and Computer Graphics*, 10(6):683–694.

[Greß and Klein, 2004] Greß, A. and Klein, R. (2004). Efficient representation and extraction of 2-manifold isosurfaces using kd-trees. *Graphical Models*, 66(6):370–397.

[Guo, 1995] Guo, B. (1995). Interval set: A volume rendering technique generalizing isosurface extraction. In *Proceedings of IEEE Visualization 1995*, pages 3–10, Los Alamitos, CA. IEEE Computer Society.

[Hall and Warren, 1990] Hall, M. and Warren, J. (1990). Adaptive polygonalization of implicitly defined surfaces. *IEEE Comput. Graph. Appl.*, 10(6):33–42.

[Hamann et al., 1997] Hamann, B., Trotts, I. J., and Farin, G. E. (1997). On approximating contours of the piecewise trilinear interpolant using triangular rational-quadratic Bézier patches. *IEEE Transactions on Visualization and Computer Graphics*, 3(3):215–227.

[Harvey et al., 2010] Harvey, W., Wang, Y., and Wenger, R. (2010). A randomized $O(m \log m)$ time algorithm for computing Reeb graphs of arbitrary simplicial complexes. In *Proceedings of the Annual Symposium on Computational Geometry*, pages 267–276, New York. ACM Press.

[He et al., 1996] He, T., Hong, L., Varshney, A., and Wang, S. W. (1996). Controlled topology simplification. *IEEE Transactions on Visualization and Computer Graphics*, 2(2):171–184.

[Hilton et al., 1996] Hilton, A., Stoddart, A., Illingworth, J., and Windeatt, T. (1996). Marching triangles: Range image fusion for complex object modeling. In *International Conference on Image Processing*, pages 381–384, Los Alamitos, CA. IEEE Computer Society.

[Ho et al., 2005] Ho, C., Wu, F., Chen, B., and Ouhyoung, M. (2005). Cubical marching squares: Adaptive feature preserving surface extraction from volume data. *Computer Graphics Forum*, 24:195–201.

[Hocking and Young, 1961] Hocking, J. and Young, G. (1961). *Topology*. Addison-Wesley series in mathematics. Dover Publications, New York.

[Hossain et al., 2011] Hossain, Z., Alim, U. R., and Möller, T. (2011). Toward high-quality gradient estimation on regular lattices. *IEEE Transactions on Visualization and Computer Graphics*, 17(4):426–439.

[Howie and Blake, 1994] Howie, C. T. and Blake, E. H. (1994). The Mesh Propagation algorithm for isosurface construction. *Computer Graphics Forum*, 13(3):65–74.

[Iske et al., 2002] Iske, A., Quak, E., and Floater, M. S., editors (2002). *Tutorials on Multiresolution in Geometric Modelling: Summer School Lectures Notes*. Springer-Verlag, Berlin.

[Itoh and Koyamada, 1994] Itoh, T. and Koyamada, K. (1994). Isosurface generation by using extrema graphs. In *Proceedings of IEEE Visualization 1994*, pages 77–84, Los Alamitos, CA. IEEE Computer Society.

[Itoh and Koyamada, 1995] Itoh, T. and Koyamada, K. (1995). Automatic isosurface propagation using an extrema graph and sorted boundary cell lists. *IEEE Transactions on Visualization and Computer Graphics*, 1(4):319–327.

[Itoh et al., 1996] Itoh, T., Yamaguchi, Y., and Koyamada, K. (1996). Volume thinning for automatic isosurface propagation. In *Proceedings of IEEE Visualization 1996*, pages 303–310, Los Alamitos, CA. IEEE Computer Society.

[Itoh et al., 2001] Itoh, T., Yamaguchi, Y., and Koyamada, K. (2001). Fast isosurface generation using the volume thinning algorithm. *IEEE Transactions on Visualization and Computer Graphics*, 7(1):32–46.

[Ju, 2004] Ju, T. (2004). Robust repair of polygonal models. *ACM Transactions on Graphics*, 23(3):888–895.

[Ju, 2009] Ju, T. (2009). Fixing geometric errors on polygonal models: a survey. *J. Comput. Sci. Technol.*, 24(1):19–29.

[Ju et al., 2002] Ju, T., Losasso, F., Schaefer, S., and Warren, J. (2002). Dual contouring of hermite data. *ACM Transactions on Graphics*, 21(3):339–346.

[Ju and Udeshi, 2006] Ju, T. and Udeshi, T. (2006). Intersection-free contouring on an octree grid. Paper presented at the 14th Pacific Conference on Computer Graphics and Applications, Taipei, Taiwan, October 11–13.

[Karron, 1992] Karron, D. (1992). SpiderWeb algorithm for surface construction in noisy volume data. In *Proceedings of the Conference on Visualization in Biomedical Computing*, pages 462–476, Bellingham, WA. SPIE.

[Karron and Cox, 1992] Karron, D. and Cox, J. (1992). Mathematical analysis of "Spi-derWeb" surface construction algorithm. In *14th Annual Conference of the IEEE Engineering in Medicine and Biology Society*, volume 3, pages 1250–1252, Los Alamitos, CA. IEEE Computer Society.

[Karron et al., 1993] Karron, D., Cox, J. L., and Mishra, B. (1993). The SpiderWeb algorithm for surface construction from medical volume data: Geometric properties of its surface. *Innovations et Technologie en Biologie et Mèdecine*, 14(6):634–656.

[Kazhdan and Hoppe, 2007] Kazhdan, M., K. A. D. K. and Hoppe, H. (2007). Uncon-strained isosurface extraction on arbitrary octrees. In *Proceedings of the Eurograph-ics Symposium on Geometry Processing*, pages 125–133, Aire-la-Ville, Switzerland. Eurographics Association.

[Khoury and Wenger, 2010] Khoury, M. and Wenger, R. (2010). On the fractal dimen-sion of isosurfaces. *IEEE Transactions on Visualization and Computer Graphics*, 16(6):1198–1205.

[Kindlmann et al., 2003] Kindlmann, G., Whitaker, R., Tasdizen, T., and Möller, T. (2003). Curvature-based transfer functions for direct volume rendering: Methods and applications. In *Proceedings of IEEE Visualization 2003*, pages 513–520, Los Alamitos, CA. IEEE Computer Society.

[Kobbelt et al., 2001] Kobbelt, L. P., Botsch, M., Schwanecke, U., and Seidel, H.-P. (2001). Feature sensitive surface extraction from volume data. In *Proceedings of the 28th Annual Conference on Computer Graphics and Interactive Techniques, SIG-GRAPH 2001*, pages 57–66, New York. ACM Press.

[Labelle and Shewchuk, 2007] Labelle, F. and Shewchuk, J. R. (2007). Isosurface stuff-ing: Fast tetrahedral meshes with good dihedral angles. *ACM Transactions on Graphics*, 26(3):57.1–57.10.

[Lachaud and Montanvert, 2000] Lachaud, J.-O. and Montanvert, A. (2000). Continu-ous analogs of digital boundaries: A topological approach to iso-surfaces. *Graphical Models*, 62:129–164.

[Lee et al., 2008] Lee, J. K., Maskey, M., Newman, T., Wood, B., and Wang, C. (2008). Evaluation of high order approximating normals for marching cubes. In *IEEE South-eastcon 2008*, pages 593–598, Los Alamitos, CA. IEEE Computer Society.

[Lewiner et al., 2003] Lewiner, T., Lopes, H., Vieira, A. W., and Tavares, G. (2003). Efficient implementation of marching cubes' cases with topological guarantees. *Jour-nal of Graphics Tools*, 8(2):1–15.

[Lin and Ching, 1997] Lin, C. and Ching, Y. (1997). A note on computing the saddle values in isosurface polygonization. *The Visual Computer*, 13(7):342–344.

[Livnat and Hansen, 1998] Livnat, Y. and Hansen, C. (1998). View dependent iso-surface extraction. In *Proceedings of IEEE Visualization 1998*, pages 175–180, Los Alamitos, CA. IEEE Computer Society.

[Livnat et al., 1996] Livnat, Y., Shen, H.-W., and Johnson, C. R. (1996). A near optimal isosurface extraction algorithm using the span space. *IEEE Transactions on Visualization and Computer Graphics*, 2:73–84.

[Lopes and Brodlie, 2003] Lopes, A. and Brodlie, K. (2003). Improving the robustness and accuracy of the marching cubes algorithm for isosurfacing. *IEEE Transactions on Visualization and Computer Graphics*, 9(1):16–29.

[Lorensen and Cline, 1987a] Lorensen, W. and Cline, H. (1987a). Marching cubes: A high resolution 3D surface construction algorithm. *Computer Graphics*, 21(4):163–170.

[Lorensen and Cline, 1987b] Lorensen, W. and Cline, H. (1987b). System and method for the display of surface structures contained within the interior region of a solid body. U.S. Patent 4,710,876.

[Luebke, 2001] Luebke, D. P. (2001). A developer's survey of polygonal simplification algorithms. *IEEE Comput. Graph. Appl.*, 21(3):24–35.

[Manson and Schaefer, 2010] Manson, J. and Schaefer, S. (2010). Isosurfaces over simplicial partitions of multiresolution grids. *Computer Graphics Forum*, 29(2):377–385.

[Marschner and Lobb, 1994] Marschner, S. R. and Lobb, R. J. (1994). An evaluation of reconstruction filters for volume rendering. In *Proceedings of IEEE Visualization 1994*, pages 100–107, Los Alamitos, CA. IEEE Computer Society.

[Matveyev, 1994] Matveyev, S. V. (1994). Approximation of isosurface in the Marching Cube: Ambiguity problem. In *Proceedings of IEEE Visualization 1994*, pages 288–292, Los Alamitos, CA. IEEE Computer Society.

[Maubach, 1995] Maubach, J. (1995). Local bisection refinement for n-simplicial grids generated by reflection. *SIAM Journal on Scientific Computing*, 16(1):210–227.

[Max, 2001] Max, N. (2001). Consistent subdivision of convex polyhedra into tetrahedra. *Journal of Graphics Tools*, 6(3):29–36.

[McCreight, 1985] McCreight, E. M. (1985). Priority search trees. *SIAM Journal on Computing*, 14(2):257–276.

[Möller et al., 1997a] Möller, T., Machiraju, R., Mueller, K., and Yagel, R. (1997a). A comparison of normal estimation schemes. In *Proceedings of IEEE Visualization 1997*, pages 19–26, Los Alamitos, CA. IEEE Computer Society.

[Möller et al., 1997b] Möller, T., Machiraju, R., Mueller, K., and Yagel, R. (1997b). Evaluation and design of filters using a Taylor series expansion. *IEEE Transactions on Visualization and Computer Graphics*, 3(2):184–199.

[Montani et al., 1994] Montani, C., Scateni, R., and Scopigno, R. (1994). A modified look-up table for implicit disambiguation of marching cubes. *The Visual Computer*, 10(6):353–355.

[Morgan, 2009] Morgan, F. (2009). *Geometric Measure Theory: A Beginner's Guide*. Academic Press, Burlington, MA.

[Müller and Stark, 1993] Müller, H. and Stark, M. (1993). Adaptive generation of surfaces in volume data. *The Visual Computer*, 9(4):182–199.

[Natarajan, 1994] Natarajan, B. K. (1994). On generating topologically consistent isosurfaces from uniform samples. *The Visual Computer*, 11(1):52–62.

[Newman and Yi, 2006] Newman, T. S. and Yi, H. (2006). A survey of the marching cubes algorithm. *Computers & Graphics*, 30(5):854–879.

[Nielson et al., 2002] Nielson, G., Huang, A., and Sylvester, S. (2002). Approximating normals for marching cubes applied to locally supported isosurfaces. In *Proceedings of IEEE Visualization 2002*, pages 459–466, Los Alamitos, CA. IEEE Computer Society.

[Nielson, 2003a] Nielson, G. M. (2003a). MC*: Star functions for Marching Cubes. In *Proceedings of IEEE Visualization 2003*, pages 67–74, Los Alamitos, CA. IEEE Computing Society.

[Nielson, 2003b] Nielson, G. M. (2003b). On Marching Cubes. *IEEE Transactions on Visualization and Computer Graphics*, 9:283–297.

[Nielson, 2004] Nielson, G. M. (2004). Dual Marching Cubes. In *Proceedings of IEEE Visualization 2004*, pages 489–496, Los Alamitos, CA. IEEE Computer Society.

[Nielson et al., 2003] Nielson, G. M., Graf, G., Holmes, R., Huang, A., and Phielipp, M. (2003). Shrouds: Optimal separating surfaces for enumerated volumes. In *Proceedings of the Symposium on Data Visualisation, VISSYM 2003*, pages 75–84, Aire-la-Ville, Switzerland. Eurographics Association.

[Nielson and Hamann, 1991] Nielson, G. M. and Hamann, B. (1991). The Asymptotic Decider: Resolving the ambiguity in Marching Cubes. In *Proceedings of IEEE Visualization 1991*, pages 83–91, Los Alamitos, CA. IEEE Computer Society.

[Nielson and Sung, 1997] Nielson, G. M. and Sung, J. (1997). Interval volume tetrahedrization. In *Proceedings of IEEE Visualization 1997*, pages 221–228, Los Alamitos, CA. IEEE Computer Society.

[Nooruddin and Turk, 2003] Nooruddin, F. S. and Turk, G. (2003). Simplification and repair of polygonal models using volumetric techniques. *IEEE Transactions on Visualization and Computer Graphics*, 9(2):191–205.

[O'Rourke, 1998] O'Rourke, J. (1998). *Computational Geometry in C*. Cambridge University Press, New York, 2nd edition.

[Parsa, 2012] Parsa, S. (2012). A deterministic O(m log m) time algorithm for the Reeb graph. In *Proceedings of the Symposium on Computational Geometry*, pages 269–276, New York. ACM Press.

[Pascucci, 2002] Pascucci, V. (2002). Slow growing subdvisions (SGS) in any dimension: Towards removing the curse of dimensionality. *Computer Graphics Forum*, 21(3):451–460.

[Pascucci and Bajaj, 2000] Pascucci, V. and Bajaj, C. L. (2000). Time critical isosurface refinement and smoothing. In *Proceedings of the 2000 IEEE Symposium on Volume Visualization*, pages 33–42, New York. ACM Press.

[Pascucci and Cole-McLaughlin, 2002] Pascucci, V. and Cole-McLaughlin, K. (2002). Efficient computation of the topology of level sets. In *Proceedings of IEEE Visualization 2002*, pages 187–194, Los Alamitos, CA. IEEE Computer Society.

[Pascucci et al., 2007] Pascucci, V., Scorzelli, G., Bremer, P.-T., and Mascarenhas, A. (2007). Robust on-line computation of Reeb graphs: simplicity and speed. *ACM Transactions on Graphics*, 26(3):58.

[Patera and Skala, 2004] Patera, J. and Skala, V. (2004). A comparison of fundamental methods for iso-surface extraction. *Machine Graphics & Vision*, 13(4):329–343.

[Payne and Toga, 1990] Payne, B. A. and Toga, A. W. (1990). Medical imaging: Surface mapping brain function on 3D models. *IEEE Computer Graphics and Applications*, 10(5):33–41.

[Pekar et al., 2001] Pekar, V., Wiemker, R., and Hempel, D. (2001). Fast detection of meaningful isosurfaces for volume data visualization. In *Proceedings of IEEE Visualization 2001*, pages 223–230, Los Alamitos, CA. IEEE Computer Society.

[Perry and Frisken, 2001] Perry, R. N. and Frisken, S. F. (2001). Kizamu: a system for sculpting digital characters. In *Proceedings of the 28th Annual Conference on Computer Graphics and Interactive Techniques, SIGGRAPH 2001*, pages 47–56, New York. ACM Press.

[Poston et al., 1998] Poston, T., Wong, T.-T., and Heng, P.-A. (1998). Multiresolution isosurface extraction with adaptive skeleton climbing. *Computer Graphics Forum*, 17(3):137–148.

[Pöthkow and Hege, 2011] Pöthkow, K. and Hege, H.-C. (2011). Positional uncertainty of isocontours: Condition analysis and probabilistic measures. *IEEE Transactions on Visualization and Computer Graphics*, 17:1393–1406.

[Preparata and Shamos, 1985] Preparata, F. P. and Shamos, M. I. (1985). *Computational Geometry: An Introduction*. Springer-Verlag, New York.

[Raman and Wenger, 2008] Raman, S. and Wenger, R. (2008). Quality isosurface mesh generation using an extended Marching Cubes lookup table. *Computer Graphics Forum*, 27:791–798.

[Reeb, 1946] Reeb, G. (1946). Sur les points singuliers d'une forme de Pfaff complement integrable ou d'une fonction numerique. *Comptes Rendus de L'Academie ses Seances, Paris*, 222:847–849.

[Roberts and Hill, 1999] Roberts, J. C. and Hill, S. (1999). Piecewise linear hypersurfaces using the Marching Cubes algorithm. In Erbacher, R. F. and Pang, A., editors, *Visual Data Exploration and Analysis VI*, volume 3643 of *Proceedings of SPIE*, pages 170–181. IS&T and SPIE.

[Samet, 1990a] Samet, H. (1990a). *Applications of Spatial Data Structures: Computer Graphics, Image Processing and GIS*. Addison-Wesley Longman Publishing Co., Inc., Boston.

[Samet, 1990b] Samet, H. (1990b). *The Design and Analysis of Spatial Data Structures*. Addison-Wesley Longman Publishing Co., Inc., Boston.

[Samet, 2005] Samet, H. (2005). *Foundations of Multidimensional and Metric Data Structures*. The Morgan Kaufmann Series in Computer Graphics and Geometric Modeling. Morgan Kaufmann Publishers Inc., San Francisco.

[Schaefer et al., 2007] Schaefer, S., Ju, T., and Warren, J. (2007). Manifold dual contouring. *IEEE Transactions on Visualization and Computer Graphics*, 13(3):610–619.

[Schaefer and Warren, 2002] Schaefer, S. and Warren, J. (2002). Dual contouring: The secret sauce. Technical Report TR 02-408, Dept. of Computer Science, Rice University.

[Schaefer and Warren, 2004] Schaefer, S. and Warren, J. (2004). Dual marching cubes: Primal contouring of dual grids. In *Proceedings of the Computer Graphics and Applications, 12th Pacific Conference*, pages 70–76, Los Alamitos, CA. IEEE Computer Society.

[Scheidegger et al., 2008] Scheidegger, C. E., Schreiner, J. M., Duffy, B., Carr, H., and Silva, C. T. (2008). Revisiting histograms and isosurface statistics. *IEEE Transactions on Visualization and Computer Graphics*, 14(6):1659–1666.

[Schreiner et al., 2006] Schreiner, J., Scheidegger, C., and Silva, C. (2006). High-quality extraction of isosurfaces from regular and irregular grids. *IEEE Transactions on Visualization and Computer Graphics*, 12(5):1205–1212.

[Shekhar et al., 1996] Shekhar, R., Fayyad, E., Yagel, R., and Cornhill, J. F. (1996). Octree-based decimation of marching cubes surfaces. In Yagel, R. and Nielson, G. M., editors, *Proceedings of IEEE Visualization 1996*, pages 335–344, Los Alamitos, CA. IEEE Computer Society.

[Shen, 1998] Shen, H.-W. (1998). Isosurface extraction in time-varying fields using a temporal hierarchical index tree. In *Proceedings of IEEE Visualization 1998*, pages 159–166, Los Alamitos, CA. IEEE Computing Society.

[Shen et al., 1999] Shen, H.-W., Chiang, L.-J., and Ma, K.-L. (1999). A fast volume rendering algorithm for time-varying fields using a time-space partitioning (TSP) tree. In *Proceedings of IEEE Visualization 1999*, pages 371–377, Los Alamitos, CA. IEEE Computer Society.

[Shen et al., 1996] Shen, H.-W., Hansen, C. D., Livnat, Y., and Johnson, C. R. (1996). Isosurfacing in span space with utmost efficiency (ISSUE). In *Proceedings of Visualization 1996*, pages 287–294, Los Alamitos, CA. IEEE Computer Society.

[Shen and Johnson, 1995] Shen, H.-W. and Johnson, C. R. (1995). Sweeping simplices: A fast iso-surface extraction algorithm for unstructured grids. In *Proceedings of IEEE Visualization 1995*, pages 143–150, Los Alamitos, CA. IEEE Computer Society.

[Shewchuk, 2002] Shewchuk, J. R. (2002). What is a good linear element? Interpolation, conditioning and quality measures. In *Proceedings of the 11th International Meshing Roundtable*, pages 115–126, Berlin. Springer-Verlag.

[Shi and JaJa, 2006] Shi, Q. and JaJa, J. (2006). Isosurface extraction and spatial filtering using persistent octree (pot). *IEEE Transactions on Visualization and Computer Graphics*, 12(5):1283–1290.

[Shinagawa and Kunii, 1991] Shinagawa, Y. and Kunii, T. L. (1991). Constructing a Reeb graph automatically from cross sections. *IEEE Comput. Graph. Appl.*, 11:44–51.

[Shirley and Tuchman, 1990] Shirley, P. and Tuchman, A. (1990). A polygonal approximation to direct scalar volume rendering. In *Proceedings of the 1990 Workshop on Volume Visualization*, pages 63–70, New York. ACM Press.

[Shu et al., 1995] Shu, R., Zhou, C., and Kankanhalli, M. S. (1995). Adaptive marching cubes. *The Visual Computer*, 11(4):202–217.

[Sutton and Hansen, 2000] Sutton, P. M. and Hansen, C. D. (2000). Accelerated isosurface extraction in time-varying fields. *IEEE Transactions on Visualization and Computer Graphics*, 6(2):98–107.

[Sutton et al., 2000] Sutton, P. M., Hansen, C. D., Shen, H.-W., and Schikore, D. (2000). A case study of isosurface extraction algorithm performance. In *Data Visualization 2000*, pages 259–268, New York. Springer.

[Szymczak and Vanderhyde, 2003] Szymczak, A. and Vanderhyde, J. (2003). Extraction of topologically simple isosurfaces from volume datasets. In *Proceedings of IEEE Visualization 2003*, pages 67–74, Los Alamitos, CA. IEEE Computer Society.

[Tarasov and Vyalyi, 1998] Tarasov, S. P. and Vyalyi, M. N. (1998). Construction of contour trees in 3D in $O(n \log n)$ steps. In *Proceedings of the Annual Symposium on Computational Geometry*, pages 68–75, New York. ACM Press.

[Tenginakai et al., 2001] Tenginakai, S., Lee, J., and Machiraju, R. (2001). Salient iso-surface detection with model-independent statistical signatures. In *Proceedings of IEEE Visualization 2001*, pages 231–238, Los Alamitos, CA. IEEE Computer Society.

[Tenginakai and Machiraju, 2002] Tenginakai, S. and Machiraju, R. (2002). Statistical computation of salient iso-values. In *Proceedings of the Symposium on Data Visualisation 2002*, pages 19–24, Aire-la-Ville, Switzerland. Eurographics Association.

[Theisel, 2002] Theisel, H. (2002). Exact isosurfaces for Marching Cubes. *Computer Graphics Forum*, 21(1):19–32.

[Theußl et al., 2001] Theußl, T., Möller, T., and Gröller, M. (2001). Optimal regular volume sampling. In *Proceedings of IEEE Visualization 2001*, pages 91–546, Los Alamitos, CA. IEEE Computer Society.

[Tierny et al., 2009] Tierny, J., Gyulassy, A., Simon, E., and Pascucci, V. (2009). Loop surgery for volumetric meshes: Reeb graphs reduced to contour trees. *IEEE Transactions on Visualization and Computer Graphics*, 15(6):1177–1184.

[Treece et al., 1999] Treece, G., Prager, R., and Gee, A. (1999). Regularised marching tetrahedra: Improved iso-surface extraction. *Computers & Graphics*, 23(4):593–598.

[Tzeng, 2004] Tzeng, L. (2004). Warping Cubes: Better triangles from Marching Cubes. European Workshop on Computational Geometry, Seville, Spain, March 24.

[van Kaick and Pedrini, 2006] van Kaick, O. M. and Pedrini, H. (2006). A comparative evaluation of metrics for fast mesh simplification. *Computer Graphics Forum*, 25(2):197–210.

[van Kreveld et al., 1997] van Kreveld, M., van Oostrum, R., Bajaj, C., Pascucci, V., and Schikore, D. (1997). Contour trees and small seed sets for isosurface traversal. In *Proceedings of the Annual Symposium on Computational Geometry*, pages 212–220, New York. ACM Press.

[Varadhan et al., 2003] Varadhan, G., Krishnan, S., Kim, Y. J., and Manocha, D. (2003). Feature-sensitive subdivision and isosurface reconstruction. In *Proceedings of IEEE Visualization 2003*, pages 99–106, Los Alamitos, CA. IEEE Computer Society.

[Vrolijk et al., 2004] Vrolijk, B., Botha, C. P., and Post, F. H. (2004). Fast time-dependent isosurface extraction and rendering. In *Proceedings of the 20th Spring Conference on Computer graphics*, pages 45–54, New York. ACM Press.

[Wang, 2011] Wang, C. (2011). Intersection-free dual contouring on uniform grids: An approach based on convex/concave analysis. Technical report, Dept. of Mechanical and Automation Engineering, The Chinese University of Hong Kong, Hong Kong.

[Wang and Chiang, 2009] Wang, C. and Chiang, Y.-J. (2009). Isosurface extraction and view-dependent filtering from time-varying fields using persistent time-octree (ptot). *IEEE Transactions on Visualization and Computer Graphics*, 15:1367–1374.

[Waters et al., 2006] Waters, K., Co, C., and Joy, K. (2006). Using difference intervals for time-varying isosurface visualization. *Visualization and Computer Graphics, IEEE Transactions on*, 12(5):1275–1282.

[Weber et al., 2001] Weber, G. H., Kreylos, O., Ligocki, T. J., Shalf, J. M., Hagen, H., Hamann, B., and Joy, K. I. (2001). Extraction of crack-free isosurfaces from adaptive mesh refinement data. In *Proceedings of the Joint Eurographics and IEEE TCVG Symposium on Visualization*, pages 25–34, Aire-la-Ville, Switzerland. Eurographics Association.

[Weigle and Banks, 1996] Weigle, C. and Banks, D. (1996). Complex-valued contour meshing. In *Proceedings of IEEE Visualization 1996*, pages 173–180, Los Alamitos, CA. IEEE Computer Society.

[Weigle and Banks, 1998] Weigle, C. and Banks, D. C. (1998). Extracting iso-valued features in 4-dimensional scalar fields. In *Proceedings of the IEEE Symposium on Volume Visualization*, pages 103–110, Los Alamitos, CA. IEEE Computing Society.

[Weiss and De Floriani, 2008] Weiss, K. and De Floriani, L. (2008). Multiresolution interval volume meshes. In *IEEE/EG Symposium on Volume and Point-Based Graphics*, pages 65–72, Aire-la-Ville, Switzerland. Eurographics Association.

[Weiss and De Floriani, 2010] Weiss, K. and De Floriani, L. (2010). Isodiamond hierarchies: An efficient multiresolution representation for isosurfaces and interval volumes. *IEEE Transactions on Visualization and Computer Graphics*, 16(4):583–598.

[Weiss and De Floriani, 2011] Weiss, K. and De Floriani, L. (2011). Simplex and diamond hierarchies: Models and applications. *Computer Graphics Forum*, 30(8):2127–2155.

[Westermann et al., 1999] Westermann, R., Kobbelt, L., and Ertl, T. (1999). Real-time exploration of regular volume data by adaptive reconstruction of iso-surfaces. *The Visual Computer*, 15:100–111.

[Wilhelms and Gelder, 1992] Wilhelms, J. and Gelder, A. V. (1992). Octrees for faster isosurface generation. *ACM Transactions on Graphics*, 11(3):201–227.

[Wood et al., 2004] Wood, Z., Hoppe, H., Desbrun, M., and Schroder, P. (2004). Removing excess topology from isosurfaces. *ACM Transactions on Graphics*, 23(2):190–208.

[Wyvill et al., 1986] Wyvill, G., McPheeters, C., and Wyvill, B. (1986). Data structure for *soft* objects. *The Visual Computer*, 2(4):227–234.

[Zhang et al., 2004] Zhang, N., Hong, W., and Kaufman, A. (2004). Dual contouring with topology-preserving simplification using enhanced cell representation. In *Proceedings of IEEE Visualization 2004*, pages 505–512, Los Alamitos, CA. IEEE Computer Society.

[Zhang and Qian, 2012] Zhang, Y. and Qian, J. (2012). Dual contouring for domains with topology ambiguity. In *Proceedings of the 20th International Meshing Roundtable*, pages 41–60, Berlin. Springer-Verlag.

[Zhou et al., 1994] Zhou, C., Shu, R., and Kankanhalli, M. S. (1994). Handling small features in isosurface generation using Marching Cubes. *Computers & Graphics*, 18(6):845–848.

[Zhou et al., 1997] Zhou, Y., Chen, B., and Kaufman, A. (1997). Multiresolution tetrahedral framework for visualizing regular volume data. In *Proceedings of IEEE Visualization 1997*, pages 135–142, Los Alamitos, CA. IEEE Computer Society.

INDEX